Proceedings of the 7th International Symposium on

FOUNDATIONS OF QUANTUM MECHANICS IN THE LIGHT OF NEW TECHNOLOGY

ISQM — TOKYO '01

Proceedings of the 7th International Symposium on

FOUNDATIONS OF QUANTUM MECHANICS IN THE LIGHT OF NEW TECHNOLOGY

ISQM — TOKYO '01

Advanced Research Laboratory
Hitachi, Ltd., Hatoyama, Saitama, Japan 27–30 August 2001

Edited by

Yoshimasa A. Ono
Advanced Research Laboratory
Hitachi, Ltd.
Saitama, Japan

Kazuo Fujikawa
Department of Physics
The University of Tokyo
Tokyo, Japan

World Scientific
New Jersey • London • Singapore • Hong Kong

Published by

World Scientific Publishing Co. Pte. Ltd.
P O Box 128, Farrer Road, Singapore 912805
USA office: Suite 1B, 1060 Main Street, River Edge, NJ 07661
UK office: 57 Shelton Street, Covent Garden, London WC2H 9HE

British Library Cataloguing-in-Publication Data
A catalogue record for this book is available from the British Library.

FOUNDATIONS OF QUANTUM MECHANICS IN THE LIGHT OF NEW TECHNOLOGY

ISBN 981-238-130-9

Printed by FuIsland Offset Printing (S) Pte Ltd, Singapore

The Seventh International Symposium on Foundations of Quantum Mechanics (ISQM-Tokyo '01)

PREFACE

The Seventh International Symposium on Foundations of Quantum Mechanics in the Light of New Technology (ISQM-Tokyo '01) was held on August 27-30, 2001 at the Advanced Research Laboratory, Hitachi, Ltd. in Hatoyama, Saitama, Japan. The symposium was organized by its own Scientific Committee under the auspices of the Physical Society of Japan, the Japan Society of Applied Physics, and the Advanced Research Laboratory, Hitachi, Ltd. A total of 126 participants (26 from abroad) attended the symposium, and 23 invited oral papers, 18 contributed oral papers, and 30 poster papers were presented.

Just as in the previous six symposia, the aim of this symposium was to link the recent advances in technology with fundamental problems in quantum mechanics. It provided a unique interdisciplinary forum where scientists from the very different disciplines, who would otherwise never meet each other, convened to discuss basic problems of common interest in quantum science and technology from various aspects and "in the light of new technology."

Quantum Coherence and Decoherence was chosen as the main theme for the present symposium because of its importance in quantum science and technology. This topic was reexamined from all aspects, not only in terms of atom optics, quantum optics, quantum computing, quantum information, and mesoscopic physics, but also in terms of the physics of precise measurement and other fundamental problems in quantum physics. We were delighted that many active and well-known researchers in these fields accepted our invitation.

We are now very happy to offer the fruits of the symposium in the form of the proceedings to a wider audience. As shown in the table of contents, the proceedings include 66 refereed papers in ten sections: quantum computing; quantum information, quantum teleportation, and entanglement; quantum optics; Bose-Einstein condensation and atom interferometry; mesoscopic magnets; single electronics and superconductors; nanoscale physics and atomics; quantum transport; precise measurements; fundamental problems in quantum physics. Here we will just mention some of the important key words to give the flavor of the proceedings: quantum computation, qubits, quantum dots, Bose-Einstein condensates, single-electron transistor, mesoscopic spins, magnetic domain wall, nanowire, Josephson junctions, and dynamics of vortices in high-temperature superconductors. We hope that the proceedings will not only serve as a good introductory book on quantum coherence and decoherence for newcomers in this field, but also as a reference book for experts.

In conclusion, we thank the participants for their contribution to the symposium's success. Thanks are also due to all the authors who prepared manuscripts, and to the referees who kindly reviewed the papers. We also thank the members of the Advisory Committee and Organizing Committee, without whose kind cooperation the symposium would not have been a success. Finally, we would like to express our deepest gratitude to the Advanced Research Laboratory, Hitachi, Ltd. and its General Manager, Dr. Nobuyuki Osakabe, for providing us with financial support and an environment ideal for lively discussion, and to his staff members, in particular Naoyuki Chino and Akemi Tsuchida, for their efforts in making the symposium enjoyable as well as fruitful.

May 2002

Yoshimasa A. Ono
Kazuo Fujikawa

COMMITTEES

Chair:

H. Fukuyama, Institute for Solid State Physics, University of Tokyo

Advisory Committee:

H. Ezawa, Gakushuin University
Y. Imry, Weizmann Institute
S. Kobayashi, The Institute of Physical and Chemical Research (RIKEN)
A.J. Leggett, University of Illinois at Urbana-Champaign
J.E. Mooij, Delft University of Technology
N. Osakabe, Advanced Research Laboratory, Hitachi, Ltd.
Y. Yamamoto, Stanford University and NTT
C.N. Yang, State University of New York at Stony Brook
A. Zeilinger, University of Vienna

Organizing Committee:

K. Fujikawa, University of Tokyo
Y. Iye, Institute for Solid State Physics, University of Tokyo
S. Nakajima, Superconductivity Research Laboratory,
International Superconductivity Technology Center
F. Shimizu, University of Electro-Communications
Y.A. Ono, Advanced Research Laboratory, Hitachi, Ltd.

Sponsors:

The Physical Society of Japan
The Japan Society of Applied Physics
Advanced Research Laboratory, Hitachi, Ltd.

CONTENTS

BOSE-EINSTEIN CONDENSATION AND ATOM INTERFEROMETRY

MESOSCOPIC MAGNETS

SINGLE ELECTRONICS AND SUPERCONDUCTORS

NANOSCALE PHYSICS AND ATOMICS

QUANTUM TRANSPORT

PRECISE MEASUREMENTS

FUNDAMENTAL PROBLEMS IN QUANTUM PHYSICS

OPENING ADDRESS

HIDETOSHI FUKUYAMA
Chair of the Organizing Committee, ISQM-Tokyo '01
Institute for Solid State Physics, the University of Tokyo
Kashiwa, Chiba 277-8581, Japan

Good morning, ladies and gentlemen, and dear friends. On behalf of the Organizing Committee, I would like to welcome you to this International Symposium on the Foundations of Quantum Mechanics in the Light of New Technology (ISQM). This is the 7th Symposium of the series, which orignally began in 1983 under the leadership of Professor Sadao Nakajima, Director of the Institute for Solid State Physics at the University of Tokyo, and Dr. Yasutsugu Takeda, General Manager of the Central Research Laboratory, Hitachi, Ltd.

It is an honor and a great pleasure for me to open this Symposium, which hosts more than 126 scientists. In particular, I want to thank the 31 participants from abroad, who flew long distance from 11 different countries despite the vacation season.

As you know, this Symposium is unique in the sense that scientists from the very different disciplines, who would otherwise never meet each other, are convening to exchange information and cultivate common scientific interests in order to deepen their understanding of quantum mechanics. This kind of collaborative spirit has been a hallmark of this Symposium from the very beginning, as exemplified by the initial achievements of Dr. Akira Tomomura, who obtained experimental verification of the Aharonov-Bohm effect, and by the strong scientific support provided to the first Symposium by Professor Chen Nin Yang, who unfortunately is not be able to attend this time. We miss him very much.

As you all know, quantum mechanics, together with the theory of relativity, has changed the entire realm of physics dramatically and has led to many great developments in science and technology over the last century. This is particularly true with respect to modern technologies. A typical example is the semiconductor, which is the main component of computers and is truly one of the essential yet invisible constituents of our daily life. Nowadays, activities in almost any branch of science–not to mention of IT–are supported by computers: for example, consider the sequence analyzers used in genomics, the sensitive detectors used in cosmic telescopes, and the instrumentation used for making all kinds of precise measurements. The understanding of such semiconductor devices has been possible only through the use of quantum mechanics.

That close link between basic science and technology will escalate in this new century, and I hope that forums like this one will continue to contribute to the development of such a connection. I think both Professor Nakajima and Dr. Takeda had good foresight and made an excellent choice in their decision to host the first Synmposium 18 years ago.

Last but not least, we should thank the Advanced Research Laboratory, Hitachi, Ltd., headed by Dr. Nobuyuki Osakabe, for their supporty in enabling this forum to happen. We also owe a great debt to Dr. Yoshimasa A. Ono for his painstaking efforts over the past few months.

I hope you enjoy the presentations and discussions.

Thank you for your attention.

WELCOMING ADDRESS

NOBUYUKI OSAKABE
Advanced Research Laboratory, Hitachi, Ltd.
Hatoyama, Saitama 350-0395, Japan

Good morning, distinguished guests, ladies and gentlemen. On behalf of our laboratory, let me welcome you to Hatoyama and ISQM-Tokyo '01. It is a great pleasure for us to host this event for you all here today.

This symposium was initiated in 1983. The motivation for Hitachi to host and sponsor the first ISQM came from Dr. Akira Tonomura's successful verification of the Aharonov-Bohm effect. In his study, new micro-fabrication technology developed in the semiconductor industry and an electron microscope technology developed at Hitachi were successfully combined to verify the important concept of the gauge field. In addition, through the strong thrust of Professor Chen Ning Yang, Hitachi decided to hold the first ISQM, i.e., the International Symposium on Foundations of Quantum Mechanics in the Light of New Technology. The aim of the symposium has been to discuss the foundation of quantum mechanics achieved by using new technologies based on industrial innovation and to contribute to the scientific community in general.

Eighteen years have past since then. The site of the symposium has moved from the Central Research Laboratory to the Advanced Research Laboratory, Hitachi, Ltd., here at Hatoyama. Many new areas in physics have been discussed over the years. Today those fundamental quantum phenomena are now being used in the world of technology: Single electron charges can now be controlled to make a new semiconductor memories to breakthrough the barrier of modern device performance. Quantum entanglement will be used to secure future communication. Macroscopic quantum tunneling and coherence will be the basis for quantum computing. The foundation of quantum mechanics will leverage industrial companies. Thus, because of all these intriguing possibilities, we found enormous value in hosting and sponsoring the symposium here again.

I hope all of you find the symposium rewarding. Before concluding, I would like to thank all the members of the Organizing and Advisory Committees for putting together such an exciting program. I also would like to thank all the invited speakers for coming to share their latest findings with the participants here. I very much look forward to hearing them.

Thank you very much for your attention.

DECOHERENCE AND DEPHASING IN SPIN-BASED SOLID STATE QUANTUM COMPUTERS

XUEDONG HU, ROGERIO DE SOUSA, AND S. DAS SARMA
Department of Physics, University of Maryland, College Park, MD 20742-4111

We discuss the operational definition of decoherence in various solid state systems. In particular, we review, in the context of spin-based solid state quantum computation, the introduction of T_1 and T_2 to describe decoherence in a two level system. We provide a perspective on recent experiments involving the manipulation of spin coherence in semiconductors, and discuss specific decoherence and dephasing issues in electron spin-based quantum dot quantum computer architectures.[1]

1 Introduction

A quantum eigenstate of a particular Hamiltonian is by definition a stationary state, in which the wavefunction might vary spatially, but does not decay in time. A quantum system can be in a superposition of its eigenstates with definite phase and amplitude relationships among the basis states. Such a superposition of states with definite phase relationships is called quantum coherence. Decoherence refers loosely to how a system loses these quantum coherence features. For example, it can refer to the amplitude decay (often exponential) and the associated disappearance of a quantum eigenstate in time (by virtue of it interacting with a surrounding bath, for instance). Or it may refer to the loss of electron phase coherence, because the definite phase relationship between the superposing states disappears over time leading to dephasing.

In an isolated quantum system decoherence could only arise from the dynamical degrees of freedom neglected in the original Hamiltonian used to define the quantum state. In a system coupled to an external bath, decoherence could arise naturally from the coupling between the system and the bath as the quantum eigenstate (presumably slowly) leaks to the environment (the bath) due to energy exchange between the system and the bath. Note that the bath or the environment does not need to be physically separated from the "system"—it is a standard practice in physics to divide a large system into sub-systems which are "reasonably isolated" (in some well-defined operational sense) from each other, i.e. the interaction Hamiltonian coupling the various subsystems is "weak" in a precisely defined manner. In situations like this (which are commonplace in condensed matter physics) the "system" and the "bath" can be two components of the same system (such as electrons and phonons of a crystal lattice).

Decoherence in quantum mechanics has received a great deal of recent attention in the context of current interest in quantum computation and information processing.[2] In a quantum computer (QC), a computation is typically performed by applying unitary operations on an array of two level systems (qubits) that carry quantum information.[2] These qubits must be isolated from the other degrees of freedom, i.e. decoherence in a QC must be much slower than a typical quantum gate operation for successful quantum computation. The ratio between gate time and decoherence time needs to be smaller than $10^{-3} \sim 10^{-6}$, which is the current quantum error correction limit.[2] Thus the control of decoherence is a crucial aspect of quantum information processing.

In condensed matter physics, terminologies such as decoherence, dephasing, relaxation, scattering, etc, are often used quite liberally (and somewhat confusingly) due to the diversity of physical systems and phenomena studied. With the many existing proposals for solid state QC architectures (e.g. references 3 and 4), there is obviously a need to accurately understand decoherence issues in each particular scheme. Most QC proposals involve quantum two level systems (TLS) serving the role of qubits. The basic QC algorithm involves dynamic manipulations of these TLS using external means to perform one and two qubit operations. It is therefore imperative that the decoherence time in the dynamics of these TLS is much larger than the qubit operation times. Because of the two-level nature of these systems, it is possible to describe their decoherence using just two dephasing times (T_1 and $T_2(\leq T_1)$), which give a phenomenological description of the population and phase relaxation in these systems. For an ensemble of TLS, another time scale $T_2^* \leq T_2$ should also be defined, since some spins may rotate faster than others leading to loss of coherence between them. The two time scales T_2^* and T_2 should be carefully distinguished, as the current QC proposals mostly involve single TLS so that T_2 is the relevant quantity, while in macroscopic measurements the observed quantity is often T_2^*.

In this paper we first review and discuss the definitions and significance of relaxation times T_1, T_2, and T_2^* in the context of a TLS. We then review measurements of these relaxation times in semiconductors, which are projected to be of potential use in several proposed solid state QC architectures. We also discuss some specific features of decoherence in a solid state QC, particularly aspects of gate induced decoherence in the form of non-adiabaticity, electrical circuit noise, and field inhomogeneity. We conclude with a brief discussion of electronic decoherence in mesoscopic systems.

2 Decoherence in a two level system

Two relaxation time scales, T_1 and T_2, were introduced and used extensively in the fields of NMR,[5] ESR,[6] and quantum optics,[7] in which either the applied static magnetic field (which causes Zeeman splitting along the field direction) or the natural TLS (e.g. for photons, where the longitudinal and transverse polarization occurs naturally) defines a longitudinal and a transverse direction. T_1 and T_2 are then respectively the longitudinal and transverse relaxation times for magnetizations in NMR and ESR, or the population difference and polarization in quantum optics. Note that using T_1 and T_2 to characterize decoherence applies only to TLS dynamics.

The definition of T_1 and T_2 is quite system specific—in fact, the strict definition of T_1 and T_2 applies specifically to magnetic resonance measurements. An arbitrary decoherence phenomenon might require more or less parameters to describe the dephasing process. In general, in the absence of magnetic field and in isotropic systems, $T_1 = T_2$. In weak localization related mesoscopic experiments, a single electron dephasing time, which parametrizes the loss of quantum interference, is sufficient to characterize the results. We would also like to point out that T_1 and T_2 are purely phenomenological parameters (characterizing longitudinal and transverse relaxation respectively), to which many different decoherence mechanisms could, in principle, contribute. Although two parameters may not completely describe TLS decoherence, experience (particularly in NMR, ESR, and optical pumping experiments) suggests that T_1 and T_2 are often quite sufficient in characterizing TLS decoherence in many diverse situations and are therefore extremely important TLS parameters.

We first discuss how T_1 and T_2 are introduced for a particular TLS—an electron spin in an external magnetic field. Consider an electron subjected to a constant field in the z direction $\mathbf{B}_{\|} = B_{\|}\ \hat{\mathbf{z}}$, and a rotating field in the xy plane with frequency ν, $\mathbf{B}_{\perp} = B_{\perp}[\cos(\nu t)\hat{\mathbf{x}}+\sin(\nu t)\hat{\mathbf{y}}]$. The spin density matrix elements satisfy

$$i\dot{\rho}_{\uparrow\uparrow} = \Delta(\rho^*_{\uparrow\downarrow} - \rho_{\uparrow\downarrow}), \tag{1}$$

$$i\dot{\rho}_{\uparrow\downarrow} = \delta\rho_{\uparrow\downarrow} + \Delta(\rho_{\downarrow\downarrow} - \rho_{\uparrow\uparrow}). \tag{2}$$

where $\delta = \omega - \nu$ is the detuning frequency, $\omega = \mu_e B_{\|}/\hbar$ is the Zeeman splitting, and $2\Delta = 2\mu_e B_{\perp}/\hbar$ is the Rabi frequency. The evolution of this rotating or precessing electron spin is unitary since we are considering a single isolated spin without any dephasing or decoherence. However, an electron spin is never isolated in a solid. It couples to the electron orbital degrees of freedom, the surrounding nuclear spins, the crystal lattice, the magnetic impurities, and to other electron spins. All these "environmental" degrees of freedom need to be included in the equations for the spin density matrix. A simple approach to tackle this problem is to add exponential decay terms to the right hand sides of the two equations above: $-i\rho_{\uparrow\uparrow}/T_1$ to Eq. (1) and $-i\rho_{\uparrow\downarrow}/T_2$ to Eq. (2) to mimic decoherence phenomenologically. This is similar to adding a friction term proportional to velocity in the classical Newton's equation. The two time constants can be calculated if sufficient information about the bath is available. The corresponding equations for the macroscopic magnetization are the Bloch equations,[5] which are quite successful in describing many experiments, ranging from NMR and ESR to quantum optics, although actual explicit calculations of T_1 and T_2 are generally quite difficult.

To describe an ensemble of spins, which may in general possess different Zeeman splittings $\hbar\omega$ (for example, by virtue of inhomogeneities in the applied magnetic field and/or in the electron g-factor) and thus having different detunings δ in the rotating field, additional ensemble averaging needs to be performed. This averaging leads to a different time constant $T_2^*(\leq T_2)$ to describe the width of the magnetic resonance signal, but it does not affect the longitudinal direction. Note that T_2 (or T_2^*) describes the dephasing process (T_2 is often called the dephasing time), and T_1 ($\geq T_2$) is the inelastic spin-flip or spin-lattice relaxation time. Often T_2 is also called the spin-spin relaxation time for reasons to be discussed below.

Spin-flip processes cause both population relaxation and dephasing, contributing to both rates $1/T_1$ and $1/T_2$. However, there exist pure dephasing processes which affect only T_2 but not T_1. One example is the molecules in an optically active gaseous medium. The molecules constantly collide with each other, most of the time elastically. These collisions lead to random shifts in the molecular energy levels, and cause a pure dephasing effect that only contributes to T_2, but not to T_1. Another well-known example of pure dephasing is the dipolar spin-spin interaction in NMR, which produces effective local magnetic field fluctuations and hence contributes essentially only to T_2 (the corresponding effect on T_1 is extremely small).

What is important for dephasing is that some change in the state of the environment must occur due to its interaction with the system—dephasing does not necessarily require an explicit inelastic scattering process for the system, although all inelastic scatterings necessarily produce dephasing. In fact, as mentioned before, T_2 in the context of ESR and NMR is often called the spin-spin relaxation time because the most important intrinsic effect contributing to $1/T_2$ is the dipolar interaction among various spins in the system, which, while transferring energy among the spins themselves, does not lead to overall energy relaxation from the total spin system. By contrast, spin-lattice interactions lead to energy relaxation (via spin-flip processes) from the spin system to the lattice, and thus contribute to T_1^{-1}, the spin-lattice relaxation rate. We note in this context that T_2 sets the time scale for the spin system to achieve equilibrium within itself whereas T_1 sets the time scale for the global thermodynamic equilibrium between the spin system and the lattice. It should be emphasized that all inelastic processes contributing to T_1 also automatically lead to dephasing, but in many circumstances there may be additional dephasing processes (e.g. dipolar spin-spin coupling in NMR and ESR) which contribute only to T_2 (and *not* to T_1), therefore $T_1 \geq T_2$ in general.

3 Measuring T_1, T_2, and T_2^* of electron spins in semiconductors

In metals and doped semiconductors there are three major spin relaxation mechanisms for conduction electrons:[8,9] the Elliot-Yafet mechanism, the Dyakonov-Perel' mechanism, and the Bir-Aronov-Pikus mechanism. At the lowest temperatures, where relaxation times are extremely long, dipolar and nuclear coupling become comparatively important. A detailed review of these mechanisms and their experimental observations in metals and semiconductors have been discussed in reference 9. We only mention here that in general electron spin relaxation in GaAs is weak due to its relatively weak conduction band spin-orbit coupling—a simple matrix element estimate indicates that the spin relaxation time is of the order of 10-100 ns at low temperatures ($T \sim 4$ K), and one should approximately have $T_1 \sim T_2$ in high quality GaAs. Such a "long" relaxation time (T_2) has recently been directly measured in ESR experiments.[10] We should emphasize that this relaxation time ($\sim$ 10-100 ns) is "long" only in a relative sense compared with electron-electron scattering times ($\sim$ fs) or momentum relaxation times ($\sim$ ps)—electron spin relaxation in GaAs should be rather slow due to very weak conduction band spin-orbit coupling, and the observed relaxation times are not long in any absolute sense. In fact, low temperature electron spin relaxation times in metals are also very long ($\sim$ μs).[9]

Coherent manipulation of electron spins and the study of electron spin relaxation have a long history going back to the first ESR experiments half a century ago,[6] and ESR remains a key technique to determine electron spin relaxation and study electron spin dynamics.[10] Below we focus on some of the recent experiments that employ alternative optical approaches to study electron spins in semiconductors, especially on optical orientation and related measurements of spin relaxation, which may have particular relevance to solid state quantum computation, where externally controlled optical pulses may be used to coherently manipulate electron spin dynamics in zinc-blende semiconductor structures.

3.1 *III-V semiconductors*

The method of optical orientation can be used to measure spin relaxation (T_1) in bulk III-V semiconductors like GaAs[8]. Typical values for T_1 in p-doped GaAs are 10 ps to 1 ns, depending on the sample temperature and impurity concentration. The hole spin relaxation rate is relatively fast in GaAs, consistent with its strong valence band spin-orbit coupling. Optical measurements have recently been used to measure T_2^* in n-doped bulk GaAs, using pump-probe and Faraday rotation techniques.[11] At $B = 1$ T, $T = 5$ K, and doping concentration $n = 10^{16} cm^{-3}$, a spin decoherence time T_2^* of approximately 10ns was measured[11]. Furthermore, $T_2^* = 130$ ns is obtained at $B = 0$, which is one order of magnitude larger than the value at B=1 T. These values of the relaxation time are consistent with the direct ESR measurement of the linewidth ($\sim$ 50 MHz) in GaAs heterostructures.[10]

Time-resolved Faraday rotation was also used to study spin precession in chemically synthesized CdSe quantum dots[12]. T_2^{Inh} (which contains averaging not only for the many electrons in one quantum dot (QD), but also over many QDs) was measured ranging from $3ns$ at $B = 0$ to less than 100 ps at $B = 4$ T. It was suggested that this strong field dependence comes from the fact that the QDs have varying g factors ranging between 1.1 and 1.7. To measure the decoherence time T_2^* of the many electrons in a single QD one could in principle use the spin echo technique to constantly realign the spins and observe the time decay of the transverse magnetization, hence extracting T_2^* from the data of an ensemble of QDs. This experiment has not yet been performed, although the tipping technique required by spin echo type of experiments has recently been demonstrated.[13]

3.2 Single spin in a quantum dot

Since a spin-based quantum dot quantum computer (QDQC) has been proposed,[3] it is desirable to have estimates of T_1 and T_2 for single electron spins in a QD, and to confirm the estimates experimentally. It is important to mention in this context that the single spin decoherence times should be longer than the macroscopic (averaged over many spins) values of T_1 and T_2, which should thus serve as an upper limit.

Phonon-assisted spin flip rates due to spin-orbit coupling in a single electron GaAs QD have been calculated.[14] It was pointed out that the spin-orbit relaxation mechanisms mentioned above are strongly suppressed in a QD, leading to a long spin-flip time: $T_1 \approx 1$ ms for $B = 1$ T and $T = 0$ K. This result is consistent with recent transport measurements,[15] which indicate that for $T = 150$ mK and $B = 0$-2 T, spin relaxation times (T_1) in a many-electron QD (less than 50 electrons) are longer than at least a few μs. This is encouraging from the perspective of the spin-based solid state QC architecture where spin relaxation times of μs or longer are most likely necessary for large scale QC operation.

ESR combined with transport techniques in principle could be used to probe T_2 in a QD in the Coulomb Blockade regime. It has recently been proposed[16] that magnetic resonance will lead to a peak in the stationary current through a single electron QD. The peak width will yield a lower bound on T_2. This is in essence an ESR measurement with a transport readout.

It is desirable that T_2 for an electron in a single QD is a factor of 10^4 or so greater than the typical gating time in a QDQC.[2] For $B = 1$ T, the Zeeman splitting in a QD is about 0.03 meV, which yields 100 ps for the precession time of one spin, which can be used as the one qubit gate (the two qubit gate time is shorter, $\hbar/J \sim 50$ ps for $J \sim 0.1$ meV). Therefore for quantum error correction to be performed reliably, T_2 for the trapped electron spin needs to be on the μs time scale, which may very well be the case at low enough temperatures in a single electron QD. We note that the existing experimental estimates of free electron spin relaxation time T_2 (or T_2^*) in GaAs (for $T = 1$-4 K) is around 10-100 ns, which is obviously a lower bound since one expects on rather general ground the spin relaxation time of individual electrons confined in QD structures to be longer. Much of the current optimism in the feasibility of spin-based QDQC architecture arises from the expected long electron spin relaxation times (many μs or longer at low temperatures) of GaAs electrons confined in high quality QDs.

4 Decoherence in a quantum dot quantum computer

In a spin-based QC, the spin up and down states are the TLS for a qubit. It is then crucial to explore all relevant degrees of freedom in the solid state environment that couple to the electron spins and determine whether they are sufficiently weak (or can be made so by suitable means) for a QC to work.

Trapped in an ideal QD in a perfect heterostructure, an electron is sitting at the bottom of the conduction band. The major environmental influences come from spin-orbit coupling (therefore phonons) and nuclear spins in the material, assuming dipolar coupling to other electron spins to be weak; otherwise dipolar coupling to other spins has to be accounted for as well.

When we bring two QDs close to each other in order to perform exchange gate operations required for quantum computing,[3] there are additional external influences on the system. One concern is whether Heisenberg exchange Hamiltonian is a complete description of the low energy dynamics of a double dot. Indeed, inherent spin-orbit coupling in 2D QDs leads to an anisotropic exchange, whose effects have to be controlled and, if necessary, corrected.[17,18] It is self-evident that the environmental influences for both single and double QD structures will have to be carefully studied before a functional QDQC (even with just a few qubits) can be fabricated.

To operate a QC, we inevitably need to manipulate the spins (or other form of qubits) using external means: gate voltages, applied magnetic fields, light, microwave, etc. When we introduce these external influences, we inevitably introduce unwanted perturbations. These can be called gate errors, and can be regarded as a form of decoherence. For example, in performing exchange gates in spin-based QDQC, the potential barrier between two neighboring QDs needs to be lowered to allow the exchange interaction to take effect. However, the mixing of the two electrons naturally leads to the possibility of exciting the previously frozen orbital degrees of freedom. It has been pointed out that adiabatic operation of the exchange gate can help suppress the errors caused by the state mixing.[19,20] Of course, the adiabatic condition can never be satisfied exactly in any real situation, therefore the important issue is to estimate the amount of mixing of the higher energy states under realistic conditions. We have recently done a calculation[21] of the time evolution of a two-electron double dot system when the central barrier between the dots is varied, so that we can quantitatively determine the adiabatic condition that a spin-based QDQC has to satisfy. Our results show[21] that for a typical configuration[22,23] of the double dot system for quantum computing, the leakage rate would be reasonably small ($< 10^{-6}$) for gate operation times longer than 50 ps. Thus adiabatic condition is not overly stringent and should not hamper the operation of a QDQC. Note that the optimum value of the gate operation time is constrained from below by the adiabatic condition (i.e. gates should not be too "fast") and from above by the spin relaxation time (i.e. gates should be much faster than the typical spin relaxation time). Our most optimistic reasonable estimate of the applicable range of gating time for spin-based QDQC operations is 50 ps to 1 ns, which is not an unrealistic operation regime.

As the exchange coupling J (the singlet-triplet splitting) is tuned[22] by changing external gate voltage in a QDQC, fluctuations in the gate voltage V will lead to fluctuations in J, thus causing phase errors in the exchange-based swap gate which is crucial for two-qubit operations. We have estimated this error by assuming a simple thermal (white) noise.[22] Under reasonable experimental conditions, the phase error accrued during a swap gate is about 0.01%. This is at the threshold of the currently available quantum error correction codes. To further lower this error rate, one can lower the experimental temperature and set up the gate architecture in such a way as to decrease the sensitivity of the exchange coupling J on the gate voltage V in the operation of a QDQC.

Another possible error in the two-qubit operations of the QDQC architecture is caused by inhomogeneous magnetic fields.[24,25] Magnetic field directly affects spin through Zeeman coupling. In an inhomogeneous field, the Zeeman terms do not commute with the exchange term in the spin Hamiltonian. We have done a detailed analysis[24,25] on how to achieve swap with such a Hamiltonian, and found that there is at the minimum an error proportional to the square of field inhomogeneity in the swap. We have estimated[24] that in GaAs a Bohr magneton can lead to an error in the order of 10^{-6}, which is within the capability of currently available quantum error correction schemes.

In concluding this section, we mention that many of the techniques developed over the last fifty years in the context of NMR and ESR studies should be useful in controlling decoherence (and carrying out error corrections) in spin-based QDQC operations. These include the spin echo and refocusing techniques. In fact, our best estimates for the electron spin relaxation time in GaAs QDs ($\sim \mu$s), the exchange coupling ($\sim$ 0.1 meV), and the gating time (ps to ns) make us guardedly optimistic that a spin-based QDQC architecture may very well be developed in the future. It should, however, be kept in mind in this context that QC architectures[2] based on atomic physics (e.g. trapped ions) and liquid state NMR have extremely long natural decoherence times ($T_2 \sim \mu$s-ms; $T_1 \sim$ minutes-hours) because of the extremely weak environmental coupling in these systems. The main problem in these architectures is not decoherence, but scaling up to more than a few (2-10) qubits which should be relatively easy in semiconductor solid state systems.[4]

5 Mesoscopic decoherence

Finally, we provide a very brief discussion on the issue of electronic decoherence in mesoscopic electronic materials (e.g. metals, semiconductors). Our discussion is necessarily brief (done only in the context of quantum computing), and certainly will not do justice to the vast literature and the great deal of current activity in the subject. We provide this discussion only for the sake of completeness, restricting ourselves entirely to rather elementary considerations.

Electronic decoherence in mesoscopic materials is often characterized by dephasing (or the phase relaxation) time τ_ϕ which shows up in electronic phase coherent processes (i.e. quantum interference phenomena) such as weak localization, conductance fluctuations, and various quantum interference oscillations (i.e. h/e and $h/2e$ oscillations in connected structures). In general, each quantum coherent phenomenon may involve a slightly different (but closely related) definition of τ_ϕ, but in all cases τ_ϕ indicates the typical dephasing time over which quantum interference memory is lost in the system.[26] The dephasing time τ_ϕ in the many-body electronic system is similar in spirit to the dephasing time T_2 in the TLS case although, in contrast to TLS dynamics, the concept of a transverse relaxation time (as T_2 is in the TLS case) is not germane to the mesoscopic problem. The inelastic scattering (or relaxation) time τ_i (sometimes also called quasiparticle lifetime) in the electronic many-body system corresponds loosely to the inelastic spin-flip relaxation time T_1 in the TLS problem. Actually, the loose analogy between T_2 and τ_ϕ on the one hand and T_1 and τ_i on the other hand could be further elucidated. Just as one of the main contributions to $1/T_2$ at low temperatures is the spin-spin (dipolar) interaction, the main contribution to $1/\tau_\phi$ at low temperatures is the electron-electron interaction. Similarly, if one restricts to energy loss from the whole electron system (and not just the individual electrons), then electron-phonon interaction is the main inelastic scattering mechanism in electronic systems, just as spin-lattice relaxation is the inelastic spin-flipping mechanism in T_1.

While τ_i is an inelastic lifetime related to the inverse of the one-electron self-energy, τ_ϕ relates directly to the dephasing process. They are closely connected, but *not* necessarily the same, particularly at low temperatures and in low dimensional systems. In general, elastic scattering by impurities (where the impurities are considered static and immobile) does not *directly* contribute to either τ_i or τ_ϕ. The presence of impurity scattering does, however, have strong *indirect* influence on both τ_i and τ_ϕ, since the electronic motion in mesoscopic systems is diffusive (with a finite transport mean free path) in the presence of impurity scattering. For example, in a 3D metallic system the electron-electron scattering contribution (which is expected to dominate at low temperatures where the phonons are frozen out) to τ_i is $\tau_i^{-1} \sim T^2$ for a ballistic system, and $\tau_\phi^{-1}, \tau_i^{-1} \sim T^{3/2}$ in a diffusive system. Note that in a diffusive 3D system $\tau_\phi \sim \tau_i$ for electron-electron scattering (this is also true for electron-phonon scattering which is important at higher temperatures). This approximate equality of τ_i and τ_ϕ for 3D systems follows primarily from the fact that large energy transfer ($\sim k_B T$) scatterings dominate both τ_i and τ_ϕ in 3D systems. In lower dimensional systems, however, this is not true in general, and we can have $\tau_i \gg \tau_\phi$ with τ_ϕ^{-1} being dominated by very small energy scattering processes which in general do not much affect τ_i^{-1}. The calculated temperature dependence of τ_i and τ_ϕ due to electron-electron scattering in 2D diffusive systems are, however, the same up to a logarithmic correction: $\tau_i^{-1} \sim T \ln T$, $\tau_\phi^{-1} \sim T$. In 1D, on the other hand, $\tau_\phi^{-1} \sim T^{2/3}$ and $\tau_i^{-1} \sim T^{1/2}$. For electron-phonon scattering one typically finds[27] $\tau_{e-ph} \approx \tau_\phi \sim T^{-p}$ with $p \approx$ 1-4 depending on systems and dimensionalities. Usually electron-electron scattering dominates τ_ϕ for $T < 10$ K and electron-phonon scattering dominates at higher temperatures. It may be worthwhile to point out that just as the existence of T_2 (with $1/T_2 \neq 0$) is essential in defining a spin temperature T_s (through spin-spin interaction) in ESR and NMR measurements,[5,6] electron-electron scattering is essential in defining an electron temperature T_e in an excited electron gas. The final equilibrium to the lattice temperature T_L (i.e. $T_s \to T_L$ or $T_e \to T_L$) is achieved in both cases through the interaction with the lattice (i.e. $1/T_1 \neq 0$ or $1/\tau_{e-ph} \neq 0$).

The strong temperature dependence of τ_ϕ ($\sim T^{-3/2}$ in 3D, T^{-1} in 2D, and $T^{-2/3}$ in 1D) at low temperatures is only observed over a rather limited range of temperature in experiments where the measured τ_ϕ in many different (particularly, low dimensional mesoscopic) systems always seems to saturate at low enough temperature and cross over to a constant value of the order of 1 ns (with the crossover temperature of the order of 100 mK).[28] Neither the saturation temperature nor the saturated value of τ_ϕ seems to have any obvious universal behavior. Whether this low temperature τ_ϕ saturation has any fundamental significance or not is currently being debated in the literature. We have nothing to add to this controversy other than to point out that even in the much simpler problem of NMR, actual quantitative calculations of T_2 are rather difficult because so many different processes could cause dephasing at low temperatures. In the τ_ϕ problem, such dephasing processes include, for example, electron heating, magnetic impurities, background radiation, unknown inelastic channels at very low energies, finite system sizes, and actual slight movements of the impurity atoms which are considered to be static in the theory. Indeed, at low enough temperatures, where the dephasing time is already relatively long, any weak coupling to the environment will cause (weak) decoherence and lead to apparent saturation of τ_ϕ at finite effective electron temperatures.

Acknowledgments

This work is supported by ARDA, DARPA, and ONR.

References

1. An extended version of this review is at e-print cond-mat/0108339.
2. M.A. Nielsen and I.L. Chuang, *Quantum Computation and Quantum Information* (Cambridge University Press, New York, 2000).
3. D. Loss and D.P. DiVincenzo, Phys. Rev. A **57**, 120 (1998).
4. B.E. Kane, Nature **393**, 133 (1998); Fortschr. Phys. **48** (2000), 1023.
5. C.P. Slichter, *Principles of Magnetic Resonance* (Springer-Verlag, Berlin, 1990).
6. C.P. Poole Jr, *Electron Spin Resonance* (Interscience, New York, 1967).
7. P. Meystre and M. Sargent III, *Elements of Quantum Optics* (Springer-Verlag, Berlin, 1999).
8. *Optical Orientation*, Modern Problems in Condensed Matter Science, Vol. 8, edited by F. Meier and B.P. Zakharchenya (North Holland, Amsterdam, 1984).
9. J. Fabian and S. Das Sarma, J. Vac. Sci. Technol. B **17**, 1708 (1999).
10. M. Seck, M. Potemski, and P. Wyder, Phys. Rev. B **56**, 7422 (1997); H.W. Jiang and E. Yablonovitch, Phys. Rev. B **64**, R041307 (2001). See also, D. Stein, K. von Klitzing, and G. Weimann, Phys. Rev. Lett. **51**, 130 (1983) for an early ESR measurement in GaAs heterostructures.
11. J.M. Kikkawa and D.D. Awschalom, Phys. Rev. Lett. **80**, 4313 (1998); A.P. Heberle, W.W. Rühle, and K. Ploog, Phys. Rev. Lett. **72**, 3887 (1994).
12. J.A. Gupta, D.D. Awschalom, X. Peng, and A.P. Alivisatos, Phys. Rev. B **59**, R10421 (1999).
13. J.A. Gupta, R. Knobel, N. Samarth, and D.D. Awschalom, Science **292**, 2458 (2001).
14. A.V. Khaetskii, Physica E **10**, 27 (2001); S.I. Erlingsson, Y.V. Nazarov, and V.I. Fal'ko, e-print cond-mat/0104148.
15. T. Fujisawa, Y. Tokura, and Y. Hirayama, Phys. Rev. B **63**, 081304 (2001).
16. H.-A. Engel and D. Loss, Phys. Rev. Lett. **86**, 4648 (2001).
17. K.V. Kavokin, Phys. Rev. B **64**, 075305 (2001).
18. N.E. Bonesteel, D. Stepanenko, D.P. DiVincenzo, e-print cond-mat/0106161; G. Burkard and D. Loss, e-print cond-mat/0108101; R. Sriharsha, X. Hu, and S. Das Sarma, unpublished.
19. G. Burkard, D. Loss, and D.P. DiVincenzo, Phys. Rev. B **59**, 2070 (1999).
20. J. Schliemann, D. Loss, A. MacDonald, Phys. Rev. B **63**, 085311 (2001).
21. X. Hu and S. Das Sarma, to be published.
22. X. Hu and S. Das Sarma, Phys. Rev. A **61**, 2301 (2000).
23. X. Hu and S. Das Sarma, Phys. Rev. A (2001, in press), e-print cond-mat/0101102; cond-mat/0102019.
24. X. Hu, R. de Sousa, and S. Das Sarma, Phys. Rev. Lett. **86**, 918 (2001).
25. R. de Sousa, X. Hu, and S. Das Sarma, Phys. Rev. A (2001, in press), e-print cond-mat/0103410.
26. B.L. Altshuler, A.G. Aronov, and D.E. Khmel'nitskii, Solid State Commun. **39**, 619 (1981).
27. D. Belitz and S. Das Sarma, Phys. Rev. B **36**, 7702 (1987).
28. P. Mohanty, E.M.Q. Jariwala, and R.A. Webb, Phys. Rev. Lett. **78**, 3366 (1997).

QUANTUM-STATE MANIPULATIONS IN A COOPER-PAIR BOX

Y. NAKAMURA, T. YAMAMOTO AND J. S. TSAI
NEC Fundamental Research Laboratories, Tsukuba, Ibaraki 305-8501, Japan
E-mail: yasunobu@frl.cl.nec.co.jp

YU. A. PASHKIN
CREST, Japan Science and Technology Corporation, Kawaguchi, Saitama, 332-0012, Japan
E-mail: pashkin@frl.cl.nec.co.jp

Quantum coherent properties of an artificial two-level system in a Josephson-junction circuit were studied experimentally. Responses of the quantum state to a sequence of gate-voltage pulses were measured in time-ensemble measurements, and free-induction decay and "charge echo" signals were obtained. It was found that the decoherence of the two-level system is dominated by the dephasing due to low-frequency charge fluctuations.

1 Introduction

Quantum coherence has turned out to be an important concept not only in fundamental physics but also in applications such as quantum computation and quantum communication, which are expected to have much more power compared to their classical counterparts.[1] However, the apparent difficulty for their implementations is the fact that quantum information processing should be completed coherently. Even with an aid of quantum error correction scheme,[2,3] the requirement for the coherence in the quantum bit (qubit) would be very strict. Therefore, regardless of the physical systems for the implementation of a qubit, it is important to know their coherent properties and to find a way to preserve the coherence for a longer time. Two-level systems such as those found in atomic states and nuclear spin states, have a long coherence time and have been used to demonstrate quantum-gate operations among a few to several qubits.[4,5] On the other hand, many kinds of solid-state implementations and integrations of qubits have been proposed with emphasis on the possible scalability to a large-scale processor.[6] But generally those solid-state qubits are coupled to their environment rather strongly, and we have to check their coherence in real devices.

Josephson-junction qubit is one of the most promising candidates for a solid-state qubit.[7] There are two types of qubit depending on which, either charge or phase, degree of freedom is used in the two-level system. Both of them can be fabricated with the present nanotechnology and are expected to have relatively long coherence time. The charge qubit, a Cooper-pair box, uses two distinct charge states in a small superconducting electrode connected to a reservoir via a Josephson junction,[8,9,10] while the phase qubit uses two phase states in a small superconducting loop interrupted by Josephson junction(s).[11,12,13] Coherent superposition of the two charge states[14,15] and that of two phase states[16,17] have been observed. Recently, coherent control of the charge qubit has been also demonstrated.[18] However, the decoherence time and the decoherence mechanism are not yet fully understood and need to be studied in detail.

To determine the decoherence time, an ensemble of quantum-state measurements is necessary. Because a single measurement on a single quantum state gives us only a probabilistic answer, we have to take the statistics. For example, in nuclear magnetic resonance (NMR) experiments,[19] a large number of identical spins are used, and an average magnetization signal of the ensemble is measured. However, as an expense of the ensemble measurement, there could be an additional contribution of inhomogeneities to the observed coherence signal; because of the spatial inhomogeneities in the magnetic field, spins precess with different velocities and the ensemble is dephased. This "inhomogeneous dephasing" does not directly mean the decoherence of each spin. However, even if it were possible to use only one spin for a quantum operation, the result would be effectively decohered in a probabilistic sense, if we are not able to precisely calibrate the external parameters (like the magnetic field) in advance of the quantum operation.

On the other hand, in our experiments we utilize a single two-level system in a Cooper-pair box. For an ensemble, identical quantum operations are repeated on identical initial states, and read-out signals of the final states are averaged over the repetition. So, this is a time-ensemble measurement, and thus, in addition to the intrinsic decoherence, there could be dephasing due to temporal inhomogeneities, that is, time-dependent fluctuations of the parameters. Even if the fluctuation is very slow, it results in uncertainty in the parameters and thus gives rise to an error in the quantum operation, unless the parameters are calibrated just before the operation. Therefore, this inhomogeneous dephasing is relevant in the context of quantum computing and should not be ignored.

In NMR, a clever technique called spin echo was invented by Hahn half a century ago. [20] In this technique, the dephasing due to inhomogeneities is cancelled out by using a phase-flip operation, and the intrinsic decoherence time is observed. In the present work, we adapt the idea to our two-level system involving two charge states and call it "charge echo".[22] We find that the dephasing in the time ensemble is also cancelled, but not perfectly. The possible origins of the fluctuations and the prospects for implementations of quantum computing will be discussed.

2 Experiment

2.1 Cooper-Pair Box

A Cooper-pair box device[18] consists of a small superconducting "box" electrode connected to a reservoir electrode via a Josephson junction and a gate electrode capacitively coupled to the box. This device works as a qubit.[8] The relevant two states are the two lowest-energy charge-number states, say, $|n=0\rangle$ and $|1\rangle$, which differ by one Cooper pair in the box. Here n represents the excess number of Cooper pairs in the box. Other charge-number states with different number of Cooper pairs have higher energies and thus can be neglected as long as the Josephson energy E_J and the thermal energy k_BT are much smaller than the single-electron charging energy E_C of the box. Moreover, quasiparticle tunneling across the Josephson junction is suppressed because of the superconducting gap in the electrodes.

An effective Hamiltonian of the system is given as

$$H = \frac{1}{2}\delta E(Q_0)\sigma_z - \frac{1}{2}E_J\sigma_x, \qquad (1)$$

where $\delta E(Q_0) \equiv 4E_C\,(Q_0/e-1)$, and σ_z and σ_x are Pauli matrices. The energy dis-

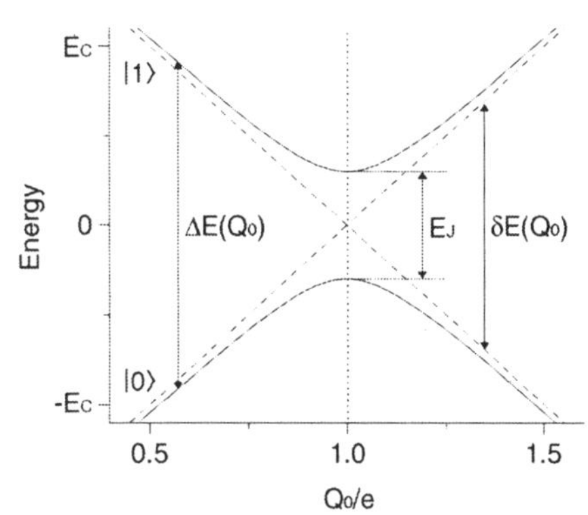

Figure 1. Energy-level diagram of a Cooper-pair box.

persion of the two-level system $\Delta E(Q_0) \equiv \sqrt{\delta E(Q_0)^2 + E_J^2}$ is shown in Fig. 1 as a function of the gate-induced charge Q_0 which is given by a product of the gate capacitance and the applied gate voltage.

2.2 Initialization

Initialization of the quantum states is realized by applying a gate charge far away from the degeneracy point at $Q_0 = e$. The upper state $|1\rangle$ eventually relaxes into the ground state $|0\rangle$ via an inelastic Cooper-pair tunneling process or, more efficiently, via the measurement process described below.

2.3 Quantum-State Manipulation

The quantum state is nonadiabatically manipulated by applying a fast gate-voltage pulse.[18] As it is clear from Eq. (1), evolution of the quantum state can be mapped on the motion of a fictitious spin-$\frac{1}{2}$ in an effective magnetic field $\vec{B} = (E_J, 0, -\delta E(Q_0))$. For example, if the gate charge Q_0 is suddenly brought to e, i.e., if $\delta E(Q_0) = 0$, then the spin precesses around x-axis, while, if $|\delta E(Q_0)| \gg E_J$, the spin rotates approximately around z-axis. By combining a sequence of gate-voltage pulses and delay times between them, complicated quantum-state manipulations can be realized.

2.4 Read-out

The quantum state after the manipulation is measured by using an additional probe electrode attached to the box via a highly resistive tunnel junction. The probe electrode is voltage-biased appropriately so that two quasiparticles tunnel out from the box only from the $|1\rangle$ state but not from the $|0\rangle$ state. Therefore, the two charge states can be distinguished. Moreover, after the read-out process, the quantum state is initialized to the $|0\rangle$ state automatically. The probe is always active even during the manipulation of the quantum state. So, in order to avoid too much decoherence due to the read-out, the quasiparticle tunneling rate through the junction should be small enough.

The read-out signal, only two electrons that tunnel through the junction, is too small for a single-shot measurement. Hence, the read-out process has to be repeated on identically prepared quantum states in order to accumulate the tunneling electrons and to obtain a measurable dc current through the probe junction. Therefore, the measured result is an average over a time-ensemble of quantum-state manipulations, although our experiment is on a single two-level system. The averaging number is typically 10^5-10^6.[21]

3 Results and Discussions

3.1 Coherent Oscillations

First we measured a response to a single-pulse manipulation.[18] The pulse was adjusted to bring the system to the point where $\delta E(Q_0) = 0$ during the pulse width Δt. The final state is expected to be a superposition of $|0\rangle$ and $|1\rangle$ states, and the measured current signal proportional to the population of the $|1\rangle$ state, or to $\langle\sigma_z\rangle+1$ in the spin representation, shows oscillating behavior as a function of Δt (Fig. 2), reflecting coherent oscillations between the two charge states. The oscillations are visible up to nearly 5 ns, although the decay envelope is not clearly seen.

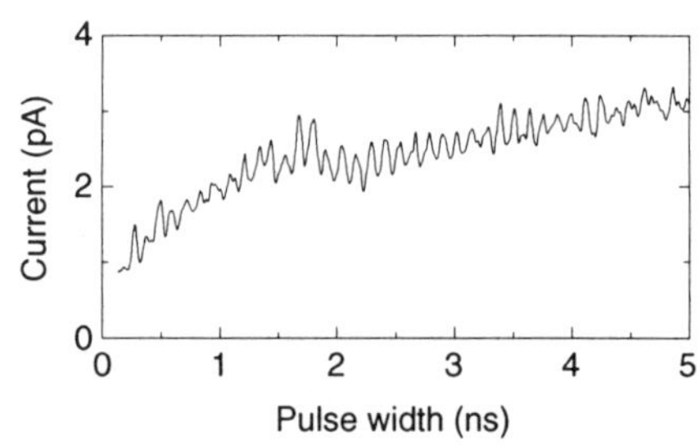

Figure 2. Pulse-induced current as a function of the pulse width in the single-pulse experiment.

3.2 *Free-Induction Decay*

In contrast to the above experiment, a free-induction decay (FID) experiment provides information about the quantum-state evolution at gate charges Q_0 far from the degeneracy point. In the FID experiment, two pulses are used. The first pulse prepares a fifty-fifty superposition of the two charge states. During the delay time t_d between the pulses, the quantum state acquires phase $\Delta E(Q_0)t_d/\hbar$ due to the energy difference between the two eigenstates. The second pulse projects the phase information on the measurement basis, i.e., σ_z-basis.

Figure 3 shows the FID signal as a function of the delay time. The oscillation period is about 15 ps, which agrees with $h/\Delta E(Q_0)$. The signal decays quite rapidly within a few hundred picoseconds. This time scale is much shorter than that observed in the above experiment. The large difference can be explained if dephasing due to charge fluctuations is a dominant origin of the decay. At the degeneracy point, the energy dispersion is flat (Fig. 1), and the phase evolution velocity is immune to the charge fluctuations. On the other hand, at the point far from the degeneracy, $\Delta E(Q_0)$ is sensitive to the charge fluctuations, and thus the system is strongly dephased by the fluctuations. Because our measurement is a time-ensemble measurement averaged over a long data-acquisition time (20 ms in the present experiment), even very low-frequency fluctuations down to about 50 Hz can contribute to the dephasing as temporal inhomogeneities in the time-ensemble.

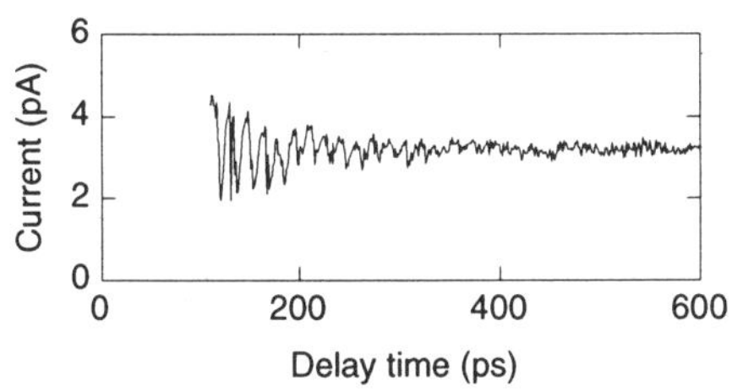

Figure 3. Free-induction decay signal.

3.3 *Charge Echo*

In order to suppress the spurious dephasing in the ensemble, we performed echo experiment by using a pulse sequence consisting of three pulses. In between the two pulses we used in the FID experient, we insert a pulse which reverses the phase accumulated in the first half of the delay time. The phase evolution is cancelled by the one during the second half of the delay time as long as the charge fluctuation during the delay time is constant. Thus, we can eleminate the dephasing due to low-frequency fluctuations.

In Fig. 4(a), a charge-echo signal is shown as a function of the temporal position of the second pulse. As the position is shifted from the middle of the delay time between the first and the third pulses, the compensation of the phase becomes incomplete, and the signal oscillates and decays (in a few-hundred picoseconds, data not shown). The amplitude of the echo signal is plotted in Fig. 4(b) as a function of the delay time. The decay time is lengthened compared to that of the FID experiment, indicating that the echo technique works successfully and that the effect of a low-frequency part of the fluctuations is cancelled. In other words, these results clearly show that the dephasing of the present two-level system is dominated by low-frequency charge fluctuations.

The echo signal decays within about 5 ns. It is worth comparing the result with estimations of decay due to several possible reasons.[22] Decoherence of the system due to the read-out process is expected to take place at the quasiparticle tunneling rate through the probe junction. This rate is estimated as about $(8\ \text{ns})^{-1}$, which is longer than the observed time scale. Estimation of decoherence due to electromagnetic environment of the device gives a decay time of about

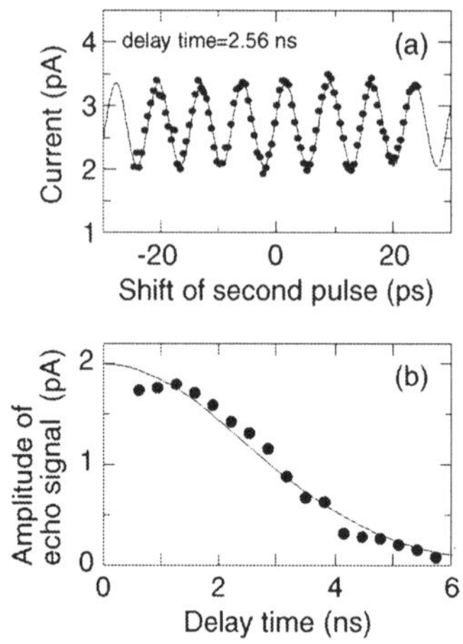

Figure 4. (a) Charge-echo signal as a function of the shift of the second pulse. (b) Decay of the amplitude of the charge-echo signal as a function of the delay time. The fit is a gaussian curve.

100 ns. In addition, the well-known $1/f$ background charge noise,[23,24] which is believed to be caused by fluctuations of charges in the substrate and/or the tunnel barrier, may give rise to the dephasing. Although reported measurements of the $1/f$ noise have been restricted to the low-frequency range, we assume that the $1/f$ spectrum extends to the infinite frequency and calculate the dephasing properties.[25] In the case of a $1/f$ spectrum, the coherence decays in a gaussian manner similar to the observed decay, and a typical amplitude of the $1/f$ spectrum leads to a time scale consistent with the experimental result. These observations suggest that the $1/f$ charge noise is a dominant dephasing source in the present two-level system.

3.4 Prospects for Two-Qubit Gates and others

In the echo experiment, the charge qubit is effectively decoupled from the external fluctuations by the second pulse which reverses the phase evolution. This is a primitive demonstration of a technique known as the decoupling in NMR[26] or as the bang-bang control proposed by Viola *et al.*[27] Although the decoupling is not perfect, the decoherence time of the qubit is elongated by a large amount. Two-qubit gates, which are key elements for universal quantum information processing and seemed to be very difficult because of the short dephasing time observed in the FID experiment, may be successfully demonstrated if the echo technique is combined with the pulse sequence for the two-qubit gate.

On the other hand, the echo technique is not so effective for high-frequency fluctuations. If the $1/f$ noise really extends to the high frequency, the dephasing time even after applying the echo technique cannot be so long. The only way to achieve long coherence time required for quantum computing would be to reduce the amount of the $1/f$ noise. It is important to clarify the exact origin of the $1/f$ noise and find a way to eliminate the noise.

The decoherence due to electromagnetic environment is relatively easy to be dealt with. By designing the environment appropriately, the decoherence time could be as long as microseconds.[7] In addition, noise from the biasing circuit should be completely filtered. Attention should be paid not only to the high-frequency filtering to suppress relaxation processes but also to the low-frequency filtering to avoid additional dephasing.

The measurement probe we used is not ideal because it cannot be switched off during the quantum-state manipulations and decoheres the system eventually. A detector which can be switched on and off at will is needed and is under development.[28,29,30] Such a detector is also expected to have a good resolution for a single-shot read-out, which is important for measuring entangled states.

4 Conclusion

We have demonstrated quantum-state manipulations in a Cooper-pair box and investigated decoherence in the two-level system. In the charge-echo experiment, coherence was recovered largely compared to the case in the FID experiment, indicating that low-frequency energy-level fluctuations contribute to the dephasing in the time-ensemble measurement. It is suggested that $1/f$ background charge noise is a source of the fluctuations.

Acknowledgments

This work has been supported by the CREST project of Japan Science and Technology Corporation (JST).

References

1. M. A. Nielsen and I. L. Chuang, *Quantum computation and quantum information* (Cambridge University Press, Cambridge, 2000).
2. P. Shor, Phys. Rev. A **52**, 2493 (1995).
3. A. Steane, Proc. Roy. Soc. Lond. A **452**, 2551 (1996).
4. C. A. Sackett *et al.*, Nature **404**, 256 (2000).
5. L. K. Vandersypen *et al.*, Phys. Rev. Lett. **85**, 5452 (2000).
6. See articles in Fortschr. Phys. **48**, No. 9-11 (2000).
7. For review, Yu. Makhlin, G. Schön, and A. Shnirman, Rev. Mod. Phys. **73**, 357 (2001).
8. A. Shnirman, G. Schön, and Z. Hermon, Phys. Rev. Lett. **79**, 2371 (1997).
9. D. V. Averin, Solid State Commun. **105**, 659 (1998).
10. Yu. Makhlin, G. Schön, and A. Shnirman, Nature **386**, 305 (1999).
11. L. B. Ioffe *et al.*, Nature **398**, 133 (1999).
12. J. E. Mooij *et al.*, Science **285**, 1036 (1999).
13. G. Blatter, V. B. Geshkenbein, and L. B. Ioffe, Phys. Rev. B **63** 174511 (2001).
14. V. Bouchiat *et al.*, Phys. Scr. **T76**, 165 (1998).
15. Y. Nakamura, C. D. Chen, and J. S. Tsai, Phys. Rev. Lett. **79**, 2328 (1997).
16. J. R. Friedman *et al.*, Nature **406**, 43 (2000).
17. C. H. van der Wal *et al.*, Science **290**, 773 (2000).
18. Y. Nakamura, Yu. A. Pashkin, and J. S. Tsai, Nature **398**, 786 (1999); Physica B **280**, 405 (2000).
19. See *e.g.*, C. P. Slichter, *Principles of Magnetic Resonance, 3rd ed.* (Springer-Verlag, New York, 1990).
20. E. L. Hahn, Phys. Rev. **80**, 580 (1950).
21. For experimental details, Y. Nakamura and J. S. Tsai, J. Low Temp. Phys. **118**, 765 (2000).
22. Y. Nakamura *et al.*, in preparation.
23. G. Zimmerli *et al.*, Appl. Phys. Lett. **61**, 237 (1992).
24. A. B. Zorin *et al.*, Phys. Rev. B **53**, 13682 (1996).
25. Existence of the $1/f$ tail at about 100 GHz is supported by a recent experiment on a single-electron pump: M. Covington *et al.*, Phys. Rev. Lett. **84**, 5192 (2000).
26. D. W. Leung *et al.*, Phys. Rev. A **61**, 042310 (2000).
27. L. Viola and S. Lloyd, Phys. Rev. A **58**, 2733 (1998); L. Viola, E. Knill, and S. Lloyd, Phys. Rev. Lett. **82**, 2417 (1999).
28. M. H. Devoret and R. J. Schoelkopf, Nature **406**, 1039 (2000).
29. A. Cottet *et al.*, to be published in the Proceedings of MQC2 (Naples, 2000).
30. A. Aassime *et al.*, Phys. Rev. Lett. **86**, 3376 (2001).

QUANTUM STATE ENGINEERING AND JOSEPHSON JUNCTIONS: CHARGE AND FLUX DETECTORS

YU. MAKHLIN[1,2], G. SCHÖN[1,3], AND A.SHNIRMAN[1]

[1] *Institut für Theoretische Festkörperphysik, Universität Karlsruhe, D-76128 Karlsruhe, Germany*
[2] *Landau Institute for Theoretical Physics, Kosygin st. 2, 117940 Moscow, Russia*
[3] *Forschungszentrum Karlsruhe, Institut für Nanotechnologie, D-76021 Karlsruhe, Germany*
E-mail: makhlin,schoen,shnirman@tfp.physik.uni-karlsruhe.de

We discuss parameters that quantify efficiency of devices suggested as detectors of the quantum state of Josephson-junction charge and flux qubits. These parameters are used to compare various detectors. We estimate the dephasing rate by a turned-off detector and show that for a dc-SQUID and a SET it vanishes very fast at low temperatures, suggesting their possible use in experiments.

1 Introduction

Low-capacitance Josephson junctions offer a particularly promising way to realize quantum bits [1] for quantum information processing. They can be embedded in electronic circuits and scaled up to large numbers of qubits. Two kinds of devices, which exploit the coherence of the superconducting state, have been proposed: they use either charge or flux/phase macroscopic quantum degree of freedom. Single- and two-qubit quantum manipulations can be controlled by gate voltages or magnetic fields, by methods established for single-charge devices or the SQUID technology, respectively. In flux qubit devices an important milestone, the observation of superpositions of different flux states in the system eigenstates, has been achieved. [2,3] In charge qubits even coherent oscillations between the eigenstates have been demonstrated in the time domain. [4] Further work concentrated on understanding the decoherence mechanisms in these devices and on the design of Josephson circuits that diminish the dephasing effect of the environment, enable manipulations of many qubits and further release requirements on the circuit parameters.

A crucial issue is the development of quantum detectors, the devices that perform the measurement of the qubit's state. [1] They are needed to read out the final state after a quantum computation or in the process of quantum error correction. Moreover, even in first experiments with simple systems high-quality detectors are required to demonstrate the result of manipulations. The detectors used so far had the advantage of being easy to realize but they lacked some properties needed for reliable and accurate quantum measurement. In this paper we discuss the relevant figures of merit that can be used to assess the quality of quantum detectors and analyze several proposed devices in terms of these parameters.

Fig. 1 presents typical flux and charge detectors. In Fig. 1a a dc-SQUID is shown, which is widely used as a sensitive magnetometer. The critical current of the SQUID is controlled by the magnetic flux through its loop. Hence it is sensitive to the state of a flux qubit coupled inductively to the SQUID as in Fig. 1a. Applying a current above the critical value allows one to read the qubit's state from the voltage response.

As an example of a quantum charge detector consider a single-electron transistor (SET; see Fig. 1b). When a transport voltage above the Coulomb blockade threshold is applied, a dissipative current starts to flow. Its value is sensitive to the state of a charge qubit, capacitively coupled to the SET's central island, thus allowing the measurement.

In the next section we discuss the de-

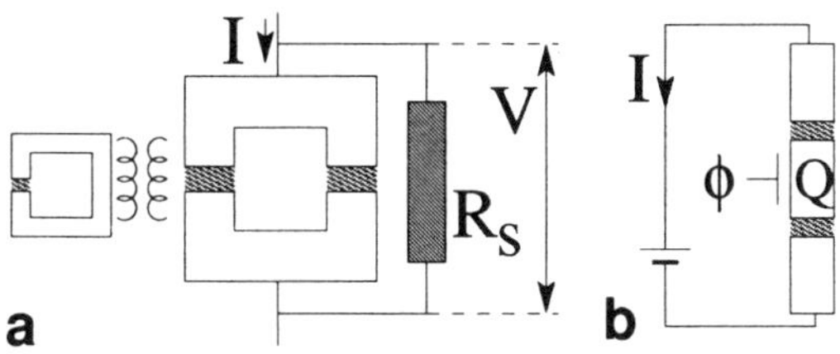

Figure 1. Quantum detectors of (a) flux (dc-SQUID) and (b) charge (SET).

scription of the quantum measurement based on the linear amplifier theory and the master equation for the density matrix. Using this description we consider several charge and flux readout devices and compare their properties. In particular, we analyze the effect of the detectors on the qubit state during manipulations, when in real devices the qubit-detector coupling cannot be switched off completely.

2 Linear amplifiers and the master equation

Three time scales characterize a quantum measurement process. [1] During a quantum measurement a detector is coupled to a qubit in such a way that the value of a macroscopic observable of the detector is influenced by the qubit's state (the charge or flux influences the output current or voltage). Monitoring the output signal for a sufficiently long *measurement* time τ_{meas} allows one to distinguish between two qubit's states. Further, the coupling introduces an additional noise from the detector into the qubit's dynamics which disturbs its quantum coherence. A proper quantum measurement requires the existence of a preferred qubit basis in which the measurement is performed. [5] In this basis the off-diagonal entries of the qubit's density matrix (DM) vanish after a short dephasing time τ_φ. The occupations of the basis states are changed by detector-induced transitions only after a longer mixing time $\tau_{\mathrm{mix}} \geq \tau_\varphi$. Quantum mechanics requires that the readout destroys coherence of the quantum state, $\tau_{\mathrm{meas}} \geq \tau_\varphi$. On the other hand the readout should be performed before the information about the initial occupations of the basis states is lost. Hence a good measurement is achieved if

$$\tau_\varphi \leq \tau_{\mathrm{meas}} \ll \tau_{\mathrm{mix}} \,.$$

2.1 Linear amplifiers

One of the approaches used to analyze the mutual influence of the qubit and detector treats the latter as linear amplifiers. This approach was developed in the context of quantum optics and applied to nanodevices only later. [6,7] It links the relevant characteristics of a detector to its noise properties.

Consider an amplifier with the output signal I and input signal ϕ, which is coupled to an observable Q of the detector via a term ϕQ (the notations are motivated by the case of a SET, see Fig. 1b). The linear response of the output signal to variations at the input is characterized by the gain coefficient $\lambda \equiv d\langle I\rangle/d\phi$. When used as a detector, the device is usually operated in a dissipative, nonequilibrium regime. With the coupling to a qubit turned on, the input signal is $\phi \propto \sigma_z$, and the coupling is $c\sigma_z Q$ (with c being a coupling constant). First, consider the case when the tunneling between the basis states is suppressed, $\mathcal{H}_{\mathrm{qb}} = -\frac{1}{2}\Delta E \sigma_z$. In this case the fluctuations of Q dephase the qubit with rate

$$\tau_\varphi^{-1} = \frac{c^2}{\hbar^2} S_Q \,. \tag{1}$$

The symbols S_Q and S_I (introduced below) stand for the noise power of the corresponding observable if the amplifier is decoupled from the qubit: $S_Q = 2\,\langle Q^2_\omega\rangle$ and $S_I = 2\,\langle I^2_\omega\rangle$. Here we assume a white-noise spectrum at the relevant frequencies. The two basis states of the qubit produce output signals $I^{0/1} = \bar{I} \pm \Delta I/2$, differing by

$$\Delta I = 2c\lambda \,. \tag{2}$$

They can be distinguished after the measurement time

$$\tau_{\text{meas}}^{-1} = \frac{(\Delta I)^2}{4S_I} \; . \tag{3}$$

Hence, the two times are related by

$$\frac{\tau_{\text{meas}}}{\tau_\varphi} = \frac{S_Q S_I}{\hbar^2 \lambda^2} = \frac{S_Q S_\phi}{\hbar^2} \; . \tag{4}$$

In the last form of Eq. (4) $S_\phi \equiv S_I/\lambda^2$ is the output noise in terms of the input, i.e., the noise which should be applied to the input to produce the noise S_I at the output.

If the tunneling is turned on, $\mathcal{H}_{\text{qb}} = -\frac{1}{2}\Delta E(\cos\eta\,\sigma_x + \sin\eta\,\sigma_z)$, both the measurement rate (3) and the rate of pure dephasing (1) acquire an additional factor $\cos^2\eta$, while their ratio (4) persists. Apart from that, the finite tunneling introduces mixing, with rate

$$\tau_{\text{mix}}^{-1} = \frac{c^2}{\hbar^2} S_Q \sin^2\eta \; . \tag{5}$$

The quantity $\sqrt{S_Q S_\phi}$ is proportional to the "noise energy", discussed in Ref. [7], measured in units of the energy quanta at the given frequency. For the noise spectra S_Q and S_ϕ one can obtain an inequality, similar to the Heisenberg uncertainty principle: [8] $S_Q S_\phi \geq \hbar^2$. By virtue of Eq. (4) this relation coincides with the constraint

$$\tau_\varphi \leq \tau_{\text{meas}} \; . \tag{6}$$

The quantum limit is (nearly) reached in several measurement devices: For a quantum point contact, an overdamped dc-SQUID, a resistively shunted superconducting SET, or a normal SET in the cotunneling regime [9] the following two relations hold under certain conditions:

$$S_I S_Q \approx |S_{IQ}|^2 \tag{7}$$

and

$$\lambda \approx \text{Im}\, S_{IQ} \; . \tag{8}$$

These requirements for a quantum limited measurement device are not valid in general. E.g., the $\approx$ sign in (7) should in general be replaced by the $\geq$ sign. Even for the detectors where they were found they break down, e.g., at finite temperatures or at bias voltages close to the Coulomb blockade threshold [9].

Eq. (8) implies a vanishing reciprocal response coefficient λ', i.e., no response of the input to the output. While in equilibrium Onsager's relations imply $\lambda = -\lambda'$, such an asymmetry can arise in a nonequilibrium stationary state. It is a characteristic feature of linear measuring devices. [8]

2.2 Master equation

The application of the above formulae requires a calculation of the noise power of the detector. It can be performed in the linear-response approximation [6] as long as the latter applies. Apart from the requirement of the operation in the linear regime, the linear amplifier theory needs that the response time of the detector is much shorter than other relevant time scales. [6] In a more general situation careful analysis of the time evolution of the density matrix of the coupled system (qubit+detector) is needed. This can be carried out through the derivation of the master equation for this time evolution. Applications to the measurement by a quantum point contact (QPC) or a SET demonstrate that one can go beyond the linear approximation (consider measurements near a threshold bias) and to time scales shorter than the tunneling time of electrons in the QPC or SET (its response time) and still describe the evolution by compact equations. Furthermore, the derivation of the master equation and its solution allow one to understand how the states of the qubit and the detector become entangled and the information is transfered between them.

For instance, in the case of a QPC (and a SET in certain limits) tracing out microscopic degrees of freedom one arrives [10,5] at a closed set of equations for the entries $\rho^{ij}(m)$ of the reduced DM, which are diagonal in

m, the charge that has tunneled through the contact (i,j refer to the qubit's charge basis). If $\Gamma^{0/1} = \bar{\Gamma} \pm \Delta\Gamma/2$ are the tunneling rates through the QPC for two qubit's states, the equation for the Fourier transform $\hat{\rho}(k) = \sum_m e^{-ikm}\hat{\rho}(m)$ reads

$$\frac{d}{dt}\hat{\rho} + \frac{i}{\hbar}[\mathcal{H}_{\rm qb}, \hat{\rho}] = \left[\bar{\Gamma}\hat{\rho} + \frac{1}{4}\Delta\Gamma\{\sigma_z, \hat{\rho}\}\right] \times \\ \times (e^{-ik} - 1) - \frac{1}{4}\gamma_\varphi e^{-ik}\{\sigma_z, [\sigma_z, \hat{\rho}]\}, \quad (9)$$

where $\gamma_\varphi \equiv \frac{1}{2}(\sqrt{\Gamma^0} - \sqrt{\Gamma^1})^2$. From a solution $\hat{\rho}(m,t)$ we can obtain by further reduction the 2×2 DM of the qubit, $\sum_m \hat{\rho}(m,t)$, the charge distribution $P(m,t) = \mathrm{tr}\,\hat{\rho}(m,t)$, and other statistical characteristics of the current. [5]

An additional advantage of the master-equation approach is that it allows one to derive [11,1] the so called conditional master equation. [12] This approach aims at the description of the qubit's dynamics conditioned on the values of the output signal $I(t)$ at earlier times. As a consequence this approach allows one to produce typical, fluctuating current-time patterns that one can identify in a given experiment.

3 Efficiency of a quantum detector

Recently, several devices performing quantum measurements have been analyzed. Apart from SETs in the sequential tunneling regime [13] they include SETs in the co-tunneling regime, [9] superconducting SETs (SSETs) and dc-SQUIDS, [6] as well as quantum point contacts (QPC). [14] All these devices have a common characteristic feature: they are dissipative systems whose response (conductance, resistance) depends on the state of a qubit coupled to them.

The efficiency of a quantum detector has several aspects. From a practical point of view the most important is the ability to perform a strong, single-shot measurement which requires that the mixing is slower than the read-out, $\tau_{\rm mix} \gg \tau_{\rm meas}$. In the SET this limit can be reached as was demonstrated in recent experiments with an rf-SET in Chalmers. [15] At the same time for flux qubits the currently discussed detectors require many measurements ("shots"). [18]

A further important figure of merit is the ratio of the dephasing and measurement times. Indeed, the ratio $\tau_{\rm mix}/\tau_\varphi$ is to a large extent fixed by qubit's properties. Hence to achieve an accurate measurement ($\tau_{\rm meas} \ll \tau_{\rm mix}$) the detector should be optimized to be close to the quantum limit $\tau_{\rm meas} = \tau_\varphi$. For a SET coupled to a charge qubit the dephasing time is (much) shorter than the measurement time. This means that the information becomes available later than it would be possible in principle. In this sense the efficiency of the SET in the sequential tunneling regime is less than 100%. [13,12] The reason for the delay is an entanglement of the qubit not only with the output degree of freedom (current) but also with microscopic degrees of freedom in the SET. To illustrate this point consider a situation where the initial state of the system $(a|0\rangle + b|1\rangle)\,|\chi\rangle\,|m=0\rangle$ evolves into $a|0\rangle\,|\chi_0\rangle\,|m_0\rangle + b|1\rangle\,|\chi_1\rangle\,|m_1\rangle$, where $|\chi\rangle$ stands for the quantum state of the uncontrolled environment. One can imagine a situation when $m_0 = m_1$, but $|\chi_0\rangle$ and $|\chi_1\rangle$ are orthogonal. Then the dephasing has occurred but no measurement has been performed. It is interesting to note that the ratio $\tau_\varphi/\tau_{\rm meas}$ grows if the SET is biased in an asymmetric way, creating a strong asymmetry in the tunneling rates, e.g. $\Gamma_{\rm L} \ll \Gamma_{\rm R}$. In other measurement devices the ratio $\tau_\varphi/\tau_{\rm meas}$ may be close to 1. This includes quantum point contacts [14] as well as SETs in the co-tunneling regime. [9] The common feature of these three examples is that the device consists of one junction or effectively reduces to it. It should be kept in mind, however, that a large ratio of dephasing and measurement times is not the only figure of merit. For instance, in a SET in the cotunneling regime the current is low and more difficult to detect as compared to a SET in the sequential tunneling regime.

Another desired property is a low back-action noise of the meter in the off-state. The corresponding dephasing rate $\tau_\varphi^{\mathrm{off}}$ is discussed in the next section.

4 Dephasing in the off-state

For nano-electronic realizations of quantum bits, it is hard to realize a coupling between the qubit and the detector that can be turned on and off when needed. Rather, one can switch between the dissipative and equilibrium states of the detector. When the detector is in a non-dissipative state it only weakly affects the qubit whereas in a dissipative state (i.e., during the measurement) the dephasing of the qubit's state is enhanced. Still, even when the transport voltage or current, which controls the detector, is switched off, the noise of the meter dephases the qubit. In this section we analyze this characteristic of the device.

One example, where the estimate of the corresponding dephasing rate is important, is the dc-SQUID used as a quantum detector of the flux state. Two approaches have been discussed, which use either an underdamped or an overdamped SQUID. In an underdamped SQUID the shunt (see Fig. 1a) is absent or has a high resistance, $R_s \gg \sqrt{E_C/E_J}\,R_K$, where E_J and E_C are the characteristic Josephson and charging energies of the SQUID. The higher the resistance R_s, the lower is the noise of the SQUID in the superconducting regime (when no measurement is performed). Therefore, in spite of the permanent presence of the SQUID, the coherent dynamics of the qubit suffers only weak dephasing. To read out the state of the qubit, the current in the SQUID is ramped, and the value of current is recorded where the SQUID switches to the dissipative regime. As this switching current depends on the flux through the SQUID, information about the state of the qubit is obtained. Unfortunately, the current switching is a random process, fluctuating even when the external flux in the SQUID is fixed. For currently available system parameters this spread is larger than the difference in switching currents corresponding to the two states of the qubit. Therefore only statistical (weak repeated) measurements are possible in this regime. [3,18]

The second strategy is to use overdamped SQUIDs, with $R_s < \sqrt{E_C/E_J}\,R_K$. When the bias current exceeds the critical value, the voltage which develops across the shunt resistor depends on the external flux in the SQUID. Thus by measuring this voltage one learns about the state of the qubit. Recently Averin [6] analyzed continuous measurements in this regime and obtained the input and output noise characteristics which determine the relevant time scales τ_φ, τ_{meas}, τ_{mix}. The main disadvantage of this strategy is that the SQUID induces dephasing during the periods of coherent manipulations when no measurement is performed. The question still remains to be settled whether a reasonable compromise between the underdamped and the overdamped limits can be found. Hence the estimate of the dephasing effect of the overdamped SQUID in the off-state is an important issue. Taking into account the equilibrium fluctuations of the SQUID's flux degree of freedom coupled to the environment, we obtain [16] the dephasing rate $(\tau_\varphi^{\mathrm{off}})^{-1} = (12/\pi^2)((I_{c,\uparrow} - I_{c,\downarrow})^2/eI_c)(R_K/R)^2(T/8\pi E_J)^3$ at low temperatures. Here $I_{c,\uparrow}$, $I_{c,\downarrow}$ are the critical currents for two qubit states.

The rate scales as the third power of temperature and therefore assumes very low values as $T \to 0$. The same temperature dependence of the dephasing time can be found for a charge qubit, which is capacitively coupled to a SET [17,9], where $\tau_\varphi^{-1} = (2E_{\mathrm{int}}^2 T^3/3\pi^2 E_C^4)(R_K/R_T)^2$. Here E_{int} is the coupling energy, R_T is the resistance of the SET's junctions in the normal state, and E_C is the charging energy needed to increase the charge Q of the central island by e. This expression assumes that decreasing Q e requires a much higher energy. In the opposite limit, when this energy is equal to E_C, the dephasing rate scales as T^5 (cf. Ref. [19]).

This should be contrasted with a qubit coupled to an Ohmic environment which enforces the dephasing rate linear in T. For instance, the tunneling in a QPC can be controlled by the qubit, $\mathcal{H}_{\mathrm{T}} = (1 + \kappa\sigma_z)\sum_{kk'}[a_k^\dagger b_{k'} + \text{h.c.}]$, and the dephasing rate in the off-state is $\tau_\varphi^{-1} = (2\kappa^2/\pi)(R_{\mathrm{K}}/R_{\mathrm{T}})T$.

We argue that the cases of the dc-SQUID and the SET are similar: a qubit is coupled nonlinearly to an Ohmic oscillator bath. In the case of the SQUID this is realized by quadratically coupling the qubit to weak fluctuations of the SQUID's flux, which in turn are coupled (linearly) to the Ohmic electromagnetic environment. In the SET the two levels of the central island, $Q = 0, e$, can be viewed as two lowest levels of an oscillator (the effect of the higher levels is of higher order in the qubit-detector coupling). This fictitious oscillator is coupled quadratically (capacitively) to the qubit and linearly (via the tunneling in the SET's junctions) to the bath of quasiparticle excitations. The system can be described by the Hamiltonian

$$\mathcal{H} = \varepsilon\sigma_z + \mathcal{H}_{\mathrm{bath}} + \frac{\sigma_z}{2E_0}\left(\sum\nolimits_i \lambda_i x_i\right)^2 . \quad (10)$$

Here the first term describes the qubit (we neglect the tunneling between the basis states), the bath $\mathcal{H}_{\mathrm{bath}} = \sum_i (m_i\omega_i^2 x_i^2 + p_i^2/m_i)/2$ has the Ohmic spectrum $J(\omega) \equiv \sum_i \frac{\pi\lambda_i^2}{2m_i\omega_i}\delta(\omega - \omega_i) = \gamma\omega$, and the last term is the coupling between the qubit and the bath. The dephasing rate for this system can be estimated as

$$\tau_\varphi^{-1} = \frac{1}{E_0^2}\int\frac{d\omega}{2\pi}\langle(\sum_i \lambda_i x_i)_\omega^2\rangle\langle(\sum_i \lambda_i x_i)_{-\omega}^2\rangle ,$$

where $\langle(\sum_i \lambda_i x_i)_\omega^2\rangle = J(\omega)[1+\coth(\hbar\omega/2T)]$. At low temperatures this rate vanishes as $\tau_\varphi^{-1} = (4\pi/3)\gamma^2 T^3/E_0^2$, i.e., faster than in a system coupled to an Ohmic bath linearly.

References

1. Yu. Makhlin, G. Schön, and A. Shnirman, Rev. Mod. Phys. **73**, 357 (2001).
2. J.R. Friedman et al., Nature **406**, 43 (2000).
3. C.H. van der Wal et al., Science **290**, 773 (2000).
4. Y. Nakamura, Yu.A. Pashkin, and J.S. Tsai, Nature **398**, 786 (1999).
5. Yu. Makhlin, G. Schön, and A. Shnirman, Phys. Rev. Lett. **85**, 4578 (2000).
6. D.V. Averin, e-print cond-mat/0004364.
7. M.H. Devoret and R.J. Schoelkopf, Nature **406**, 1039 (2000).
8. V.B. Braginsky and F.Y. Khalili, *Quantum Measurement*, Cambridge University Press, Cambridge, 1992.
9. D.V. Averin, e-print cond-mat/0010052.
10. S.A. Gurvitz, Phys. Rev. B **56**, 15215 (1997).
11. A.N. Korotkov, Phys. Rev. B **63**, 115403 (2001).
12. A.N. Korotkov, Phys. Rev. B **60**, 5737 (1999).
13. A. Shnirman and G. Schön, Phys. Rev. B **57**, 15400 (1998).
14. A.N. Korotkov and D.V. Averin, e-print cond-mat/0002203.
15. A. Aassime et al., Phys. Rev. Lett. **86**, 3376 (2001).
16. Yu. Makhlin, G. Schön, and A. Shnirman, Proceedings of SQUID2001.
17. Yu. Makhlin, G. Schön, and A. Shnirman. In: *Macroscopic Quantum Coherence and Computing*. Eds. D.V. Averin, B.Ruggiero, P. Silvestrini, Plenum, 2001.
18. See, however, H. Tanaka, Y. Sekine, S. Saito, H. Takayanagi, e-print cond-mat/0108231.
19. Yu.V. Nazarov, Sov. J. Low Temp. Phys. **16**, 422 (1990).

MULTIPLE SIGE QUANTUM DOTS FOR QUANTUM COMPUTATION

P. A. CAIN AND H. AHMED
Microelectronics Research Centre, Cavendish Laboratory, University of Cambridge, Madingley Road, Cambridge, CB3 0HE United Kingdom
E-mail: pac36@cam.ac.uk

D. A. WILLIAMS
Hitachi Cambridge Laboratory, Hitachi Europe Ltd., Cavendish Laboratory, Madingley Road, Cambridge, CB3 0HE United Kingdom
E-mail: williams@phy.cam.ac.uk

We have fabricated coupled double quantum dots, each with a diameter of around 40nm, using trench isolation. A range of transport measurements have been performed on various double quantum dot systems at temperatures down to 20 mK. Peak splitting in the Coulomb oscillations has been clearly observed at 4.2 K due to the capacitative coupling of the two dots. An advantage of the trench isolation approach to fabricating coupled quantum dots is that secondary structures, such as single electron electrometers can be fabricated close to the dots quite easily. We present measurements showing that an electrometer 50 nm away from the double dot can detect a single hole added to one of the dots.

1 Introduction

It is widely believed that useful quantum computers must be constructed in the solid state. Multiple quantum dot structures are a candidate system where the ability to manipulate and measure the states of each quantum dot is possible[1,2,3]. Most studies of laterally coupled double quantum dots have been performed in GaAs:AlGaAs heterostructures using surface gates selectively to deplete a two-dimensional electron gas [4,5,6,7,8]. These structures have shown charge delocalisation and interdot Coulomb interaction manifested as peak splitting in the Coulomb oscillations. The ability to control the size of the tunnel barrier between the dots in principle allows charge superposition states to be set up across the dots, a requirement for charge-state quantum computation. However, in these surface gated structures it would be difficult to measure the charge state of the quantum dots (another requirement), since the surface gates must surround the dots for good definition, leaving little or no room for further surface gates to define a single electron electrometer.

We have fabricated coupled quantum dots using trench isolation to define the structure. The advantage of this approach is that secondary structures such as extra gates or single electron electrometers can be fabricated close to the dots. This allows the charge state of the double-dot structure to be measured.

2 Device Layout

Figure 1 shows a double quantum dot device, fabricated using electron beam lithography followed by trench isolation. The material used consists of a 30 nm $Si_{0.9}Ge_{0.1}$ layer grown on a silicon-on-insulator substrate. The SiGe layer is doped to $10^{19}cm^{-3}$ with boron (p-type). The quantum dots shown in figure 1 have a lithographic diameter close to 50nm, but due to surface depletion the conducting diameter of the dots is closer to 30 nm. The constrictions between the dots are less than 20nm wide lithographically, and so form tunnel barriers due to the

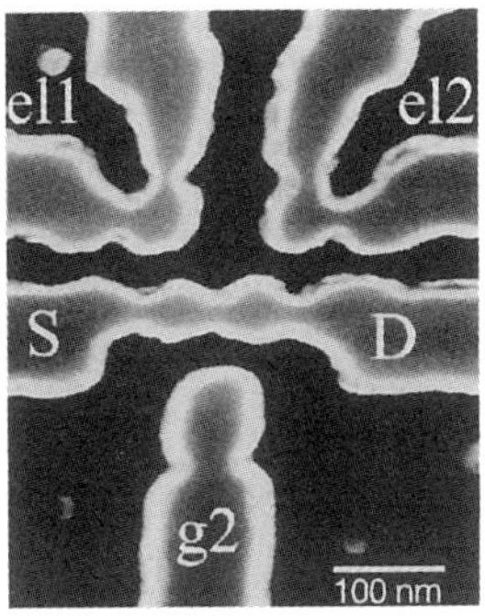

Figure 1. A double quantum dot device fabricated by trench isolation in a 30nm SiGe layer grown on an SOI wafer. The dark areas are where the SiGe has been etched away, revealing the oxide layer below. Adjacent to the double dot structure are two single quantum dot structures (el1, el2) which act as single-hole electrometers to measure the charge on the double-dot.

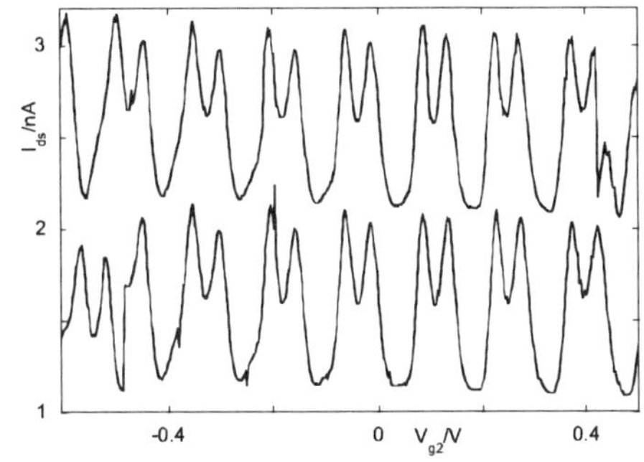

Figure 2. Two sets of Coulomb oscillation measurements taken consecutively and for the same source-drain bias of 1 mV, but offset from each other for clarity. The degree of peak splitting suggests that the inter-dot capacitance is similar to the dot-lead capacitances. The two sets of data overlap very well, except at the far right of the upper curve, and the far left of the lower curve where two random telegraph switching events have occurred.

surface depletion.

Adjacent to the double-dot structure are two single quantum dot structures intended to be used as single hole electrometers (labelled el1 and el2). They can also be used as gates for the double quantum dot structure by applying an offset voltage to their source and drain. The third gate (g2) is positioned between the dots to vary the size of the tunnel barrier by depletion. Previous measurements on a similar double dot structure in this material have shown that the tunnel barriers can be raised and lowered if sufficient voltage is applied (although this has not always been possible if the tunnel barrier is naturally very large or very small) [9].

3 Measurements at 4.2 K

Preliminary measurements at 4.2K on a structure similar to that in figure 1 revealed that the device operated in the Coulomb blockade regime. The device on which these measurements were performed differs slightly from that in figure 1 in that it had two more gates instead of the electrometers. Figure 2 shows measurements of the Coulomb oscillations as the voltage applied to gate g2 is swept. The oscillations clearly split into pairs of peaks, as expected if the two dots are coupled [10]. This effect has only previously been seen at millikelvin temperatures in surface gated GaAs:AlGaAs structures [5,7,8]. The degree of peak splitting is related to the strength of inter-dot coupling (parameterised as the inter-dot capacitance, $C_{int} \approx 22aF$), and in this case the splitting is in an intermediate regime, where the inter-dot capacitance is similar to the capacitance between the dots and the source/drain regions. Measurements on other similar structures have sometimes shown a very high level of interdot coupling, as well the ability to adjust the size of the interdot tunnel barrier with the applied gate voltage[9].

4 Electrometer measurements at 20 mK

Measurements on the device in Figure 1 were made in a dilution refrigerator at a base temperature of 20 mK. Using gate g2, single period Coulomb oscillations were observed in current between the source (S) and drain (D). The period of the oscillations was twice the average period of those observed in the first device shown in Figure 2. This indicates that one of the outer constrictions defining the double dot is not fully depleted, so that only one quantum dot is present. Further measurements using el1 and el2 as gates indicated that only the left quantum dot is present. Single period Coulomb oscillations were also observed in el1 by sweeping the voltage applied to gate g2.

In order to measure the level of interaction between the electrometer and the 'double dot' structure, the Coulomb oscillations in each were measured simultaneously while the voltage to gate g2 was swept. Figure 3 shows both measurements for source-drain bias of 1mV and 3mV on the 'double dot' and el1 respectively. Every time an extra hole is added to the 'double dot' structure (by passing through an oscillation) a kink in the electrometer current is observed due to the change in the local electrostatic field. The size of the kinks is equivalent to a 30 mV change in the gate voltage V_{g2}. This is calculated to be a change in the electrochemical potential on the electrometer quantum dot of $0.1meV$. Although this change is quite small, it is easily resolved at 20 mK in a dilution refrigerator (where $k_B T \approx 0.002meV$).

The measurements show that this system may be useful for silicon-based quantum computation. A superposition charge state across the two dots could in principle be formed by first introducing a single hole to one of the two dots using the peak-split Coulomb oscillation as a guide. A Hadamard gate could be then be realised by pulsing the gate voltage to the central gate (g2) in order to lower the interdot tunnel barrier for a specific time. A subsequent measurement using the electrometers should find the extra charge on either dot with equal probability[9]. The decoherence time for the charge states may be short due in part to the electromagnetic environment,

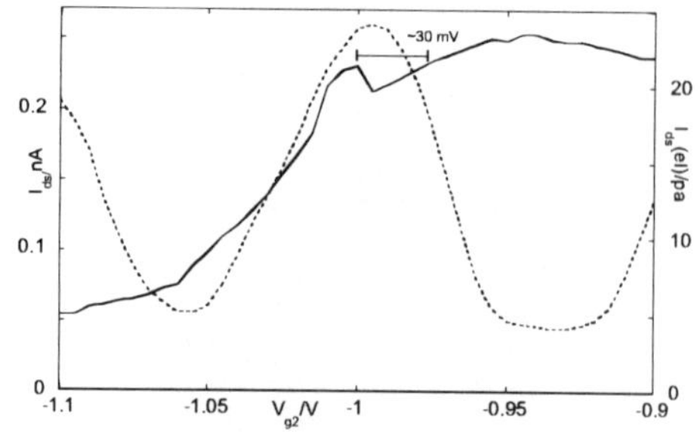

Figure 3. Measurements of the electrometer current for el1 (solid) and the double dot current (dotted) as the voltage to gate g2 is swept. $V_{ds} = 3mV$ and $V_{ds}(el1) = 4mV$. The field from the extra hole induced onto the double dot after passing through an oscillation affects the potential on the electrometer, reducing the effective gate voltage.

although no measurements of this have been performed. A modified version of this system could also be useful in a spin-based qubit approach, where the ability to adjust the inter-dot coupling allows the interaction between neighbouring spins to be controlled[9].

We have shown that silicon based coupled quantum dots formed by trench isolation can exhibit similar coupling effects previously observed in surface gated GaAs:AlGaAs based structures. We have also shown that electrometers can be fabricated very close to the double quantum dot and can measure an extra hole on the island, and the strength of the interaction is likely to be enough to determine on which of the two dots the hole resides. This approach therefore provides a practically realistic route to fabricating a single qubit in silicon.

Acknowledgments

P.A.C. acknowledges financial support given by an industrial CASE studentship from EPSRC and Hitachi Europe Ltd.

References

1. Y. Nakamura, Yu. A. Pashkin and J. S. Tsai, *Nature* **39** *8786 (1999).*
2. *B. E. Kane, Nature* **393**, *133 (1999).*
3. *D. Loss and D. P. DiVincenzo, Phys. Rev. A* **57**, *120 (1998).*
4. *R. H. Blick, R. J. Haug, J. Weis, D. Pfannkuche, K. v. Klitzing and K. Eberl, Phys. Rev. B* **53**, *7899 (1996).*
5. *D.Dixon, L. P. Kouwenhoven, P.L.McEuen and Y. Nagamune, J. Motohisa and H. Sakaki, Phys. Rev. B* **53**, *12625 (1996).*
6. *C. H. Crouch, C. Livermore, R. M. Westervelt, K. L. Campman, A. C. Gossard, Appl. Phys. Lett* **71**, *817 (1997).*
7. *F. R. Waugh, M. J. Berry, C. H. Crouch, C. Livermore, D. J. Mar, R. M. Westervelt, K. L. Campman and A. C. Gossard, Phys. Rev. B* **53**, *1413 (1996).*
8. *N. C. Van der Vaart, L. P. Kouwenhoven, S. F. Godijn, Y. V. Nazarov, C. J. P. M. Harmans and J. E. Mooij, Superlatt. and Micro.* **24**, *435 (1998).*
9. *P. A. Cain, D. A. Williams, J. M. Bonar and H. Ahmed, Appl. Phys. Lett* **77**, *3415 (2000).*
10. *P. A. Cain, D. A. Williams and H. Ahmed, Appl. Phys. Lett* **78**, *3624 (2001).*

DECOHERENCES IN STRONGLY INTERACTING QUBITS

HAYATO NAKANO AND HIDEAKI TAKAYANAGI

NTT Basic Research Laboratories,
3-1, Morinosato-Wakamiya, Atsugi-shi, Kanagawa 243-0198, Japan
E-mail: nakano@will.brl.ntt.co.jp

A strategy to obtain a quantum bit (qubit) having a longer coherence time is discussed. If the main origin of the decoherence is the weak and continuous interaction with a macroscopic bath, we can make one qubit decoherence time longer with plural two-level systems and the appropriate interaction between them. The "appropriate" interactions can be chosen from the policies of usual Quantum Error Correction Codes (QECC). The redundancy introduced by using plural two-level systems gives a Hilbert space that is much larger than that necessary for a single qubit. The two states of lower energies are assigned as the two states of the qubit ("coding space"). The interaction is chosen so as to bring the other states to higher energies. The "errors" correspond to the escapes from the lower energy states to higher energy states. Then the escapes are strongly suppressed by the interaction energy, which results in a longer coherence time when the plural two-level systems are used as a single quantum bit.

1 Introduction

After the quantum Fourier transformation algorithm was reported[1], quantum computations became very hot topics, in physics, information science, and many other disciplines. Before we can realize a quantum computer, we must first make a quantum bit (qubit) having a long coherence time because all quantum computation processes should be completed during the coherence time.

The ideas for overcoming this problem proposed are based on the quantum error correction methods, such as, error-correction code (QECC), error-avoiding code, automatic correction, and Decoherence-Free-Subspace (DFS)[2]. However, the error correction processes must be complicated if the intrinsic coherence time of one qubit is short because the correction process should be accomplished with the time interval shorter than the coherence time.

In this paper, we propose another way to get a qubit having a longer coherence time. It involves two two-level systems (two qubits) interacting with each other. Artificially applying an interaction between them makes it possible to get a longer coherence time of the coupled system as one qubit.

2 Quantum error correction and Decoherence-Free Subspace

When $n(>k)$ qubits are used for keeping the quantum information of k qubits, the Hilbert space spanned by the n qubit states is much larger than that of the k qubits. A part of the larger Hilbert space is allocated for the "coding subspace", which is the size of the k qubit Hilbert space. The rest of the subspace corresponds to the erred states. If an error occurs, the state of the n qubit escapes from the coding subspace. The escape can be detected by the syndrome measurement. When one uses enough redundant qubits, the error can be corrected by the correcting projections, and the erred state comes back into the coding subspace. This is QECC.

For example, consider a two-qubit error detection code. Suppose that an original qubit in the state

$$|\psi_{or}\rangle = a_0|0\rangle + b_0|1\rangle \tag{1}$$

is given. This qubit information is distributed to two-qubit system as

$$|\psi_{code}\rangle = \frac{1}{\sqrt{2}}a_0(|00\rangle+|11\rangle)+\frac{1}{\sqrt{2}}b_0(|01\rangle-|10\rangle), \tag{2}$$

through the quantum circuit in Fig. 1. This

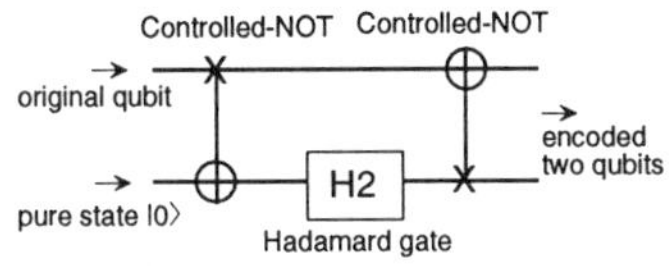

Figure 1. Coding of a two-qubit error detection code.

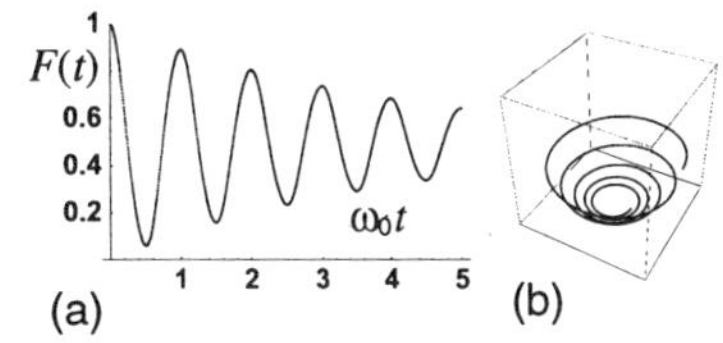

Figure 2. Decoherence in a single qubit interacting with a Boson bath. (a) Time evolution of the fidelity $F(t)$. (b) Time evolution of the state vector on the Bloch sphere.

code can detect any one bit-flip error ($|0\rangle \leftrightarrow |1\rangle$), because any erred state is orthogonal to any correct code. One can detect and correct various errors by using more redundant qubits. However, this method is not very useful for the decoherences caused by a weak interaction between the qubits and environment because such an interaction invokes small but many errors.

The idea of DFS is to find an appropriate "coding subspace" in a given larger Hilbert space. For a given environment, some states of the large qubit system cannot be entangled with the environment by the qubit-environment interaction. If it is possible to make a subspace spanned only by such states, that space will be decoherence-free. Using the subspace as the coding space, one can realize a decoherence-free qubit system. However, it is often difficult to find the DFS for a given physical system.

3 Decoherence caused by the environment

We adopt a model in which decoherence is caused by the unwanted weak interactions between the qubit system and the environment. The Hamiltonian of the system and the bath is given by

$$H_{tot} = H_S + H_B + H_{dis}, \qquad (3)$$

$$H_B = \sum_k \hbar\omega_k b_k^\dagger b_k, H_{dis} = \sum_k \hbar\sigma_x(g_k b_k^\dagger + g_k^* b_k), \qquad (4)$$

where H_S is the one qubit Hamiltonian $(\hbar\omega_0/2)\sigma_z$, g_k, g_k^* are the coupling constants, and $b_k(b_k^\dagger)$ is the annihilation (creation) operator of the Boson of the mode k. Eq. (4) means the decoherence is produced through an operator σ_x of the qubit. This interaction creates an entangled state between the system and the bath. Tracing out the microscopic degrees of freedom of the bath creates a mixed state of the qubit system resulting in the decoherence.

The result of a numerical calculation for single qubit case based on this model is shown in Fig. 2. The reduced density operator of the qubit at the time t, $\rho_S(t)$, is calculated by solving a perturbative master equation[3] with a frequency-independent $\sum_k |g_k|^2\delta(\omega - \omega_k)$. The fidelity is given by

$$F(t) = \langle\psi_{in}|\rho_S(t)|\psi_{in}\rangle, \qquad (5)$$

where $|\psi_{in}\rangle \equiv (\uparrow + \downarrow)/\sqrt{2}$ is the initial wave function of the qubit. The decrease of the amplitude of the oscillation in $F(t)$ corresponds to the decoherence.

4 Decoherences in strongly interacting coupled qubits

We propose in this section, a strategy to obtain one qubit from a coupled qubits strongly interacting with each other. The method is a modification of the ideas of QECC, and DFS. Our idea is to make a pseudo-DFS by artificially introducing an appropriate and strong interaction between two qubits.

Figure 3 shows examples of the interaction between two qubits and the environments. In Fig. 3(a), two qubits interact with

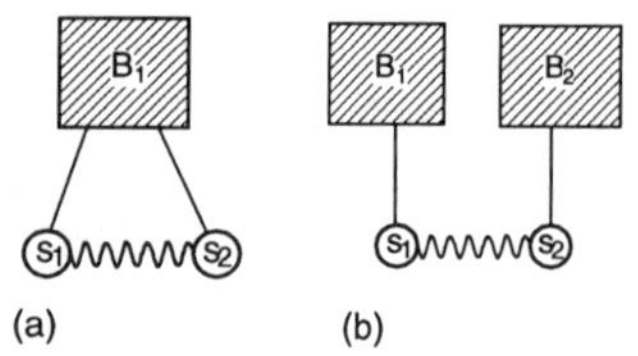

Figure 3. Models of the interaction between qubits and environments. (a) Collective interaction. (b) Local interactions.

one environment through a collective variable of the two qubit system. In Fig. 3(b), each qubit interacts with a different environment.

Protecting qubit information under the collective interaction model in Fig. 3(a) has been discussed, and the use of a logical basis in which the collective variable is not given a large amplitude, has been suggested. We found a concrete example, which realizes making such a basis. Introducing a qubit-qubit interaction that "squeezes" the collective variable is very effective[3].

A more interesting situation is the local interaction model in Fig. 3(b). Usually introducing more redundant qubits induces stronger decoherence because additional qubits are affected by their own environments. All environments additively destroy the coherence of the system of plural qubits. This situation corresponds to the local interaction model in Fig. 3(b). Therefore, it might seem that using two-qubits for one-qubit information is not promising. However, a given interaction with the environments does not destroy all of the coherence in the two-qubit system. As described below, a part of the coherence is robust under such a condition.

The Hamiltonian of the system is given by

$$H_{tot} = H_{S1} + H_{S2} + H_{int} + H_B + H_{dis}, \quad (6)$$

where

$$H_{Sn} = \frac{\hbar\omega_n}{2}\sigma_{nz}, \qquad (n = 1, 2) \quad (7)$$

$$H_B = \sum_{k1} \hbar\omega_{k1} b_{k1}^\dagger b_{k1} + \sum_{k2} \hbar\omega_{k2} b_{k2}^\dagger b_{k2}, \quad (8)$$

$$H_{dis} = \sum_{k1} \hbar\sigma_{1x}\left(g_{k1}b_{k1}^\dagger + g_{k1}^* b_{k1}\right) + \sum_{k2} \hbar\sigma_{2x}\left(g_{k2}b_{k2}^\dagger + g_{k2}^* b_{k2}\right), \quad (9)$$

Since the two baths are not correlated, their microscopic degrees of freedom are traced out separately. This gives a quite different decoherence from the single bath model.

One appropriate interaction for protecting the qubit information from the unwanted interaction Eq. (9) is

$$H_{int} = -\hbar U_{int}(\sigma_{1x}\sigma_{2x} - 2\sigma_{1y}\sigma_{2y}). \quad (10)$$

In this case, taking $\omega_1 = \omega_2 \equiv \omega_0$, the eigenenergies of the coupled qubits are

$$\varepsilon_0 = -\sqrt{{\omega_0}^2 + 9{U_{int}}^2}, \varepsilon_1 = -U_{int}, \quad \varepsilon_2 = U_{int}, \varepsilon_3 = \sqrt{{\omega_0}^2 + 9{U_{int}}^2}. \quad (11)$$

And the corresponding eigenstates are

$$|e_0\rangle = \frac{-\omega_0 + \sqrt{{\omega_0}^2 + 9{U_{int}}^2}}{3U_{int}} \uparrow\uparrow + \downarrow\downarrow,$$
$$|e_1\rangle = (\uparrow\downarrow - \downarrow\uparrow)/\sqrt{2}, |e_2\rangle = (\uparrow\downarrow + \downarrow\uparrow)/\sqrt{2},$$
$$|e_3\rangle = \frac{-\omega_0 - \sqrt{{\omega_0}^2 + 9{U_{int}}^2}}{3U_{int}} \uparrow\uparrow + \downarrow\downarrow, \quad (12)$$

where, for example, $\downarrow\uparrow$ means that the qubit 1 is in its down-spin state and qubit 2 is in its up-spin state in their spin representations.

We take the states $|e_0\rangle, |e_1\rangle$ as the logical basis. Then, under the condition $U_{int} \gg \omega_0$, the erred states induced by the interaction with the environments are

$$\sigma_{1x}|e_0\rangle \sim |e_2\rangle, \sigma_{1x}|e_1\rangle \sim |e_3\rangle, \quad \sigma_{2x}|e_0\rangle \sim |e_2\rangle, \sigma_{2x}|e_1\rangle \sim |e_3\rangle. \quad (13)$$

The erred states coincide with the two highest energy states: $|e_2\rangle$ or $|e_3\rangle$. Therefore, the error process should make a state of higher energy and that is orthogonal to the correct states. This is strongly suppressed by the system Hamiltonian $H_S + H_{int}$ itself. The introduction of the qubit-qubit interaction makes two of the four eigenstates of the two-qubit

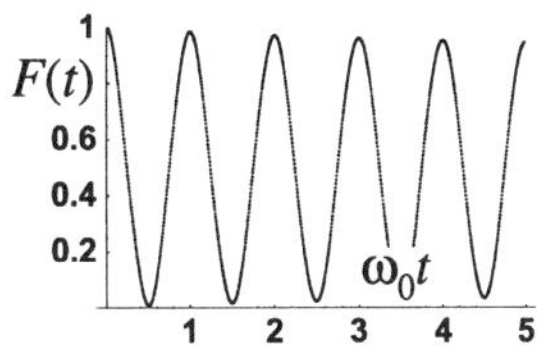

Figure 4. Time-evolution of the two-qubit system in the local interaction model. Fidelity of the decoded wave function.

system robust for the decoherence caused by the environment.

Using the state $|e_0\rangle, |e_1\rangle$ as the logical basis means that the coded two-qubit wavefunction is a superposition of

$$|\psi_{code}\rangle = \frac{1}{\sqrt{2}}a_0(\uparrow\uparrow + \downarrow\downarrow) + \frac{1}{\sqrt{2}}b_0(\uparrow\downarrow - \downarrow\uparrow). \tag{14}$$

If we put the basis as

$$|00\rangle = \uparrow\uparrow, |01\rangle = \uparrow\downarrow, |10\rangle = \downarrow\uparrow, |11\rangle = \downarrow\downarrow, \tag{15}$$

the state in Eq. (14) is almost the same as the state for the two-qubit error detection code, Eq. (2). Therefore, the quantum circuit in Fig. 1 can be used to make an initial state for our two-qubit system.

After this coding, the two-qubit system interacts with the two Boson baths. The time evolution of the reduced density operator of the two-qubit system is calculated by solving the perturbative master equation[3]. The fidelity of the decoded state is shown in Fig. 4. All of the parameters of the strength of the interactions with the environments are the same as in the single-qubit case in Fig. 2. The Larmor frequencies of the qubits are renormalized so that the Rabi frequency is the same as the single-qubit calculation. The strength of the qubit-qubit interaction U_{int} is set to be the same as the Larmor frequencies of the qubits. The decrease of the oscillation amplitude is very slow compared to the single-qubit result.

This is the quantitative demonstration of our idea for getting a qubit with a longer coherence time.

5 Conclusion

We investigated decoherences on one-qubit information put on two interacting qubits. The decoherences in the models are caused by the unwanted weak interactions between the coupled qubits and the macroscopic Bosonic heat baths. When an appropriate qubit-qubit interaction whose magnitude is comparable to other relevant energies is introduced, two of the four states of the two-qubit system become robust in their interaction with the environments, we get the result that the fidelity of the quantum information is kept against the deocoherence. Although it is difficult to give such a spin-spin interaction if the qubits are real spins (for example, electron spins or nuclear spins), we expect that it is possible for superconducting flux qubits or charge qubits.

Acknowledgments

The authors are grateful to Dr. S. Ishihara for his encouragement throughout this work. This work has been partially supported by the Core Research for Evolutional Science and Technology (CREST).

References

1. P. Shor, in *Proceedings, 35th Annual Symposium on Foundations of Computer Science*, edited by S. Goldwasser (IEEE Press, New York, 1994), 56.
2. For a review, see A. Steane, *Rep. Prog. Phys.*, **61**, 177 (1998); P. Zanardi and M. Rosetti, *Phys. Rev. Lett.* **79**, 3306 (1997); J. Barnes and W. Warren, *Phys. Rev. Lett.***85**,856 (2000); G. M. Palma *et al.*, *Proc. R. Soc. London* **A 452**, 567 (1996); D. A. Lidar *et al.*, *Phys. Rev. Lett.* **82**, 4556 (1999).
3. H. Nakano and H. Takayanagi, in preparation.

DEPHASING OF A QUBIT COUPLED WITH A POINT-CONTACT DETECTOR

Y. RIKITAKE, H. IMAMURA, Y. UTSUMI, M. HAYASHI, H. EBISAWA

Graduate School of Information Sciences, Tohoku University, Sendai 980-8579, Japan

E-mail: yoshiaki@cmt.is.tohoku.ac.jp

The dephasing of a qubit coupled with a point-contact detector is theoretically studied. We calculate the time evolution of the reduced density matrix of qubit by using the perturbation expansion. We show that the dephasing rate is proportional to the temperature at zero bias-voltage, while it is proportional to the bias-voltage when the bias-voltage is large. We also evaluate the dephasing rate by using the real time renormalization group method and show that the higher order processes of the particle-hole excitation enhances dephasing of qubit.

Much attention has been devoted in recent years to quantum computation and quantum information, which are the new technology of information processing based on quantum mechanical principles[1]. The dephasing[2] of a qubit is one of the fundamental problems in quantum information sciences. Gruvitz has studied the dephasing and collapse of a qubit in a double-dot(DD) caused by the continuous measurement by a point-contact(PC) detector[3,4]. He showed that the collapse and the role of the observer in quantum mechanics can be resolved experimentally via a non-destructive continuous monitoring of a single quantum system. However, extensive theoretical studies on the dephasing of qubit, such as the temperature dependence and bias-voltage dependence, are needed to construct the reliable quantum information processing system.

In this paper, we study the dephasing of a qubit coupled with a PC detector. Using the lowest order approximation, we show that the dephasing rate is proportional to the temperature at zero bias-voltage while it is proportional to the bias-voltage in the limit of large bias-voltage. We also evaluate the dephasing rate by using the real time renormalization group method developed by Shoeller[5,6] and show that the dephasing rate is enhanced by the higher order processes of the particle-hole excitation.

The system we consider is a DD coupled with a PC detector. The PC is placed near the upper dot as shown in Fig. 1[3,4]. The barrier height of the PC, therefore the tunneling current through PC, is modified by the electron state of DD. We assume that current can flow if and only if the electron in the DD occupies the lower dot. In our system, the dephasing of qubit is caused by the interaction between the qubit and the PC.

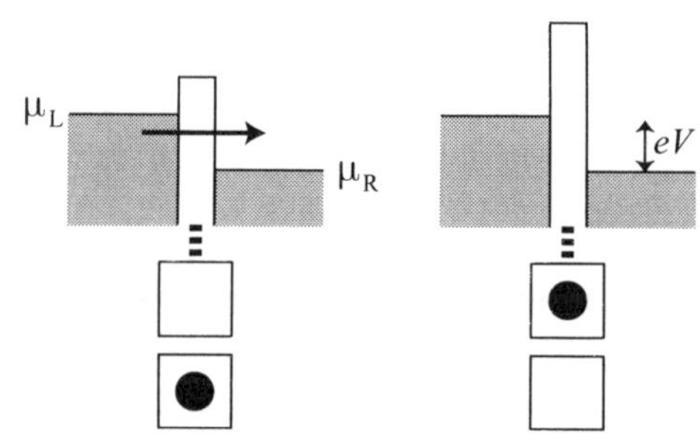

Figure 1. The double-dot (qubit) coupled with a point-contact detector. μ_L and μ_R are the chemical potential of the left and right reservoirs of point-contact and eV is the difference between them. The filled circle represents the electron in a double-dot. The height of the potential barrier of the point-contact is modified by the electron state of the double-dot. The current through the point contact flows only when the electron occupies the lower dot.

We can map the electron state of the DD into that of the equivalent two-level system and the Hamiltonian of the DD can be expressed in terms of Pauli spin matrices $\boldsymbol{\sigma}$. The electron state with upper(lower) dot occupied corresponds to the eigenstates of σ_z $|\uparrow\rangle$ ($\sigma_z = 1$) ($|\downarrow\rangle$ ($\sigma_z = -1$)). The total Hamiltonian of the system is given by

$$H = H_{qb} + H_L + H_R + H_{int}, \tag{1}$$

where

$$H_{qb} = \frac{\epsilon_C}{2}\sigma_z, \tag{2}$$

$$H_L = \sum_l \epsilon_l c_l^\dagger c_l, \quad H_R = \sum_r \epsilon_r c_r^\dagger c_r, \tag{3}$$

$$H_{int} = \frac{1}{2}(\sigma_z + 1)\sum_{l,r}(\Omega c_l^\dagger c_r + \text{h.c.}). \tag{4}$$

Here H_{qb}, $H_{L(R)}$ and H_{int} are the Hamiltonians of the qubit, the left(right) reservoir of PC and qubit-PC interaction, respectively. $\epsilon_{l(r)}$ are the energy levels in the left(right) reservoir, and Ω is the tunneling matrix element of the PC. When the upper dot is occupied ($\sigma_z = -1$), the tunneling current does not flow through the PC. Since our interest is in the dephasing of the qubit, we neglect the tunneling between upper and lower dots.

We consider the following reduced density matrix of the qubit:

$$\rho = Tr_{\rm PC}\ \rho_{\rm tot} = \begin{pmatrix} \rho_{\uparrow\uparrow} & \rho_{\uparrow\downarrow} \\ \rho_{\downarrow\uparrow} & \rho_{\downarrow\downarrow} \end{pmatrix}, \tag{5}$$

where $\rho_{\rm tot}$ is the density matrix of the total system consisting of the qubit and PC, $Tr_{\rm PC}$ denotes tracing out the degrees of reservoirs.

The time evolution of the total density matrix $\rho_{\rm tot}(t)$ is described by the von Neumann equation, $i\hbar\dot{\rho} = [H, \rho]$. Due to interaction with the detector, the time evolution of the reduced density matrix $\rho(t)$ is not unitary and it approaches the statistical mixture represented by the diagonal matrix in the limit of $t \to \infty$. Therefore, we can evaluate the dephasing of the qubit by calculating the time evolution of the off-diagonal element of reduced density matrix $\rho_{\uparrow\downarrow}(t)$.

We assume that the density matrix of the total system at the initial time $t = 0$ is in product form and the qubit and reservoirs of PC are decoupled, and the reservoirs are in thermal equilibrium[2,7]. We then have

$$\rho_{\rm tot}(0) = \rho(0) \otimes \rho_{L0} \otimes \rho_{R0}, \tag{6}$$

where

$$\rho_{L0} = \frac{{\rm e}^{-\sum_l(\epsilon_l - \mu_{\rm L})c_l^\dagger c_l/k_{\rm B}T}}{{\rm Tr}_L {\rm e}^{-\sum_l(\epsilon_l - \mu_{\rm L})c_l^\dagger c_l/k_{\rm B}T}}, \tag{7}$$

$$\rho_{R0} = \frac{{\rm e}^{-\sum_r(\epsilon_r - \mu_{\rm R})c_r^\dagger c_r/k_{\rm B}T}}{{\rm Tr}_R {\rm e}^{-\sum_r(\epsilon_r - \mu_{\rm R})c_r^\dagger c_r/k_{\rm B}T}}. \tag{8}$$

The time evolution of the off-diagonal element $\rho_{\uparrow\downarrow}(t)$ obeys the generalized master equation defined by

$$\left[\frac{\rm d}{{\rm d}t} + i\frac{\epsilon_C}{\hbar}\right]\rho_{\uparrow\downarrow}(t) = \int_0^t {\rm d}t' \rho_{\uparrow\downarrow}(t')\Sigma(t-t'), \tag{9}$$

where Σ is a 'self-energy' which describes the qubit-detector interactions[8,9].

We first calculate the self-energy $\Sigma(t - t')$ by using the perturbation expansion with respect to the interaction term H_{int}. In the lowest order approximation, the self-energy is given by

$$\Sigma(t - t') = \exp[-i\epsilon_C(t - t')]\gamma(t - t'), \tag{10}$$

where $\gamma(t - t')$ is the propagator of the particle-hole excitation defined as

$$\gamma(t - t') = \alpha\left(\frac{\pi k_{\rm B}T}{\hbar}\right)^2 \times \frac{\cos[eV(t - t')/\hbar]}{\sinh^2[\pi k_{\rm B}T(t - t' - i/D)/\hbar]}. \tag{11}$$

Here D is the high frequency cutoff and α is the dimensionless conductance of the PC defined as $\alpha = 2\Omega^2 N_L N_R$, where $N_{L(R)}$ is the density of states in the left(right) reservoir. Suppose that the reduced density matrix varies very slowly compared with the time scale of the life time of the particle-hole excitation $\hbar/k_{\rm B}T$, $\rho_{\downarrow\uparrow}(t)$ is written as

$$\rho_{\downarrow\uparrow}(t) = \rho_{\downarrow\uparrow}(0){\rm e}^{-i\epsilon_C t/\hbar}{\rm e}^{-\Gamma t}, \tag{12}$$

where

$$\Gamma = \int_0^\infty {\rm d}t\Sigma(t). \tag{13}$$

The dephasing rate, or the decay rate of $\rho_{\uparrow\downarrow}(t)$, is given by $\mathrm{Re}\Gamma$. Since the dephasing is caused by the particle hole excitation in reservoirs, it depends on the temperature T and bias-voltage V as shown in Fig. 2.

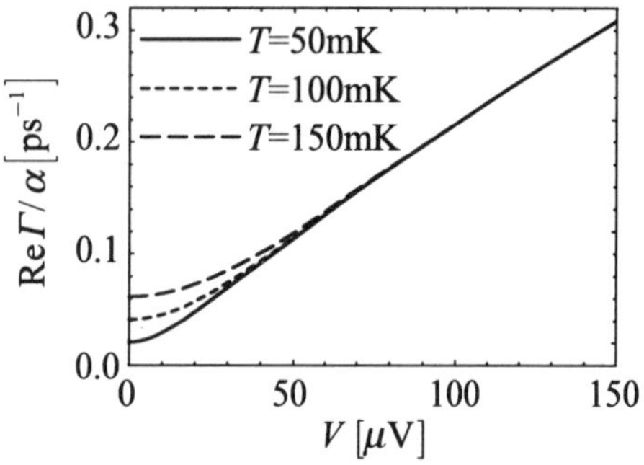

Figure 2. Dephasing rate of the qubit as a function of bias-voltage V applied to the point contact. Solid, dotted and dashed lines represent the dephasing rate for $T = 50, 100$ and 150mK, respectively. The high frequency cutoff is assumed to be $\hbar D = 20$K.

It is easy to show that, in the lowest order approximation for the self-energy, the dephasing rate at $V = 0$ is given by

$$\mathrm{Re}\,\Gamma = \alpha\frac{\pi k_{\mathrm{B}}T}{\hbar} \tag{14}$$

and is proportional to the temperature T as shown in Fig. 2. This can be explained as follows. The dephasing rate is determined by the number of particle-hole excitations in the reservoirs which interact with the qubit. The number of particle-hole excitations at $V = 0$ is proportional to the temperature T. On the other hand, the number of particle-hole excitations for $eV \gg k_{\mathrm{B}}T$ is proportional to the bias-voltage V. Indeed, the dephasing rate for $eV \gg k_{\mathrm{B}}T$ is given by

$$\mathrm{Re}\,\Gamma = \alpha\frac{1}{2}\frac{eV}{\hbar} \tag{15}$$

and is proportional to V as shown in Fig. 2.

Next we go beyond the lowest order approximation. We employ the real time renormalization group (RTRG) method developed by Schoeller[5,6]. Following Shoeller, we introduce the short-time cutoff t_c in the propagator of the particle-hole excitation by

$$\gamma_{t_c}(\tau) = \gamma(\tau)\theta(\tau - t_c), \tag{16}$$

where $\theta(t)$ is a step function. In Laplace space, the self-energy $\Sigma(z)$ is expressed by

$$\Sigma(z) = \Sigma_{t_c}(z) + \mathcal{F}_{t_c}, \tag{17}$$

where $\Sigma_{t_c}(z)$ includes only the time scales which are precisely smaller than t_c and $\mathcal{F}_{t_c}$ consisting of the other time scales is the functional of renormalized correlation function γ_{t_c}, vertex g_{t_c} and frequency ω_{t_c}

Let us increase the cutoff by an infinitesimal amount $t_c \to t_c + dt_c$. The change of $\gamma_{t_c}(\tau)$ is given by

$$\begin{aligned} d\gamma(\tau) &= \gamma_{t_c}(\tau) - \gamma_{t_c+dt_c}(\tau) \\ &= \gamma(t_c)\delta(\tau - t_c)dt_c. \end{aligned} \tag{18}$$

The change of the self-energy $\Sigma_{t_c+dt_c}(z)$ caused by $d\gamma(\tau)$ can be expressed by $d\gamma$, g_{t_c} and ω_{t_c}. The change of g_{t_c} and ω_{t_c} due to $d\gamma(\tau)$ can also be represented by $d\gamma$. As we increase t_c, the long time scale of γ is renormalized. Since $\gamma(\tau)$ is a decreasing function of τ with the life time $\hbar/k_{\mathrm{B}}T$, $\mathcal{F}_{t_c}$ is zero for $t_c \to \infty$. Therefore, the self-energy $\Sigma(z)$ in Laplace space is expressed as

$$\Sigma(z) = \lim_{t_c\to\infty}\Sigma_{t_c}(z). \tag{19}$$

The renormalization group (RG) equations is obtained by calculating the renormalization of ω_{t_c}, g_{t_c} and Σ_{t_c} due to the infinitesimal deviation $d\gamma_{t_c}$. The straightforward calculation leads to the following RG equations:

$$\frac{\mathrm{d}\Sigma_{t_c}(z)}{\mathrm{d}t_c} = \int_0^\infty \mathrm{d}t\left(\frac{\mathrm{d}\gamma_{t_c}}{\mathrm{d}t_c}\right)(t)a_{t_c}(t)b_{t_c}(0) \tag{20}$$

$$\frac{\mathrm{d}\omega_{t_c}}{\mathrm{d}t_c} = i\int_0^\infty \mathrm{d}t\left(\frac{\mathrm{d}\gamma_{t_c}}{\mathrm{d}t_c}\right)(t)g_{t_c}(t)g_{t_c}(0) \tag{21}$$

$$\frac{\mathrm{d}g}{\mathrm{d}t_c} = 0 \tag{22}$$

$$\begin{aligned} \frac{\mathrm{d}a_{t_c}}{\mathrm{d}t_c} &= \int_0^\infty \mathrm{d}t\int_{-\infty}^0 \mathrm{d}t'\left(\frac{\mathrm{d}\gamma_{t_c}}{\mathrm{d}t_c}\right)(t-t') \\ &\times [a_{t_c}(t)g_{t_c}(0)g_{t_c}(t') - g_{t_c}(t)a_{t_c}(0)g_{t_c}(t')] \end{aligned} \tag{23}$$

$$\begin{aligned} \frac{\mathrm{d}b_{t_c}}{\mathrm{d}t_c} &= -\int_0^\infty \mathrm{d}t\int_{-\infty}^0 \mathrm{d}t'\left(\frac{\mathrm{d}\gamma_{t_c}}{\mathrm{d}t_c}\right)(t-t') \\ &\times g_{t_c}(t)g_{t_c}(0)b_{t_c}(t'), \end{aligned} \tag{24}$$

where

$$\left(\frac{d\gamma_{t_c}}{dt_c}\right)(\tau) = \gamma(t_c)\delta(\tau - t_c). \qquad (25)$$

We numerically solve the above RG equations with the initial condition $\omega = \epsilon_C/\hbar$ and $g = a = b = 1$ at $t_c = 0$. Once we obtain the self-energy in Laplace space, the off-diagonal element of reduced density matrix in Laplace space is given by

$$\rho_{\downarrow\uparrow}(z) = \frac{\rho_{\downarrow\uparrow}(0)}{z - i\epsilon_C - \Sigma(z)}. \qquad (26)$$

Finally, we obtain $\rho_{\downarrow\uparrow}(t)$ by performing an inverse Laplace transformation.

We define the dephasing time τ_{dep} as the time when the absolute value $|\rho_{\downarrow\uparrow}(t)|$ becomes one-half of its initial value $|\rho_{\downarrow\uparrow}(0)|$. In Fig. 3 we plot the dephasing rate defined as $1/\tau_{dep}$ at $V = 0$ as a function of α. For small α, the result agrees well with that obtaind by the lowest order approximation. For $\alpha \gtrsim 0.2$, however, the dephasing rate becomes larger than that calculated in the lowest order approximation. As shown in Fig. 3, the higher order processes of the particle-hole excitation enhances the dephasing rate, since the number of particle-hole excitations increases due to the quantum fluctuation described by those higher order processes.

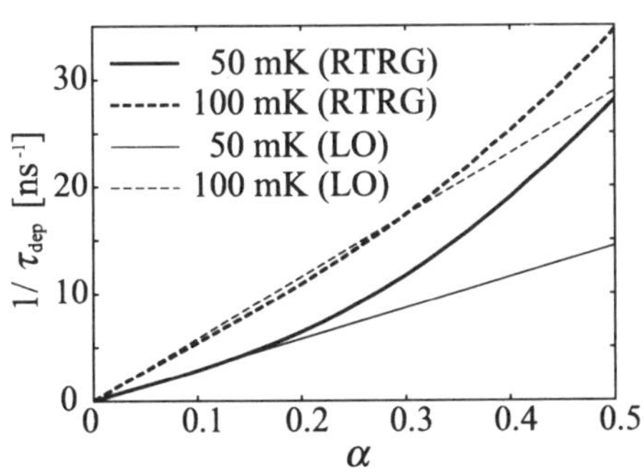

Figure 3. Dephasing rate calculated by using the real time renormalization group (RTRG) method at $k_BT = 50(100)$mK is plotted by the solid(dotted) line against the dimensionless conductance α. Thin solid(dotted) line represents the dephasing rate obtained by using the lowest order (LO) approximation at $k_BT = 50(100)$mK. The high frequency cutoff is assumed to be $\hbar D = 4$K.

In conclusion, we have studied the dephasing of a double-dot qubit coupled with a point-contact detector. The time evolution of the reduced density matrix of the qubit is calculated by using the perturbation expansion. In the lowest order approximation of the self-energy, we show that the dephasing rate is proportional to the temperature at $V = 0$, while it is proportional to the bias-voltage at large bias-voltage, $eV \gg k_BT$. The real time renormalization group method is also applied to evaluate the dephasing rate of the qubit. We show that the dephasing rate is enhanced by the higher order processes of the particle-hole excitation.

One of the author H. I. is supported by MEXT, Grant-in-Aid for Encouragement of Young Scientists, 13740197, 2001.

References

1. M. A. Nielsen and I. L. Chuang, *Quantum computation and quantum information* (Camblidge University Press, Camblidge, UK, 2000), and references therein.
2. U. Weiss, *QUANTUM DISSIPATIVE SYSTEMS* (World Scientific, Singapore, 2000).
3. S. A. Gurvitz, Phys. Rev. B **56**, 15215 (1997).
4. S. A. Gurvitz, 1998, con-dmat/9808058.
5. H. Schoeller, Lect. Notes Phys. **544**, 137 (2000).
6. M. Keil and H. Schoeller, 2000, con-dmat/0011051.
7. R. P. Feynman and F. L. Vernon, Ann. Phys (N.Y.) **24**, 118 (1963).
8. G. Shön, in *Quantum Transport and Dissipation* (WILEY-VCH, Winheim, 1998), Chap. 3.
9. A. Shnirman and G. Schön, Phys. Rev. B **57**, 15400 (1998).

QUANTUM GATES IN CAPACITIVELY COUPLED QUANTUM DOTS AND MEASUREMENT PROCEDURE

TETSUFUMI TANAMOTO

Corporate R & D Center, Toshiba Corporation, 1, Komukai Toshiba-cho, Saiwai-ku, Kawasaki 212-8582, Japan

E-mail: tetsufumi.tanamoto@toshiba.co.jp

Coulomb blockade effects in capacitively coupled quantum dots can be utilized for constructing an N-qubit system with antiferromagnetic Ising interactions. Starting from the tunneling Hamiltonian, we theoretically show that the Hamiltonian for a weakly coupled quantum-dot array is reduced to that for nuclear magnetic resonance (NMR) spectroscopy. Quantum operations are carried out by applying only electrical pulse sequences. A possible measurement scheme in an N-qubit system is quantitatively discussed.

Quantum computers have been widely investigated from many perspectives [1]. A quantum-dot array is a promising candidate for the basic element of a quantum computer from the viewpoint of technological feasibility[2,3,4,5]. Although a spin-based quantum dot computer has been intensively discussed[2], it seems that the quantum dot system using charged states is more accessible because the latter will be able to be constructed of various materials other than III-V group materials such as GaAs.

In this paper we advance the analysis of the quantum computer based on charged states and show a general N-qubit scheme including measurement procedure using field effect. Starting from tunneling Hamiltonian in the Coulomb blockade regime, we demonstrate that the Hamiltonians for one- and two-dimensional arrays of weakly coupled quantum dots are reduced to those for standard nuclear magnetic resonance (NMR) spectroscopy[6]. This enables any quantum computation to be described by electric pulse sequences. We set e=1 and k_B = 1.

A qubit based on charged states is composed of two quantum dots coupled via a thin tunneling barrier and a gate electrode that is attached on a thick insulating material(Fig. 1). The quantum dots are assumed to be sufficiently small for charging effects to be observed[7,8,9]. $N_{\alpha i}$ and $N_{\beta i}$ are the numbers of excess electrons from the neutral states in two quantum dots. One excess charge is assumed to be inserted from a substrate first and to stay in the two-coupled dots ($C_{Bi} > C_{Ci}$). When the excess charge exists in the upper dot and lower dot, we call them $|0\rangle = |\uparrow\rangle$ state and $|1\rangle = |\downarrow\rangle$ state, respectively.

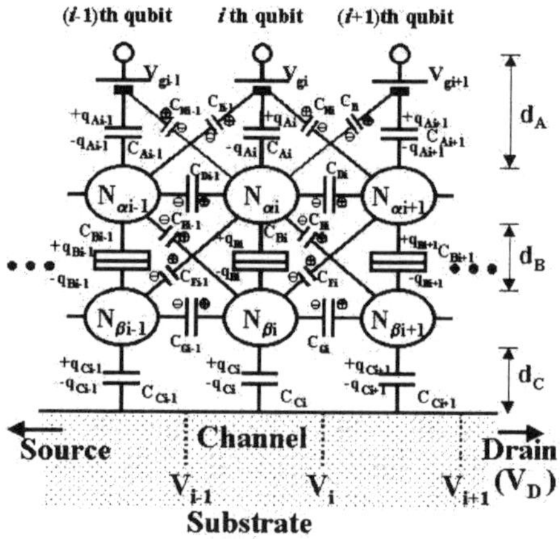

Figure 1. Schematic of the capacitance network of a one-dimensionally coupled quantum-dot array. The capacitances C_{Ai}, C_{Bi}, and C_{Ci}, the gate V_{gi} and the two quantum dots constitute the ith qubit. The electron transfer between qubits is prohibited.

Here we show the Hamiltonian of the one-dimensionally arrayed coupled quantum dots for quantum computation by starting from the tunneling Hamiltonian:

$$H=\sum_{i=1}^{N}(t\hat{a}_i^\dagger\hat{b}_i+t^*\hat{b}_i^\dagger\hat{a}_i+\epsilon_{\alpha i}\hat{a}_i^\dagger\hat{a}_i+\epsilon_{\beta i}\hat{b}_i^\dagger\hat{b}_i)+H_{\rm ch}, \tag{1}$$

where $\hat{a}_i$ ($\hat{b}_i$) describes the annihilation operator when the excess electron exists in the upper (lower) dot, and $\epsilon_{\alpha i}$ ($\epsilon_{\beta i}$) shows the electronic energy of the upper (lower) dot. $H_{\rm ch}$ is the charging energy that includes the interaction between qubits and analyzed by minimizing the general formulation of the charging energy:

$$H_{\rm ch}=\sum_{i=1}^{N}\left\{\frac{q_{{\rm A}i}^2}{2C_{{\rm A}i}}+\frac{q_{{\rm B}i}^2}{2C_{{\rm B}i}}+\frac{q_{{\rm C}i}^2}{2C_{{\rm C}i}}-q_{{\rm A}i}V_{gi}\right\}$$
$$+\sum_{i=1}^{N-1}\left\{\frac{q_{{\rm D}i}^2}{2C_{{\rm D}i}}+\frac{q_{{\rm E}i}^2}{2C_{{\rm E}i}}+\frac{q_{{\rm F}i}^2}{2C_{{\rm F}i}}+\frac{q_{{\rm G}i}^2}{2C_{{\rm G}i}}\right\}$$
$$+\sum_{i=2}^{N}\left\{\frac{q_{{\rm H}i}^2}{2C_{{\rm H}i}}-q_{{\rm H}i}V_{{\rm g}i-1}\right\}+\sum_{i=1}^{N-1}\left\{\frac{q_{{\rm I}i}^2}{2C_{{\rm I}i}}-q_{{\rm I}i}V_{{\rm g}i+1}\right\},$$

with constraints ($i=1,..,N$) $-N_{\alpha i}=q_{{\rm A}i}-q_{{\rm B}i}+q_{{\rm D}i}-q_{{\rm D}i-1}+q_{{\rm E}i}-q_{{\rm F}i-1}+q_{{\rm H}i-1}+q_{{\rm I}i}$ and $-N_{\beta i}=q_{{\rm B}i}-q_{{\rm C}i}+q_{{\rm G}i}-q_{{\rm G}i-1}-q_{{\rm E}i-1}+q_{{\rm F}i}$. It is assumed that the coupling between qubits is smaller than that within a qubit and we neglect higher order terms than $(C_{di}^2/D_i)^2(\ll 1)$. We consider the gate voltage region, where the $n_i\equiv N_{\alpha_i}-N_{\beta_i}$=-1 state and n_i=1 state are near-degenerate as in References[3,4,10], and we have:

$$H=\sum_{i=1}^{N}[t\hat{a}_i^\dagger\hat{b}_i+t^*\hat{b}_i^\dagger\hat{a}_i+\Omega_i\hat{I}_{iz}]+\sum_{i=1}^{N-1}J_{i,i+1}\hat{I}_{iz}\hat{I}_{i+1z}, \tag{2}$$

where $|\uparrow_i\rangle=|n_i=1\rangle$ and $|\downarrow_i\rangle=|n_i=-1\rangle$, $\hat{I}_{iz}=(\hat{a}_i^\dagger\hat{a}_i-\hat{b}_i^\dagger\hat{b}_i)/2$, $\Omega_i=(4C_{Ci}/D_i)[Q_V-Q_V^{\rm res}]$ and

$$J_{i,i+1}=\frac{C_{bi}C_{bi+1}C_{ei+1}+C_{ci}C_{ci+1}C_{di+1}}{2D_iD_{i+1}}. \tag{3}$$

with $C_{ai}\equiv C_{Ai}+C_{Ci}+4C_{Bi}+2(C_{Di}+C_{Ei})+C_{Hi}+C_{Ii}$, $C_{bi}\equiv C_{Ai}+C_{Ci}+2(C_{Di}+C_{Ei})+2(C_{Di-1}+C_{Ei-1})+C_{Hi}+C_{Ii}$, $C_{ci}\equiv C_{Ci}-C_{Ai}+C_{Hi}+C_{Ii}$, $C_{di}\equiv C_{Di}+C_{Ei}$, $C_{ei}\equiv C_{Di}-C_{Ei}$ ($C_{Di}=C_{Gi}$, $C_{Ei}=C_{Fi}$) and $D_i\equiv C_{ai}C_{bi}-C_{ci}^2$ and $Q_V\equiv C_{Ai}V_{gi}+C_{Hi-1}V_{gi-1}+C_{Ii}V_{gi+1}$. $Q_V^{\rm res}$ includes $\epsilon_{\alpha i}$ and $\epsilon_{\beta i}$ and shows the gate voltage when the $|0\rangle$ and $|1\rangle$ degenerate (on resonance). From the definition of Ω_i, we define the characteristic charging energy of the system as $E_C\equiv C_b/(2D)$. We control the time-dependent quantum states of the qubits in the vicinity of on resonant gate bias by applying a gate voltage such as $V_i(\tau)=V_i^{\rm res}+v_i(\tau)$ where $V_i^{\rm res}$ is the gate bias of on resonance. In this on resonant region, a transformation of the coordinate:

$$\begin{pmatrix}\hat{\alpha}_{+i}\\ \hat{\alpha}_{-i}\end{pmatrix}\equiv U_0\begin{pmatrix}\hat{a}_i\\ \hat{b}_i\end{pmatrix},\ U_0\equiv\frac{1}{\sqrt{2}}\begin{pmatrix}1&1\\1&-1\end{pmatrix}. \tag{4}$$

is convenient (we neglect the phase of t_i for simplicity). Then the Hamiltonian given by Eq.(2) can be described as

$$H(\tau)=\sum_{i=1}^{N}[2t_i\hat{I}'_{zi}-\Delta_i(\tau)\hat{I}'_{xi}]+\sum_{i=1}^{N-1}J_{i,i+1}\hat{I}'_{xi}\hat{I}'_{xi+1} \tag{5}$$

where $\hat{I}'_{xi}\equiv(\hat{\alpha}_{+i}^\dagger\hat{\alpha}_{-i}+\hat{\alpha}_{-i}^\dagger\hat{\alpha}_{+i})/2$ and $\hat{I}'_{zi}\equiv(\hat{\alpha}_{+i}^\dagger\hat{\alpha}_{+i}-\hat{\alpha}_{-i}^\dagger\hat{\alpha}_{-i})/2$. and

$$\Delta_i(\tau)=\frac{4C_{Ai}C_{Ci}}{D_i}[v_i(\tau)+\delta v_{i,{\rm cr}}(\tau)]. \tag{6}$$

Here $\delta v_{i{\rm cr}}(\tau)\equiv[C_{Hi-1}\,v_{i-1}(\tau)+C_{Ii}\,v_{i+1}(\tau)]/C_{Ai}$ is a cross-talk term and indicates an effect of other gate electrodes. The Hamiltonian given by (5) is an NMR Hamiltonian when we regard $\omega_i=2t_i$ as a Zeeman energy and $\Delta_i(\tau)=\Delta_{0i}\cos(\omega_i\tau+\delta_i)$ as a transverse magnetic field, if $\omega>\Delta_0\gg J$. Thus, quantum operations of the coupled quantum dot system can be carried out in a manner similar to that of operations in NMR quantum computers. Similarly, we can discuss the Hamiltonian of the two-dimensionally coupled dots.

Here we summarize the criterion for realizing the above-mentioned scheme. In view of time-dependent operation, the excess charge is assumed to be affected by the electronic potential generated by the capacitance network. Therefore, all quantities concerning time evolution should be smaller than the CR constant of the network. Here we take

$CR = C_{\rm int}R_{\rm int}$ where $C_{\rm int}(\equiv D/C_b)$ and $R_{\rm int}$ are capacitance and resistance of the inter-dot tunneling barrier in a qubit, respectively. Thus, we have the following condition for the operation:

$$T \ll J \ll \Delta_0 < t \ll (CR)^{-1}. \quad (7)$$

We can roughly estimate this criterion by taking typical values, r_0 =2.5 nm (radius of a quantum dot), d_A=8 nm, d_B= 1.5 nm, d_C= 2.5 nm, and the distance between qubits d_D is 12 nm ($\epsilon_{\rm ox}$ =4 (SiO_2) and $\epsilon_{\rm Si}$=12), reflecting several experimental data [8,9]. Using relations, $C_{A,C}$=$2\pi\epsilon_{\rm ox}r_0^2/(d_{A,C}+(\epsilon_{\rm ox}/\epsilon_{\rm Si})r_0)$ and $C_B = 2\pi\epsilon_{\rm ox}r_0^2/(d_B+2(\epsilon_{\rm ox}/\epsilon_{\rm Si})r_0)$, we obtain $E_C \sim$ 13meV(150K), $t \sim$ 0.4meV and $J \sim 0.1$ meV. If $R_{\rm int}$ is of the order of MΩ, we obtain $(C_{\rm int}R_{\rm int})^{-1}$ is about 3.1THz and greater than t($\sim$ 100GHz). Thus, the condition (7) is satisfied if the operational temperature should be much smaller than 1K. If we could prepare r_0 = 0.5nm, d_B = 1.2nm and d_D = 2nm, we obtain t ~120 K and $J \sim$ 90K and quantum calculations can be expected to be carried out at around liquid nitrogen temperature. The effects of cross-talk, which are of the order of $C_{Hi}/C_{Ai} \sim d_{Ai}/d_{Di}$, can be controlled by adjusting the δv_{icr} term.

At the near-degeneracy point, the system becomes a two-state system and the estimation of the decoherence (in order of μsec) discussed in Ref.[5] may be applicable. Although we do not have any corresponding experimental data at present, we can say that various methods developed in NMR spectroscopy, such as the composite pulse method[6], can be utilized to reduce the imperfections of the pulse and coherence transfer. In addition, if the speed of quantum computations can be increased to more than the shortest decoherence time, $\tau_c \sim \omega_c^{-1}(\sim 10^{-14}$ s), group-theoretic approaches [12] for decreasing the decoherence will be effective. Thus, the effects of imperfections and decoherence of the system will be reduced by developing above-mentioned contrivances.

Next, we quantitatively illustrate the *reading out* process based on an FET structure of the one-dimensionally arrayed qubits(Fig.1). Measurement is carried out, after quantum calculations, by applying a finite bias $V_{\rm D}$ between the source and drain[5]. The detection mechanism is such that the change of the charge distribution in a qubit induces a threshold voltage shift $\Delta V_{\rm th}$ of the gate voltage above which the channel current flows in the substrate. The $\Delta V_{\rm th}$ is of the order of $ed_q/\epsilon_{\rm ox}$(d_q is a distance between the centers of two quantum dots in a qubit). The effect of $\Delta V_{\rm th}$ differs depending on the position of the qubit, because the width of the depletion region in the substrate changes gradually from source to drain. A model of a metal-oxide semiconductor field-effect-transistor (MOSFET)[11] is used for the current that flows under the ith qubit

$$I_{\rm D}^{(i)}=\Lambda\frac{[V_{gi}-V_{{\rm th}i}](V_i-V_{i-1})-(1/2)\eta_i(V_i^2-V_{i-1}^2)}{1+\Theta(V_i-V_{i-1})}, \quad (8)$$

where $\Lambda \equiv Z\mu_0C_0/L_0$ (Z is the channel width, μ_0 is the mobility, L_0 is the channel length of one qubit, and C_0 is the capacitance of the gate insulator), $\eta_i \equiv 1+\zeta_i$ where ζ_i is determined by the charge of the surface depletion region, Θ indicates a velocity saturation effect, and V_i is the voltage of ith qubit to be determined from $V_{\rm N} = V_{\rm D}$ and $I_{\rm D}^{(1)} = I_{\rm D}^{(2)} - \cdots - I_{\rm D}^{(N)}$. The threshold voltage of ith qubit $V_{{\rm th}i}$ is given by $V_{{\rm th}i} = V_{\rm th}(\zeta_i)+\Delta V_{{\rm th}i}$. Figure 2 shows the ratio of the current change $|I_i$-$I_0|/I_0$ as a function of V_D in 8 qubits with $\zeta_i = 0$, where I_0 is the initial current and I_i is the current when V_{gi}-$V_{{\rm th}i}$ of ith qubit changes by 10%. This ratio is largest for a qubit near the drain (i=8) because it has the narrowest inversion layer. To show how to distinguish qubits, we compare the current where only the ith qubit shifts its threshold voltage with that where only the (i+1)th qubit shifts its threshold voltage. Figure 3 shows the results for (i)i=1, (ii)i=$N/2$, and (iii)i=$N-1$, with the same voltage shift as in Fig.2. The maxi-

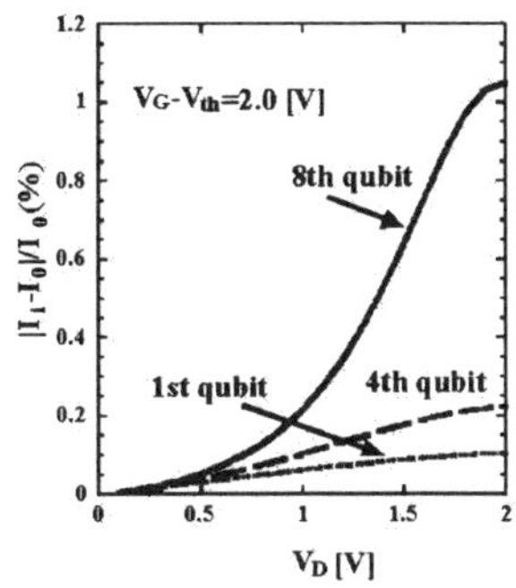

Figure 2. The ratio of current change $|I_i - I_0|/I_0$ as a function of V_D in an 8-qubit quantum computer ($i = 1$, $i = 4$, and $i = 8$). I_i represents the current where only the ith qubit shifts its threshold voltage $V_{\text{th}}(0) + \Delta V_{\text{th}i}$. I_0 represents the current where all qubits have the same threshold voltage $V_{\text{th}}(0)$. $V_g - V_{\text{th}}$=2V and Θ=0.3 V^{-1}. The threshold shift is 10% of $V_g - V_{\text{th}}$.

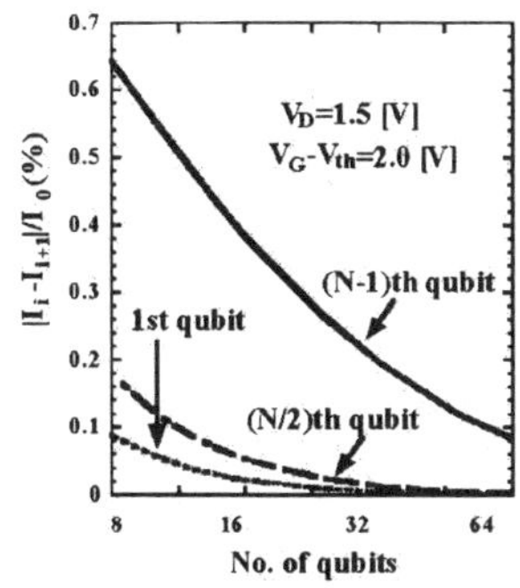

Figure 3. The ratio of change $|I_i - I_{i+1}|/I_0$ as a function of the number of qubits, in the case $i = 1$(near source), $i = N/2$ (middle) and $i = N - 1$(near drain) at $V_D = 1.5$V. The same notation and parameter region as Fig.2.

mum allowable number of arrayed qubits depends on the sensitivity of the external circuit to the channel current. To construct a large qubit array, additional dummy qubits are needed over the source and drain so that separated qubits on different FETs are connected.

In conclusion, we have theoretically shown that the Hamiltonian for a weakly coupled quantum-dot array in the Coulomb blockade regime is reduced to that for NMR spectroscopy. The flexible quantum information processing developed in the NMR quantum computer and a variety of error-correction methods in conventional NMR are applicable to the quantum-dot system that is considered to be the most feasible system in view of the present technology. Recently, nanofabrication of two-dimensionally distributed self-aligned Si doubly-stacked dots has been successfully realized in the form of a non-volatile memory device[9] and the detailed analysis of the behavior of electrons, such as an artificial antiferromagnet, is expected to be performed.

References

1. D. P. DiVincenzo, Fortschr. Phys. **48**, 771 (2000).
2. D. Loss and D. P. DiVincenzo, Phys. Rev. A **57**, 120 (1998).
3. D. V. Averin, Solid State Comm. **105**, 659 (1998).
4. Y. Makhlin, G. Schön and A. Shnirman, Nature **398**, 305 (1999).
5. T. Tanamoto, Phys. Rev. A **61**, 022305 (2000); *ibid* **64**, 062306 (2001).
6. R. R. Ernst, G. Bodenhausen and A. Wokaun, *Principles of Nuclear Magnetic Resonance in One and Two Dimensions* (Clarendon Press, Oxford, 1987).
7. C. Livermore, *et al.* Science **274**, 1332 (1996).
8. S. Tiwari *et al.*, Appl. Phys. Lett. **68**,1377 (1996).
9. R. Ohba *et al.*, *2000 IEEE International Electron Devices Meeting* (San Francisco, 2000), p313.
10. Y. Nakamura, C. D. Chen and J. S. Tsai, Phys. Rev. Lett. **79**, 2328 (1997).
11. B. Hoefflinger, IEEE. Trans. Electron. Devices **ED-28**, 971 (1981).
12. L. Viola, S. Lloyd and E. Knill, Phys. Rev. Lett. **82**, 2417 (1999).

STATISTICAL ESTIMATION OF A QUANTUM CHANNEL

AKIO FUJIWARA

Department of Mathematics, Osaka University, Toyonaka, Osaka 560-0043, Japan
E-mail: fujiwara@math.wani.osaka-u.ac.jp

This paper addresses a quantum channel identification problem: given a parametric family $\{\Gamma_\theta\}_\theta$ of quantum channels, find the best strategy of estimating the true value of the parameter θ. As illustrative examples, we present statistical estimation of a depolarizing channel and a unitary channel acting on a two-level quantum system.

1 Introduction

Let $\mathcal{H}$ be a Hilbert space that represents the physical system of interest and let $\mathcal{S}(\mathcal{H})$ be the set of density operators on $\mathcal{H}$. It is well known that a dynamical change $\Gamma : \mathcal{S}(\mathcal{H}) \to \mathcal{S}(\mathcal{H})$ of the physical system, called a *quantum channel*, is represented by a trace preserving completely positive map[1]. But how can we identify the quantum channel that we have in a laboratory? A general scheme may be as follows: input a well prepared state σ to the quantum channel and estimate the dynamical change $\sigma \mapsto \Gamma(\sigma)$ by performing a certain measurement on the output state $\Gamma(\sigma)$. It is then natural to inquire what is the best strategy of estimating a quantum channel. The purpose of this paper is to study this problem from a noncommutative statistical point of view. For mathematical simplicity, we restrict ourselves to the case when the quantum channel to be identified lies in a smooth parametric family $\{\Gamma_\theta; \theta = (\theta_1, ..., \theta_n) \in \Theta\}$ of quantum channels. When $\mathcal{H}$ is finite dimensional, this is not an essential restriction[2].

Once an input state σ for the channel is fixed, we have a parametric family $\{\Gamma_\theta(\sigma)\}_{\theta\in\Theta}$ of output states, and as long as the parametrization $\theta \mapsto \Gamma_\theta(\sigma)$ is nondegenerate, the problem of estimating the quantum channel is reduced to a parameter estimation problem for the noncommutative statistical model $\{\Gamma_\theta(\sigma)\}_{\theta\in\Theta}$ of density operators. As a consequence, the parameter estimation problem for a family $\{\Gamma_\theta\}_{\theta\in\Theta}$ of quantum channels amounts to finding an optimal input state σ for the channel and an optimal estimator for the parametric family $\{\Gamma_\theta(\sigma)\}_\theta$ of output states.

Now that each channel Γ_θ is completely positive, it can be extended to the composite quantum system $\mathcal{H}\otimes\mathcal{H}$. In view of parameter estimation, there are two essentially different extensions that have the same parametrization θ as Γ_θ: one is $\Gamma_\theta \otimes \mathrm{Id} : \mathcal{S}(\mathcal{H}\otimes\mathcal{H}) \to \mathcal{S}(\mathcal{H}\otimes\mathcal{H})$, where Id denotes the identity channel, and the other is $\Gamma_\theta \otimes \Gamma_\theta : \mathcal{S}(\mathcal{H}\otimes\mathcal{H}) \to \mathcal{S}(\mathcal{H}\otimes\mathcal{H})$. A question arises naturally: what happens when we use an entangled state as an input to the extended channel? In the subsequent sections, we demonstrate nontrivial aspects of this problem.

This article is a summary of our recent works. All the proofs omitted here are found in Fujiwara[3], where more detailed discussions are also presented.

2 Depolarizing channel

Let $\mathcal{H} := \mathbb{C}^2$ and let the channel $\Gamma_\theta : \mathcal{S}(\mathcal{H}) \to \mathcal{S}(\mathcal{H})$ be a depolarizing channel having a parameter θ that represents the magnitude of isotropic depolarization. To ensure that Γ_θ is completely positive, the parameter θ must lie in the closed interval $\Theta := [-1/3, 1]$. We thus have a one parameter family $\{\Gamma_\theta; \theta \in \Theta\}$ of quantum channels, and our task is to estimate the true value of θ. In view of the Cramér-Rao inequality for a one-dimensional quantum statistical model, our task is reduced to finding an optimal input for the channel that maximizes the symmetric logarithmic derivative (SLD) Fisher information of the corresponding parametric family of output states.

Let us start with the maximization of the SLD Fisher information of the family $\{\Gamma_\theta(\sigma)\}_\theta$ with respect to the input state $\sigma \in \mathcal{S}(\mathcal{H})$. An important observation is that the maximum is attained by a pure state. Since our channel Γ_θ is unitarily invariant (i.e., isotropic in the Stokes parameter space), we can take without loss of generality the optimal input to be $\sigma = |e\rangle\langle e|$ where $\langle e| = (1,0)$. The corresponding output state is $\rho_\theta := \Gamma_\theta(\sigma) = \frac{1}{2}\begin{bmatrix} 1+\theta & 0 \\ 0 & 1-\theta \end{bmatrix}$, and the SLD Fisher information becomes $J_\theta = 1/(1-\theta^2)$.

We next study the extended channel $\Gamma_\theta \otimes \mathrm{Id} : \mathcal{S}(\mathcal{H}\otimes\mathcal{H}) \to \mathcal{S}(\mathcal{H}\otimes\mathcal{H})$. In this case we can use a possibly entangled state as the input. For the same reason as above, we can take the input to be a pure state: $\hat{\sigma} = |\psi\rangle\langle\psi|$ where $\psi \in \mathcal{H}\otimes\mathcal{H}$. By the Schmidt decomposition, the vector ψ is represented as

$$|\psi\rangle = \sqrt{1-x}\,|e_1\rangle|f_1\rangle + \sqrt{x}\,|e_2\rangle|f_2\rangle, \quad (1)$$

where x is a real number between 0 and 1, and $\{e_1, e_2\}$ and $\{f_1, f_2\}$ are orthonormal bases of $\mathcal{H} = \mathbb{C}^2$. Since the channels Γ_θ and Id are both unitarily invariant, we can assume without loss of generality that the optimal input takes the form (1) with $\langle e_1| = \langle f_1| = (0,1)$ and $\langle e_2| = \langle f_2| = (1,0)$. The constant x remains to be determined. By a direct computation, the SLD Fisher information of the corresponding output state $\hat{\rho}_\theta := \Gamma_\theta \otimes \mathrm{Id}(\hat{\sigma})$ becomes

$$\hat{J}_\theta = \frac{1+3\theta+8x(1-x)}{(1-\theta^2)(1+3\theta)}.$$

When $x = 0$ or 1, the above SLD Fisher information $\hat{J}_\theta$ is identical to J_θ. This is a matter of course: the input state is disentangled in this case and no information about the parameter θ is available via the independent channel Id. When $x \neq 0$ and $\neq 1$, the SLD Fisher information $\hat{J}_\theta$ diverges at $\theta = 1$ and $-1/3$. This is because the complete positivity of the channel Γ_θ breakes across these values. Now let us specify the optimal input state. For every θ, the SLD Fisher information $\hat{J}_\theta$ takes the maximum $3/(1-\theta)(1+3\theta)$ at $x = 1/2$. Therefore the optimal input for the channel $\Gamma_\theta \otimes \mathrm{Id}$ is the maximally entangled state. The implication of this result is deep: although we use the channel Γ_θ only once, extra information about the channel is obtained via entanglement of the input state. In particular, the use of entanglement improves exceedingly the performance of estimation as θ approaches $-1/3$.

Let us proceed to the analysis of the other extension $\Gamma_\theta \otimes \Gamma_\theta : \mathcal{S}(\mathcal{H}\otimes\mathcal{H}) \to \mathcal{S}(\mathcal{H}\otimes\mathcal{H})$. As before, we can take the input to be a pure state $\check{\sigma} = |\psi\rangle\langle\psi|$ where ψ is given by Eq. (1) with $\langle e_1| = \langle f_1| = (0,1)$ and $\langle e_2| = \langle f_2| = (1,0)$. The SLD Fisher information of the corresponding output state $\check{\rho}_\theta := \Gamma_\theta \otimes \Gamma_\theta(\check{\sigma})$ becomes

$$\check{J}_\theta = \frac{4\theta^4+5\theta^2-1}{2\theta^2(1-\theta^4)} + \frac{8\theta^2\,x(1-x)}{1-\theta^4} + \frac{1-\theta^2}{2\theta^2(1+\theta^2)(1-\theta^2+16\theta^2\,x(1-x))}.$$

When $x = 0$ or 1, the above SLD Fisher information $\check{J}_\theta$ becomes $2/(1-\theta^2)$, which precisely doubles the J_θ. Again this is a matter of course: the input state is disentangled in this case and the same amount of information about the parameter θ is obtained per independent use of the channel Γ_θ. When $x \neq 0$ and $\neq 1$, the SLD Fisher information $\check{J}_\theta$ diverges at $\theta = 1$ but does not at $\theta = -1/3$. This is because the requirement of positivity for the channel $\Gamma_\theta \otimes \Gamma_\theta$ is strictly weaker than that for the channel $\Gamma_\theta \otimes \mathrm{Id}$ (i.e., the complete positivity for Γ_θ). Now we examine a rather unexpected behavior of the optimal input state. For $1/\sqrt{3} \leq \theta < 1$, the SLD Fisher information $\check{J}_\theta$ takes the max-

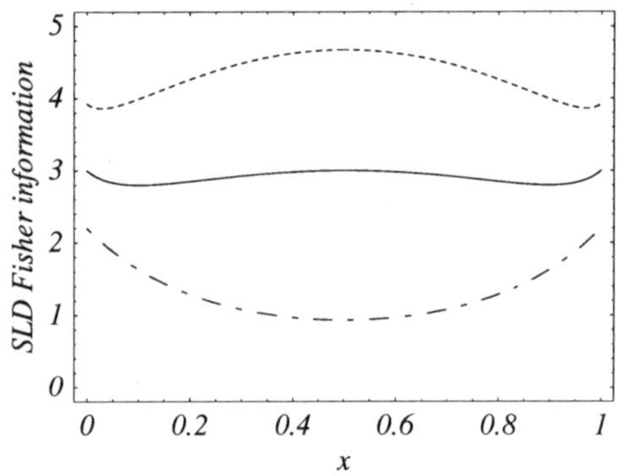

Figure 1. SLD Fisher information $\check{J}_\theta$ versus x, for $\theta = 0.7$ (dashed), $\theta = 1/\sqrt{3}$ (solid), and $\theta = 0.3$ (chained).

imum $12\theta^2/(1-\theta^2)(1+3\theta^2)$ at $x = 1/2$, while for $-1/3 \le \theta \le 1/\sqrt{3}$, it takes the maximum $2/(1-\theta^2)$ at $x = 0$ and 1. (See Figure 1.) Namely, the optimal input state "jumps" from the maximally entangled state to a disentangled state at $\theta = 1/\sqrt{3}$. It is surprising that the seemingly homogeneous family $\{\Gamma_\theta\}_\theta$ of depolarizing channels involves a transitionlike behavior.

3 $SU(2)$ channel

Suppose an unknown operation Γ acting on $\mathcal{S}(\mathbb{C}^2)$ is noiseless, in that there is a unitary operator $U \in SU(2)$ such that $\Gamma(=: \Gamma_U) : \rho \mapsto U\rho U^*$, and our problem is to estimate the unknown U. The group $SU(2)$ is a 3-dimensional manifold and is parametrized, for example, as

$$U = U_{(\phi,\alpha,\beta)} := \begin{bmatrix} e^{i\alpha}\cos\phi & -e^{i\beta}\sin\phi \\ e^{-i\beta}\sin\phi & e^{-i\alpha}\cos\phi \end{bmatrix},$$

where $0 \le \phi \le \pi/2$, and $0 \le \alpha, \beta < 2\pi$. Since, for any $\rho \in \mathcal{S}(\mathbb{C}^2)$, the family $\{\Gamma_U(\rho); U \in SU(2)\}$ of output states is at most 2-dimensional, we must extend Γ_U on an enlarged Hilbert space $(\mathbb{C}^2)^{\otimes n}$, $(n \ge 2)$, in order for the parametrization of output states to be nondegenerate. In this section, we study the extension $\Lambda_U := \Gamma_U \otimes \mathrm{Id}$, i.e.,

$$\Lambda_{U_\theta} : \sigma \longmapsto (U_\theta \otimes I)\sigma(U_\theta \otimes I)^* \qquad (2)$$

where $\theta := (\theta^1, \theta^2, \theta^3) := (\phi, \alpha, \beta)$. Obviously this is regarded as a continuous analogue of the dense coding. Since $\Lambda_{-U} = \Lambda_U$, we might as well express that our problem is to estimate the parameter θ of $SU(2)/\{\pm I\} \simeq SO(3)$. Consequently, the estimation of $SU(2)$ operation must be a local one: the domain Θ of the parameter θ to be estimated forms a local chart of $SU(2)$ on which the parametrization $\theta \mapsto \Lambda_{U_\theta}(\sigma)$ is one-to-one.

Letting $J_\theta(\sigma)$ be the SLD Fisher information matrix for the output family $\{\Lambda_{U_\theta}(\sigma)\}_\theta$, it is shown that the Cramér-Rao lower bound $J_\theta(\sigma)^{-1}$ takes the minimum if and only if σ is a maximally entangled state. This fact hints that the optimal input is a maximally entangled state. However, it alone does not lead to a decisive conclusion, because the Cramér-Rao lower bound $J_\theta(\sigma)^{-1}$ is not always achievable for a multi parameter quantum statistical model. Here we say that the Cramér-Rao lower bound is *achievable* at θ if there is a locally unbiased estimator M that satisfies $V_\theta[M] = J_\theta(\sigma)^{-1}$, where $V_\theta[M]$ denotes the covariance matrix. In this sense the next result is the key to the conclusion that a maximally entangled state is in fact the optimal one: Letting $\sigma = |\psi\rangle\langle\psi|$ where ψ is given by Eq. (1), the Cramér-Rao lower bound $J_\theta(\sigma)^{-1}$ is achievable if and only if $x = 1/2$, i.e., if and only if σ is a maximally entangled state. The implication of this result is profound. The existence of an estimator that achieves the Cramér-Rao lower bound implies the existence of compatible observables that correspond to the parameters of $SU(2)$. The above result thus asserts that the noncommutative nature of the $SU(2)$ parameters is "suppressible" (at least locally) by using a maximally entangled input.

In order to get a deeper insight into the role of entanglement, we next explore differential geometry of the manifold of output states, i.e., the orbit of $SU(2)$ action $\psi \mapsto (U \otimes I)\psi$ on the 3-dimensional complex projective space $\mathbb{C}P^3$. We regard $\mathbb{C}P^3$ as a 6-dimensional real Riemannian manifold equipped with the SLD Fisher metric g, which is identical to the Fubini-Study metric.

The volume element of the $SU(2)$ orbit that passes through the vector (1) is

$$\omega = 8\sqrt{x(1-x)}\sin 2\phi \; d\phi d\alpha d\beta.$$

This simple formula already offers some information about the relation between entanglement and the geometry of orbits: an orbit maximally inflates at $x = 1/2$, and collapses as $x \to 0$ and $\to 1$. Note that the scaling factor $\sqrt{x(1-x)}$ is identical, up to a constant factor, to the concurrence.

The Riemannian curvature R of the Levi-Civita connection is completely determined by the following six components:

$$\begin{aligned} R_{1212} &= -4\cos^2\phi\,[1+(1-2x)^2 \\ &\qquad \times\{3-4(1+x-x^2)\cos^2\phi\}] \\ R_{1313} &= -4\sin^2\phi\,[1+(1-2x)^2 \\ &\qquad \times\{3-4(1+x-x^2)\sin^2\phi\}] \\ R_{2323} &= -16x^2(1-x)^2\sin^2 2\phi \\ R_{1213} &= -4(1-2x)(1+x-x^2)\sin^2 2\phi \\ R_{1223} &= R_{1323} = 0 \end{aligned}$$

where $R_{ijkl} := g(R(\partial_i, \partial_j)\partial_k, \partial_l)$ with $\partial_i = \partial/\partial\theta^i$. It is shown that the orbit is Einstein if and only if $x = 1/2$ or $x = 0, 1$. The scalar curvature $\rho = 2(1-x+x^2)$ indicates that the closer to $1/2$ the parameter x is, the "flatter" the orbit becomes on average. Let us take a closer look at this point.

The sectional curvature with respect to the subspace spanned by $\{\partial_1, \partial_2\}$ is given by

$$\frac{R_{1212}}{(g_{12})^2 - g_{11}g_{22}} = 1 + x - x^2 - \frac{8x(1-x)}{1+4x(1-x)-(1-2x)^2\cos 2\phi}.$$

This is independent of ϕ if and only if $x = 1/2$ or $x = 0, 1$. When $x = 1/2$, the orbit turns out to be a space of constant positive curvature $1/4$, in that $R_{ijkl} = \frac{1}{4}(g_{jk}\, g_{il} - g_{ik}\, g_{jl})$, for all $i, j, k, l = 1, 2, 3$. Since the fundamental group of the orbit is $\mathbb{Z}_2$, it is the 3-dimensional real projective space $\mathbb{R}P^3(2)$ of radius 2, i.e., $S^3(2)/\{\pm I\}$. It is also important to observe that for $x \neq 1/2$, (and $\neq 0, 1$), the orbit is *not isometric* (though diffeomorphic) to $\mathbb{R}P^3$. Since the manifold $SU(2)$ equipped with the Cartan-Killing metric is isometric to S^3, these facts could be paraphrased by saying that the "shape" of the Riemannian manifold $SU(2)/\{\pm I\} \cong SO(3)$, the coordinates of which are to be estimated, comes into full view only through the $SU(2)$ action $\psi \mapsto (U \otimes I)\psi$ on a maximally entangled ψ.

When $x = 0$ or 1, on the other hand, the orbit collapses to a lower dimensional manifold in which $\partial_2 = \partial_3$. In this case, the only independent component R_{1212} of the Riemannian curvature tensor satisfies $R_{1212} = (g_{12})^2 - g_{11}g_{22}$. Since the collapsed manifold is simply connected, it is the 2-dimensional sphere S^2 of unit radius. This is, of course, in accordance with the known isomorphism between $\mathbb{C}P^1$ and S^2.

Acknowledgment

The author acknowledges support of CREST of JST (Japan Science and Technology Corporation).

References

1. K. Kraus, *States, Effects, and Operations.* Lecture Notes in Physics Vol. 190 (Springer, Berlin, 1983).
2. A. Fujiwara and P. Algoet, Phys. Rev. A **59**, 3290 (1999).
3. A. Fujiwara, Phys. Rev. A **63**, 042304 (2001); A. Fujiwara, "Estimation of $SU(2)$ operation and dense coding: An information geometric approach," preprint (2001).

HIGH-FIDELITY EXPERIMENTAL QUANTUM TELEPORTATION AND ENTANGLEMENT SWAPPING

T. JENNEWEIN, J.-W. PAN, S. GASPARONI, G. WEIHS, AND A. ZEILINGER
Institut für Experimentalphysik, Universität Wien,
Boltzmanngasse 5, 1090 Wien, Austria.

Entanglement swapping in its most fundamental form is the teleportation of a quantum state that is itself entangled to another quantum system. This connection to teleportation provides us with a tool to demonstrate and verify the non-local nature of the teleportation procedure. In this work the non-locality is experimentally confirmed by a violation of Bell's inequality using teleported entanglement.

Within the field of quantum information and quantum communication a very prominent position is taken by the quantum teleportation protocol. It is the only known way in which quantum information could be sent to a different place without actually sending a physical object that carries the quantum information. Not only does this open a completely new way of communication for future quantum computers, but also it can enhance the efficiency of potential linear optics realizations of quantum information processing.

If one were to assess the quality of a certain teleportation[1,2,3,4] device there are two fundamental criteria. One is the efficiency of the apparatus, that is the percentage of cases in which the procedure is completed successfully, after an input has been provided to the machine. The other important parameter is the fidelity, which is defined as the overlap between output and input states.[a] If we restrict ourselves to pure input states the fidelity F can be calculated as follows.

$$F = \langle \Psi_{\rm in} | \, \rho_{\rm out} \, | \Psi_{\rm in} \rangle \,, \qquad (1)$$

where $|\Psi_{\rm in}\rangle$ is the (pure) input and $\rho_{\rm out}$ is the output density operator. A fidelity of 0.5 corresponds to random output, whereas a value 1 indicates perfect operation.

There are two other important thresholds for the teleportation fidelity, the first one being the limit of classically teleporting an unknown quantum state. Its maximum possible fidelity of 2/3 is the state estimation limit for a single copy of an arbitrary pure qubit state.[7] A possible realization, is to measure in an arbitrary basis, communicate the result and prepare the state on the other side corresponding to the measurement result.

Even though this limit cannot be surpassed by classical physics, a local hidden variable theory could in principle do better. Yet, such a theory in turn has to obey the limit of $(1+\sqrt{2})/\sqrt{8} \approx 0.854$ according to Bell's inequality. If hidden variables are considered to be a part of classical physics, then this value is the boundary above which a teleportation device performs provably non-classical.

The fidelity could be measured for a set of possible input states by comparing input and output for a large number of sample transmissions. Another method is to use entanglement swapping[8] to probe the input state space uniformly. Here a quantum state that is itself entangled with an ancilla is teleported and the entanglement of the output with the ancilla is observed. Measuring this "teleported entanglement"[b] we can calculate

[a]From a particular teleportation point of view this fidelity applies to our experiment only conditioned on the actual detection on the receiver's side due to present technical limitations.[5,6] In the entanglement swapping experiment presented here, however, the particles must be detected in order to test the entanglement. Therefore the fidelity will naturally be conditioned on detections of particles 0 and 3 (see Fig. 1).

[b]Of course one does not actually teleport an *entangled* state but only the relation of the input to the ancilla.

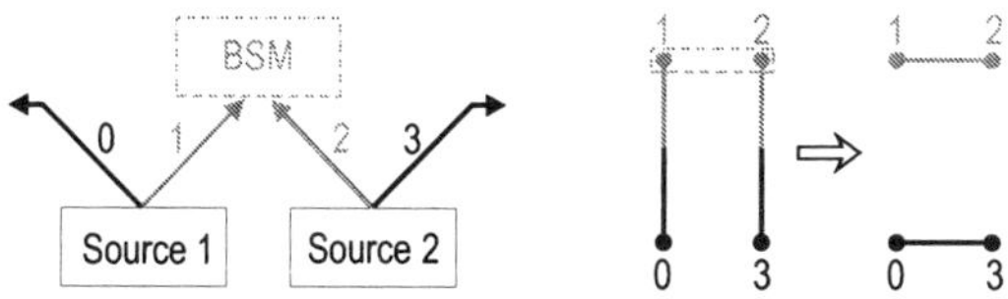

Figure 1. Entanglement swapping or equivalently teleportation of entanglement. The quantum state "1" is teleported to mode "3", (except for the missing application of a unitary transformation) or, by symmetry, state "2" is teleported to mode "0". It turns out, that if the outputs of "0" and "3" are selected according to the result of the Bell-state measurement (BSM), it is not necessary to actually perform the unitary transformation that is part of the teleportation protocol.

the fidelity without varying the input state. Depending on the quality of the original entanglement the whole input state space is probed at once. In the following we will therefore only consider entanglement swapping, which was demonstrated first by Pan et al.[9]

As shown in Fig. 1 (right) we can think of entanglement swapping as a general procedure to redistribute entanglement within a collection of quantum systems. This is completed by entangling a subset of previously separable particles via a projection onto an entangled state basis. In Fig. 1 only the most basic protocol for two two-particle entangled states is shown, which can be generalized to higher numbers of particles.[10]

If we restrict ourselves to start out with two entangled two-particle systems, then entanglement swapping reduces to a Bell-state measurement between two initially separable particles. Consider Fig. 1(left): here we start out with the two entangled systems 0-1 and 2-3. Let their quantum states be $|\Psi^-_{01}\rangle$ and $|\Psi^-_{23}\rangle$, where $|\Psi^\pm\rangle$, $|\Phi^\pm\rangle$ denote the maximally entangled two qubit basis (Bell-basis, Bell-states). The joint state of the four particles is consequently $|\Psi^-_{01}\rangle\otimes|\Psi^-_{23}\rangle$. The Bell-state measurement on particles 1 and 2 will then yield one of the four results $|\Psi^\pm_{12}\rangle$, $|\Phi^\pm_{12}\rangle$ with equal probabilities. The action of the Bell-state measurement is most easily seen in a basis of entangled states between particles 2 and 3. Then the state reads

$$\Psi_{\text{total}} = \frac{1}{2}\left[|\Psi^+_{03}\rangle|\Psi^+_{12}\rangle - |\Psi^-_{03}\rangle|\Psi^-_{12}\rangle - \right.$$
$$\left. - |\Phi^+_{03}\rangle|\Phi^+_{12}\rangle + |\Phi^-_{03}\rangle|\Phi^-_{12}\rangle\right]. \quad (2)$$

Whenever we measure a specific Bell-state for particles 1 and 2 particles 0 and 3 will end up in exactly the same state.

To subsequently assess the produced entanglement in our experiment we chose to do a Bell-inequality test. Another possibility would be to do a tomography of the two-particle space to recover the full two-particle density matrix.

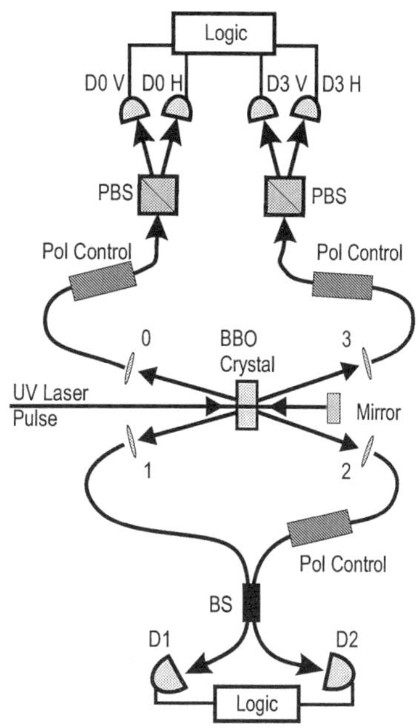

Figure 2. Experimental apparatus to perform teleportation of entanglement created by spontaneous parametric down-conversion from a pulsed laser. Bell-state measurement is performed with the help of a beam-splitter and detectors D1 and D2, while polarization is analyzed on the other side with polarizing beam-splitters followed by detectors D0 H/V and D3 H/V. Due to spurious birefringence in the single-mode optical fibers polarization controllers have to be used to align the various analyzer bases.

Our experimental apparatus is depicted in Fig. 2. We use spontaneous parametric down-conversion from a fs UV laser in a double-pass configuration inside a nonlinear optical crystal to create two $|\Psi^-\rangle$ entangled photon pairs. The photons are coupled to single mode optical fibers. Photons 1 and 2 are overlapped on a fiber beam-splitter. Coincidence detection after the beam-splitter projects destructively on to the state $|\Psi^-_{12}\rangle$. In turn particles 0 and 3 are also projected onto a $|\Psi^-_{12}\rangle$ state whenever the Bell-state measurement was successful.

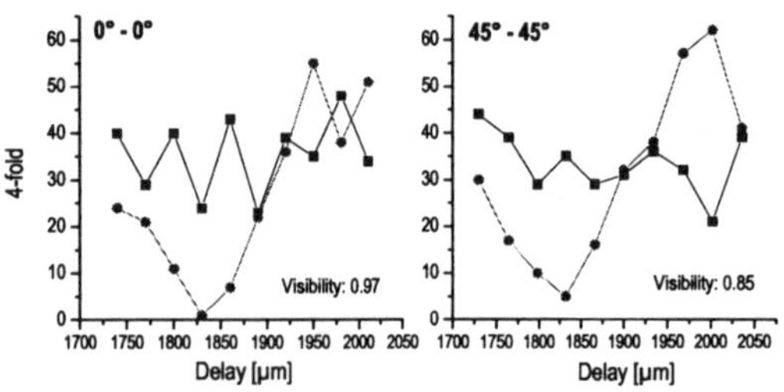

Figure 3. Two photon interferometry of independent photons conditioned on the detection of four photons as used for the Bell-state measurement of $|\Psi^-\rangle$. Polarization anticorrelation shows up as a dip in the coincidence rate for parallel polarizer setting **(circles)** as the relative delay to the beam-splitter is scanned across the point of indistinguishability. For orthogonal polarizers the rate stays constant **(squares)**.

Projection on to the mentioned state of particles 1 and 2 can only be achieved if their indistinguishability is assured. In the experiment we vary the difference between the optical paths of either photon in order to continuously vary the degree of distinguishability. Typical interferograms can be seen in Fig. 3. They are already verifications of entanglement swapping. At matching delay (zero difference) four fold coincidence between all outputs show a clear dip in the curve when we look at identical polarizer outputs for particles 0 and 3. In a $|\Psi^-\rangle$ state we expect the coincidences to go to zero.

The finite experimental visibility V ($V = 2F - 1$) reflects two separate imperfections. On the one hand the initial entanglement between particles 0-1 and 2-3 will never be perfect and also the Bell-state measurement in practice falls short of perfection. The nature of our source makes the quality of the initial entanglement basis dependent. In the vertical-horizontal basis we observe higher contrast than in any other basis. The critical visibility is therefore measured at 45° off horizontal (or at circular polarization alternatively).

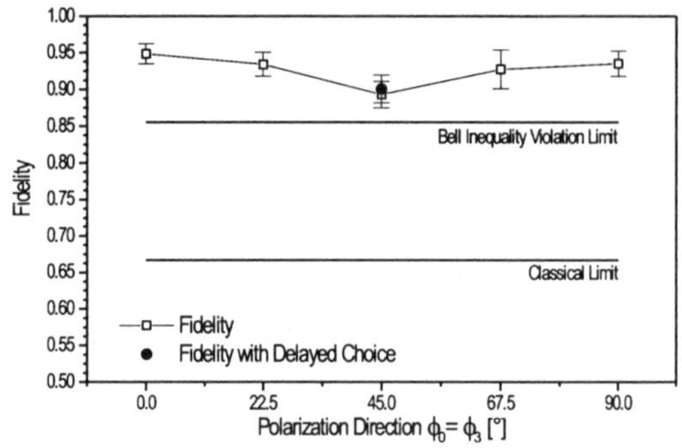

Figure 4. The experimentally achieved fidelity as a function of the absolute analyzer angle. The solid dot shows a measurement for which the Bell-state measurement was delayed until after the other two photons had been detected. Within experimental errors there is no difference in fidelity. The photon pairs are originally produced as HV or VH pairs in the parametric down-conversion crystal. Only by careful compensation are the two components made coherent. The perfection of this compensation is, however, limited by group dispersion and transverse ("walk-off") effects. These imperfections show up in measurements at 45°, where we observe reduced correlation for the individual pairs, which transfer to a reduced correlation for the teleported entanglement if measured at 45°. Comparing the correlation measurements taken for individual pairs and the above data we conclude that the fidelity of our Bell-state measurement must be better than 97%.

In our experiment we achieved a fidelity that was higher than 0.89 for any basis. The results are shown in Fig. 4. This, for the first time made it possible to violate Bell's inequality by a teleportation experiment, i.e. by particles that did not interact locally. For this purpose the correlation is evaluated at the following linear polarization analyzer angles (particles 0, 3): $(0°, 22.5°)$, $(0°, 67.5°)$, $(45°, 22.5°)$, and $(45°, 67.5°)$. The values of the correlation function were determined to $E_{11} = -0.6281 \pm 0.0461$, $E_{12} = +0.6766 \pm$

0.0423, $E_{21} = -0.5407 \pm 0.0454$, and $E_{22} = -0.5748 \pm 0.0472$ respectively. When we evaluate the Bell inequality[11]

$$S = |E_{11} - E12| + |E_{21} + E_{22}| \leq 2 \quad (3)$$

with this data we get $\mathbf{S = 2.421 \pm 0.091}$, which is 4.6 standard deviations above the limit for any local realistic theory, provided the fair sampling assumption is obeyed by the theory. Also, the values within reasonable experimental error agree with the quantum physical predictions when the imperfections of the apparatus are taken into account.

Because entanglement swapping and teleportation are two aspects of a fundamental quantum communication protocol the appear in various situations. They are part of quantum repeaters, enable basic forms of entanglement purification and lately it has been discovered that they can also enhance quantum operations with linear optics to obtain scalability.[12]

Acknowledgments

This work was supported by the Austrian Science Fund (FWF) project no. F1506, by the TMR network "The physics of quantum information", and by the IST-FET project "QuComm" of the European commission. G.W. wishes to acknowledge the warm hospitality of the ERATO project on Quantum Computation and Information (Imai project) while preparing this manuscript.

References

1. C.H. Bennett, G. Brassard, C. Crépeau, R. Jozsa, A. Peres, and W.K. Wootters, Teleporting an unknown quantum state via dual classical and Einstein-Podolsky-Rosen channels, *Phys. Rev. Lett.*, **70**, 1895 (1993).
2. D. Bouwmeester, J.-W. Pan, K. Mattle, M. Eibl, H. Weinfurter, and A. Zeilinger, Experimental quantum teleportation, *Nature*, **390**, 575 (1997).
3. A. Furusawa, J.L. Sorensen, S.L. Braunstein, C.A. Fuchs, H.J. Kimble, and E.S. Polzik, Unconditional quantum teleportation, *Science*, **282**, 706 (1998).
4. D. Boschi, S. Branca, F. De-Martini, L. Hardy, and S. Popescu, Experimental realization of teleporting an unknown pure quantum state via dual classical and Einstein-Podolsky-Rosen channels, *Phys. Rev. Lett.*, **80**, 1121 (1998).
5. S.L. Braunstein and H.J. Kimble, A posteriori teleportation, *Nature*, **394**, 840 (1998).
6. D. Bouwmeester, J.-W. Pan, M. Daniell, H. Weinfurter, M. Zukowski, and A. Zeiling, A posteriori teleportation - Bouwmeester al. reply, *Nature*, **394**, 841 (1998).
7. S. Massar and S. Popescu, Optimal extraction of information form finite quantum ensembles, *Phys. Rev. Lett.*, **74**, 1259 (1995).
8. M. Żukowski, A. Zeilinger, M.A. Horne, and A.K. Ekert, "Event-ready-detectors" Bell experiment via entanglement swapping, *Phys. Rev. Lett.*, **71**, 4287 (1993).
9. J.W. Pan, D. Bouwmeester, H. Weinfurter, and A. Zeilinger, Experimental entanglement swapping: Entangling photons that never interacted, *Phys. Rev. Lett.*, **80**, 3891 (1998).
10. S. Bose, V. Vedral, and P.L. Knight, Multiparticle generalization of entanglement swapping, *Phys. Rev. A*, **57**, 822 (1998).
11. J.F. Clauser, M.A. Horne, A. Shimony, and R.A. Holt, Proposed experiment to test local hidden-variable theories, *Phys. Rev. Lett.*, **23**, 880 (1969).
12. E. Knill, R. Laflamme, and G. Milburn, A scheme for efficient quantum computation with linear optics, *Nature*, **409**, 46 (2000).

ENTANGLEMENT AND BELL'S INEQUALITY EXPERIMENTS USING TRAPPED IONS

C. A. SACKETT, D. KIELPINSKI, V. MEYER, M. A. ROWE, W. M. ITANO, C. MONROE, AND D. J. WINELAND

Time and Frequency Division, National Institute of Standards and Technology
325 Broadway, Boulder, CO 80305 USA
E-mail: david.wineland@boulder.nist.gov

Recent improvements in the ability to generate entangled states of trapped ions have permitted the application of entanglement to several problems. First, entangled ions were used to demonstrate a violation of Bell's inequality, marking the first such demonstration with high detection efficiency. Second, a pair of ions was used to store a quantum bit of information in a decoherence-free subspace, allowing the qubit to resist environmental decoherence. Third, an entangled state was used to demonstrate improved spectroscopic resolution. These results are summarized here.

1 Introduction

Quantum mechanics allows for many-particle wave functions that cannot be factorized into a product of single-particle states, even when the constituent particles are entirely distinct. Such entangled states explicitly demonstrate the non-local character of quantum theory[1] and have potential applications in quantum communication and computation. Entanglement has been explored in a variety of physical systems, ranging from gas to liquid to solid state.[2] One promising system consists of atomic ions confined in a trap. Strong coupling of the ions is obtained through the Coulomb potential, but the ions remain well-isolated from their environment, allowing a high degree of coherence and control.

The experiments described here were performed using a pair of $^9Be^+$ ions confined along the axis of a linear radio-frequency trap. The axial center-of-mass oscillation frequency was $\nu = 2\pi \times 5$ MHz. Two ground hyperfine states, $|0\rangle \equiv |F=2, m_F=-2\rangle$ and $|1\rangle \equiv |F=1, m_F=-1\rangle$ were coupled by a Raman transition, and formed an effective two-level system. The Raman beams were tuned near the $2S \leftrightarrow 2P$ transition at $\lambda =$ 313 nm, and were perpendicularly oriented with their wave vector difference lying along the trap axis. The beams therefore only coupled to ion motion along the axis. This motion was cooled to near its ground state.[3]

The linewidth of the Raman transition was small compared to the oscillation frequency, so that the vibrational states of the motion were resolved. Thus by tuning the difference frequency of the Raman beams to the hyperfine splitting $\omega_0 \approx 2\pi \times 1.25$ GHz, the carrier transition $|0, n_C, n_S\rangle \leftrightarrow |1, n_C, n_S\rangle$ could be driven, where n_C and n_S are respectively the number of quanta in the symmetric center-of-mass and the antisymmetric stretch vibrational modes. Alternatively, if the difference frequency was tuned to $\omega_0 \pm \nu$, the blue or red sideband transitions $|0, n_C, n_S\rangle \leftrightarrow |0, n_C \pm 1, n_S\rangle$ were driven. Similar transitions occur at the stretch mode frequency of $\nu_S = \sqrt{3}\nu$.

The carrier transition corresponds to individual rotations of the effective spin-1/2 particles, described by

$$R(\theta, \phi) = \begin{bmatrix} \cos(\theta/2) & e^{i\phi}\sin(\theta/2) \\ -e^{-i\phi}\sin(\theta/2) & \cos(\theta/2) \end{bmatrix} \quad (1)$$

in the interaction picture. The rotation angle θ is given by Ωt, where t is the duration of the laser pulse and Ω is the Rabi frequency $2\langle 0|V|1\rangle/\hbar$ for interaction V. Typically Ω was $2\pi \times 500$ kHz.

The angle ϕ appearing in (1) is the phase

of the beat note of the Raman beams at the position of the ion, and is given by

$$\phi = \phi_0 + 2^{1/2}kz \qquad (2)$$

where ϕ_0 is the phase at an arbitrary reference position, z is the axial displacement of the ion from this reference, and $k = 2\pi/\lambda$. The uniform part ϕ_0 can be controlled by adjusting a microwave oscillator from which the difference frequency is derived. The spatial dependence of ϕ can be used to perform independent rotations of the two ions. Independent rotations are most easily achieved by applying separate laser beams to each ion,[4] but this is technically difficult here because the ions are separated by only 3 μm. Instead, rotation of a single ion is achieved in three steps. First, a laser pulse is used to rotate both ions by $\theta/2$. The trap electrode voltages are then adjusted so that one ion remains fixed but the other is displaced by an amount $d = \pi/(\sqrt{2}k)$, changing ϕ by π. A second laser pulse identical to the first then completes the rotation of the first ion but reverses the rotation of the second ion. Using this technique, operation fidelities of about 0.95 can be obtained.

Entanglement of the ions is achieved by the technique of Mølmer and Sørensen.[5,6,7] Three laser beams are applied, with difference frequencies of $\omega_0 + \nu_S + \delta$ and $\omega_0 - \nu_S - \delta$. This drives the red and blue stretch sidebands simultaneously, but the detuning δ prevents the $|0\rangle \leftrightarrow |1\rangle$ transitions from strongly occurring. However, the double transition $|00\rangle \leftrightarrow |11\rangle$ is resonant, with Rabi frequency

$$\tilde{\Omega} = \frac{\eta^2 \Omega^2}{\delta} \qquad (3)$$

for Lamb-Dicke parameter $\eta \approx 0.16$. By applying the lasers for a time $\pi/(2\tilde{\Omega})$, the entangling operator

$$E = \frac{e^{i\pi/4}}{\sqrt{2}} \begin{bmatrix} 1 & 0 & 0 & i \\ 0 & 1 & i & 0 \\ 0 & i & 1 & 0 \\ i & 0 & 0 & 1 \end{bmatrix} \qquad (4)$$

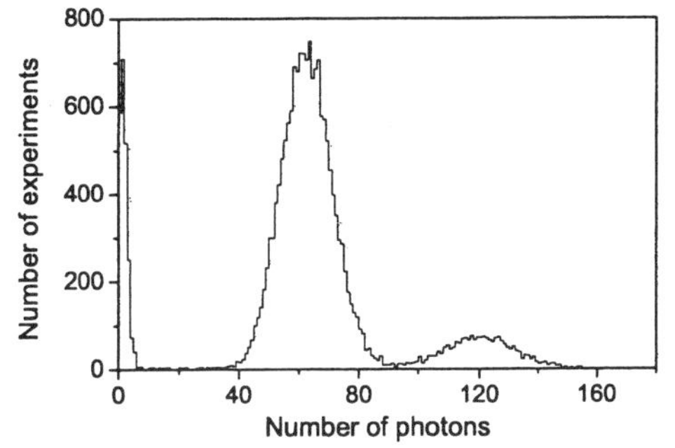

Figure 1. Histogram showing detection of two $^9Be^+$ ions. When the ions repeated prepared in state ψ are illuminated with the detection laser beam, ion i fluoresces brightly with probability $|\langle\psi\,|0\rangle_i\,|^2$. (In order to suppress fluorescence from the $|1\rangle$ state, the population of the $|1\rangle \equiv |1,-1\rangle$ state is transferred to the $|1,1\rangle$ hyperfine state before detection.) On average, approximately 60 photons are detected from each fluorescing ion, but shot noise causes this number to fluctuate according to Poissonian statistics. The histogram shows the results obtained when the ions are identically prepared and detected 2×10^4 times in succession. The three peaks correspond to the possibilities that 0, 1 or 2 ions were found in state $|0\rangle$, and from the areas under the peaks, the respective probabilities can be determined.

is realized, in the $\{|00\rangle, |10\rangle, |01\rangle, |11\rangle\}$ basis. A phase convention is chosen such that the laser phase at each ion is zero. The experimental fidelity of the operation is about 0.9, and is limited by spontaneous emission, laser technical noise, and heating of the ions.[8] Effects of heating are minimized by using the stretch mode.[3] The Mølmer and Sørensen technique has also been used to produce a four-ion entangled state.[7]

After manipulating the ions, their states can be measured by applying a laser tuned to the $2S\,|F=2, m_F=-2\rangle \leftrightarrow 2P\,|F=3, m_F=-3\rangle$ cycling transition. An ion in state $|0\rangle$ will fluoresce brightly in response to this light, while an ion in $|1\rangle$ remains dark. This allows detection of an individual ion with efficiency of about 99%. When two ions are equally illuminated, it is impossible to distinguish the states $|01\rangle$ and $|10\rangle$, as seen in Fig. 1. However, this distinction is unnecessary for many applications.

2 Bell's Inequality

Bell's inequalities have long been used to contradict the notion of local realism. Loosely, this is the idea that objects have definite properties whether or not they are measured, and that these properties are not affected by events taking place sufficiently far away. Bell and others showed that all local realistic measurement predictions must obey certain inequalities, whereas quantum mechanics allows the inequalities to be violated.[9,10,11] Many experiments have since demonstrated violations of the inequalities.[11,12]

Experiments to date, however, have been subject to one or more significant loopholes, allowing local realism to retain some viability. For instance, if the required measurements are not performed in a relativistically separate way, then it is possible for an unknown subluminal signal to affect the observed results. This locality loophole has been the subject of much experimental effort using entangled photon sources, starting with Aspect *et al.*[13] and more recently in the Geneva[14] and Innsbruck[15] experiments. Whether these experiments have conclusively closed the locality loophole is still a matter of debate,[16] but it seems clear that they have made local realistic theories increasingly implausible.

The second main loophole is due to the low detection efficiency of most experiments, which makes it possible that the totality of the events satisfies Bell's inequality even though the subensemble of detected events violates it.[11] Because ion states can be measured with near unit efficiency, the trapped-ion experiment closes this detection loophole for the first time. However, no experiment yet has simultaneously closed both loopholes, so again all that can be made is a plausibility argument.

The ion experiment tested the CHSH form of Bell's inequality.[10] Generically, the argument applies to a pair of spin-1/2 particles, prepared in an arbitrary state. A classical rotation of ϕ_1 is applied to particle 1, and ϕ_2 to particle 2. The spin projection of each particle along a fixed axis is then measured, and a correlation function q is assigned the value $+1$ if the two results agree, and -1 if not. This procedure is repeated many times for four combinations of rotation angles. Bell's inequality then states that the quantity

$$B \equiv |\langle q(\delta_1,\gamma_2)\rangle - \langle q(\alpha_1,\gamma_2)\rangle| \\ +|\langle q(\delta_1,\beta_2)\rangle + \langle q(\alpha_1,\beta_2)\rangle| \quad (5)$$

must satisfy $B \leq 2$. Here α_1, δ_1, β_2, and γ_2 are rotation angles, and $\langle q(\phi_1,\phi_2)\rangle$ is the average correlation obtained using angles ϕ_1 and ϕ_2.

In the experiment, the entangled state

$$\frac{1}{\sqrt{2}}(|00\rangle + i\,|11\rangle) \quad (6)$$

was generated by the Mølmer and Sørensen technique. Rotations were applied using (1) with θ fixed at $\pi/2$. The phases ϕ_i played the role of the angles α, β, γ and δ, and were independently varied by adjusting the ion positions. Detection was performed as in Fig. 1, with simple discriminator levels used to determine the number of ions in the bright $|0\rangle$ state for each experimental run. This number specified the value of q according to whether it was even or odd, since there is no need to distinguish which particle is in which state.

Quantum theory predicts Bell's inequality to be maximally violated for $\alpha_1 = \beta_2 = -\pi/8$ and $\delta_1 = \gamma_2 = 3\pi/8$, with $B = 2\sqrt{2}$. The experimentally obtained value was $B = 2.25 \pm 0.03$, which is consistent with the quantum prediction given experimental imperfections in state preparation, rotations, and detection.

3 Decoherence Free Subspace

Ultimate applications of quantum logic will likely rely on a variety of error correction and error avoidance methods to protect quantum data from weak interactions with a noisy environment.[17] One such method is to encode a qubit of information into a decoherence-free subspace (DFS) of several particles.[18,19] This protects the information from an environment which couples to each of the physical particles in the same way. For example, trapped-ion quantum states suffer decoherence due to fluctuating ambient magnetic fields, which introduce an uncontrolled phase shift between $|0\rangle$ and $|1\rangle$. In this case, a DFS exists for two ions, spanned by the states $|01\rangle$ and $|10\rangle$, since any superposition of these states is unaffected by a uniform phase shift $|1\rangle \to e^{i\alpha}|1\rangle$.

In order for a DFS scheme to be useful, it must include a way to encode a qubit from the natural basis to the DFS. This can be accomplished for two ions using the Mølmer-Sørensen operation E. A two-qubit state of the form $|0\rangle(a|0\rangle + b|1\rangle)$ was first prepared using the individual addressing technique of Section 1. Here the second ion holds the arbitrary qubit to be encoded, and the first serves as an ancilla qubit. Encoding was performed by applying E^{-1}, and then the operator $R_1(\pi/2, \pi/2) \otimes R_2(\pi/2, 0)$, where R_i is the rotation operator (1) applied to ion i. The net effect is to transform the test state to the encoded state

$$\frac{a}{\sqrt{2}}(|01\rangle + i|10\rangle) + \frac{b}{\sqrt{2}}(|01\rangle - i|10\rangle), \quad (7)$$

which is in the DFS. The E^{-1} operator was implemented by applying E three times, since $E^4 = 1$. Decoding is achieved by reversing these steps, applying $R_1(\pi/2, -\pi/2) \otimes R_2(\pi/2, \pi)$ and then E.

The encoding procedure was tested using Ramsey's method of separated fields. First, in a control experiment, the second ion was rotated with $R_2(\pi/2, \beta)$, held for a fixed time, and then rotated again with a different phase, $R_2(\pi/2, \beta')$. The first ion remained in $|0\rangle$ throughout. The probability for the second ion to ultimately be in state $|0\rangle$ then oscillates as $\sin(\beta - \beta')$, and the contrast of this oscillation gives the coherence of the operations. Decoherence was introduced by illuminating both ions with a noisy laser beam that induced random ac-Stark shifts to the states. As seen in Fig. 2, the dephasing that resulted caused the contrast to rapidly decay as the duration of the noisy pulse was increased.

To see the effect of the DFS, the same experiment was performed, but the state of the second ion was encoded into the DFS before applying the noise, and decoded after. As the figure shows, in this case the noise had negligible effect. The unscaled contrast for the test state was 0.69 in the absence of noise, and for the encoding state it was 0.43. The encoding/decoding fidelity was therefore about 0.6, roughly consistent with the entangling fidelity of 0.9 since E was applied four times.

Resistance of the encoded state to naturally occurring decoherence was also observed, by performing the same two experiments but without the noisy pulse and with a variable time between the initial and final rotations. The unprotected state was found to decay with a time constant of 120 ± 20 μs, consistent with ambient magnetic field noise of several mG. The encoded state decay time was 450 ± 60 μs, which was due to degradation of the decoding fidelity as the ions' motion was heated by the environment.

4 Spectroscopy

An important motivation for studying entanglement is the possibility to increase the sensitivity of quantum-limited measurements.[20] This is a very general effect: if a measurement O can be made on a single particle with accuracy $\Delta O = (\langle O^2\rangle - \langle O\rangle^2)^{1/2}$, then typically a measurement using N uncorrelated particles can reduce the uncertainty to $\Delta O/\sqrt{N}$. If the N particles are entangled, however, in some cases an uncertainty of $\Delta O/N$ can be achieved. This is the minimum possible value, termed the Heisenberg limit.

A demonstration of this effect was made

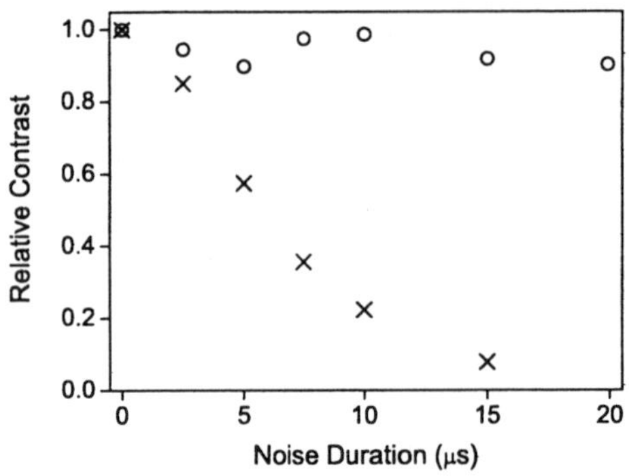

Figure 2. Decay of the test state (crosses) and DFS encoded state (circles) under applied noise. Noise is applied for a fraction of the 25 μs delay between encoding and decoding, and the contrast of the Ramsey interference is measured. Contrasts are normalized to their values for no applied noise, to remove the effects of imperfect encoding operations.

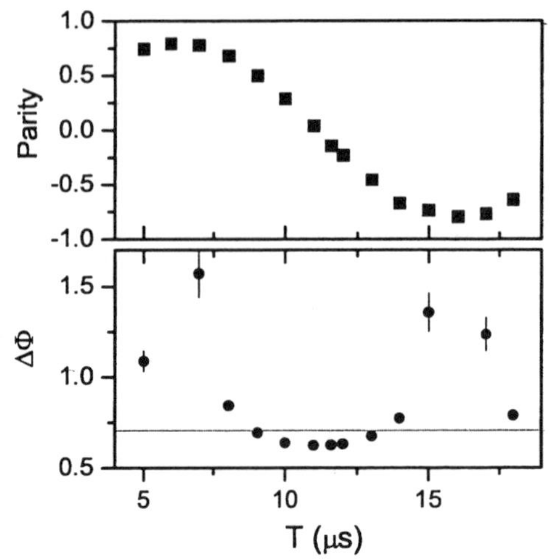

Figure 3. Scheme for measurement of $^9Be^+$ hyperfine frequency using entanglement. A Ramsey experiment is performed using the initial state $(|01\rangle + |10\rangle)/\sqrt{2}$. (a) The parity Π is detected, and oscillates as $\cos 2\Phi$, where Φ is equal to $(\omega - \omega_0)T$ for drive frequency ω and resonant frequency ω_0. (b) The phase accuracy $\delta\Phi$, determined from the measured parity variance $\Delta\Pi$ and measured sensitivity $d\Pi/d\Phi$. When $\delta\Phi < 2^{-1/2}$ (the grey line shown), the frequency detuning $\omega - \omega_0$ is determined with an accuracy better than $(T\sqrt{2})^{-1}$, which is the best accuracy that can possibly be achieved using unentangled ions. At the optimum time, $\delta\Phi = 0.62 \pm 0.01$.

using two $^9Be^+$ ions. A variety of methods for reducing measurement noise have been proposed, and several were investigated by Meyer *et al.*[21] We discuss here one used to demonstrate improved spectroscopic resolution of the $^9Be^+$ hyperfine frequency.[22]

The experiment is essentially a Ramsey experiment of the type described in the previous section. If the initial and final rotation pulses are driven at a fixed frequency ω which is detuned from the carrier frequency ω_0, then a phase $\Phi = (\omega - \omega_0)T$ accumulates during the variable time T between the two pulses. The final state oscillates with this phase, and by observing this oscillation, the detuning and thus ω_0 can be determined. For one measurement of a single ion, the phase can be determined with an accuracy of 1 rad, giving a frequency accuracy of $\delta\omega = T^{-1}$. For two unentangled ions, this is reduced by $\sqrt{2}$.

To improve this resolution, a state of the form $(|01\rangle+|10\rangle)/\sqrt{2}$ was produced by applying E to $|00\rangle$ followed by the uniform carrier rotation $R(\pi/2, \pi/4)$. The first Ramsey pulse $R(\pi/2, 0)$ then yields the state

$$\frac{1}{\sqrt{2}}(|00\rangle - |11\rangle) \tag{8}$$

which precesses to

$$\frac{1}{\sqrt{2}}\left(|00\rangle - e^{-2i\Phi} |11\rangle \right) \tag{9}$$

The final Ramsey pulse is again $R(\pi/2, 0)$, and the resulting state is

$$\frac{1}{\sqrt{2}}\Big[- \cos\Phi\left(|01\rangle + |10\rangle \right) + i\sin\Phi\left(|00\rangle + |11\rangle \right)\Big]. \tag{10}$$

Although the average number of ions in state $|0\rangle$ is constant, the parity

$$\Pi \equiv \sum_{j=0}^{2}(-1)^j P_j \tag{11}$$

oscillates as $\cos 2\Phi$. Here P_j is the probability for j ions to be detected in state $|0\rangle$. Analogous to the single-ion signal, the phase of the parity oscillation can be determined with an accuracy of 1 rad, but since the parity oscillates twice as fast, the resulting frequency accuracy is $\delta\omega = (2T)^{-1}$.

The experimental result is shown in Fig. 3. The top graph shows the oscillation of the parity, while the bottom shows the phase accuracy

$$\delta\Phi = \frac{\Delta\Pi}{d\Pi/d\Phi} \tag{12}$$

As can be seen, a region exists where the observed accuracy is below the limit achievable without entanglement. This technique may ultimately be directly applicable to atomic clocks based on trapped ions.

5 Conclusions

As the above experiments demonstrate, trapped ion entanglement experiments have reached a stage where basic operation can be performed with some degree of reliability and repeatability. These tools have been sufficient to demonstrate applications including closure of the detection loophole in Bell's inequality, a decoherence-free subspace encoding scheme, and improvements in spectroscopic resolution. This work was supported by the U.S. National Security Agency and the Advanced Research and Development Agency under contract MOD-7171, by the U.S. Office of Naval Research, and by the U. S. Army Research Office.

References

1. See for instance J. Bell, *Speakable and Unspeakable in Quantum Mechanics* (Cambridge Univ. Press, Cambridge, 1987).
2. See for instance *Fortsch. Phys.*, special issues on quatum computation, 4-6 **46** (1998) and 9-11 **48** (2000).
3. B. E. King *et al.*, *Phys. Rev. Lett.* **81**, 1525 (1998).
4. H. Rohde *et al.*, *J. Opt. B* **3**, S34 (2001).
5. A. Sørensen and K. Mølmer, *Phys. Rev. Lett.* **82**, 1804 (1999); K. Mølmer and A. Sørensen, *ibid* 1835; A. Sørensen and K. Mølmer, *Phys. Rev. A* **62**, 022311 (2000).
6. E. Solano, R. L. de Matos Filho, and N. Zagury, *Phys. Rev. A* **59**, R2539 (1999).
7. C. A. Sackett *et al.*, *Nature* **404**, 256 (2000).
8. Q. A. Turchette *et al.*, *Phys. Rev. A* **61**, 063418 (2000).
9. J. S. Bell, *Physics* **1**, 195 (1965).
10. J. F. Clauser, M. A. Horne, A. Shimony, and R. A. Holt, *Phys. Rev. Lett.* **23**, 880 (1969).
11. J. F. Clauser and A. Shimony, *Rep. Prog. Phys.* **41**, 1883 (1978).
12. A. Aspect, *Nature* **398**, 189 (1999).
13. A. Aspect, P. Grangier, and G. Roger, *Phys. Rev. Lett.* **49**, 91 (1982).
14. W. Tittel, J. Brendel, H. Abinden, and N. Gisin, *Phys. Rev. Lett.* **81**, 3563 (1998); N. Gisin and H. Zbinden, *Phys. Lett. A* **264**, 103 (1999).
15. G. Weihs *et al.*, *Phys. Rev. Lett.* **81**, 5039 (1998).
16. L. Vaidman, quant-ph/0102139.
17. See for instance J. Preskill, *Phys. Today* **52**, 24 (1999).
18. P. Zanardi and M. Rasetti, *Phys. Rev. Lett.* **79**, 3306 (1997); D. A. Lidar, I. L. Chuang, and K. B. Whaley, *Phys. Rev. Lett.* **81**, 2594 (1998); L. M. Duan and G. C. Guo, *Phys. Rev. A* **57**, 737 (1998).
19. P. G. Kwiat, A. J. Berglund, J. B. Altepeter, and A. G. White, *Science* **290**, 498 (2000).
20. B. Yurke, *Phys. Rev. Lett.* **56**, 1515 (1986); D. J. Wineland *et al.*, *Phys. Rev. A* **46**, R6797 (1992).
21. V. Meyer *et al.*, *Phys. Rev. Lett.* **86**, 5870 (2001).
22. J. J. Bollinger, W. M. Itano, D. J. Wineland, and D. J. Heinzen, *Phys. Rev. A* **54**, R4649 (1996).

EXPERIMENTAL REALIZATION OF CONTINUOUS-VARIABLE TELEPORTATION

AKIRA FURUSAWA

Department of Applied Physics, University of Tokyo, 7-3-1 Hongo, Bunkyo-ku, Tokyo113-8656, Japan

E-mail: akiraf@ap.t.u-tokyo.ac.jp

Quantum entanglement and a teleportation experiment of continuous-variables are briefly reviewed, and the plans of our teleportation and related research are presented.

1 Introduction

Quantum teleportation is a method of quantum state transportation with a classical channel and a quantum channel. In this technique, only the "information" contained in a quantum state is transferred from a sending station (Alice) to a receiving station (Bob), with the original quantum state thereby reconstructed at Bob's place with the received information and previously shared entanglement.

Quantum teleportation was originally proposed by Bennett et al.[1], with two-dimensional systems (e.g., the states of spin $\frac{1}{2}$ particles) having received the greatest attention. On a different front, quantum teleportation with continuous variables in an infinite dimensional Hilbert space was first proposed by Vaidman.[2] His proposal was further investigated theoretically by Braunstein and Kimble, who introduced a teleportation scheme with non-singular squeezed-state entanglement.[3] This latter scheme was experimentally demonstrated by the Quantum Optics group at Caltech in 1998.[4] Somewhat remarkably, in this scheme complete Bell-state measurements can be performed by way of quadrature-phase measurements with homodyne techniques whose detection efficiency can be close to unity. The high detection efficiency of this scheme together with the EPR entanglement generated via summing of independent squeezed beams enabled the boundary between the classical and quantum teleportation to be crossed for the first time. More specifically, a teleportation fidelity of 0.58 ± 0.02 was obtained for the teleportation of coherent states, where the relevant quantum-classical boundary is 0.50 for this experiment.[5]

In this article, I will briefly review the EPR entanglement and the teleportation experiment of continuous-variables, and present our plans for the teleportation and related research.

2 Quantum entanglement

Quantum entanglement is the most important resource for quantum teleportation. In quantum optics, one of the easiest (but somehow unrealistic) way to create quantum entanglement is the usage of a half beam splitter and a single photon (a single-photon state $|1\rangle$). When a single photon is an input (state) for one of input ports of the half beam splitter and a vacuum is the other input, the output state of the half beam splitter $|\mathrm{EPR2}\rangle$ can be written as

$$|\mathrm{EPR2}\rangle = \frac{1}{\sqrt{2}}(|0\rangle_{\mathrm{A}}|1\rangle_{\mathrm{B}} + |1\rangle_{\mathrm{A}}|0\rangle_{\mathrm{B}}), \quad (1)$$

where the subscripts A and B denote the two output modes. This output state $|\mathrm{EPR2}\rangle$ is fully entangled and can be a good resource for two-dimensional teleportation.

By using the similar way with a half beam splitter, can we make some entanglement of continuous variables? The answer is "Yes", but it is a rather complicated story.

When a coherent state $|\alpha\rangle$ is used instead of a single photon in the above scheme, the output state $|\phi\rangle$ can be written as

$$|\phi\rangle = \left|\frac{\alpha}{\sqrt{2}}\right\rangle_{\rm A} \otimes \left|\frac{\alpha}{\sqrt{2}}\right\rangle_{\rm B}. \quad (2)$$

The output state $|\phi\rangle$ is separable even when the input is in "single-photon level" or $|\alpha| = 1$. For getting some entanglement, one of the input has to be a squeezed state at least. In a simple case, both inputs are squeezed states. The two-mode squeezed state $|{\rm EPR}\rangle$ can be written as [6]

$$|{\rm EPR}\rangle = \sqrt{1-q^2}\sum_{n=0}^{\infty} q^n\,|n\rangle_{\rm A}\,|n\rangle_{\rm B}\,, \quad (3)$$

where $q = \tanh r$ and r is a squeezing parameter. Note that some amount of entanglement always exists when $r > 0$. Also note that a single-photon state $|1\rangle$ is an ultimate amplitude squeezed state in some sense and the situation for the single-photon input is very similar to the case of squeezed-state input.

The two-mode squeezed vacuum can be written with the Wigner function as well. The Wigner function $W_{\rm EPR}$ can be written as [3]

$$\begin{aligned} &W_{\rm EPR}(\alpha_{\rm A};\alpha_{\rm B}) \\ &= \frac{4}{\pi^2}\exp\{-e^{-2r}[(x_{\rm A}-x_{\rm B})^2+(p_{\rm A}+p_{\rm B})^2] \\ &\qquad -e^{+2r}[(x_{\rm A}+x_{\rm B})^2+(p_{\rm A}-p_{\rm B})^2]\}, \quad (4) \end{aligned}$$

where $\alpha_j = x_j + ip_j$ and (x_j, p_j) are canonically conjugated variables like "position" and "momentum" of A and B. When $r \to \infty$, Eq. (4) becomes $C\delta(x_{\rm A}+x_{\rm B})\delta(p_{\rm A}-p_{\rm B})$ which is precisely the EPR state.

We used this type of entanglement for the experiment of continuous-variable teleportation.

3 Teleportation experiment

We generated a two-mode squeezed vacuum or EPR beams via summing of independent squeezed beams. The EPR beams were shared with Alice and Bob. The Wigner function of the output state of the teleportation, $W_{\rm out}$, can be written with the input Wigner function $W_{\rm in}$ as [3]

$$W_{\rm out} = W_{\rm in} \circ G_\sigma, \quad (5)$$

where $\circ$ denotes convolution and G_σ is a complex Gaussian distribution of (x,p) with variance $\sigma = e^{-2r}$.

The experimental result is shown in Fig.(1) with the Wigner function. The output state with the EPR beams is more "similar" to the original input state compared to the case without the EPR beams. Thus we declared the success of teleportation.

We are now trying to generate the GHZ state of continuous variables.[7] The GHZ state enables us to do teleportation in three parties.[7]

Acknowledgments

The teleportation experiment was performed at Caltech with Prof. H. J. Kimble and his group. Our project now in University of Tokyo is partially supported by MPHPT.

References

1. C. H. Bennett, et al., *Phys. Rev. Lett.* **70**, 1895(1993).
2. L. Vaidman, *Phys. Rev. A* **49**, 1473 (1994).
3. S. L. Braunstein and H. J. Kimble, *Phys. Rev. Lett.* **80**, 869 (1998).
4. A. Furusawa, J. L. Sørensen, S. L. Braunstein, C. A. Fuchs, H. J. Kimble, and E. S. Polzik, *Science* **282**, 706 (1998).
5. S. L. Braunstein, C. A. Fuchs, H. J. Kimble, and P. van Loock, *Phys. Rev. A* **64**, 022321 (2001).
6. D. F. Walls and G. J. Milburn, *Quantum Optics*, (Springer-Verlag, Berlin, 1994).
7. P. van Loock and S. L. Braunstein, *Phys. Rev. Lett.* **84**, 3482 (2000)

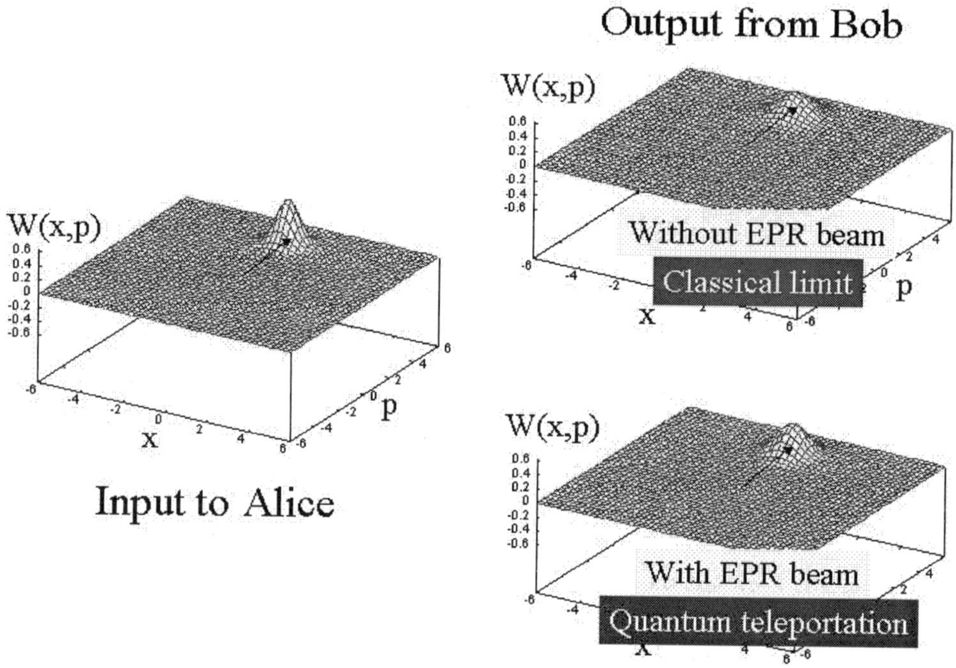

Figure 1. Experimental results of teleportation of a coherent state. The arrows represent displacements.

EXACTLY DISTILLABLE ENTANGLEMENT

FUMIAKI MORIKOSHI
NTT Basic Research Laboratories
3-1 Morinosato-Wakamiya, Atsugi-shi, Kanagawa, 243-0198, Japan
E-mail: fumiaki@will.brl.ntt.co.jp

MASATO KOASHI
CREST Research Team for Interacting Carrier Electronics
School of Advanced Sciences, The Graduate University for Advanced Studies (SOKEN)
Hayama, Kanagawa, 240-0193, Japan
E-mail: koashi@soken.ac.jp

We discuss deterministic extraction of Bell pairs from a finite number of partially entangled pairs by using local operations and classical communication. The maximum number of Bell pairs extracted with certainty is derived. It is shown that the optimal deterministic entanglement concentration can be performed by successive two-pair collective manipulations instead of manipulating all entangled pairs at once. Finally, this scheme reveals an entanglement measure that quantifies the exactly distillable entanglement.

1 Introduction

Quantum entanglement is a valuable resource in quantum information theory. In quantum teleportation[1], Alice and Bob communicate a quantum state via an entangled state,

$$|\Phi^+\rangle_{AB} = \frac{1}{\sqrt{2}}(|00\rangle_{AB} + |11\rangle_{AB}), \tag{1}$$

where qubits A and B are possessed by Alice and Bob, respectively. It is essential that they share a Bell pair, $|\Phi^+\rangle_{AB}$, in advance. If Alice and Bob initially share partially entangled pairs, they need to prepare Bell pairs from the initial pairs only by operating locally on their respective systems and communicating classically. Extraction of maximally entangled pairs from partially entangled pairs by using local operations and classical communication (LOCC) is generally called entanglement concentration. (Note that neither global operations on the whole system nor transmission of quantum systems is allowed in the process.) Various studies on its limitations and efficiency have been carried out[2,3,4,5,6,7]. All the proposed entanglement concentration processes are probabilistic for a finite number of entangled pairs.

We will discuss an entanglement concentration scheme within the framework of exact (deterministic) transformations[8]. Our scheme converts a collection of two-qubit partially entangled pairs having different amounts of entanglement into a bunch of Bell pairs with probability 1. Whereas the above probabilistic concentration processes run the risk of losing all the entanglement contained in the initial states with certain probability, our concentration scheme answers the question of how many Bell pairs can be distilled from partially entangled pairs without gambling.

2 Deterministic entanglement concentration of two pairs

We begin with the deterministic entanglement concentration of two pairs, which will turn out to be the building blocks of the n-pair concentration.

In this paper, we deal with bipartite pure entangled states, which can be written in a standard form called Schmidt decomposition:

$|\psi\rangle_{AB} = \sum_i \sqrt{a_i}|i\rangle_A|i\rangle_B$, where $\{|i\rangle_A\}$ and $\{|i\rangle_B\}$ are the orthonormal bases of respective systems.

Suppose Alice and Bob wish to extract a Bell pair form the following two partially entangled pairs shared between them:

$$\begin{aligned} |\psi\rangle &= \sqrt{a}\ |00\rangle + \sqrt{1-a}\ |11\rangle, \\ |\phi\rangle &= \sqrt{b}\ |00\rangle + \sqrt{1-b}\ |11\rangle. \end{aligned} \tag{2}$$

Without loss of generality, we can set

$$\frac{1}{2} < a \leq b < 1. \tag{3}$$

By using Nielsen's theorem[9], we can prove the following two cases in the deterministic entanglement concentration of two pairs[8,10].

Case (a): $\frac{1}{4} < ab \leq \frac{1}{2}$. In this case, Alice and Bob can extract a Bell pair from $|\psi\rangle \otimes |\phi\rangle$ and keep the residual entanglement in another entangled pair such that

$$|\omega\rangle = \sqrt{2ab}\ |00\rangle + \sqrt{1-2ab}\ |11\rangle. \tag{4}$$

Thus they perform

$$|\psi\rangle \otimes |\phi\rangle \rightarrow |\Phi^+\rangle \otimes |\omega\rangle. \tag{5}$$

Alice and Bob cannot retain more entanglement in the residual pair than in $|\omega\rangle$, which means that this is the optimal concentration in a deterministic way.

Case (b): $\frac{1}{2} < ab < 1$. It is impossible for Alice and Bob to extract a Bell pair with certainty. The best they can do is gather entanglement of two pairs into one pair such that

$$|\omega\rangle = \sqrt{ab}\ |00\rangle + \sqrt{1-ab}\ |11\rangle, \tag{6}$$

in order to make an entangled pair as close to a Bell pair as possible. Thus the concentration in this case becomes

$$|\psi\rangle \otimes |\phi\rangle \rightarrow |\omega\rangle \otimes |00\rangle. \tag{7}$$

Note that the product of the larger amplitudes of each entangled pair, $\sqrt{ab}$, is conserved before and after the transformations in both cases. This fact is the crux of the optimality of the deterministic concentration presented in Sec. 3.

3 Tournament-like concentration of n pairs

Suppose Alice and Bob share n partially entangled pairs $|\psi_1\rangle \otimes \cdots \otimes |\psi_n\rangle$, where

$$|\psi_i\rangle = \sqrt{a_i}\ |00\rangle + \sqrt{1-a_i}\ |11\rangle \quad (i = 1, \ldots, n), \tag{8}$$

with

$$\frac{1}{2} < a_i < 1, \tag{9}$$

and wish to extract as many Bell pairs as possible from these entangled pairs by LOCC with certainty.

If Alice and Bob obtain k Bell pairs and $(n-k)$ disentangled pairs, the concentration becomes

$$|\psi_1\rangle \otimes \cdots \otimes |\psi_n\rangle \rightarrow \otimes^k |\Phi^+\rangle \otimes^{n-k} |00\rangle. \tag{10}$$

With Nielsen's theorem, it can be proved that this concentration is accomplished if

$$a_1 \cdots a_n \leq \frac{1}{2^k}. \tag{11}$$

Inequality (11) gives the maximum number of Bell pairs extracted from the partially entangled pairs given by Eq. (8),

$$k_{max} = \lfloor -\log_2(a_1 \cdots a_n) \rfloor, \tag{12}$$

where $\lfloor x \rfloor$ represents the largest integer equal to or less than x.

At first glance, the optimal n pair concentration seems to require a collective manipulation of all pairs. However, with Eq. (12), we can prove that the optimal deterministic concentration can be constructed out of two-pair concentrations. The maximum number of Bell pairs, k_{max}, is determined only by the product of the lager amplitudes of each pair to be concentrated, i.e., $\sqrt{a_1 \cdots a_n}$. Furthermore, we found in Sec. 2 that the product never changes during the optimal deterministic concentration of two pairs. Therefore, we can perform the optimal deterministic concentration of n pairs simply by successive two-pair concentrations without losing any Bell pairs we are to distill from the initial state.

Let us look at the concentration procedure in detail. First Alice and Bob choose two arbitrary entangled pairs. Then they perform a two-pair concentration on them. If they obtain a Bell pair [case (a) in Sec. 2], it is put aside, and the residual entangled pair will be reused in a later pairwise concentration. If the selected two pairs are insufficient for a Bell pair [case (b) in Sec. 2], they gather the entanglement contained in the two pairs into a partially entangled pair, which will be reused in a later pairwise concentration. They repeat these operations until they achieve the maximum number of Bell pairs.

After going through the above process, Alice and Bob finally share k_{max} Bell pairs, $(n - k_{max} - 1)$ disentangled pairs, and one partially entangled pair. Since the product of the larger amplitudes of each pair, $\sqrt{a_1 \cdots a_n}$, remains unchanged throughout these concentration processes, the residual partially entangled state becomes

$$|\omega\rangle = \sqrt{2^{k_{max}} a_1 \cdots a_n}|00\rangle + \sqrt{1 - 2^{k_{max}} a_1 \cdots a_n}|11\rangle, \quad (13)$$

where $\frac{1}{2} < 2^{k_{max}} a_1 \cdots a_n \leq 1$. This residual entangled pair can be used in another pairwise concentration if Alice and Bob obtain an extra partially entangled pair to be concentrated together later. Therefore, they do not waste any potential entanglement that can be distilled in a future concentration.

As seen in the above, in each pairwise concentration, at least one of the pairs becomes a Bell pair or a disentangled pair, which does not proceed to the next concentration. Thus, we perform a pairwise concentration at most $n-1$ times in a process similar to elimination in a tournament. The pairwise nature of this *tournament-like* concentration is useful in a practical sense too. Even if we fail in the manipulation of two pairs, or if some error occurs in the concentration, only the two pairs are affected. The other pairs remain intact in spite of such unexpected effects. This contrasts sharply with other proposed concentration schemes that require collective manipulation of all pairs.

4 Entanglement measure for the deterministic concentration

We can rewrite the formula for the maximum number of Bell pairs [Eq. (12)] as follows:

$$k_{max} = \left\lfloor \sum_{i=1}^{n} - \log_2 a_i \right\rfloor \quad (14)$$

This implies that the contribution from a partially entangled state $|\psi_i\rangle = \sqrt{a_i}\,|00\rangle + \sqrt{1 - a_i}\,|11\rangle$ can be described as

$$D(\psi_i) \equiv - \log_2 a_i \qquad (\frac{1}{2} \leq a_i \leq 1). \quad (15)$$

It is easily seen that $D(\Phi^+) = 1$ and $D(|00\rangle) = 0$. Thus $D(\psi)$ does not diverge in $\frac{1}{2} \leq a_i \leq 1$.

The quantity $D(\psi)$ is an entanglement measure for the deterministic concentration. It satisfies the following conditions for entanglement measures $M(\psi)$ in deterministic transformations of pure states:

(i) $M(\psi) = 0$ iff $|\psi\rangle$ is separable.

(ii) $M(\psi)$ remains unchanged under local unitary transformations.

(iii) $M(\psi)$ cannot be increased by deterministic LOCC.

These conditions are modified versions of the conditions for entanglement measures proposed by Vedral *et al.*[11,12]. The restriction of deterministic transformations makes the third condition weaker than the counterpart in the original versions[11,12], which requires the non-increasing property of the *expected* value of the measure when the final states are not unique. Note that if we allow probabilistic transformations, the expected value of our measure can be increased in some cases.

The measure $D(\psi_i)$ quantifies the number of Bell pairs contributed by the state $|\psi_i\rangle$ in the deterministic concentration of a finite number of entangled pairs. For example, if Alice and Bob share two pairs, $|\psi_1\rangle$ and $|\psi_2\rangle$, where $D(\psi_1) = 0.7$ and $D(\psi_2) = 0.6$, then they can obtain a Bell pair and a residual state $|\omega\rangle$ with $D(\omega) = 0.3$, because $0.6+0.7 = 1 + 0.3$. On the other hand, another entanglement measure[2]

$$E(\psi_i) \equiv -a_i \log_2 a_i - (1 - a_i) \log_2(1 - a_i) \tag{16}$$

represents the amount of entanglement per one partially entangled pair in the concentration of an infinite number of identical copies of $|\psi_i\rangle$. The strong restriction of deterministic transformations makes our concentration scheme less efficient than the original Schmidt projection method[2], which attains $E(\psi)$ in the asymptotic limit. Therefore, the quantity $D(\psi)$ does not converge on $E(\psi)$ even in the asymptotic limit.

The most fascinating property of the measure $D(\psi)$ is additivity. The additivity shows that the tournament-like concentration procedure does not depend on the way of pairing entangled pairs, because we can always attain the optimality as long as the sum $\sum_i D(\psi_i)$ is conserved. That is, the initially shared partially entangled pairs are finally lead to a bunch of Bell pairs and a bunch of disentangled pairs by *any* pairwise manipulation that conserves $\sum_i D(\psi_i)$.

5 Conclusion

We have studied a deterministic entanglement concentration scheme for a finite number of partially entangled pairs consisting of two qubits. We derived the maximum number of Bell pairs extracted by LOCC, and proved that it can be attained simply by a series of two-pair manipulations. The optimality requires only tournament-like manipulations instead of collective manipulations of all entangled pairs. Furthermore, this concentration scheme revealed an entanglement measure for the deterministic concentration, which quantifies the amount of entanglement that we can use with certainty.

Acknowledgments

We are grateful to N. Imoto for valuable discussions. This work was supported by a Grant-in-Aid for Encouragement of Young Scientists (Grant No. 12740243) and a Grant-in-Aid for Scientific Research (B) (Grant No. 12440111) by the Japan Society of the Promotion of Science.

References

1. C. H. Bennett, G. Brassard, C. Crépeau, R. Jozsa, A. Peres, and W. K. Wootters, Phys. Rev. Lett. **70**, 1895 (1993).
2. C. H. Bennett, H. J. Bernstein, S. Popescu, and B. Schumacher, Phys. Rev. A **53**, 2046 (1996), quant-ph/9511030.
3. H.-K. Lo and S. Popescu, Phys. Rev. A **63**, 022301 (2001), quant-ph/9707038.
4. G. Vidal, Phys. Rev. Lett. **83**, 1046 (1999), quant-ph/9902033.
5. S. Bose, V. Vedral, and P. L. Knight, Phys. Rev. A **60**, 194 (1999), quant-ph/9812013.
6. D. Jonathan and M. B. Plenio, Phys. Rev. Lett. **83**, 83, 1455 (1999), quant-ph/9903054.
7. L. Hardy, Phys. Rev. A **60**, 1912 (1999), quant-ph/9903001.
8. F. Morikoshi and M. Koashi, Phys. Rev. A **64**, 022316 (2001), quant-ph/0107120.
9. M. A. Nielsen, Phys. Rev. Lett. **83**, 436 (1999), quant-ph/9811053.
10. F. Morikoshi, Phys. Rev. Lett. **84**, 3189 (2000), quant-ph/9911019.
11. V. Vedral, M. B. Plenio, M. A. Rippin, and P. L. Knight, Phys. Rev. Lett. **78**, 2275 (1997), quant-ph/9702027.
12. V. Vedral and M. B. Plenio, Phys. Rev. A **57**, 1619 (1998), quant-ph/9707035.

INFORMATION EXTRACTION AND QUANTUM STATE DISTORTIONS IN CONTINUOUS VARIABLE QUANTUM TELEPORTATION

HOLGER F. HOFMANN

CREST, Japan Science and Technology Corporation (JST), Research Institute for Electronic Science, Hokkaido University, Sapporo 060-0812, Japan

TOSHIKI IDE, TAKAYOSHI KOBAYASHI

Department of Physics, Faculty of Science, University of Tokyo, 7-3-1 Hongo, Bunkyo-ku, Tokyo 113-0033, Japan

AKIRA FURUSAWA

Department of Applied Physics, Faculty of Engineering, University of Tokyo, 7-3-1 Hongo, Bunkyo-ku, Tokyo 113-8656, Japan

We analyze the loss of fidelity in continuous variable teleportation due to non-maximal entanglement. It is shown that the quantum state distortions correspond to the measurement back-action of a field amplitude measurement

1 Introduction

Quantum teleportation transfers a quantum state to a remote location using shared entanglement and classical communication [1]. Ideally, this procedure does not change the transmitted state at all, even though classical information is obtained in an irreversible measurement. This is only possible if the classical information is completely independent of the teleported state. In the case of continuous variable quantum teleportation [2], only non-maximal entanglement is available. As a result, the classical information obtained in the measurement does depend on the input state, and a corresponding measurement back-action is observed in the output. In the following, this limitation of fidelity in continuous variable teleportation is discussed using the recently introduced transfer operator formalism [3].

2 Continuous variable teleportation

In continuous variable teleportation, an unknown input state $\mid \psi_{\text{in}}\rangle$ of the input field $\hat{a}$ is transfered by a precise measurement of the field difference $\hat{a} - \hat{r}^\dagger = \beta$ between $\hat{a}$ and a reference field $\hat{r}$. The reference field $\hat{r}$ is entangled with the remote field $\hat{b}$. This entanglement is obtained by squeezing the vacuum to suppress the fluctuations of $\hat{b} - \hat{r}^\dagger \approx 0$ below the standard quantum limit. Therefore, the measurement result β is approximately equal to the field difference between the unknown input field $\hat{a}$ and the output field $\hat{b}$. The original quantum state of the input field can then be restored by $\hat{b} + \beta \approx \hat{a}$.

The most serious technical limitation of this teleportation scheme is the amount of squeezed state entanglement available. At present, it seems unrealistic to assume a noise suppression of more than 10 dB. While it may be possible to raise this limit in the future, maximal entanglement would require the unrealistic limit of infinite squeezing. Non-maximal entanglement is therefore a fundamental feature of continuous variable teleportation. The quantum state of non-maximal squeezed state entanglement can be formulated in the photon number basis as

$$| EPR(q)\rangle = \sqrt{1-q^2}\sum_n q^n \mid n\rangle\otimes\mid n\rangle. \quad (1)$$

The entanglement parameter q is related to the squeezing factor s by $q = (1-s)/(1+s)$. For example, a noise suppression by one half (3 dB) would correspond to $q = 1/3$.

With this definition of the initial state, it is possible to describe how the measurement of $\hat{a} - \hat{r}^\dagger = \beta$ conditions the output state of $\hat{b}$. Using the properly normalized eigenstates of $\hat{a} - \hat{r}^\dagger$ [4], the projection reads

Initial state

$$\sqrt{1-q^2}\sum_n q^n \quad \mid \psi_{\text{in}}\rangle \quad \otimes \mid n\rangle \otimes \mid n\rangle$$

Measurement projection

$$\frac{1}{\sqrt{\pi}}\sum_n \quad \langle n \mid \hat{D}(-\beta) \otimes \langle n \mid$$

Conditional output state

$$\sqrt{\frac{1-q^2}{\pi}}\sum_n q^n \quad \langle n \mid \hat{D}(-\beta) \mid \psi_{\text{in}}\rangle \quad \mid n\rangle. \quad (2)$$

For $q \to 1$, the conditional output state is equal to the displaced input state $\hat{D}(-\beta) \mid \psi_{\text{in}}\rangle$. The effect of $q < 1$ reduces the contributions of states with high photon numbers.

In the final step of quantum teleportation, the displacement is reversed by modulating the output field $\hat{b}$. This modulation is proportial to β and can be described by a displacement operator $\hat{D}(g\beta)$, where the gain factor g permits an amplification or attenuation of the output amplitude. The process of continuous variable quantum teleportation can then be described by a transfer operator, such that both the probability distribution $P(\beta)$ of the measurement results and the normalized conditional output states $\mid \psi_{\text{out}}(\beta)\rangle$ are described by

$$\sqrt{P(\beta)} \mid \psi_{\text{out}}(\beta)\rangle = \hat{T}(\beta) \mid \psi_{\text{in}}\rangle$$

with

$$\hat{T}(\beta) = \sqrt{\frac{1-q^2}{\pi}}\sum_{n=0}^{\infty} q^n \, \hat{D}(g\beta) \mid n\rangle\langle n \mid \hat{D}(-\beta). \quad (3)$$

The transfer operator $\hat{T}(\beta)$ establishes the general relationship between the measurement information β obtained in the teleportation and the conditional quantum state distortions caused by this measurement for arbitrary input states. In particular, the information obtained about the unknown input state $\mid \psi_{\text{in}}\rangle$ is characterized by a positive operator valued measure given by

$$P(\beta) = \langle \psi_{\text{in}} \mid \hat{T}^\dagger(\beta)\hat{T}(\beta) \mid \psi_{\text{in}}\rangle$$

with

$$\hat{T}^\dagger(\beta)\hat{T}(\beta) = \frac{1-q^2}{\pi}\sum_{n=0}^{\infty} q^{2n} \, \hat{D}(\beta) \mid n\rangle\langle n \mid \hat{D}(-\beta). \quad (4)$$

The eigenvalues of this positive operator valued measure are the displaced photon number states $\hat{D}(\beta) \mid n \rangle$. In phase space, these displaced photon number states can be associated with concentric circles of radius $\sqrt{n+1/2}$ around β. The higher n, the greater the difference between the actual field value and β. $\hat{T}(\beta)$ therefore describes a finite resolution measurement of the complex field amplitude $\hat{a}$.

3 Coherent state teleportation

The properties of the transfer operator are best illustrated by applying it to typical input states. For a coherent state input $\mid \alpha\rangle$, the quantum teleportation process is characterized by

$$\hat{T}(\beta) \mid \alpha\rangle = \sqrt{\frac{1-q^2}{\pi}} \exp\left(-(1-q^2)\frac{|\alpha-\beta|^2}{2}\right) \times \exp\left((1-gq)\frac{\alpha\beta^* - \beta\alpha^*}{2}\right) \times \mid q\alpha + (g-q)\beta\rangle. \tag{5}$$

This result consists of a probability factor $\sqrt{P(\beta)}$ describing a Gaussian probability distribution centered around $\beta = \alpha$, a phase factor important only if a superposition of coherent states is considered (e.g. cat states or squeezed states), and finally the modified coherent state, with an attenuated original amplitude of $q\alpha$ and a measurement dependent displacement of $(g-q)\beta$.

Note that equation (5) may be applied to any input state if that state is written as a superposition of coherent states. The phase factor is then crucial in determining the coherence of the output. For a simple coherent state, however, the distortions of the output state are best characterized by defining the measurement fluctuation $\phi = \beta - \alpha$. The probability distribution over ϕ is then given by a Gaussian centered around $\phi = 0$, and the output statistics are described by

$$\begin{aligned} P(\phi) &= \frac{1-q^2}{\pi} \exp\left(-(1-q^2)|\phi|^2\right) \\ \mid \psi_{\text{out}}(\phi)\rangle &= \mid g\alpha + (g-q)\phi\rangle. \end{aligned} \tag{6}$$

The correlation between the measurement fluctuation ϕ and the output amplitude $g\alpha + (g-q)\phi$ is given by the gain dependent factor $g-q$. In particular, $g > q$ indicates a positive correlation between the measurement fluctuation and the output amplitude, while $g < q$ indicates a negative correlation. In the special case of $g = q$, the output amplitude does not depend on the measurement result. At this gain condition, continuous variable quantum teleportation simply attenuates the coherent state to an amplitude of $q\alpha$. As pointed out by Polkinghorne and Ralph [5], this situation corresponds to the attenuation of the signal at a beam splitter. Our formalism allows a generalization of this analogy to back action evasion measurements using feedback compensated beam splitters [6].

4 Photon number state teleportation

It is possible to identify the beam splitter analogy more directly by examining the effects of the Transfer operator on the creation operator $\hat{a}^\dagger$. For $g = q$,

$$\hat{T}_{g=q}(\beta)\hat{a}^\dagger = \left((1-q^2)\beta^* + q\hat{a}^\dagger\right)\hat{T}_{g=q}(\beta). \tag{7}$$

Effectively, $\hat{T}_{g=q}(\beta)$ attenuates $\hat{a}^\dagger$ by a factor of q and replaces the loss with a complex amplitude of $\sqrt{1-q^2}\beta^*$. The attenuated amplitude can be identified with the component transmitted by a beam splitter of reflectivity $R = 1 - q^2$ and the β^* dependent addition can be interpreted as the measurement back-action from the reflected parts of the input field. In general, the teleportation of a photon number state can then be described by

$$\hat{T}(\beta)\frac{1}{\sqrt{n!}}\left(\hat{a}^\dagger\right)^n \mid 0\rangle = \hat{D}((g-q)\beta) \times \sqrt{\frac{1-q^2}{\pi n!}} \exp\left(-(1-q^2)\frac{|\beta|^2}{2}\right) \times \left((1-q^2)\beta^* + q\hat{a}^\dagger\right)^n \mid 0\rangle. \tag{8}$$

The measurement back-action causes photon losses and introduces coherence by replacing a component of the creation operators with a complex amplitude. This corresponds to the loss of photons at a beam splitter and the measurement back-action of a projection on coherent states, e.g. by eight port homodyne detection.

5 Conclusions

The transfer operator $\hat{T}(\beta)$ provides a complete description of the measurement information extracted and the quantum state distortions in continuous variable teleportation. The correlations between the errors caused and the information obtained correspond to the back-action of a non-destructive quantum measurement of the coherent field amplitude. In particular, the distortions correspond to the attenuation of the original signal amplitude and a measurement back-action conditioned by the field information obtained from a measurement of these losses.

References

1. C.H. Bennet, G. Brassard, C.Crepeau, R. Jozsa, A. Peres, and W.K. Wootters, *Phys. Rev. Lett.* 70, 1895 (1993).
2. L. Vaidman, *Phys. Rev.* A 49, 1473 (1994), S.L. Braunstein and H.J. Kimble, *Phys. Rev. Lett.* 80, 869 (1998), A. furusawa, J.L. Sorensen, S.L. Braunstein, C.A. Fuchs, H.J. Kimble, and E.S. Polzik, *Science* 282, 706 (1998).
3. H.F. Hofmann, T. Ide, T. Kobayashi, and A. Furusawa, *Phys. Rev.* A 62, 062304 (2000).
4. Note that the proper normalization for the continuous variable β is the delta function $\langle \beta \mid \beta' \rangle = \delta(\beta - \beta')$.
5. R.E.S. Polkinghorne and T.C. Ralph, *Phys. Rev. Lett* 83, 2095 (1999).
6. H.F. Hofmann, T. Ide, T. Kobayashi, and A. Furusawa, quant-ph/0102097, to be published in *Phys. Rev.* A.

CONTINUOUS VARIABLE TELEPORTATION OF SINGLE PHOTON STATES

TOSHIKI IDE, TAKAYOSHI KOBAYASHI

Department of Physics, Faculty of Science, University of Tokyo,
7-3-1 Hongo, Bunkyo-ku, Tokyo113-0033, Japan

HOLGER F. HOFMANN

CREST, Japan Science and Technology Corporation (JST),
Research Institute for Electronic Science, Hokkaido University,
Sapporo 060-0812, Japan

AKIRA FURUSAWA

Department of Applied Physics,
Faculty of Engineering, University of Tokyo,
7-3-1 Hongo, Bunkyo-ku, Tokyo113-8656, Japan

We investigate the changes to a single photon state caused by the non-maximal entanglement in continuous variable quantum teleportation. It is shown that the teleportation measurement introduces field coherence in the output.

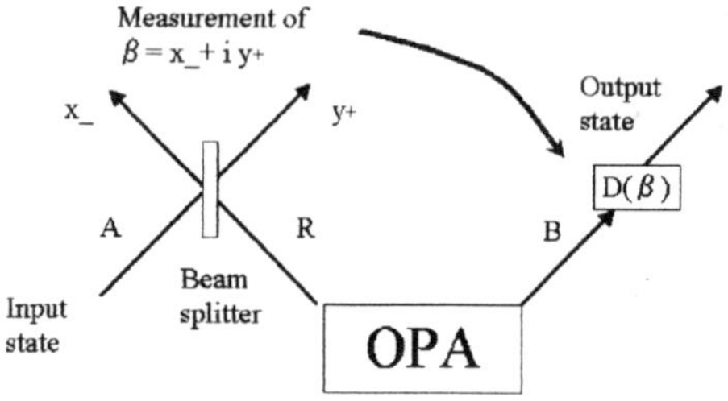

Figure 1. The schematic figure for the continuous variable teleportation.

1 Introduction

Quantum teleportation is a method for Alice (sender) to transmit an unknown quantum input state to Bob (receiver) at a distant place by sending only classical information using a shared entangled state as a resource [1]. In continuous variable quantum teleportation [2,3,4], the available entanglement is non-maximal, limited by the amount of squeezing achieved. Fig.1 shows the setup of a continuous variable quantum teleportation. Alice transmits an unknown quantum state $|\psi\rangle_A$ to Bob. Alice and Bob share EPR beams in advance. Alice mixes her input state with the reference EPR beam by a 50% beam-splitter and performs an entanglement measurement of the complex field value β. After Bob gets the information of the field measurement value β from Alice, Bob applies a displacement to the output state by mixing the coherent field of a local oscillator with the output EPR beam B.

As has been shown previously [6], the properties of this transfer process can be summarized by the transfer operator $\hat{T}_q(\beta)$ which describes both the probability distribution $P(\beta)$ of measurement results β and the normalized conditional output state $|\psi_{\text{out}}(\beta)\rangle$ for any input state $|\psi_{\text{in}}\rangle$, such that

$$\sqrt{P(\beta)}\,|\psi_{\text{out}}(\beta)\rangle = \hat{T}_q(\beta)\,|\psi_{\text{in}}\rangle. \tag{1}$$

In its diagonalized form, this transfer operator reads

$$\hat{T}_q(\beta) = \sqrt{\frac{1-q^2}{\pi}} \sum_{n=0}^{\infty} q^n \hat{D}(\beta)\,|n\rangle_{BA}\langle n|\,\hat{D}(-\beta). \tag{2}$$

The non-maximal entanglement is described by the parameter q, which is 0 for a non-entangled vacuum and 1 for maximal entanglement. In the following, this operator will be applied to characterize the teleportation of a single photon input state, with special consideration of the field coherence created in the output by the teleportation process.

2 Teleportation of a single photon state

The output of a one photon input state is characterized by

$$\hat{T}_q(\beta)\,|1\rangle = \sqrt{\frac{1-q^2}{\pi}} e^{-(1-q^2)\frac{|\beta|^2}{2}} \hat{D}((1-q)\beta)((1-q^2)\beta^*\,|0\rangle + q\,|1\rangle). \tag{3}$$

The normalized output state is then given by

$$|\psi_{\text{out}}(\beta)\rangle = \hat{D}((1-q)\beta)\frac{1}{\sqrt{q^2+(1-q^2)^2|\beta|^2}} ((1-q^2)\beta^*\,|0\rangle + q\,|1\rangle). \tag{4}$$

The output state can be described by a displaced coherent superposition of a zero photon and a one photon component. Both the displacement and the coherence depend on the complex measurement value β. They therefore represent a measurement induced coherence of the output state. We characterize this coherence by the expectation value of the complex field amplitude,

$$C_q(\beta) = \langle \psi_{\text{out}}(\beta) \mid \hat{a} \mid \psi_{\text{out}}(\beta) \rangle = \frac{q(1-q^2)}{q^2 + (1-q^2)^2|\beta|^2}\beta + (1-q)\beta. \quad (5)$$

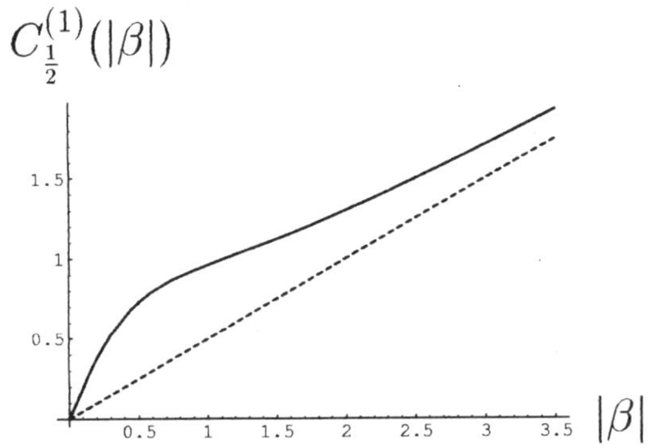

Figure 2. The absolute value of β dependence of the field amplitude $C^{(1)}_q(\beta)$ in the case of $q = 0.5$. The curve is approaching $y = 0.5q$.

Fig.2 illustrates this dependence of C_q on β for $q = 1/2$. For low values of β, C_q rises sharply, levelling off around $|\beta| \approx 0.5$. At higher values of $|\beta|$, C_q slowly approaches $(1-q)\beta$, as indicated by the dotted line. Since the input photon number state has a field expectation value of zero, the field coherence in the output is a consequence of the measurement β. As has been argued elsewere [6], β corresponds to the result of a field measurement performed on the input state. This measurement creates coherence by projection onto displaced photon number states as indicated by the diagonalized form of $\hat{T}_q(\beta)$ in equation (2). Since the photon is the quantum mechanical equivalent of field intensity, each photon represents an addition of one quantum unit to the field fluctuations. The measurement then converts the field fluctuation into an actual field. This process is responsible for the rapid increase in coherence at low values of β.

3 Conclusions

We have investigated the effects of continous variable quantum teleportation on a single photon input state. Because of the non-maximal entanglement used in the teleportation, the measurement of β introduces coherence into the output state. We have quantified this coherence, tracing its origin to the field fluctuations of the single photon input.

References

1. C.H. Bennet, G. Brassard, C. Crepeau, R.Jozsa, A. Peres, and W.K. Wootters, *Phys. Rev. Lett.* **70**, 1895 (1993).
2. L. Vaidman, *Phys. Rev. A* **49**, 1473 (1994).
3. S.L. Braunstein and H.J. Kimble, Phys. Rev. Lett. **80**, 869 (1998).
4. A. Furusawa, J. L. Sørensen, S. L. Braunstein, C. A. Fuchs, H. J. Kimble, and E. J. Polzik, *Science* **282**, 706 (1998).
5. S. J. van Enk, *Phys. Rev. A* **60**, 5095 (1999).
6. H. F. Hofmann, T. Ide, T. Kobayashi, and A. Furusawa, *Phys. Rev. A* **62**, 062304 (2000).

QUANTUM INFORMATION TRANSFER WITHOUT AN EXTERNAL CHANNEL

JOHN V. CORBETT
Department of Mathematics, Macquarie University
Sydney, N.S.W. 2109, Australia
Email: jvc@ics.mq.edu.au

DIPANKAR HOME
Department of Physics, Bose Institute
Calcutta 700 009, India
Email: dhom@boseinst.ernet.in

Retrievable, usable quantum information is transferred in a scheme which, in a striking contrast to quantum teleportation, requires no external channel and does not involve the transfer of a quantum state from one subsystem to another. The process uses a three particle system in which the information is transmitted from 3 to 1 even though 1 and 3 never interact but are both entangled with 2. That this scheme implies a previously unexplored form of quantum nonlocality is also demonstrated.

Quantum entanglement is used as a resource for transferring and processing information. Most work, starting from[1], has been on studies of quantum teleportation (QT). In QT the information transfer from one wing to another involves a maximally entangled state in conjunction with a Bellbasis measurement in one wing with an external channel through which information about the outcome is classically communicated to the other wing. The information is transferred by transporting a quantum state in which it is encoded.

We show that the quantum formalism permits a *different* protocol involving a tripartite entanglement in which the information transfer does *not* require any external channel and does *not* involve the transfer of a quantum state from one system to another. In it, the Bellbasis measurement is replaced by an interaction between two subsystems, 2 and 3, in one wing while 2 is entangled with subsystem 1 in the other wing. The external channel is replaced by a long range interaction between 1 and 2 which, along with entanglement, transports the information.

In this ipso-information transfer (IIT) scheme, particles 1 and 2 are first prepared in an entangled state. They separate but continue to interact via a long-range residual interaction, $V_R(1,2)$. While $V_R(1,2)$ acts an independent interaction, $V_N(2,3)$, between 2 and 3 may be switched on by Alice. *If* Alice switches it on then the joint state of 1, 2 and 3 becomes entangled. Then, after all the particles have separated and ceased interacting, 1 and 3 are entangled with each other because both are in turn entangled with 2. The pair of single particle states of 2 are not orthogonal. Hence this tripartite entangled state is *not* Schmidt decomposable, and so it is not asymptotically equivalent to a GHZ state [2].

We note that there are two levels of information transferred from Alice to Bob: 1.The *yes/no* decision taken by Alice whether to turn on the interaction, $V_N(2,3)$. 2. Contingent on a *yes* decision, information about the final states of 3 given by their inner product. Provided that Bob knows the initial entangled state of 1 and 2 and the form of $V_N(2,3)$, he can determine both the strength of $V_N(2,3)$ and if the initial state of 3 has been changed. Both are obtained from the expectation values of certain observables of 1. The difference in the expectation values measured by Bob when Alice has or has not turned on the interaction is proportional to the inner product of the final states of 3. For this protocol to work, as well as the tripartite entanglement, both the interactions $V_R(1,2)$ and $V_N(2,3)$ are necessary. But it does not require any extra communication between Alice and Bob as is required in QT.

The other significant feature of this example is that it implies a form of quantum nonlocality which is quite different from those previously studied.

1 Communicating through an entangled tripartite state

To illustrate the way the information is transferred from Alice to Bob, consider the following scenario: before t_1, Alice prepared the entangled state, $\Psi(1,2)$, of 1 and 2.

$$\Psi(1,2) = a\psi_+(1)\phi_+(2) + b\psi_-(1)\phi_-(2), \quad (1)$$

where each of the single particle wave functions is normalised and $\langle\psi_+(1)|\psi_-(1)\rangle = 0$, however, $\langle\phi_+(2)|\phi_-(2)\rangle = \gamma_2 \neq 0$. We take $|a|^2 + |b|^2 = 1$ so $\Psi_0(1,2)$ is also normalised.

Alice sends particle 1 off towards a distant observer Bob at some time $t > t_1$. As a result 1 and 2 separate spatially but continue to interact through $V_R(1,2)$ whose strength diminishes as the separation grows. Alice must decide whether and when to switch on an interaction, $V_N(2,3)$, between 2 and 3. If Alice switches on $V_N(2,3)$ during the period from t_2 to t_3, while $V_R(1,2)$ is still non-negligible compared to the kinetic energy, then a tripartite entangled state is formed.

The tripartite entangled state is the normalised wave function $\Psi_f(1,2,3)$, given by,

$$a\psi_+(1)\phi_+(2)\chi_+(3) + b\psi_-(1)\phi_-(2)\chi_-(3). \quad (2)$$

Each single particle wave function in it has norm 1, but only $\langle\psi_+(1)|\psi_-(1)\rangle = 0$, whereas, $\langle\phi_+(2)|\phi_-(2)\rangle = \gamma_2 \neq 0$ and $\langle\chi_+(3)|\chi_-(3)\rangle = \gamma_3$ may be non-zero.

At times $t > t_3$, 2 and 3 cease interacting and move apart. At a later time t_4, when Bob receives 1, $V_R(1,2)$ is also negligible in comparison with the kinetic energy terms and the particles 1, 2 and 3 are all non-interacting.

At a time $t > t_4$, Bob measures an observable $A(1)$ of 1, that satisfies $\langle\psi_+(1)|A(1)|\psi_-(1)\rangle = \alpha \neq 0$. The expectation value of $A(1)$ in $\Psi_f(1,2,3)$ is

$$\begin{aligned} &|a|^2\langle\psi_+(1)|A(1)|\psi_+(1)\rangle \\ &+|b|^2\langle\psi_-(1)|A(1)|\psi_-(1)\rangle \\ &+2\Re e\, \bar{a}b\gamma_2\gamma_3\alpha. \end{aligned} \quad (3)$$

If particles 2 and 3 had *not* interacted the tripartite system would have a final state $\Psi'_f(1,2,3) = \Psi(1,2)\chi_0(3)$ where $\chi_0(3)$ is normalised and $\Psi(1,2)$ is given by equation (1). The expectation value of $A(1)$ in $\Psi'_f(1,2,3)$ differs from that in $\Psi_f(1,2,3)$, ie (3), by δ

$$\delta = 2\Re e\, \bar{a}b\gamma_2(1 - \gamma_3)\alpha. \quad (4)$$

Once Bob knows α and $\Psi(1,2)$, ie a, b and γ_2, he can determine γ_3 from δ.

Hence Bob can determine which choice Alice had made by measuring the expectation value of $A(1)$. Therefore, information of whether Alice switched on $V_N(2,3)$ and, if she had, about the value of γ_3 is transferred to Bob without 1 ever interacting with 3 and after 1 has ceased interacting with 2. Alice and Bob need no communication channel external to the tripartite system.

2 The preparation of the entangled states

The explicit dynamical model of the IIT scheme that we employ is based upon the von Neumann model of measurement[3] and uses time-dependent coordinate wave functions. The position coordinate space is one dimensional to simplify the notation.

The initial state of particle 1 is

$$\zeta(q_1,t) = a\psi_+(q_1,t) + b\psi_-(q_1,t), \quad (5)$$

is normalised when $|a|^2 + |b|^2 = 1$. The support of $\psi_+(q_1,t)$ is the interval I_+ while

$\psi_-(q_1,t)$'s support is the interval I_-, with a distance $d > 0$ between the intervals. 2 has the initial wave function $\phi_0(q_2,t)$.

The von Neumann interaction, $V_N(1,2)$ is turned on at time t_0.

$$V_N(1,2) = g(1,2)(q_1 \cdot p_2). \qquad (6)$$

1 and 2 also interact through a weaker, long range potential $V_R(1,2)$. The unitary evolution of 1 and 2 is governed by the Hamiltonian H,

$$H = \frac{p_1^2}{2m_1} + \frac{p_2^2}{2m_2} + V_R(1,2) + V_N(1,2). \quad (7)$$

We assume that $g(1,2)$ is so large that both $V_R(1,2)$ and the kinetic energy terms can be neglected in comparison with $V_N(1,2)$ so that the total Hamiltonian is effectively

$$H = V_N(1,2). \qquad (8)$$

Given an initial wave function $\Psi(q_1,q_2,t_0) = \zeta(q_1,t_0)\phi_0(q_2,t_0)$ the solution of the Schrodinger equation with Hamiltonian H (8), is an entangled bipartite state,

$$\Psi(q_1,q_2,t_1) = a\psi_+(q_1,t_0)\phi_+(q_2,t_0) + b\psi_-(q_1,t_0)\phi_-(q_2,t_0) \quad (9)$$

where $\phi_s(q_2,t_0)$, for $s = \pm$, are given by weighted shifts in q_2 of $\phi_0(q_2,t_0)$,

$$\phi_s(q_2,t_0) = \int |\psi_s(q_1,t_0)|^2 \phi_0(q_2 - q_1', t_0) dq_1. \qquad (10)$$

with $q_1' = g(1,2)(t_1 - t_0)q_1$[4]. The interaction $V_N(1,2)$ is impulsive.

At a time $t_2 > t_1$, while 1 and 2 are still interacting through $V_R(1,2)$, 2 and 3 interact through $V_N(2,3)$. The Hamiltonian for the system is ,

$$H = \frac{p_1^2}{2m_1} + \frac{p_2^2}{2m_2} + \frac{p_3^2}{2m_3} + V_N(2,3) + V_R(1,2). \qquad (11)$$

$V_N(2,3)$ is given by

$$V_N(2,3) = g(2,3)(q_2 \cdot p_3). \qquad (12)$$

$g(2,3)$ is so large that while $V_N(2,3)$ is acting the total Hamiltonian is effectively

$$H = V_N(2,3). \qquad (13)$$

$V_N(2,3)$ acts impulsively and prepares an entangled wave function $\Psi(q_1,q_2,q_3,t_3)$ [4].

After t_3, the wave function evolves under the Hamiltonian H,

$$H = \frac{p_1^2}{2m_1} + \frac{p_2^2}{2m_2} + \frac{p_3^2}{2m_3} + V_R(1,2) \quad (14)$$

until a time when $V_R(1,2)$ becomes negligible in comparison to the kinetic energy terms in H then Bob can measure the observable $A(1)$.

3 Summary of the IIT protocol

It is crucial that 1 and 2 continued to interact while 2 and 3 were interacting. If $V_R(1,2)$ was negligible compared to the kinetic energy terms when $V_N(2,3)$ was acting, the final entangled state, $\Psi_f(1,2,3)$, given by (2), could not be formed by a unitary dynamics because the states of 2 are not orthogonal [5]. The magnitude of $V_R(1,2)$ is *irrelevant* as long as it is not negligible compared to the kinetic energy.

The IIT scheme is robust in that parameters like $g(1,2)$ and $g(2,3)$ can be varied and information is still tranferred. It is secure against eavesdropping by Alice. By local measurements on particle 2, Alice cannot detect any change *after* or even *while* the information is transported from Carol to Bob because the reduced density matrix of 2 remains unchanged.

Furthermore, the two levels of transferred information in the IIT scheme exhibit different responses to variations of the parameters. The first level is *not* sensitive to variations provided that γ_2 is non-zero but the second level is. The information contained in γ_3 can be analysed in terms of binary information, for example using a binary tree.

4 A novel form of quantum nonlocality

Consider the following scenario; After the entangled state (1) has been prepared, particle 1 goes towards Bob and particle 2 goes to Alice. As well, Alice receives particle 3 from

Carol who has prepared 3 in a certain initial state. Then Alice switches on the interaction $V_N(2,3)$ forming the tripartite entanglement.

Subsequently, Bob measures an observable $A(1)$. If Carol varies the initial state of 3 but Alice keeps $V_N(2,3)$ unchanged, then the value measured by Bob will change.

Thus a local change made by Carol at a region spatially distant from Bob affects the statistical result of a measurement performed by Bob. The key point is that although the particle 3 send by Carol eventually interacts with 2, which in turn interacts with 1, the information from Carol's initial state preparation sector is *not* simply transported in a classical way by the interactions $V_R(1,2)$ and $V_N(2,3)$. The reduced density operator of 2 remains *unchanged* during the process so that Alice has *no* access to the information that is being transported. Moreover there is no direct channel between Carol and Bob.

Such an effect is therefore nonlocal, because according to any local model, the results of local measurement on particle 1 are determined entirely by the initial entangled state (1) of 1,2 and their interaction $V_R(1,2)$ because the states of 2 in the tripartite state (2) are unaffected by changes in the initial state of 3. This requirement is not satisfied in the IIT example, thereby implying a new form of quantum violation of Einstein's locality condition, viz."...the real factual situation of the system $S2$ is independent of what is done with the system $S1$ which is spatially separated from the former"[6]. This form of non-locality is independent of how "realism" is specified or what form of "hidden variables" is used in a local model.

In IIT, nonlocality is discernible through statistical measurements in one of the wings without needing to use correlations between observations in the different wings. This contrasts with the usual arguments for quantum nonlocality (whether or not a Bell-type inequality is used)[7]. Long range interactions in conjunction with kinematic entanglement produce this type of non-locality in non-Schmidt decomposable states. Further studies of the properties of the interactions needed to implement the scheme are planned.

5 Acknowledgements

The authors acknowledge the financial support for the visit of D.H. to Macquarie University via a Research Grant to J.V.C. The research of D.H. is supported by the Dept. of Science and Technology, Govt. of India. Thanks are due to Sougato Bose of Claredon Laboratory, Oxford and to Dale Woodside of Macquarie University for helpful suggestions.

References

1. C. H. Bennett, G. Brassard, C. Crepeau, R. Jozsa, A. Peres, and W. K. Wooters, Phys.Rev.Lett.**70**,1895 (1993);
2. A. V. Thapliyal. Phys. Rev. A **59**, 3336 (1999);
3. J. von Neumann, *Mathematical Foundations of Quantum Mechanics*, (Princeton University, Princeton, 1955), p. 443.
4. J.V.Corbett,and D.Home, *Ipso-Information-transfer* quant-ph/0103146.
5. J. V. Corbett, and D. Home, Phys. Rev. A **62**, 062103 (2000).
6. A.Einstein, in Albert Einstein: Philosopher - Scientist, edited by P.A.Schilpp (Open Court, La Salle, 1949), p.85.
7. J.S.Bell, Physics (Long Island, NY)1, 195 (1964); D.M.Greenberger, M.A.Horne and A.Zeilinger, in Bell's Theorem, Quantum Theory, and Conceptions of the Universe, edited by M.Kafatos (Kluwer, Dordrecht, 1989), p. 69; N.D.Mermin, Phys.Rev.Lett. 65,3373 (1990); L.Hardy, Phys.Rev.Lett. 71,1665 (1993); A.Cabello, Phys.Rev.Lett. (to be published) quant-ph/0008085; quant-ph/0101108.

QUANTUM INFORMATION TRANSFER USING A TIME-DEPENDENT BOUNDARY CONDITION

A. S. MAJUMDAR

S. N. Bose National Centre for Basic Sciences, Block JD, Sector III, Salt Lake, Calcutta 700098, India

E-mail:archan@boson.bose.res.in

DIPANKAR HOME

Department of Physics, Bose Institute 93/1 A. P. C Road, Calcutta 700009, India

E-mail:dhom@boseinst.ernet.in

We compute the time evolving reflection probability of a wave packet incident on a potential barrier while its height is reduced. A time interval is found during which the reflection probability is larger (superarrivals) compared to the unperturbed case. To explain this essestially nonclassical effect, a wave function acts as a "field" through which a disturbance resulting from the boundary condition being perturbed propagates at a speed depending upon the rate of reducing the barrier height. This phenomenon suggests a hitherto unexplored application for secure information transfer.

1 Introduction

Several interesting applications of wave-packet dynamics have been pointed out in recent years[1]. In this paper we investigate a new effect revealed in a time-dependent situation. Generally, the reflection/transmission probabilities for the scattering of wave packets by potential barriers are calculated after a complete time-evolution when asymptotic values have been attained. Here we study a phenomenon that occurs during the time evolution. For this purpose we consider the dynamics of wave packet scattered from a barrier while its height is reduced to zero before the asymptotic value of reflection probability is reached.

If an initially localysed wave packet $\psi(x, t=0)$ moves to the right and is scattered from a rectangular potential barrier of finite height and width, the time evolving reflection probability obtained at the left of a point x' is given by

$$|R(t)|^2 = \int_{-\infty}^{x\prime} |\psi(x,t)|^2 \, dx \qquad (1)$$

Note that $x\prime$ lies at the left of the initial profile of the wave packet such that $\int_{-\infty}^{x\prime} |\psi(x,t=0)|^2 \, dx$ is negligible. Let us now consider the case if during the time evolution of this wave packet the barrier is perturbed by reducing its height to zero. Let this height reduction take place within a small interval of time compared to the time taken by the reflection probability to attain its asymptotic value $|R_0|^2$. We compute effects of this perturbation on $|R(t)|^2$.

We find that there is a time interval during which $|R(t)|^2$ is remarkably larger in the perturbed case even though the barrier height is reduced. We call this effect "superarrivals"[2]. We showed in [2] that this essentially nonclassical phenonenon can be explained in terms of the dynamical kick imparted on the wave packet by the time evolving barrier. The time interval and the magnitude of superarrivals depend on the rate at which the barrier height is made zero. Information about barrier perturbation reaches the detector with a finite speed (signal velocity, v_e) which is also proportional to the rate at which the barrier is reduced. Here we suggest a scheme for secure information transfer using superarrivals.

2 Superarrivals through barrier perturbation

Let us consider an initial wave packet given by

$$\psi(x,t=0) = \frac{1}{\left[2\pi(\sigma_0)^2\right]^{1/4}} exp\left[-\frac{(x-x_0)^2}{4\sigma_0^2} + ip_0x\right] \quad (2)$$

with width σ_0 centered around $x = x_0$ and its peak moving with a group velocity $v_g = \frac{\langle p \rangle}{m}$ towards a rectangular potential barrier centered at a point x_c. (We set $\hbar = 1$ and $m = 1/2$. The numerical values of all quantities are evaluated in terms of these units.) The point x_0 is chosen such that $\psi(x,t=0)$ has a negligible overlap with the barrier. For computing $|R(t)|^2$ given by Eq. (1) the time dependent Schrodinger equation is solved numerically. Height of the barrier (V) before perturbation is so chosen that the asymptotic value of reflection probability is close to 1 for the static case. $|R(t)|^2$ is computed according to Eq. (1) by taking various values of x' satisfying the condition $x' \leq x_0 - 3\sigma_0/\sqrt{2}$. The computed evolution of $|R(t)|^2$ corresponds to the building up of reflected particles with time. It means that a detector located within the region $-\infty < x < x'$ measures $|R(t)|^2$ by registering the reflected particles arriving in that region up to various instants. First, we compute $|R_s(t)|^2$ for the wave packet scattered from a static barrier. Next we compute the time evolution of $|R_p(t)|^2$ in the perturbed case by choosing different rates of barrier reduction. In all cases the barrier reduction starts at a time t_p and the barrier height is reduced to zero linearly in time. t_p is chosen such that at that instant the overlap of the wave packet with the barrier is significant. The time interval during which the barrier is perturbed is denoted by ϵ. We choose values for which $\epsilon \ll t_0$, t_0 being the time required for $|R(t)|^2$ to attain the asymptotic value $|R_0|^2$. Figure 1 shows the evolution of $|R(t)|^2$ for various values of ϵ. One sees that

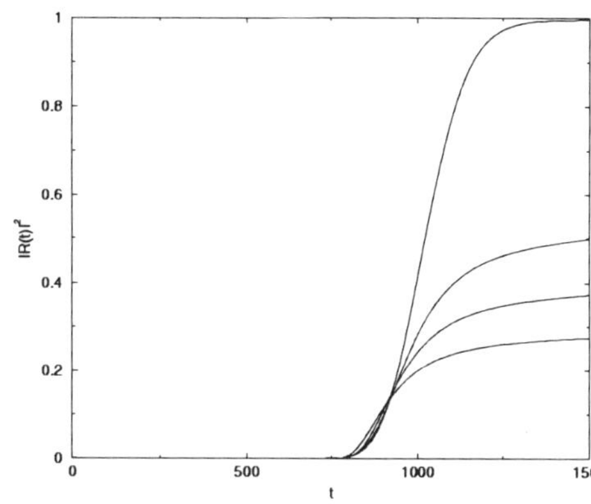

Figure 1. $|R(t)|^2$ is plotted for various values of ϵ. The top curve corresponds to the static case. The other curves correspond to the perturbed cases with different rates of reduction. As one increases ϵ, the magnitude of superarrivals decreases.

during the time interval $t_d < t < t_c$,

$$|R_p(t)|^2 > |R_s(t)|^2 \quad (3)$$

t_c is the instant when the two curves cross each other, and t_d is the time from which the curve corresponding to the perturbed case starts deviating from that in the unperturbed case. A detector placed in the region $x < x'$ would therefore register more counts during this time interval $\Delta t = t_c - t_d$ even though the barrier height had been reduced to zero prior to that.

In order to demonstrate the quantum mechanical nature of superarrivals, one could consider a probabilistic distribution of particles given by the initial wave packet in terms of the spreads in both position and momentum. Solving the classical Liouville equation for the same time-varying potential and obtaining $|R_s(t)|^2$ and $|R_p(t)|^2$ for the classical case shows that there are no superarrivals[2].

The magnitude of superarrivals can be defined by a quantity η, given by

$$\eta = \frac{I_p - I_s}{I_s} \quad (4)$$

where the quantities I_p and I_s are given by

$$I_p = \int_{\Delta t} |R_p(t)|^2 \, dt \quad (5)$$

$$I_s = \int_{\Delta t} |R_s(t)|^2 \, dt \tag{6}$$

Now we consider the question as to how fast the influence of barrier perturbation travels across the wave packet. Note that the information content of a wave packet does not always propagate with the group velocity v_g of a wave packet which is usually identified with the velocity of the peak of a wave packet. A local change in potential affects a wave packet globally, the global effect being manifested through time evolution of the packet. The action due to a local perturbation (barrier height reduction) propagates across the wave packet at a finite speed say, v_e affecting the time evolving reflection probability. Thus a distant observer who records the growth of reflection probability becomes aware of perturbation of the barrier (starting at an instant t_p) from the instant t_d when the time varying reflection probability starts deviating from that measured in the unperturbed case. Thus v_e is given by

$$v_e = \frac{D}{t_d - t_p} \tag{7}$$

It has been shown[2] that the duration of superarrivals Δt, magnitude of superarrivals η, and the signal velocity v_e all decrease monotonically with increasing ϵ (or decreasing rate of barrier reduction). The reducing barrier imparts a kick (the magnitude of which is proportional to the rate of reduction) on the wave packet. This disturbance is propagated accross the wave packet to reach the detector.

3 A scheme of secure information transfer

We have seen from our above analysis that a perturbation in the boundary condition (potential barrier) affects the dynamics of a wave packet reflected from it. The reduction of barrier height provides a kick to the wave packet, breaking it up into a reflected and a transmitted portion. An analysis of the profile of the wave packet shows that the reflected part has a secondary peak which is shifted towards the detector[2]. We have also seen that information about barrier perturbation taking place at the location x_c and starting at the time t_p propagates through the wave packet and reaches the detector located at x' at the time t_d with a finite velocity v_e. It is interesting to investigate the details of this information transfer occuring through quantum superarrivals.

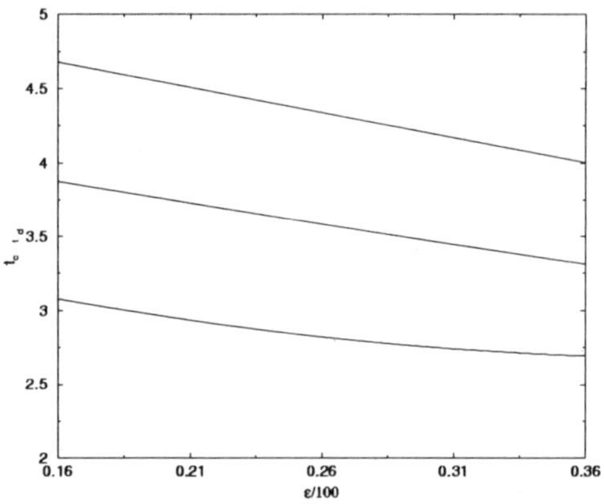

Figure 2. The duration of superarrivals Δt is plotted versus ϵ. The three different curves denote three different values of the detector position.

A lot of interest is being currently devoted to study and develop new schemes of quamtum information transfer (see, for instance Alber *et al*[3]). Let us see, how the present scheme of superarrivals could, in principle, be used for such a purpose. In order to do so, it is important to focus on the variation of $\Delta t \equiv t_c - t_d$ (the duration of superarrivals) as a function of ϵ (the time taken for barrier reduction). This is plotted in Figure 2 for three different values of the detector position x'. Note that Δt decreases monotonically with increasing ϵ for a wide range of values of ϵ. Now suppose a particular curve in Figure 2 (functional relation between Δt and ϵ for a fixed value of detector position x') is chosen as a key which is shared by two persons Alice and Bob who want to exchange information. Alice is at the barrier and receives a continuous inflow of particles whose wave function is given by the initial Gaussian. Alice has the choice of reducing the barrier height at completely random different rates. She chooses one particular value of ϵ for a single run of the experiment, and she wants Bob who is at the detector to be able to decipher this value of ϵ. Bob monitors the time evolution of $|R_p(t)|^2$ through the detector counts and is able to decipher t_c, t_d, and hence Δt by comparison with the curve $|R_s(t)|^2$ for the static case. He then uses his key to infer the exact value of ϵ corresponding to the particular value of Δt he has measured. The whole procedure can be repeated with different rates of barrier reduction as many times as required by Alice and Bob. In this way an exchange of information takes place between Alice and Bob. This exchange is secure because it would not be possible for any eavesdropper to decipher Alice's chosen value of ϵ without having access to the key. It is important to note that information transfer takes place in this scheme without any shared entanglement between the two players Alice and Bob. Also, the variable ϵ can vary continuously in the allowed parameter range.

In conclusion, we have suggested a scheme of secure transfer of continuous information by the use of the phenomenon of quantum superarrivals recently pointed out by us[2]. Our method is inherently nonclassical in the sense that the quantum wave function acts as a medium or "carrier" through which information about perturbation of the boundary condition propagates. The objective "field-like" characteristic of the quantum wave function is exploited in this scheme of information transfer, which makes it rather distinct from the other schemes available in the literature. The most natural and widely used form of a localised wave function is of a Gaussian type. Since our work is of the first type in exploiting a time-varying boundary condition and also the objective "field-like" property of the wave function for the purpose of information transfer, we have chosen an initially localised Gaussian wave packet having negligible overlap with the barrier. Of course, the experimental viability of our scheme is contingent upon the robustness of the key (Figure 2) against distortion of the initial wave packet. A comprehensive study is required to check thoroughly this feature. Nevertheless, our preliminary results indicate that the key is quite stable against small distortions of the width and the location of the initial wave packet, provided its Gaussian profile is maintained. Further investigations with more varied sets of parameters and different types of perturbation in boundary conditions are necessary to test the feasability of single particle experiments[4] for realizing such a scheme.

References

1. D. M. Greenberger, *Physica B* **151**, 374 (1988); M.V. Berry, *J. Phys.* A **29**, 6617 (1996); A. Venugopalan and G.S. Agarwal, *Phys. Rev.* A **59**, 1413 (1999).
2. S. Bandyopadhyay, A. S. Majumdar and D. Home, *quant-ph*/0103001.
3. G. Alber, T. Beth, M. P. & R. Horodecki, M. Rotteler, H. Weinferter, R. F. Werner and A. Zeilinger, *Quantum information – an introduction to basic theoretical concepts and experiments*, (Springer Tracts in Modern Physics, 2001).
4. A. Tonomura, J. Endo, T. Matsuda, T. Kawasai and H. Ezawa, *Am. J. Phys.*, **57** (2), 117 (1989); S. A. Werner in *Fundamental Problems in Quantum Theory,* eds. D. M. Greenberger and A. Zeilinger (N. Y. Academy of Science, 1995), p. 241.

A DETAILED ANALYSIS ON THE FIDELITY OF QUANTUM TELEPORTATION USING PHOTONS

KENJI TSUJINO
Research Institute for Electronic Science, Hokkaido University, Kita-12 Nishi-6, Kita-ku, Sapporo, Hokkaido 060-0812 Japan, E-mail: tsujino@es.hokudai.ac.jp

SHIGEKI TAKEUCHI
Research Institute for Electronic Science, Hokkaido University and JST-CREST project Kita-12 Nishi-6, Kita-ku, Sapporo, Hokkaido 060-0812 Japan, E-mail: takeuchi@es.hokudai.ac.jp

KEIJI SASAKI
Research Institute for Electronic Science, Hokkaido University, Kita-12 Nishi-6, Kita-ku, Sapporo, Hokkaido 060-0812 Japan, E-mail: sasaki@es.hokudai.ac.jp

We investigate the condition in order to perform a high fidelity teleportation from experimental viewpoint. Our scheme referred to the analysis suggested by P. Kok *et al* [4]. and added our ideas to their analysis. As a result, it is possible to demonstrate an experiment of high fidelity teleportation without receiver's feedback using a visible light photon counter and high visibility entangled photon sources.

1 Introduction

Quantum teleportation using photons is the cornerstone of the quantum communication and information. The protocol of the quantum teleportation uses a pair of two entangled photons, each of which is shared by Alice and Bob. In order to teleport the state of the original photon to Bob, Alice performs Bell state measurement on the original photon and the one of the Einstein-Podolsky-Rosen (EPR) pair shared by Alice. She tells the result of the measurement to Bob. According to result, Bob performs the appropriate unitary transformation on the other of the EPR pair, and then the original state is reconstructed.

In 1997, D. Bouwmeester *et al.* demonstrated an experimental quantum teleportation of polarized photons [1], and reported high fidelity in the experiment. Later, it was pointed out by Braunstein and Kimble that the fidelity might be lower than $1/2$ when the output of vacuum states was taken into account [2]. They also suggested that the employment of a so-called detector cascade will improve the fidelity including the effect of the vacuum states.

In our opinion, the definition of the fidelity may change according to the experimental setup and the supposed usage of the output state. In this sense, we think both of the "fidelities" are reasonable according to circumstances. In a sense, it may be said that Bouwmeester *et al.* improved the fidelity *using the receiver's feedback.*

For some applications of the quantum teleportation, we have to decrease the vacuum state in the output. In this sense, an experimental demonstration of "high fidelity quantum teleportation without receiver's feedback" is a challenging theme.

In this paper, we clarify the required parameters to perform the demonstration using a "visible light photon counter (VLPC)" in-

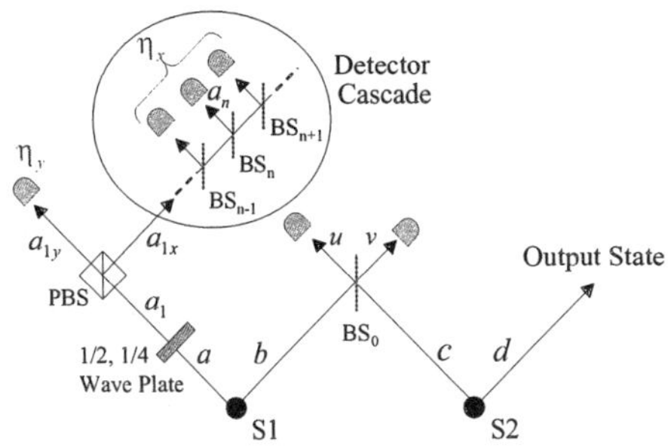

Figure 1. Scheme of a teleportation experimental setup.

stead of a detector cascade. This detector has a high quantum efficiency of 88%, and the ability to distinguish the number of incident photon [3]. Following the previously reported analysis by P. Kok *et al.*[4], we considered the situations where the EPR twins are in the mixed state and the combination of different type of detectors.

2 Background

2.1 Innsbruck Experiment

Figure 1 shows the scheme of an experimental setup. The EPR source S_1 and the EPR source S_2 are parametric down-converters [1,5]. One of the photons of an EPR pair emitted into the mode b from the source S_1 carries the original state. The original state is determined by the measurement of the other photon in mode a. The EPR pair created from the source S_2 corresponds to the EPR pair in the original proposal of Bennett [6].

The parametric down-converter generates not only one EPR pair, but also two EPR pairs simultaneously. Unfortunately, in Innsbruck experiment the detectors couldn't distinguish one photon incidence from several-photon incidences. Therefore, teleportation fidelity was smaller than 2/3 [4] when the vacuum state was also included in the "teleported" density matrix.

2.2 P. Kok's analysis

P. Kok *et al.* [4] reported the calculation of the fidelity of the teleportation when a detector cascade is used in the experiment. the calculation was performed as follows.

In Fig.1, the state of the photons before the detectors is given by

$$|\chi\rangle\langle\chi| = U_{cas}U_\theta U_{BS}U_{src2}U_{src1}|0\rangle \times\langle 0|U^\dagger_{src1}U^\dagger_{src2}U^\dagger_{BS}U^\dagger_\theta U^\dagger_{cas}, \quad (1)$$

where $|0\rangle$ is a vacuum state. The U_{src1}, U_{src2} are unitary transformations to generate a wavepacket states of the EPR pairs from the sources S_1 and S_2, respectively. The U_{BS}, U_θ and U_{cas} are the unitary transformations of the beam splitter(BS_0), the polarization rotator and the detector cascade, respectively. The detection of the photon is described by the projection operator valued measures (POVM's). Taking the partial trace over the detected modes, the outgoing state is given by

$$\rho_{out} = \mathrm{Tr}_{a_1,\ldots,a_n,u,v}[\sqrt{E_{cas}E_u E_v}|\chi\rangle \times\langle\chi|\sqrt{E_v E_u E_{cas}}], \quad (2)$$

where E_{cas} is the POVM that corresponds to the detection of a x-polarized photon in the detector cascade and no y-polarize photon. The E_u and E_v are the POVMs which correspond to the detection event of the detectors in the mode u and v.

3 Teleportation fidelity using the VLPC with EPR pairs in the pure states

The Visible Light Photon Counters are the alternative unique detectors. S.Takeuchi *et al.* [3] showed that VLPCs have high quantum efficiencies (88.2%±5%) and can distinguish between a single-photon incidence and two-photon incidence [3]. With our estimation, a VLPC is equivalent to a detector cascade with several hundred detectors. Here we calculate the fidelity of our future experiment using the VLPCs.

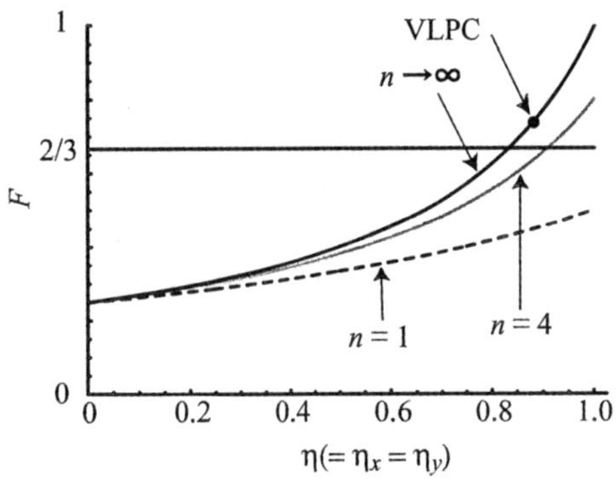

Figure 2. Theoretical curves of the teleportation fidelity F of Eq.(4) for cascading detector of n =1, 4 and ∞, $p_1 = p_2$.

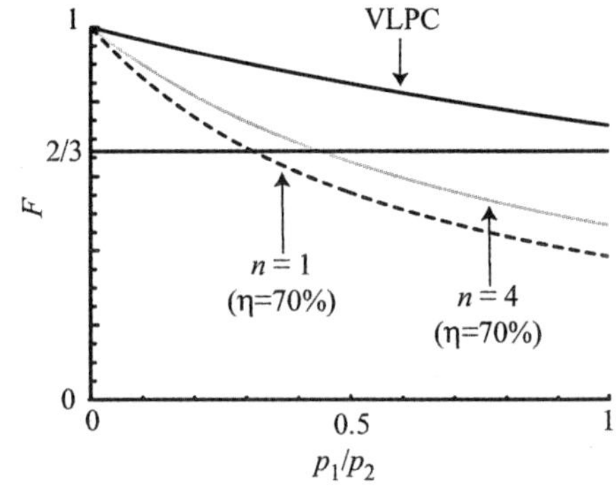

Figure 3. Curves of teleportation fidelity F of Eq.(4) for the ratio probability of creating one entangled photon pair from two source, p_1/p_2.

In addition to the Kok's analysis, we added the phase shifter between x-polarization and y-polarization in Eq.(1) because we can decrease the critical value of the fidelity from 3/4 to 2/3. According to the actual setup developing, we considered the combination of detectors that have different quantum efficiencies between x-polarized photons and y-polarized photons.

The probability of one entangled photon pair creation of the parametric down-conversion at $S_1(S_2)$ is $p_1(p_2)$. We calculated the output state up to the order p^2 (i.e., p_1^2 or p_1p_2) and found

$$\rho_{out} \propto \frac{p_1}{n}[1 + n(1-\eta_y) + (2n-1)(1-\eta_x)] \times|0\rangle\langle 0| + p_2|\phi\rangle\langle\phi|, \quad (3)$$

where $|\phi\rangle = \cos\theta|0,1\rangle + e^{i\varphi}\sin\theta|1,0\rangle$ is initial state send to Bob. Quantum teleportation fidelity F is given by

$$F = \frac{1}{\frac{p_1}{p_2}[\frac{1}{n} + (1-\eta_y) + (2-\frac{1}{n})(1-\eta_x)] + 1}, \quad (4)$$

where η_x and η_y are quantum efficiencies of the x-polarized photon on mode a and the y-polarized photon on mode a, respectively.

The functional behavior of the fidelity F is shown in figure 2 for the case $p_1 = p_2$. Even in this condition, it is possible to achieve high fidelity teleportation without receiver's feedback using the VLPC; otherwise, it is impossible to achieve high fidelity using detector cascade with conventional detectors (i.e. SPCMs) or a photon number indistinguishable detector with the quantum efficiency $\eta = 1$. This result shows that both of the features (high quantum efficiency and multi-photon distinguishability) contribute to the fidelity.

Figure 3 shows the teleportation fidelity F when p_1/p_2 is changed. In this situation the rate of the teleportation is sacrificed to improve the fidelity. It is shown that the rate is much larger when the VLPC is adopted.

4 Teleportation fidelity using the VLPC with EPR pairs in the mixed states

In this section, we calculate the fidelity for the case where the EPR pairs are in the mixed states. We consider the states which is a simply mixture of the pure EPR pair state and the state of the randomly polarized photons. Within the order up to p_2 (i.e., p_1^2 or p_1p_2), the density matrix of the pair is given as follows :

$$\rho_{src} = \gamma U_{src}|0\rangle\langle 0|U_{src}^{\dagger} + (1-\gamma)[|0\rangle\langle 0| + p\frac{1}{4}\boldsymbol{I} + p^2\frac{1}{16}|2,0\rangle|2,0\rangle\langle 2,0|\langle 2,0| + \cdots], \quad (5)$$

where $\boldsymbol{I}$ is the identity operator and γ is a mixing parameter.

Under these conditions, we calculated the output state following the calculation shown in the previous section; the density matrix of the output state ρ'_{out} is given by

$$\rho'_{out} \propto 2p_1[\frac{1}{n} + (1-\eta_y) + (2-\frac{1}{n})(1-\eta_x)] \times (3 - 6\gamma + 4\gamma^2)|0\rangle\langle 0| + 2p_2\gamma^2|\phi\rangle\langle\phi| + 2p_2(1-\gamma^2)\frac{1}{2}\mathbf{1}. \quad (6)$$

Therefore, the teleportation fidelity is calculated as follows.

$$F' = \frac{(1+\gamma^2)/2}{\frac{p_1}{p_2}P_v(3 - 6\gamma + 4\gamma^2) + 1}, \quad (7)$$

where

$$P_v = [\frac{1}{n} + (1-\eta_y) + (2-\frac{1}{n})(1-\eta_x)]. \quad (8)$$

Figure 4 shows the required parameters to performe the "quantum" teleportation. The weakly shaded area shows the region where VLPC (infinite detector cascade, η=88%) is fully utilized, and dark area shows the region where the effective efficiency is reduced to 80% by the loss at the filters.

This result shows that it is possible to perform a "receiver's feedback free quantum teleportation" using the VLPCs with the reported EPR source ($\gamma > 85\%$).

5 Conclusion

We expanded the experimental parameters to perform the high fidelity teleportation without a receiver's feedback. In addition to the previously reported analysis, we considered the combination of the different detectors and also the EPR pairs in mixed states. The VLPC can be distinguished between one- and several photons.

We succeeded to clarity the experimental parameters to perform the "receiver's feedback free quantum teleportation" using the VLPC with a reasonable experimental time when the effective quantum efficiency of the VLPC is reduced to 80% due to some losses in the actual setup. We can perform the experiment using the reported EPR sources (γ=85%).

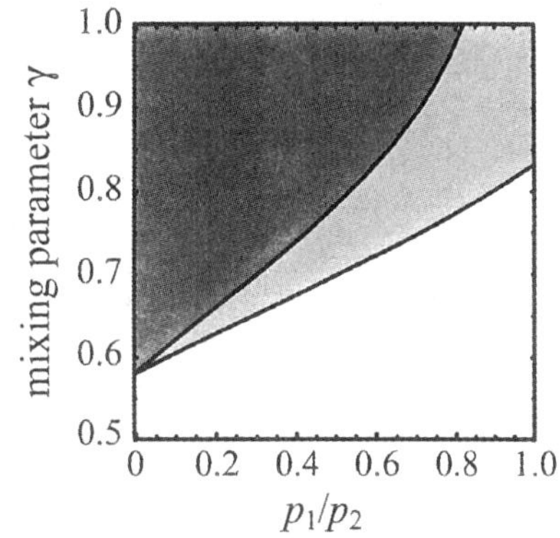

Figure 4. Teleportation fidelity F' of Eq.(7) under the condition that mixed states are generated from sources.

Acknowledgments

We would like to thank Peter Kok for useful discussions and correspondence. This work is supported by the JST CREST project.

References

1. D. Bouwmeester, J.-W. Pan, K. Mattle, M. Eibl, H. Weinfurter, and A. Zeilinger, *Nature* (London) **390**, 575 (1997).
2. S.L. Braunstein and H.J. Kimble, *Nature* (London) **394**, 840 (1998).
3. S. Takeuchi, J. Kim, Y. Yamamoto, and H.H. Hogue, *Appl. Phys. Lett.* **74**, 1063 (1999); J. Kim, S. Takeuchi, Y. Yamamoto, and H.H. Hogue, *ibid.* **74**, 902 (1999).
4. P. Kok and S.L. Braunstein, *Phys. Rev. A.* **61**, 042304 (2000).
5. P.G. Kwiat, K. Mattle, H. Weinfurter, and A. Zeilinger, *Phys. Rev. Lett.* **75**, 4337 (1995).
6. C.H. Bennett, G. Brassard, C. Crepeau, R. Jozsa, A. Peres, and W.K. Wootoers, *Phys. Rev. Lett.* **70**, 1895 (1993).

HOLEVO BOUND FOR CLASSICAL INFORMATION TRANSMITTED THROUGH ENTANGLED STATES

L.C. KWEK[A,B] AND C.H. OH[A]

[A] *Physics Department, National University of Singapore, Kent Ridge, Singapore, 119260*

[B] *Nanyang Technological University, National Institute of Education, 1, Nanyang Walk, Singapore 637616*

Recently, seemingly conflicting results were obtained for the efficiency of the transmission of classical signals over quantum channels. In one study, it was shown that entangled states can improve the transmission of classical information in a two-Pauli channel. However, in another study, Bruss et al showed that entanglement cannot possibly improve transmission of classical information in a depolarizing channel. By studying the Holevo information in a general quantum channel, we analyze the efficiency of such transmission of classical states.

1 Introduction

Quantum entanglement[1] has always been an important and useful tool in quantum information and communication theory. Indeed, a good understanding of quantum entanglement can vastly improve our knowledge regarding optimal capabilities of quantum channels. A quantum channel permits quantum information to be transferred from a sender to a receiver. If such transfer of information is free from noise or disturbance due to decoherence effects, the quantum information remains faithfully intact. However, most practical channels do not conform to this class of ideal channels. Real channels are always plagued by some form of noise, typically through depolarization, amplitude damping and dephasing.

For noisy channels, the efficiency of the transmission of information is an important issue in quantum information theory. Analogous to the classical Shannon coding theorem, a quantum version has been formulated[2,3] several years ago. The quantum coding theorem concerns the transmission of an unknown quantum state from a sender to a receiver. On the other hand, it is also possible to envisage the transmission of classical information as a sequences of zeros and ones using quantum states. Such consideration has given rise to several rich and interesting information-theoretic notions based on the idea of von Neumann entropy[4]. It is interesting to note that quantum states involved in classical information need not be orthogonal to each other.

It is always possible to transmit classical information through a noiseless quantum communication channel using orthogonal signal states. However, the situation becomes slightly more complicating for non-orthogonal states. In this case, it is often instructive to look at the maximum amount of information that a message can be recovered in a measurement performed on the system. One such measure of the recovered information is the notion of mutual information[4,5,7] or Holevo information.

Recently, it has been pointed out that entangled states can improve the transmission of classical information in a noisy channel, specifically the two-Pauli channel[8,9]. However, a seemingly conflicting result was reported for the case of a depolarizing channel[10]. Specifically, it has been shown that such mutual information cannot be enhanced in a memoryless depolarizing channel.

In a previous work[11], by generalizing prototype channels so that the new modified channel can straddle between the two-Pauli channel and the depolarizing channel, we see why these apparent conflicts arise within the parameter space. Since the two-Pauli channel permits the transmission of classical information via entangled states while the depolarizing channel forbids such transmission efficiently, our modified channel provides a formalism for studying the transition from one channel to the other using a tuning parameter.

In section 2, we first introduce the generalized channel and compute the Holevo information for equal probable states associated with the channel. In section 3, we briefly consider parametrizations of the modified channel with one single parameter, q, so that in the limit $q = 0$, we get the depolarizing channel whereas in the other limit of $q = 1$, we get the two-Pauli channel (Sect. 3.1). This result has been reported elsewhere [11]. Moreover, for any arbitrary one-parameter parametrization, it is also very instructive to consider $\eta - q$ plots for the channels where η is a parameter for the noise in the channel. In section 4 we consider more generally a possible multi-parameter extension of the previous model. Finally, in section 5, we summarize some of our main points in the paper.

2 Modifying the channel

In this section, our generalized channel is described by the following Krauss operators

$$M_0 = \sqrt{1-\eta}\mathbf{1} \tag{1}$$

$$M_i = \sqrt{\frac{\eta}{3}(1+\eta_i)}\sigma_i \tag{2}$$

where σ_i, $(i = 1 \cdots 3)$ are Pauli matrices and $\eta_1 + \eta_2 + \eta_3 = 0$. Also, the parameter η measures the amount of noise within the channel. It is easy to check that the operators satisfy the relation $\sum_i M_i^\dagger M_i = 1$. Notice that if $\eta_1 = \eta_2 = \eta_3 = 0$, we get the usual depolarizing channel and if $\eta_1 = \eta_2 = \frac{1}{2}, \eta_3 = -1$, we get the two-Pauli channel.

The action of a memoryless channel on n arbitrary signals π_i is described by[10]

$$\phi(\pi_i) = \sum_{k_1,k_2,\cdots k_n} (M_{k_1} \otimes \cdots M_{k_n}) \pi_i \left(M_{k_1}^\dagger \otimes \cdots M_{k_n}^\dagger\right). \tag{3}$$

Further, if the input ensemble of states is the set $\mathcal{E} = \{p_i, \pi_i\}$ with $p_i \geq 0$ and $\sum p_i = 1$, the mutual information, $I_n(\mathcal{E})$ is defined as

$$I_n(\mathcal{E}) = S(\rho) - \sum_{i=1}^{n} p_i S(\phi(\pi_i)) \tag{4}$$

where the index n denotes the number of channels, $\rho = \sum_i p_i \phi(\pi_i)$ and

$$S(\chi) = -\mathrm{tr}\ \ (\chi \log_2 \chi) \tag{5}$$

is the von Neumann entropy.

We next consider the following set of equally probable input states

$$|\pi_1\rangle = \cos\theta\ |00\rangle + \sin\theta\ |11\rangle \tag{6}$$

$$|\pi_2\rangle = \sin\theta\ |00\rangle - \cos\theta\ |11\rangle \tag{7}$$

$$|\pi_3\rangle = \cos\beta\ |01\rangle + \sin\beta\ |10\rangle \tag{8}$$

$$|\pi_4\rangle = \sin\beta\ |01\rangle - \cos\beta\ |10\rangle \tag{9}$$

which becomes maximally entangled for $\theta = \beta = \frac{\pi}{4}$ and becomes completely un-entangled for $\theta = \beta = 0$.

For the quantum state $|\pi_1\rangle$, the corresponding output density matrix, in the $\{|00\rangle, |01\rangle, |10\rangle, |11\rangle\}$ is given by

$$\phi(\pi_1) = \frac{1}{4}\begin{pmatrix} \xi_1^+ & 0 & 0 & \xi_2 \\ 0 & \xi_3 & \xi_4 & 0 \\ 0 & \xi_4 & \xi_3 & 0 \\ \xi_2 & 0 & 0 & \xi_1^- \end{pmatrix} \tag{10}$$

where $\xi_1^\pm = 1 \pm \gamma\cos 2\theta + \gamma^2$, $\xi_2 = \sin 2\theta(\alpha^2 + \delta^2)$, $\xi_3 = 1 - \gamma^2$ and $\xi_4 = \sin 2\theta(\alpha^2 - \delta^2)$, $\alpha = 1 - 4/3\eta + 2/3\eta\eta_1$, $\delta = 1 - 4/3\eta + 2/3\eta\eta_2$ and $\gamma = 1 - 4/3\eta + 2/3\eta\eta_3$. The eigenvalues of the density matrix $\phi(\pi_1)$ can be easily found to be

$$\lambda_{1,2} = \frac{1}{4}\left(1 - \gamma^2 \pm (\alpha^2 - \delta^2)\sin 2\theta\right) \tag{11}$$

$$\lambda_{3,4} = (1 + \gamma^2 \pm \kappa). \tag{12}$$

where $\kappa = \sqrt{4\gamma^2 \cos^2 2\theta + (\alpha^2 + \delta^2 \sin^2 2\theta)}$. A similar computation can also be done on the other input states, namely $\phi(\pi_i), i = 2,3,4$. Suppose the input states π_i occur with equal probability in the set $\mathcal{E}$. It is then possible to calculate the mutual information, $I_2(\mathcal{E};\theta,\beta)$ for the set, $\mathcal{E}' \equiv \{1/4, \pi_i\}$, $i = 1,\cdots 4$. For a fixed values of η and η_i, it turns out that the mutual information exhibits periodic extremal values at various values of θ and β. By computing $\frac{\partial I_2}{\partial \theta} = 0$ and $\frac{\partial I_2}{\partial \beta} = 0$,we see that the only solution to these equations occurs when $\sin 2\theta \cos 2\theta = 0$, $\sin 2\beta \cos 2\beta = 0$ yielding $\theta = m\frac{\pi}{4}$, $\beta = n\frac{\pi}{4}$ for some integers, m and n.

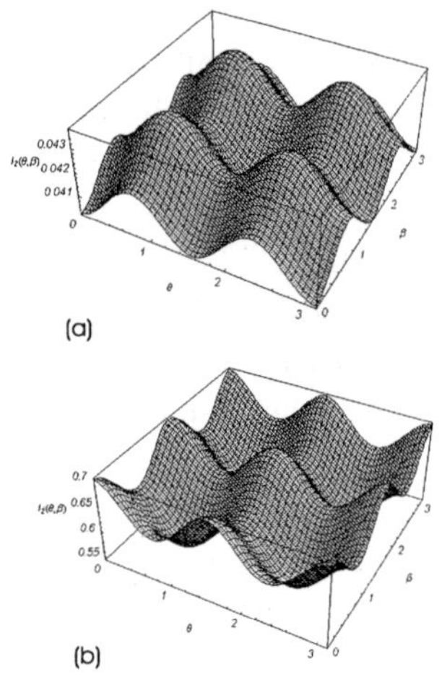

Figure 1. Variation of the mutual information, $I_2(\theta,\beta)$ for (a) $q = 0.5 = \eta$ and (b) $q = 0.5, \eta = 0.2$ as a function of θ and β.

3 One-Parameter Parametrizations

3.1 *A special parametrization*

For simplicity and convenience, we confine ourselves to the simplest non-trivial case of two channels and compute the mutual information $I_2(\mathcal{E})$ and choose the parameters, $\eta_i = \frac{q}{2}, i = 1,2$ and $\eta_3 = -q$, so that we get the depolarizing channel for $q = 0$ and the two-Pauli channel for $q = 1$.

The variation of the mutual information, $I_2(\theta,\beta)$ for $q = 1/2 = \eta$ as a function of θ and β is plotted in Fig. 1. Notice that the mutual information is maximal or minimal for $\theta = m\frac{\pi}{4}$ and $\beta = n\frac{\pi}{4}$ for some integers, m, n as shown earlier for the general case. In particular, we see that for $q = \eta = 0.5$, the mutual information for maximally entangled states ($\theta = \beta = \pi/4$) is greater than that for un-entangled states ($\theta = \beta = 0$) whereas for $q = 0.5$ and $\eta = 0.2$, the behavior is reversed so that the mutual information for maximally entangled state is now less than that for un-entangled ones.

It is also useful to consider the difference between the mutual information for maximally entangled states and un-entangled states, ΔI_2, defined by

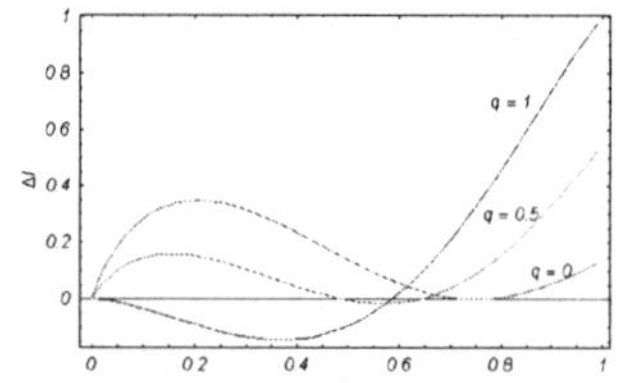

Figure 2. Difference between mutual information for unentangled and maximally entangled states ($\theta = \pi/4$ and $\beta = \pi/4$) as a function of η for $q = 0$ (depolarizing), $q = 0.5$ and $q = 1$ (two-Pauli channel).

$$\Delta I_2(q,\eta) = (I_2)\,|_{\theta=0=\beta} - (I_2)\,|_{\theta=\pi/4=\beta}. \tag{13}$$

In this way, maximally entangled states can only enhance to classical communication and transmission if $\Delta I \equiv \Delta_2 I < 0$. In particular, we have plotted the variations of ΔI as a function of η for various values of q ($q = 0$, $q = 0.5$ and $q = 1$) in Fig. 2. The figure clearly shows that for $q = 0$, $\Delta I \geq 0$ but as $q \to 1$, there is an increasing dip in the curve with a maximal dip of about -0.143443 at $\eta = .368807$ for $q = 1$. The three-dimensional variation of ΔI as a function of both q and η is shown in Fig. 3.

Thus, as η and q vary, the region on the $\eta - q$ plot in which $\Delta I < 0$ provides an in-

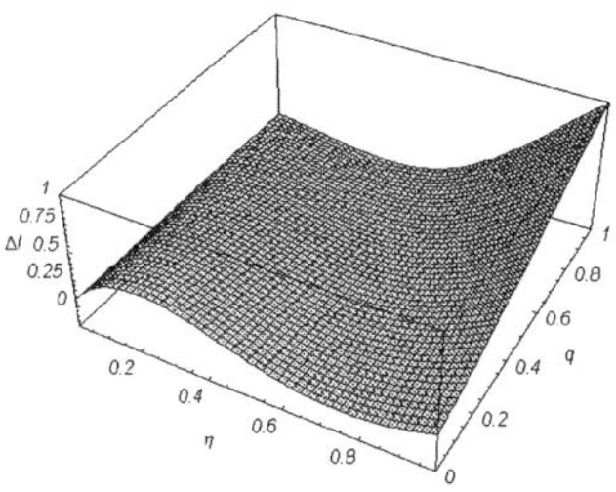

Figure 3. Difference between mutual information for unentangled and maximally entangled states as a function of η and q.

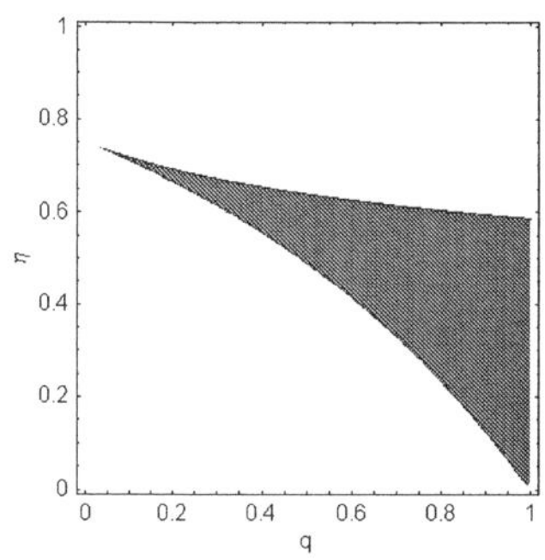

Figure 4. Contour plot of the difference in mutual information as a function of η and q. The shaded portion denotes the region in which entangled quantum states has a distinct advantage over classical transmission.

dication of the possibility of classical transmission using maximally entangled quantum states. Such a plot is shown in Fig. 4. Clearly for $q = 0$ (depolarizing channel), it is never possible to increase mutual information using quantum entangled states, but for $q = 1$ (two-Pauli channel), it is possible to do so for certain values of η, namely $0 < \eta < \eta_0$, $\eta_0 \approx 0.585795$ (see Fig. 2). It is also intriguing to note that for arbitrary q, (say $q = 0.5$), some noise seems to aid rather than impede classical transmission of information.

3.2 Other parametrizations

In the previous section, we have considered a particular modification of the depolarizing-two-Pauli channel. Other parametrizations

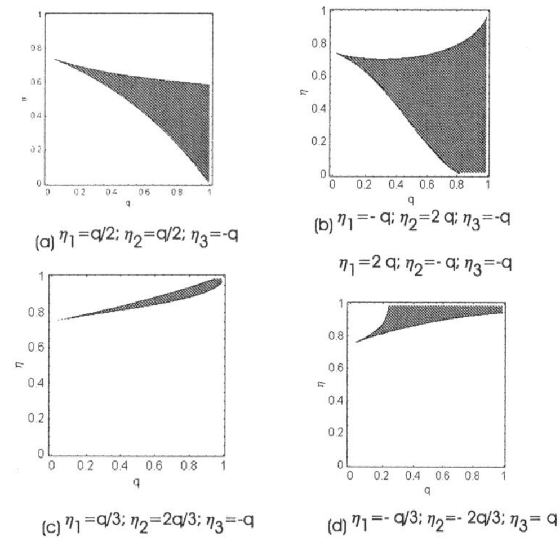

Figure 5. Contour plots of ΔI as a function of η and q ($\eta - q$ plots) for various parametrizations. Note that except for (a) (included for comparison) which is the case discussed in this paper, the other $\eta - q$ plots do not straddle between the depolarizing and usual two-Pauli channel. In the limits $q = 1$, case (b) and case (c) become asymmetric two-Pauli channels.

with different $\eta_i, i = 1, 2, 3$ are also possible and the $\eta - q$ plots for these parametrizations are shown in Fig. 5. Note also that, except for $\eta_1 = \eta_2 = q/2$ and $\eta_3 = -q$ as in the case discussed earlier, other parametrizations do not in general straddle between the depolarizing and two-Pauli channel. It is interesting to note the changes in the $\eta - q$ plots for case (c) and (d) since $\eta_i^{(c)} = -\eta_i^{(d)}$ for each $i = 1, 2, 3$. In the limit $q \to 1$, although case (c) becomes an asymmetric two-Pauli channel (and similarly for case(b)), case (d) continues as an asymmetric depolarizing channel. Note that in our plots, we recover the depolarizing channel in the limit $q \to 0$.

4 Extension with two or more parameters

It is easy to extend the previous analysis to a more general channel, for instance by setting $\eta_i = q_i, i = 1, 2$ and $\eta_3 = -q_1 - q_2$. Fig. 6 shows the graph of the difference between mutual information for unentangled

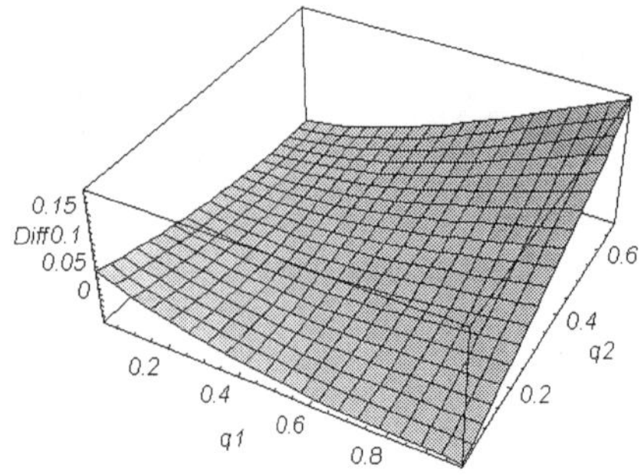

Figure 6. Difference between mutual information for unentangled and maximally entangled states as a function of q_1 and q_2 for $\eta = 0.6$.

and maximally entangled states for $\eta = 0.6$ as a function of q_1 and q_2. Despite the additional complications due to the introduction of extra parameters, the behavior appears similar.

5 Concluding remarks

In summary, we note that by modifying the depolarizing channel so that it can smoothly interpolate between the two different channels, we study and understand how mutual information changes from one type of channel to the another. By considering the mutual information, we find that the mutual informations in all modified channels are extremized for either completely un-entangled states or maximally entangled states. Moreover, the modified channel shows that there exists a large class of possible channels which can provide transmission of classical information via quantum entanglement.

Acknowledgments

We would also like to thank A. Ekert for stimulating and lively discussion as well as Y. Liu (ISN) and X.B. Wang (ERATOS) in the initial phase of this research. This work was supported by NUS Research Grant No. R144-000-054-112.

References

1. C.H. Bennett, P.W. Shor, J.A. Smolin and A.V. Thalpliyal, *Phys. Rev. Lett.* **83** (1999)3081, LANL eprint quant-ph/9904023; S. Bose, M. Plenio and V. Vedral, *J. Mod. Opt.* **47** (2000) 291, LANL eprint quant-ph/9810023; C.H. Bennett, D.P. DiVincenzo and J.A. Smolin, *Phys. Rev. Lett.* **78**, 3217 (1997); H. Barnum, J.A. Smolin and B. Terhal, *Phys. Rev. A* **58** (1998) 3496.
2. B. Schumacher, *Phys. Rev A* 51 (1995) 2738.
3. R. Jozsa and B. Schumacher, *J. Mod. Opt.* 41 (1994) 2343.
4. P. Hausladen, R. Jozsa, B. Schumacher, M. Westmoreland and W. Wootters, *Phys. Rev. A* 54 (1996) 1869.
5. N.J. Cerf and C. Adami, *Physica D* 120 (1998) 62.
6. M.J.W. Hall, *Phys. Rev. A* 55 (1997) 100.
7. Vidiello-Barranco, *Phys. Lett. A* 260 (1999) 335.
8. C.H. Bennett and S.J. Wiesner, *Phys. Rev. Lett.* **69** (1992) 2881.
9. C.H. Bennett, C. Fuchs and J. Smolin *Quantum Communication, Computing and Measurement*, edited by Hirota O., A.S. Holevo and C.M. Caves (New York, Plenum Press, p. 79 (1997); quant-ph/9611006.
10. D. Bruss, L. Faoro and C. Macchiavello, *J. Mod. Opt.* **47** (2000) 325 .
11. L.C. Kwek, L. Yong, C.H. Oh and X.B. Wang, *J. Mod. Opt.* Forthcoming issue, 2001.

ENTANGLEMENT MANIPULATION WITH ATOMS AND PHOTONS IN A CAVITY

S. HAROCHE

Ecole Normale Supérieure et Collège de France, Paris, France

E-mail: haroche@lkb.ens.fr

We review experiments in which entanglement is engineered and manipulated in a system made of circular Rydberg atoms interacting with microwave photons in a superconducting cavity.

1 Introduction

Engineering entanglement in systems made of several particles acting as quantum bits has become a very active research field at the border between physics and information science [1]. In this paper, I briefly review experiments performed with a cavity QED set-up involving Rydberg atoms and microwave photons in interaction. These experiments demonstrate various aspects of entanglement, as well as the deep relationship between entanglement and complementarity. We give here only a sketchy description of experiments published elsewhere in more details. A recent review article [2] provides more information on our experimental procedures.

2 Experimental set-up

Our apparatus manipulates individual Rydberg atoms and microwave photons interacting in a controlled and coherent way in a high Q superconducting cavity. The set-up, sketched in Fig. 1, comprises a cavity C made of two spherical superconducting mirrors facing each other. It stores microwave photons (51.1 GHz frequency, 6 mm wavelength) for up to 1 millisecond and let them interact with velocity selected atoms prepared by a pulsed process into a circular Rydberg state (in box B). The atoms, excited according to a Poissonian process with an average atom number per pulse of the order of 0.1, cross one at a time the cavity. The set-up is cooled to about 1 K to minimize thermal radiation noise. The atom are detected by the field ionization detector D after they have crossed the cavity. Some experiments are performed by directly detecting the atomic final energy state. In other experiments, atomic state superpositions are used. These superpositions are prepared, before the atoms cross C, by applying to them an auxiliary microwave classical field pulse (produced in zone R_1, see Fig. 1). Another pulse, produced in R_2, is used to mix again the atomic energy states after the atom has interacted with C. The successive application of these two pulses constitutes a Ramsey interferometer. Interference fringes are obtained in the probability for finding the atom in a given final state either when the frequency of the $R_1 - R_2$ fields is swept across a transition between two Rydberg levels, or when the energy gap between these levels is tuned by a Stark effect induced by a variable electric field applied across the cavity mirrors. We study how the fringe phase and amplitude is affected by the presence of photons in the cavity and we gain in this way useful information on the atom-photon interaction process.

3 Two-particle entanglement via resonant and non-resonant processes

Atom-field entanglement is achieved by exploiting the Rabi oscillation at frequency Ω which occurs when an atom interacts resonantly with a cavity mode containing 0 or 1 photon. Let us call e and g the upper and

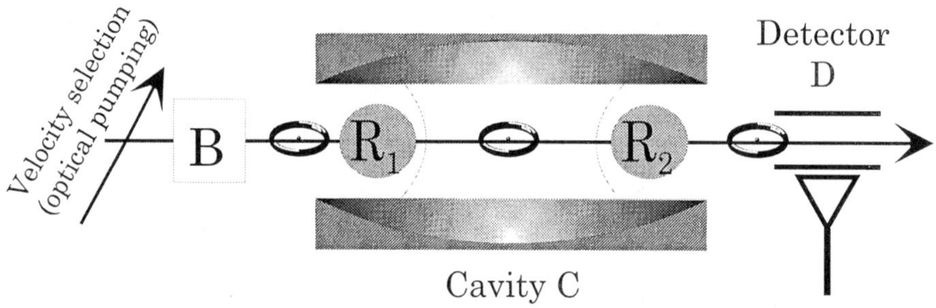

Figure 1. Scheme of Cavity QED experimental set-up

lower circular Rydberg states of the transition resonant with the cavity mode (corresponding respectively to electronic principal quantum numbers 51 and 50). The Rabi oscillation brings, after an interaction time t, the atom-field system in the linear superposition $\cos(\Omega t/2)|e,0\rangle + \sin(\Omega t/2)|g,1\rangle$, where the two states correspond to an atom in level e (g) with 0 (1) photon respectively in the cavity. In our set-up, $\Omega/2\pi = 50$ kHz, so that the Rabi period is much shorter than the cavity field relaxation time. The interaction time t is adjusted by Stark switching the $e-g$ atomic transition out of resonance while the atom flies across the cavity, thus freezing the system evolution when the desired value of the "pulse area" Ωt has been achieved. When $\Omega t = \pi/2$, maximum atom-field entanglement is obtained [3]. When the pulse area is π, the atom and the field fully exchange their energies [4]. This process can be used to swap excitation between the atom and the field and to transform atom-field entanglement into an atom-atom one. The experiment involves two atoms crossing the cavity one after the other, the first initially in level e, the second in g. The first atom undergoes a $\pi/2$ Rabi pulse and gets entangled with the cavity mode. The second is submitted to a π pulse, copying the cavity state and getting entangled in the process with the first atom. This entanglement has been checked by performing various measurements on the final states of the two-atoms and analyzing their correlations [3]. This massive pair of particles, of the Einstein-Podolsky-Rosen (EPR) type, could be used for Bell's inequality tests.

The above atom-atom entangling procedure relies on the transient real emission of a photon in the cavity. It is also possible to entangle two atoms directly, via a collision process assisted by non-resonant cavity modes [5]. The first atom (A_1) is again initially in e and the second (A_2) in g. The atoms have now different velocities, so that the second catches up the first at cavity center, before exiting first from C. The two closest cavity modes are now detuned from the $e-g$ transition at frequency ω by amounts $\delta_1, \delta_2 > \Omega$. Due to energy non-conservation, no real photon emission can occur in this case. Atom A_1 can however virtually emit a photon immediately reabsorbed by A_2. This leads, as in the resonant case, to atom-atom entanglement. The system ends up in a superposition of the $|e,g\rangle$ and $|g,e\rangle$ states. The quantum amplitudes associated to these states are periodic functions of the collision duration (which depends on the atoms velocities). The oscillation frequency associated to this second order collision process is $(\Omega^2/4)(1/\delta_1 + 1/\delta_2)$. By repeating the experiment, we reconstruct the probabilities P_{eg} and P_{ge} for finding finally the atom pair in states $|e,g\rangle$ and $|g,e\rangle$. We plot these probabilities versus the dimensionless parameter $\eta = \omega(1/\delta_1 + 1/\delta_2)$ (see Fig. 2). The oscillations of P_{eg} and P_{ge} as a function of η are well accounted for by theo-

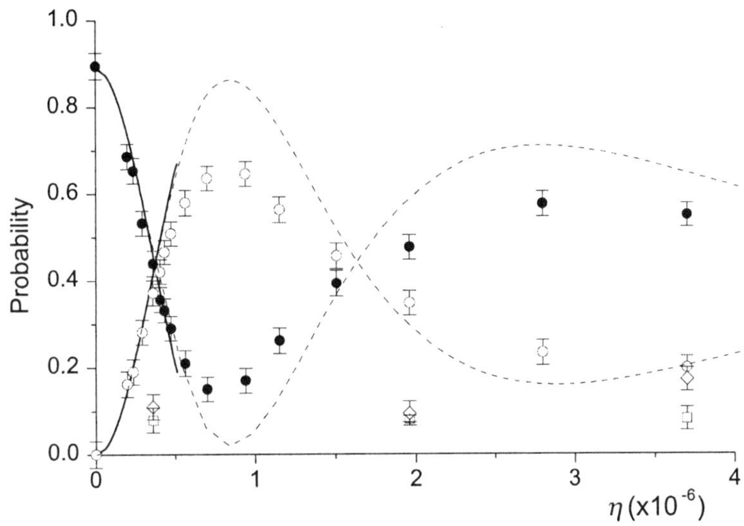

Figure 2. Joint detection probabilities P_{eg} and P_{ge} versus the parameter η. Points are experimental. Solid lines for small η values correspond to a simple analytical model based on second order perturbation theory. The dashed lines (large η) present the results of a numerical integration of the system evolution (adapted from [5])

retical models (solid and dashed lines in Fig. 2). We have realized the situation of maximum entanglement by adjusting η to the value corresponding to $P_{eg} = P_{ge} = 0.5$. We have also checked the coherent nature of the two-atom state prepared in this way by performing measurements of observables whose eigenstates are superpositions of energy states. This new method to entangle atoms might be used to demonstrate elementary steps of quantum logic, essentially immune from noise due to cavity photon decay or thermal field background.

4 Quantum gates and non destructive measurement of photons

We now come back to the resonant atom-cavity case. After a full cycle of Rabi oscillation (2π pulse), the atom-field system comes back to its initial state, with a π phase shift of its wave function. If the system is initially in the $|g,1\rangle$ state, it ends up in the state $-|g,1\rangle = \exp(i\pi)|g,1\rangle$. This is reminiscent of the sign change of a spin state undergoing a 2π rotation. When the cavity is initially empty, the sign of the state is unaltered, the $|g,0\rangle$ state remaining unchanged. If we view the atom and the field as qubits, the 2π Rabi pulse couples them according to the dynamics of a quantum phase gate. We have shown that this gate works in a coherent and reversible way [6]. We have also applied this gate to perform a non-destructive measurement of a single photon [7]. The phase change of the atomic wave function when it undergoes a 2π Rabi pulse in C can be translated into an inversion of the phase of the fringe pattern of the Ramsey interferometer sandwiching the cavity (see Fig. 1). By setting the interferometer at a fringe extremum, we correlate the photon number (0 or 1) to the final state of the atom. The atom is then a "meter" measuring, without destroying it, a single photon in C. The same photon can be measured repeatedly by successive atoms, without being absorbed. This is quite different from ordinary absorbing photon counting procedures.

5 Multiparticle entanglement

By combining quantum Rabi pulses of various durations and auxiliary Ramsey pulses on successive atoms, one can generate and analyse entangled states involving more than two particles. By applying a $\pi/2$ Rabi pulse on a first atom, one entangles it to a 0/1 photon field. A second atom then undergoes a 2π Rabi pulse combined to Ramsey pulses, in order to measure this field in a non destructive way. Before this atom is detected, a three-part entanglement involving the two atoms and the photon field is generated. The field state is finally copied on a third atom, initially in the lower state of the transition resonant with the cavity mode (a π Rabi pulse is used for this copying procedure). The characteristics of this three-particle entangled state are analyzed by performing various measurements on the three atoms [8]. These measurements involve the application of auxiliary Ramsey pulses after the atoms have interacted with the cavity. The procedure could be generalized to situations of increasing complexity, with larger numbers of atoms.

6 Complementarity and entanglement at the quantum-classical boundary.

We have also used our set-up to perform a complementarity test very closely related in its principle to the double slit experiment described by Bohr in his 1927 discussions with Einstein. Bohr had analyzed a situation where particles are crossing a Young interferometer in which one slit is carried by a light assembly, free to move independently of the other. In an ordinary Young design with fixed slits, interference fringes reveal the wave nature of the particle. In this design, the momentum imparted to the moving slit by the deflected particle provides a "which path" information, suppressing the fringes and revealing the corpuscular aspect. In this experiment, the quantum moving slit and the particle are in fact the two components of a correlated EPR system. The trajectory of the particle gets entangled to the motion of the slit. Observing (really or virtually) this motion lifts the ambiguity of the particle path and suppresses the interferences. Intermediate situations can be considered, by varying for example the mass of the movable slit. There is a continuous transition from the quantum slit case (very small mass) to the classical one (infinite mass). In between, we expect fringes with a limited contrast, reflecting the partial degree of entanglement between the slit and the particle.

A Young double slit experiment with an ultra-light quantum slit would be very difficult to realize. This gedanken experiment can however be translated into an easier to perform cavity QED experiment, presenting the same conceptual features [9]. We use, instead of a Young device, our double oscillatory field Ramsey interferometer, in a slightly modified version. The field pulses, resonant on the $e-g$ transition act as atomic beam splitters transforming these states into superpositions. The first resonant field, instead of being produced by R_1, is now a small coherent field injected through a wave guide inside the long lived cavity mode C. Depending upon the average photon number n in this field, it can be considered either as a microscopic or macroscopic "beam splitter". The second pulse is produced downstream in R_2. This latter pulse, produced by photons fed into a low Q field mode and recycled at a very fast rate, can always be considered as macroscopic and classical. The first and second pulses are thus equivalent to the movable and fixed slits of Bohr gedanken experiment. If n is large, Ramsey fringes with maximum contrast are expected. When n gets smaller, the one photon change produced by the atomic transition from e to g leads to atom-field entanglement, in the same way as the recoil of the slit in

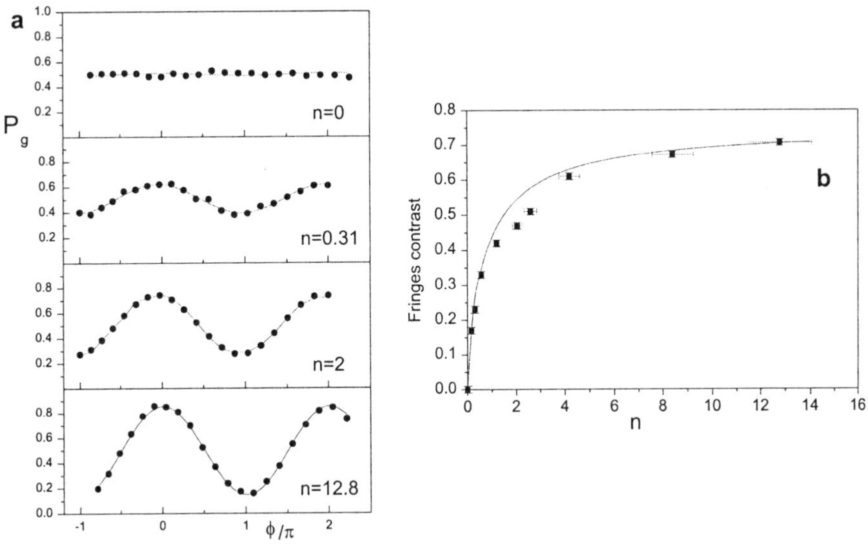

Figure 3. a) Ramsey fringes obtained for different mean photon numbers n in the first coherent field pulse. The phase Φ is swept by Stark tuning the atomic energy levels. b) Fringe contrast as a function of n. The points are experimental. The line represents the theoretical predictions corrected for the imperfections of the interferometer (adapted from [9])

the Young apparatus leads to slit-particle entanglement. Fringe contrast is then expected to be reduced. The experimental signals are shown in Fig. 3a. We observe that the fringe contrast becomes smaller and smaller as n decreases. At the limit where the field in C is the vacuum, the fringes vanish, revealing maximum atom-field entanglement in C. Fig. 3b shows the fringe contrast versus n. The points are experimental and the curve is a theoretical fit.

We have performed other entanglement experiments, based on the use of dispersive non-resonant atom-field interaction. Schrodinger cat like states of the field have been produced in this way and their decoherence studied [10]. All these experiments illustrate fundamental concepts of quantum theory and demonstrate the feasibility of elementary logical steps in quantum information processing. We are considering various improvements of our set-up and procedures, in order to be able to increase the number of atoms and photons involved in these experiments.

Acknowledgements

This research is supported by the CNRS, the European community and the Japan Science and technology Corporation (International Cooperation Research Project, Quantum Entanglement Project).

References

1. Bouwemeester D., Ekert A. and Zeilinger A. editors, The Physics of Quantum Information, (2000), Springer Verlag, Berlin.
2. Raimond J.M., Brune M. and Haroche S., Rev. Mod. Phys.,73 (2001)pp.565-582.
3. Hagley E., Matre X., Nogues G., Wunderlich C., Brune M., Raimond J.M. and Haroche S, Phys.Rev.Lett. 79, (1997) pp. 1-4.
4. Maître X., Hagley E., Nogues G., Wun-

derlich C., Goy P., Brune M., Raimond J.M. and Haroche S. Phys.Rev.Lett.79 (1997)pp.769-772.

5. Osnaghi S., Bertet P., Auffeves A., Maioli P., Brune M., Raimond J.M. and Haroche S, Phys.Rev.Lett. 87 (2001)pp.037902-037905.
6. Rauschenbeutel A., Nogues G., Osnaghi S., Bertet P., Brune M, Raimond J.M. and Haroche S, Phys.Rev.Lett. 83 (1999)pp.5166-5169.
7. Nogues G., Rauschenbeutel A.,Osnaghi S., Brune M., Raimond J.M. and Haroche S., Nature, 400,(1999)pp.239-242
8. Rauschenbeutel A., Nogues G.,Osnaghi S.,Bertet P., Brune M., Raimond J.M. and Haroche S., Science, 288(2000)pp.2024-2027.
9. Bertet P., Osnaghi S., Rauschenbeutel A., Nogues G., Auffeves A., Brune M., Raimond J.M. and Haroche S., Nature 411(2001) pp.166-170.
10. Brune M., Hagley E., Dreyer J., Matre X., Maali A., Wunderlich C., Raimond J.M. and Haroche S., Phys.Rev.Lett. 77 (1996)pp.4887-4890.

GENERATION OF SINGLE PHOTONS AND ENTANGLED PHOTON PAIRS FROM A QUANTUM DOT

Y. YAMAMOTO

Quantum Entanglement Project, ICORP, JST,
E. L. Ginzton Laboratory, Stanford University, Stanford, California, U.S.A.,
NTT Basic Research Laboratories, Atsugishi, Kanagawa, Japan
E-mail: yamamoto@loki.stanford.edu

M. PELTON, C. SANTORI AND G. S. SOLOMON

Quantum Entanglement Project, ICORP, JST,
E. L. Ginzton Laboratory, Stanford University, Stanford, California, U.S.A.

Current quantum cryptography systems are limited by the Poissonian photon statistics of a standard light source: a security loophole is opened up by the possibility of multiple-photon pulses. By replacing the source with a single-photon emitter, transmission rates of secure information can be improved. A single photon source is also essential to implement a linear optics quantum computer. We have investigated the use of single self-assembled InAs/GaAs quantum dots as such single-photon sources, and have seen a hundred-fold reduction in the multi-photon probability as compared to Poissonian pulses. An extension of our experiment should also allow for the generation of triggered, polarization-entangled photon pairs.

1 Introduction

Quantum cryptography has emerged as a significant field of study over the last fifteen years, because it offers the promise of private communication whose security is assured by the laws of quantum mechanics. Most implementations of quantum cryptography so far have used a protocol introduced by Bennet and Brassard, generally known as BB84 [1]. The message can be encoded on the polarization state of single photons, with a random choice between two non-othogonal polarization bases when the photons are sent and recieved.

However, sources of single photons have not been generally available. Experimental implemetations of BB84 have used pulses from lasers or light-emitting diodes, attenuated to the point where the average photon number per pulse is significantly less than one [2]. However, the number of photons in these pulses is described by Poissonian statistics, so that there is always a possibility of more than one photon being sent in a given pulse. Such pulses are vulnerable to a photon-splitting attack, where the eavesdropper removes one photon from the pulse, leaving the remaining photons undisturbed [3].

A pulse stream with reduced multi-photon probability compared to the Poissonian case is said to be *antibunched,* and can only be described only quantum mechanically. Mathematically, such non-classical photon statistics can be described using the second-order correlation function $g^{(2)}(\tau)$, defined as follows:

$$g^{(2)}(\tau) = \frac{\langle \hat{a}^\dagger(t)\hat{a}^\dagger(t+\tau)\hat{a}(t+\tau)\hat{a}(t)\rangle}{\langle \hat{a}^\dagger \hat{a}\rangle^2}, \quad (1)$$

A pulsed source with Poissonian statistics will have a $g^{(2)}(\tau)$ function consisting of a series of peaks with unit area, when normalized by the pulse repetition period. This can be seen in Fig. 1, which shows a measured second-order correlation function for attenuated pulses from a mode-locked $Ti{:}Al_20_3$ laser. This reflects the fact that the probability of detecting a photon in a given pulse is independent of whether a photon has already

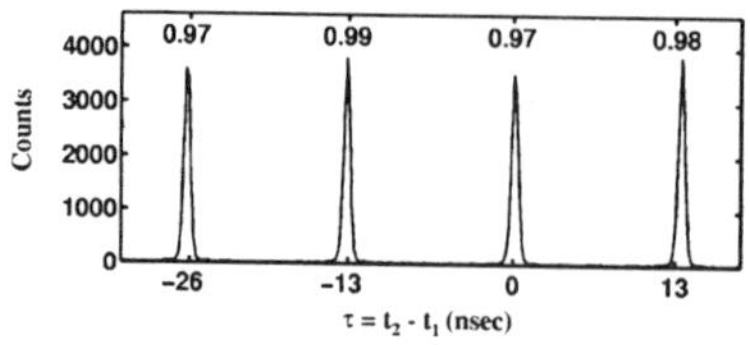

Figure 1. Measured photon-photon correlation function for pulses from a mode-locked Ti:Sapphire laser with a 13 ns repetition period. The numbers above the peaks are the normalized areas.

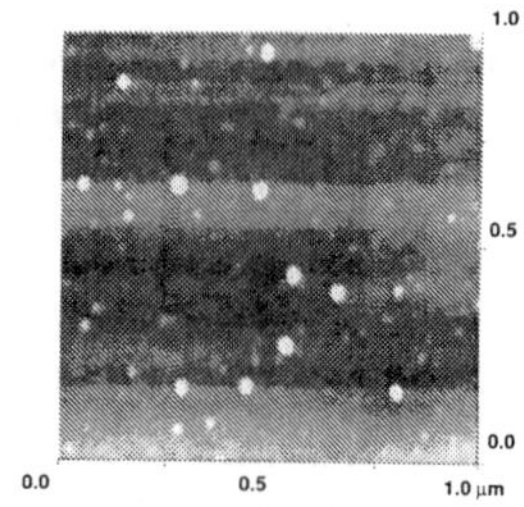

Figure 2. Atomic-force-microscope image of sparse self-assembled InAs quantum dots grown on GaAs.

been detected. For an ideal single-photon source, the central peak at $\tau = 0$ is absent, indicating that, once a single photon has been detected in a pulse, another one will never be detected. An antibunched source will have a zero-delay peak with an area between zero and one. This area gives an upper bound on the probability $P(n_j \geq 2)$ that two or more photons are present in the same pulse:

$$2P(n_j \geq 2)/\left\langle n\right\rangle^2 \leq \frac{1}{T}\int_{-\epsilon}^{\epsilon} g^{(2)}(\tau)d\tau\,. \quad (2)$$

where $\langle n\rangle$ is the mean photon number per pulse, T is the pulse repetition period.

Excitons in quantum dots are promising as single-photon sources, since they behave as single emitters. It has been shown that the fluorescence from a single quantum dot exhibits antibunching [4]. We have achieved triggered generation of antibunched photons from a single quantum dot by exciting with a pulsed laser and spectrally filtering the emission [5]. The normalized area of the $g^{(2)}(0)$ peak can be as low as 0.03. This system for generating single photons, which has also been reported by other groups [6] is stable over long periods of time, and is compatible with mature semiconductor technologies. This allows for the possibility of injecting carriers into the dot electrically instead of optically, producing arrays of sources, and integrating into larger structures [7].

The single quantum dot can serve as a source of other non-classical radiation states. For example, there may be the possibility to create triggered pairs of polarization-entangled photons [8]. Generation of single photons and entangeld photon-pairs is also a key for quantum computation and network based on photonic qubits. Optical quantum computation based on nonlinear Fredkin gates [9,10] and linear optical elements [11] requires the synchronized parallel production of many single photons.

2 Single InAs/GaAs Quantum Dots

Semiconductor quantum dots are small regions of a low-bandgap semiconductor inside a crystal of a larger-bandgap semiconductor. The bandgap difference acts as a potential barrier for carriers, confining them inside the dot. Quantum dots are small enough that the carriers form standing waves inside the confinement region, and can only occupy discrete energy levels. When InAs is deposited on GaAs, a strained planar layer, known as a wetting layer, initially forms. The strain energy that builds up in this layer is eventually partially relieved by the formation of nanometer-scale islands on the surface. The islands form without defects, and can subsequently be covered with a capping layer of GaAs.

We grew InAs/GaAs quantum dots by molecular beam epitaxy (MBE). The samples used in our experiments were grown under conditions that give relatively sparse dots, with a surface density of 11 - 75 μm^{-2}. Fig. 2 shows an atomic-force microscope image of dots similar to the dots used in our experiments, except for the absence of a GaAs capping layer. Single dots are isolated by etching mesas in the MBE-grown sample. The mesas are fabricated by electron-beam lithography and dry etching. Fig. 3 shows PL spectra from the quantum dot used to generate single photons. With continuous-wave (CW) excitation above the GaAs bandgap, the emission spectrum displays several lines. We believe that these lines all come from a single dot, because other mesas show nearly identical emission pattern (peak heights, spacings and widths), except for an overall wavelength shift, suggesting that this pattern is not random. When the laser is tuned to an absorption resonance at 857.5 nm of an excited but QD confined state, thus creating excitons directly inside the dot, emission peaks 3 and 4 almost disappear. We therefore believe that they represent emission from charged states of the dot. We identify peak 1 as ground-excitonic state emission after the capture of a single exciton, and peak 2 as "biexcitonic" emission after the capture of two excitons. A biexcitonic energy shift of 1.7 meV is due to electrostatic interactions among carriers.

Assignment of the different peaks is supported by the dependence of the emission line intensities on pump power. We can see linear growth of peak 1 and quadratic growth of peak 2 in the weak-pump limit, as expected for excitons and biexcitons, respectively. Further support for the peak indentification comes from time-dependent spectra, as collected by the streak camera. The camera produced two-dimensional images of intensity *vs.* wavelength and time after exciting with a laser pulse. Integration times were about 5 minutes, corresponding to about 20

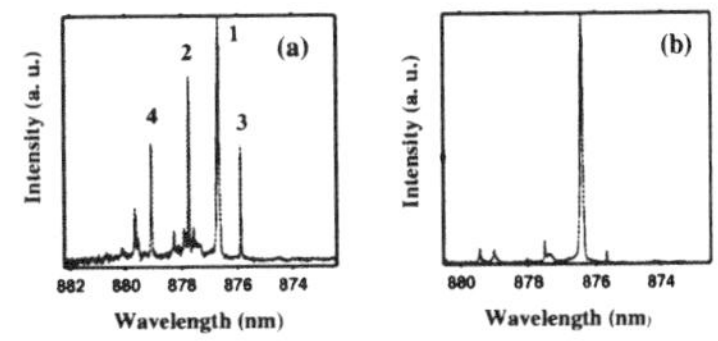

Figure 3. Photoluminescence spectra from a single InAs/GaAs self-assembled quantum dot. For **(a)**, the pump laser had energy above the GaAs bandgap. For **(b)**, the pump energy was resonant with a higher-order transition in the dot, so that excitons are created only in the dot.

billion pulses. Time resolution as determined by the spectrometer is about 25 ps. The images were corrected for background counts, non-uniform sensitivity, and a small number of cosmic ray events. By integrating intensity within frequency windows corresponding to the peaks shown in Fig. 3, time-dependent intensities are obtained for the different lines. The results are shown in Fig. 4 for different pump powers.

Under weak excitation, the single-exciton line (line 1) appears quickly after the excitation pulse, and then decays exponentially. This decay time has been measured accurately under resonant excitation to be 0.47 nsec. Under higher excitation power, however, line 1 reaches its maximum only after a long delay. Most of the emission immediately after the excitation pulse now comes from line 2. A simple explanation for this behavior is that, since the laser pulse now initially creates several exciton-hole pairs on average, some time is required before the population of the dot reduces to one electron-hole pair, and only then can the single-exciton emission occur.

In an even high excitation power, the bi-excitonic emission is delayed and the third peak (tri-excitonic emission at a slightly longer wavelength than the bi-exciton line)

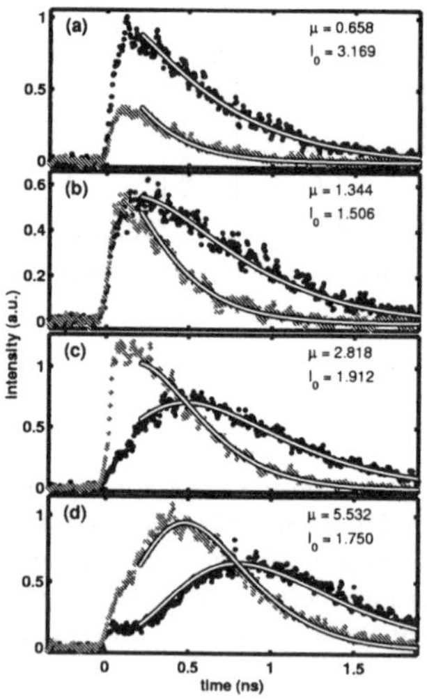

Figure 4. Time-dependence of luminescence intensities of the single-exciton line 1 (black) and the biexciton line 2 (gray) after pulsed excitation above the GaAs bandgap. Hollow lines show model fit results, with the parameter values for the fits shown in the figures. The pump powers are **(a)** 27 μW, **(b)** 54 μW, **(c)** 108 μW, and **(d)** 432 μW.

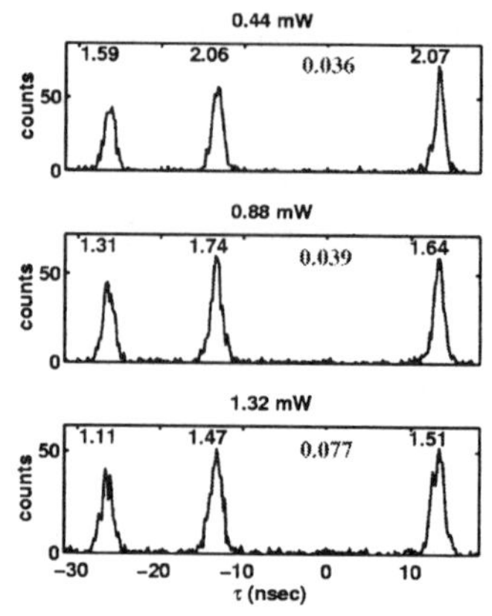

Figure 5. Photon-photon correlation functions for emission from a single InAs quantum dot under pulsed, resonant excitation. The numbers above the peaks represent the normalized peak area g^2_τ.

appears immediately after the excitation pulse.

3 Generation of Single Photons

The emission from a quantum dot containing a single and last exciton has a distinct wavelength due to multi-particle interaction effect. There should be only one photon emitted at this energy for each excitation laser pulse, regardless of the number of electron-hole pairs originally created inside the dot. Spectral filtering was used to select this last photon, resulting in only one photon per pulse.

The Hanbury Brown and Twiss-type configuration was used to measure the second-order correlation function. A beamsplitter sends photons towards one of two single-photon detectors. The detectors are EG&G "SPCM" avalanche photodiodes, which have efficiencies of 40% at 877 nm and 0.2 mm-wide active areas. The electronic pulses from the photon counters were used as start (t_1) and stop (t_2) signals for a time-interval counter, which recorded a histogram of delays $\tau = t_2 - t_1$. Normalized histograms are shown in Fig. 5.

In the limit of low collection and detection efficiency, these histograms approximate the second-order correlation function. The $\tau = 0$ peak shows a large reduction in area, indicating strong anti-bunching. The numbers printed above the peaks indicate the peak areas, properly normalized by dividing the histogram areas by the photon count rate at each detector, the laser repetition period, and the measurement time. For the numbers shown, the only background counts subtracted were those due to the known dark count rates of the photon counters (130 s^{-1} and 180 s^{-1}), which are small compared to the total count rates (19800 s^{-1} and 14000 s^{-1}) for the two counters at 0.88 mW pump power. When only counts within 2.8 ns of $\tau = 0$ were included, a normalized $g^{(2)}(\tau = 0)$ peak area of 0.03 was obtained at 0.88 mW. Subtracting the constant background floor seen in the data gave an even lower value of 0.01.

The residual non-zero probability of having more than one photon per pulse is believed to be primarily due to imperfect spec-

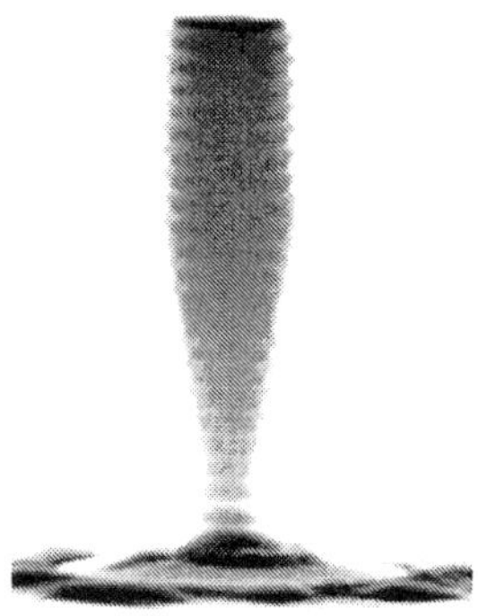

Figure 6. A new micro-post DBR microcavity structure.

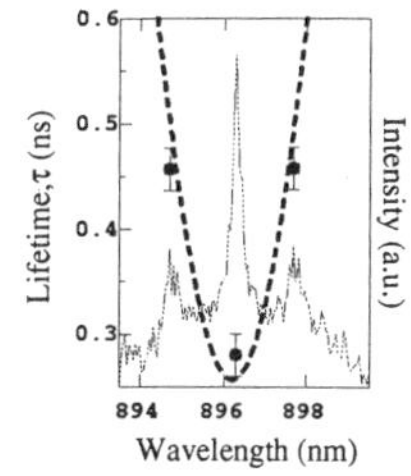

Figure 7. Photoluminesence lifetime *vs.* wavelength for isolated quantum dots in a micropost microcavity with a top diameter of 0.5 μm. The points are measured values, while the dashed line is the predicted result. The thin, solid line represents the photoluminescence intensity at the same wavelengths.

tral filtering. As well, there is a broad-band emission background that contributes some spurious photons. By reducing the filter bandwidth, we believe another fivefold reduction in the $g^{(2)}(0)$ peak should be possible.

4 Coupling Single Quantum Dots to Micropost Microcavities

A single quantum dot has been shown to be a good source of single photons. However, the usefulness of the source for quantum communication or other quantum information applications is limited by its efficiency. Only one out of approximately every 3000 photons emitted from the dot was ultimately detected by the single-photon detectors. The largest cause of this inefficiency is the fact that the dot emits primarily into the semiconductor substrate, and only 0.6% of the emission is collected by the aspheric lens in front of the cryostat window.

In order to increase an output coupling efficiency, a three-dimensional microcavity is incorporated into a single QD, as shown in Fig. 6 [12].

The expected spontaneous emission rate can be calculated according to the following equation:

$$\frac{\gamma}{\gamma_0} = \frac{Q\lambda_c^3}{2\pi^2 n_{\text{eff}}^3 V_0} \frac{\Delta\lambda_c^2}{\Delta\lambda_c^2 + 4(\lambda - \lambda_c)^2} + f\,, \quad (3)$$

where γ_0 is the (experimentally-determined) spontaneous emission rate of a quantum dot without a cavity, λ_c is the cavity resonant wavelength, $\Delta\lambda_c$ is the cavity linewidth, and $(\lambda - \lambda_c)$ is the detuning of the dot emission wavelength from the cavity resonance. $f\gamma_0$ is the spontaneous decay rate into leaky modes (i.e., emission is incident on the post edges at an angle larger than the critical angle for total internal reflection).

Fig. 7 shows the spontaneous emission lifetime for a cavity with top diameter of 0.5 μm. There is now a single quantum dot on resonance with the cavity mode. A significant reduction in the spontaneous emission time, to 0.28 ns, is seen. The dashed line is the theoretical result.

The spontaneous emission lifetime can be converted into a spontaneous emission coupling coefficient β, the fraction of light that is captured in the fundamental cavity mode:

$$\beta = \frac{\gamma - \gamma_0 - \gamma_c}{\gamma}\,, \quad (4)$$

where γ is the enhanced spontaneous emission decay rate into the fundamental mode of the cavity, γ_0 is the spontaneous emission decay rate in the absence of a cavity, and γ_c is the fractional spontaneous emission decay rate into the solid angle of the cavity mode in the limit that the mirror reflectivity, $R \to 0$. Since this solid angle is at most a few degrees in our case, $\gamma_c \ll \gamma_0$. Using this formula, we determine that 78% of the light from a single quantum dot is collected by a single mode.

5 Entangled Photon-Pairs from a Single Quantum Dot

A fundamental nonlinear effect in a QD is the saturation of a single energy level by two electrons and holes of opposite spins due to the Pauli exclusion principle. We propose a device that produces regulated pairs of entangled photons using this effect [8].

The two photons arise from the decay of the biexcitonic ground state of the QD, where the correlated electrons and holes have opposite spins. For quantum wells in direct-gap materials with a cubic lattice, any photons emitted are circularly polarized, because the $J_z = \pm 1/2$ electron recombines with the $J_z = \pm 3/2$ heavy hole [13]. In the case of a QD, the strong confinement introduces level mixing and the hole ground state may have contributions from the $J_z = \pm 1/2$ hole states. Accordingly, when a $J_z = +1/2$ electron radiatively recombines with a hole in a QD, the emitted light is predominately σ^+ polarized, but may also have a σ^- component. Thus, the two photons that arise from the decay of the biexcitonic ground state are not necessarily perfectly anticorrelated with respect to σ^+ and σ^- polarization. An asymmetric dot shape, strain, and piezoelectric effects further reduce the anticorrelation. However, there is experimental evidence from polarized photoluminescence [14] and two photon absorption measurements [15] that the anticorrelation in σ^+ and σ^- polarization is preserved in QD's.

We point out that a previous single photon turnstile device relies on the relatively small Coulomb splitting [7]. This limits the operation of this device to very low temperatures (40 mK) in order to guarantee that thermal energy fluctuations are negligible. In the proposed device, the turnstile operation is maintained up to much higher temperatures due to the very large splitting between the electron and hole ground and excited states. Electron and hole tunneling could be controlled merely by the Pauli exclusion principle, even if the Coulomb blockade effect were absent.

We now focus on the production of pairs of entangled photons at well-defined time intervals. Starting from the biexcitonic ground state of the QD, a first electron can recombine with a hole and emit a σ^+ or a σ^- photon. Then, the second electron of opposite spin recombines with a hole, and a photon of opposite polarization is emitted. This situation is very similar to a two-photon cascade decay in an atom [16]. The two-photon state has the same form in any basis and is a maximally entangled (Bell) state: $|\psi\rangle = \frac{1}{\sqrt{2}}(|\sigma^+\rangle_1|\sigma^-\rangle_2 + |\sigma^-\rangle_1|\sigma^+\rangle_2)$. Because of additional binding energy, the biexcitonic ground state has a smaller energy than twice the excitonic ground state. Therefore, the first emitted photon 1 and the second emitted photon 2 have different energies.

The advantage of the proposed structure compared to other sources of entangled photons, such as two-photon cascade decay in atoms or parametric down-conversion in non linear crystals, is that entangled photon pairs are provided one by one with a tunable repetition rate of up to 1 GHz by a compact semiconductor device. The source is electrically pumped and the photons are emitted in resonant modes of an optical resonator, which greatly improves, e.g., the efficiency of subsequent fiber coupling.

References

1. C.H. Bennett and G. Brassard, in *Proceedings of IEEE International Conference on Computers, Systems, and Signal Processing, Bangalore, India* (IEEE, New York, 1984), p. 175.

2. C.H. Bennett, F. Bessette, G. Brassard, L. Salvail, and J. Smolin, *J. Crypto.* **5**, 3 (1992).
3. N. Lütkenhaus, *Phys. Rev.* A **61**, 2304 (2000).
4. P. Michler *et al*, *Nature* **406**, 968 (2000); B. Lounis, H. A. Bechtel, D. Gerion, P. Alivisatos, and W. E. Moerner, *Chem. Phys. Lett.* **329**, 399 (2000).
5. C. Santori, M. Pelton, G .S. Solomon, Y. Dale, and Y. Yamamoto, *Phys. Rev. Lett.* **86**, 1502 (2001).
6. P. Michler *et al*, *Science* **290**, 2282 (2000); V. Zwiller *et al*, *Appl. Phys. Lett.* **78**, 2476 (2001).
7. J. Kim, O. Benson, H. Kan, and Y. Yamamoto, *Nature* **397**, 500 (1999); A. Imamoglu and Y. Yamamoto, *Phys. Rev. Lett.* **72**, 210 (1994).
8. O. Benson, C. Santori, M. Pelton, and Y. Yamamoto, *Phys. Rev. Lett.* **84**, 2513 (2000).
9. Y. Yamamoto, M. Kitagawa and K. Igeta, *Proc. 3rd Asia Pacific Phys. Conf.* (World Scientific, Singapore, 1988) p. 777.
10. G. J. Milburn, *Phys. Rev. Lett.* **62**, 2124 (1989).
11. E. Knill, R. Laflamme and G. Milburn, *Natuer* **409**, 46 (2001).
12. G. S. Solomon, M. Pelton, and Y. Yamamoto, *Phys. Rev. Lett.* **86**, 3903 (2001); Y. Yamamoto, S. Machida and G. Björk, *Phys. Rev.* A **44**, 657 (1991).
13. C. Weisbuch and B. Winter, *Quantum Semiconductor Structures* (Academic Press, San Diego, 1991).
14. Y. Toda, S. Shinomori, K. Suzuki, and Y. Arakawa, *Phys. Rev.* B **58**, R10147 (1998).
15. K. Brunner, G. Abstreiter, G. Böhm, G. Tränkle, and G. Weimann, *Phys. Rev. Lett.* **73**, 1138 (1994).
16. A. Aspect, J. Dalibard, and G. Roger, *Phys. Rev. Lett.* **49**, 1804 (1982).

TWIN PHOTON BEAMS FOR SINGLE PHOTON GENERATION

SHIGEKI TAKEUCHI
Research Institute for Electronic Science, Hokkaido University and JST-CREST project
Kita 12 Nishi 6, Kita-ku, Sapporo, 060-0812, Japan
E-mail: takeuchi@es.hokudai.ac.jp

We report a new method to generate parametric fluorescence into two small spots. Because the generated twins are concentrated to the spots, we observed a high single count rate and coincidence count rate per pump power. We also observed that the ratio of coincidence count rate to single count rate was 80%, where the loss at filters and the quantum efficiency of detectors were compensated. We also investigated the photon number distribution of the ‘ single photon pulses’ using gated parametric down conversion with several types of detectors including our high-quantum efficiency multi-photon detectors. We found that the unity fraction of the single photon pulses is achievable using a CW laser pumped PDC, where the fraction of the two photon pulses are small enough.

1 Introduction

People started to apply the basic concepts of quantum mechanics for information processing to realize some functions that can never be obtained using classical mechanics. The uncertainty principle makes it impossible for an eavesdropper to get any information without leaving any evidence of his existence. This concept is so called 'quantum cryptography'. A quantum computer, which utilizes quantum superposition states in a huge Hilbert space of entangled quantum particles for parallel processing, may solve a problem in a minute which can never be solved by a super computer in billions of years.

Our group has been working on the experimental demonstrations of such concepts and development of basic techniques to realize them. An experimental demonstration of a quantum computation algorithm using linear optics and a single photon [1,2], an experimental demonstration of quantum cryptographic system over 1km [3] are the examples.

For the practical use of these concepts using photons, single photon generators are indispensable elements. Among the candidates to realize them, the methods that utilize the twin photons generated by parametric down conversion (PDC) have the following unique features: wide wavelength tunability, controlled directivity, and so on.

In the followings, we report a new method to generate twin photon beams for single photon generation using PDC[4]. Because the generated twin photons are concentrated to the spots, we observed a high single count rate and coincidence count rate per pump power. We also observed that the ratio of coincidence count rate to single count rate was 80%, where the loss at filters and the quantum efficiency of detectors were compensated.

We have also been developing high-quantum efficiency multi-photon counting systems using Visible Light Photon Counters (VLPCs) [5,6], which can distinguish the number of incident photons with the quantum efficiency of 88 ± 5 %. In this paper, we also discuss the photon number distribution of the ‘ single photon pulses’ using the gated PDC with several types of detectors including our high-quantum efficiency multi-photon detectors[7].

2 Beamlike twin photon generation using parametric down conversion [4]

The creation of twin photons by use of spontaneous parametric down conversion (PDC)

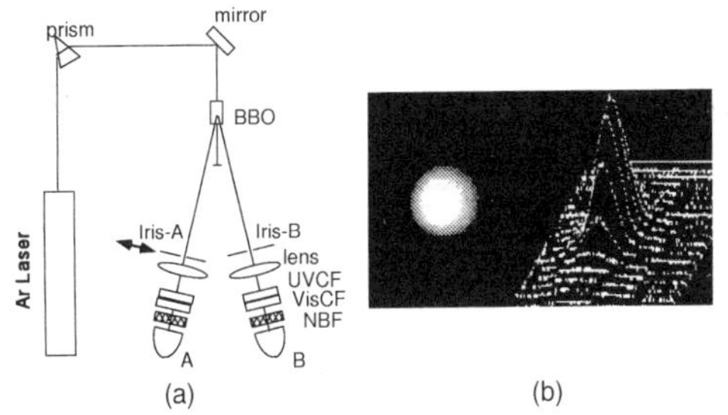

Figure 1. (a) The schematic of the experimental setup. (b) The image parametric fluorescence observed with Peltier-cooled CCD camera. The intensity distribution of the image is shown on the right side.

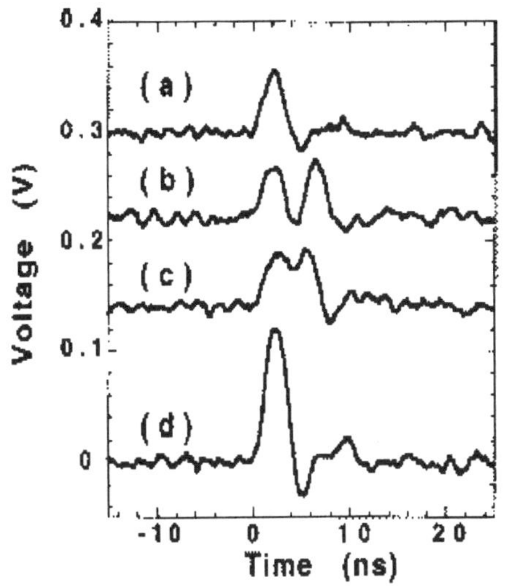

Figure 2. Real-time trace of photon detection signal recorded by a 5GS/s digitizing oscilloscope. Modifying the optical path length changes the time delay between the two beams. The trace shifted vertically for clarity. (a) Single photon detection signal. (b) 5 ns delay (c) 3ns delay (d) Zero delay.

has been a very important tool to explore the non-local feature of quantum mechanics, and recently for experiments in quantum information technology. The demonstration of quantum teleportation using photons is an example. In the field of quantum cryptography, PDC should be an indispensable technique for generation of single photon pulses, quantum repeaters and so on. The bit rate of these experiments was limited by the small conversion efficiency of the previous sources. This was mainly because the photon pairs are radiated to a widely spreading region and it was difficult to utilize all of them. In addition, the intensity distribution of the selected photons by an iris was not symmetrical, which caused difficulties in the experiments.

We developed a new method to generate parametric fluorescence into two small spots (0.9^o). We investigated the so called tuning curves, which are plots of the signal and idler wavelengths as a function of the crystal output angles. The tuning curves changes according to the angle of the nonlinear crystal to the pump, and we found that when the angle is adjusted such that the tuning curves are tangent to the line at the degenerate wave length, the generated twins are concentrated to the spots (Fig. 1). We also observed a high single count rate and coincidence count rate per pump power (8000 cps was observed with 150mW pump, 5mm BBO crystal, with 0.32nm FWMH filters). We also observed that the ratio of coincidence count rate to single count rate was 80%, where the loss at filters and the quantum efficiency of detectors were compensated. This result is important for the application of this method to the single photon generator because we may be able to find a single photon in a gated pulse with up to 80% probability.

3 High quantum efficiency multi photon counter[5,6]

A high quantum efficiency single-photon counting system has been developed. In this system, single photons were detected by a visible light photon counter operated at 6.9 K. The visible light photon counter is a solid state device that makes use of avalanches across a shallow impurity conduction band in silicon. Threefold tight shielding and viewports that worked as infrared blocking filters were used to eliminate the dark count caused by room-temperature radiation. Corrected quantum efficiencies as high as 88.2%± 5% at 694 nm were observed, which we believe is the highest reported value for a single-photon detector. The dark count increased as the exponential of the quantum efficiency with changing temperature or bias voltage, and was 2.0×10^4 cps at the highest quantum efficiency. Visible light photon counters feature noise-free avalanche multiplication and narrow pulse height distribution for single photon detection events. Such a well-defined pulse height distribution for a single photon detection event, combined with the fact that the avalanche multiplication is confined to a small area of the whole detector, opens up the possibility for the simultaneous detection of two photons. We investigated this capability using twin photons generated by PDC. Figure 2(a) shows an electrical pulse resulting from a single photon detection event. Figures 2(b) and 2(c) show the cases when the optical delay between the two beams is 5 and 3 ns, respectively. The heights of the pulses are almost identical, indicating that the number of electrons released per single photon detection event is well defined. Finally, Fig. 2(d) shows when the optical delay is reduced to zero. The two pulses resulting from the two photon detection events completely overlap in time, and the pulse height is twice that of a single detection event.

4 Single photon sources using gated PDC [7]

In this section, we discuss on the single photon source using gated parametric down conversion (GPDC). The basic concept of GPDC is shown in Fig. 3. When a pair of photons is created using PDC, both of the photons are emitted at the same time. Therefore, Detecting one of the photons make it possible to monitor the existence of the other photon and gating the emission of the photon into the output mode using a high speed shutter.

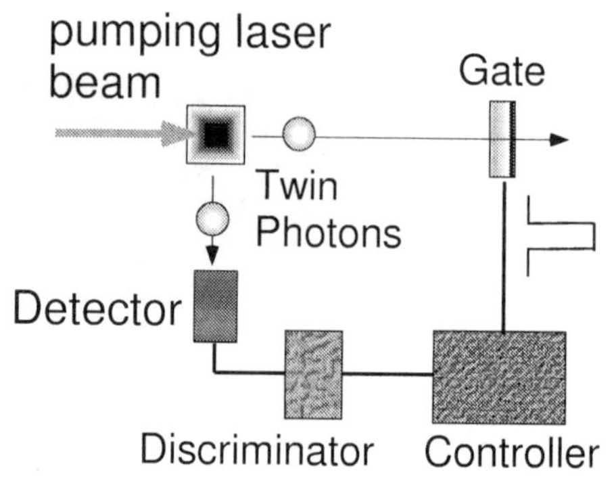

Figure 3. A schematic of the single photon source utilizing parametric down conversion.

In this paper, we analyze two GPDC methods. In one of the methods, we use a femto-second (fs) pulse laser as a PDC pumping source and monitor the number of created pairs in a pulse using a multi photon detector like the VLPCs. In the other method, we adopt a constant wave (CW) laser for pumping.

4.1 *femto-second laser pumping with multi photon detector*

Conventional fs-lasers can generate 10^8 optical pulses of 100 fs pulse width. When the nonlinear crystal is pumped with the laser, some pairs of photons are created in a pulse. The number of created pairs follows the Poisson distribution according to the pumping power. In this case, we use a detector which can distinguish the number of incident photons to monitor the number of created pairs in a pulse. Only when the signal of the detector shows the 'single photon incidence event', we open the shutter for a short time.

Figure 4 shows the ratio of the single photon pulses according to the average number of pairs created in a pulse $\bar{N}$. Let us consider the case that the quantum efficiency (QE) of the detector is unity, whose result is shown by the solid line. One interesting but natural result is that the maximum value of the ratio is 0.36, which can be obtained when $\bar{N} = 1$.

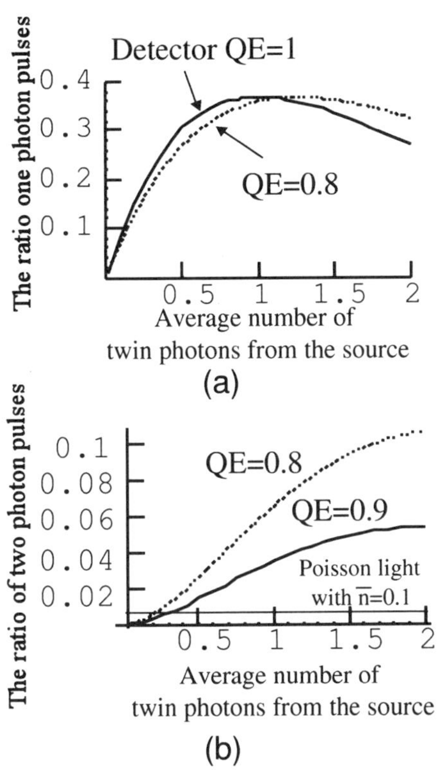

Figure 4. Photon number distributions for the case using a fs-laser and the multi photon detector. (a) The ratio of single photon pulses to the total pulses according to the average number of twin photons from the source in a pulse. The solid line is for the case where the quantum efficiency (QE) of the multi photon counter is 1, and the dotted line is for $QE = 0.8$. (b) The ratio of two photon pulses to the total pulses($P(2)$). The horizontal line shows the case where $P(2) = 0.005$.

The increase in $\bar{N}$ causes the increase in the fraction of the pulses which contains more than two photons and the decrease in those of one photon. The dotted line shows the case $QE = 0.8$. Even in this case we can achieve the maximum ratio 0.36 by increasing $\bar{N}$.

However, non-unity quantum efficiency have a bad influence on having small probability of two photon states. The fraction of two photon pulses ($P(2)$) are shown in Fig. 4(b). As shown here, $P(2)$ increases rapidly with a slight decrease of QE. In order to keep the quantum cryptographic system secure, $P(2)$ should be small. The demonstrations of quantum cryptography have been performed using weak light pulses with average photon number of 0.1, where $P(2)$ were 0.005. As shown in Fig. 4(b), It seems difficult to have some improvements in the fraction of single photon pulses with keeping $P(2)$ lower than 0.005.

Although this method is not suitable for quantum cryptography, there are some advantages. First, the single photons are emitted in a very short time (100fs) synchronized with the pump pulses. this feature is important for quantum information processing using photons, which requires good time correlation between photons. Second feature is that it will be possible to generate multi-photon state, which will lead to the artificial manipulation of the photon number distribution.

4.2 CW pumping with conventional detector

Next, let us analyze the case that a CW laser is used as a pumping source. In the previous subsection, the width and the timing of the single photon pulses are determined by the fs-laser. In this case, these parameters are determined by an inner/outer clock pulses.

The mechanism of this method is as follows. With CW pumping, the photon pairs are emitted at random in time. We adjust the experimental parameters to have some photon pairs created in a pulse. Then, according to the signal by the detector, the controller opens the shutter for a short time when the first photon created in a pulse passes the gate, and close the shutter to block the other photons in the pulse.

The photon number distribution of the output state using this method is shown in Fig. 5. Figure 5(a) shows the case the shutter opening time is almost negligible when it is compared to the pulse width. The fraction of the zero photon pulses ($P(0)$) and that of the single photon ($P(1)$) are shown by the solid and dotted lines respectively. The important result is that we can achieve the unity fraction of the single photon pulses when we increase the average number of twin photons from the source.

When we want to have a high repetition rate, the ratio of shutter opening time to the pulse width is increased. Figure 5(b) shows the case that the ratio is 0.2. As shown in the figure, probability of two photon pulses ($P(2)$) increases as the average number of photons increases. This is because the other photons may pass the gate during the shutter opening time. However, please note that we can decrease $P(2)$ as much as we need by adjusting the ratio of shutter opening time to the pulse width. These features are useful for the single photon source of the quantum cryptography.

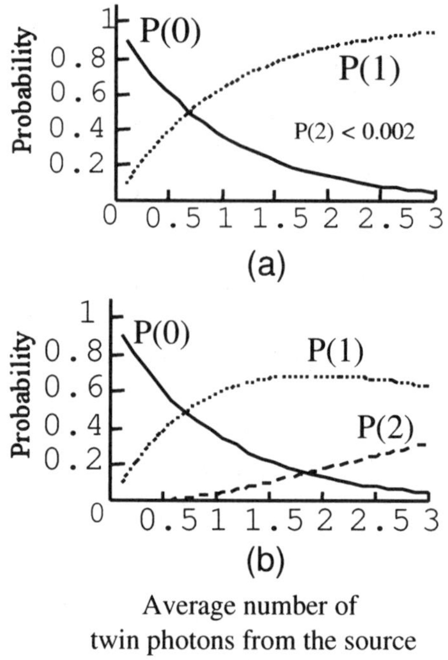

Figure 5. The solid line ($P(0)$) shows the ratio of no photon (empty) pulses to the total pulses according to the average number of twin photons from the source in a pulse. $P(1)$ is for the percentage of single photon pulses and $P(2)$ is of the two photon pulses. (a) When the ratio of the shutter opening time to the pulse width is 0.001. (b) The ratio is 0.2.

5 Conclusion

In this paper, we reported a new method to generate parametric fluorescence into two small spots . Because the generated twins are concentrated to the spots, we observed a high single count rate and coincidence count rate per pump power. We also observed that the ratio of coincidence count rate to single count rate was 80%, where the loss at filters and the quantum efficiency of detectors were compensated. This result is important for the application of this method to the single photon generator because we may be able to find a single photon in a gated pulse with up to 80% probability.

We also discussed the photon number distribution of the single photon pulses using gated parametric down conversion (G-PDC). We found that the unity fraction of the single photon pulses is achievable using the G-PDC with a CW pump laser, where the fraction of the two photon pulses are small enough. This result shows that this method is a good candidate for the photon source of quantum cryptography. We also analyzed the photon number distribution of the output state of the G-PDC with a fs laser and multi photon counter and found that this method might not be suitable for quantum cryptography but for manipulating the photon number distribution in a pulse.

Acknowledgments

This work is supported by JSPS Grant no. 12555008 and no. 12740242. This work is also partly supported by International Communications Foundation, JST-PRESTO project, and Mitsubishi Electric Corporation. The development of the VLPC photon detection system was performed in Stanford Univ. under the collaboration with ERATO Yamamoto project.

References

1. S. Takeuchi, *Phys. Rev.* A. **62**, 032301 (2000).
2. S. Takeuchi, *Phys. Rev.* A. **61**, 052302 (2000).
3. T. Hasegawa, T. Nishioka, H. Ishizuka, J. Abe, K. Shimizu, M. Matsui and S. Takeuchi, to be published in *IEICE Transactions on Fundamentals of Electronics, Communications and Computer Sciences*,(2002).
4. S. Takeuchi, *Opt. Lett.* **26**, No. 11, 843 (2001).
5. S. Takeuchi, J. Kim, Y. Yamamoto and H.H.Hogue, *Appl. Phys. Lett.* **74** 1063 (1999);
6. J. Kim, S. Takeuchi, Y. Yamamoto and H.H.Hogue, *ibid.* 902.
7. S. Takeuchi, manuscript in preparation.

MEASUREMENT OF THE PHOTONIC DE BROGLIE WAVELENGTH OF PARAMETRIC DOWN-CONVERTED PHOTONS USING A MACH-ZEHNDER INTERFEROMETER

KEIICHI EDAMATSU, RYOSUKE SHIMIZU, AND TADASHI ITOH

Division of Materials Physics, Graduate School of Engineering Science, Osaka University, Toyonaka 560-8531, Japan

E-mail: eda@mp.es.osaka-u.ac.jp

We propose and demonstrate the measurement of photonic de Broglie wavelength of an entangled photon pair state (a biphoton) generated by parametric down-conversion utilizing a normal Mach-Zehnder interferometer. The observed interference manifests the concept of the photonic de Broglie wavelength.

1 Introduction

Recently, interferometric properties of multiphoton states and their application have been attracting much attention. Jacobson *et al.*[1] proposed the concept of "photonic de Broglie wave" in multiphoton states. They argued that the photonic de Broglie wavelength of an ensemble of photons with wavelength λ and average number of photons N can be measured to be λ/N using a special interferometer with "effective beam splitters" that do not split the multiphoton states into constituent photons. Fonseca *et al.*[2] measured the photonic de Broglie wavelength of a two-photon state using a kind of Young's double slit interferometer. Boto *et al.*[3] proposed the principle of "quantum lithography", utilizing reduced interferometric diffraction of non-classical N photon states. Very recently, a proof-of-principle experiment of the quantum lithography was demonstrated.[4]

We propose and demonstrate the measurement of photonic de Broglie wavelength for N=2 in a very simple and straightforward manner, utilizing entangled photon pairs ("biphotons") generated by parametric down-conversion and a normal Mach-Zehnder (M-Z) interferometer. We discuss the nature of the biphoton interference, which is essentially governed by the frequency correlation between constituent two photons.

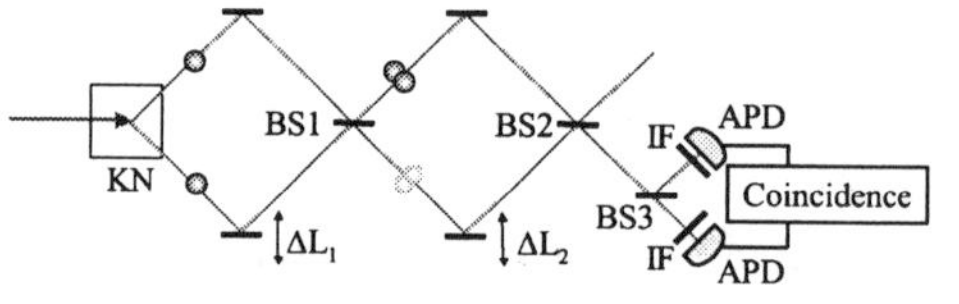

Figure 1. Schematic experimental setup of the biphoton interference. KN: $KNbO_3$ crystal, BS1$\sim$3: beam splitters, IF: interference filters, APD: avalanche photodiodes.

2 Experiment

Figure 1 shows the schematic view of our experimental setup. Pairs of entangled photons were generated by spontaneous parametric down-conversion (SPDC) in a 5 mm long $KNbO_3$ (KN) crystal, pumped by the second harmonic light of a single longitudinal mode Ti:sapphire laser (linewidth $\Delta\nu_0 \sim 40$ MHz) operating at λ_0=861.6 nm. We selected degenerate photon pairs, each wavelength of which was centered at λ_0, using two pinholes after the SPDC. The M-Z interferometer was composed of two 50/50% beam splitters (BS1 and BS2). Biphotons were generated at either arm of the interferometer when a pair of down-converted photons simultaneously entered at both input ports of a beam splitter (BS1), as a result of Hong-Ou-Mandel (HOM) interference.[5] The path-length difference between the two arms of the M-Z interferometer was controlled by a piezoelec-

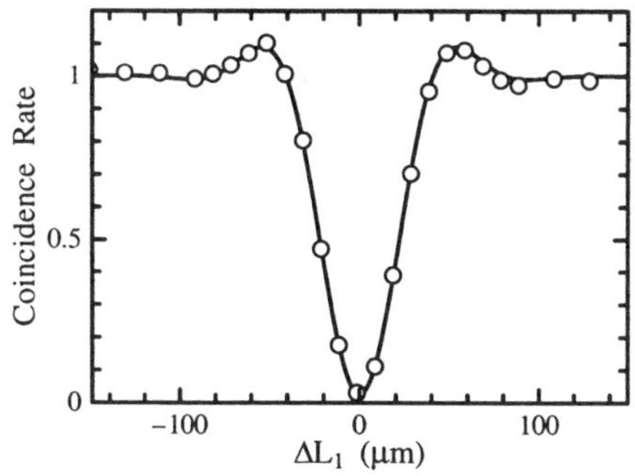

Figure 2. Coincidence counting rate detecting the photons after the two output ports of BS1 as a function of the optical path-length difference (ΔL_1) between the two paths from KN and BS1 in Fig. 1.

tric positioner. The two paths are combined together at the output beam splitter (BS2), and the biphoton interference was measured at one of the output ports of BS2. The biphoton interference pattern was recorded by a two-photon detector, consisting of a 50/50% beam splitter (BS3) and two avalanche photodiodes (APD) followed by a coincidence counter. In front of each APD, we put an interference filter (center wavelength $\lambda_c = 860$ nm, bandwidth $\Delta\lambda = 10$ nm). As a whole, the interferometer was designed to measure the photonic de Broglie wavelength of the biphoton without using any special "effective beam splitters". For comparison, we also measured the usual one-photon interference using a single detector and blocking one of the input ports.

It is worth discussing interference patterns expected in our experiment. For simplicity, we consider only single frequency (monochromatic) photons. The monochromatic treatment is adequate to predict the most distinct properties of the interference patterns, although spectral distribution should be considered to discuss more detailed phenomena such as coherent length of the interference. The one and two-photon counting rates (R_5 and R_{55}, respectively) at one of the output ports of the interferometer are

$$R_5 \propto \langle \psi_0, \psi_1 | a_5^\dagger a_5 | \psi_0, \psi_1 \rangle, \tag{1}$$

$$R_{55} \propto \langle \psi_0, \psi_1 | a_5^\dagger a_5^\dagger a_5 a_5 | \psi_0, \psi_1 \rangle, \tag{2}$$

where $a_5^\dagger$ and a_5 are photon creation and annihilation operators at the output port, ψ_0 and ψ_1 denote the quantum states of the two input ports. The photon operators at the output ports are connected to those of the input ports through the scattering matrices of the beam splitters and the optical path-length difference.[6] Thus we can calculate the counting rates (1) and (2) for arbitrary input states of light. The resultant two-photon interference pattern for the case $|\psi_0, \psi_1\rangle = |1, 1\rangle$, i.e., both inputs are $N=1$ Fock states, is

$$R_{55} \propto 1 - \cos 2\phi, \tag{3}$$

where $\phi = 2\pi\Delta L_2/\lambda$ is the optical phase difference in the two arms, ΔL_2 the path-length difference, and λ the wavelength of the input light. The corresponding one-photon interference for the case $|\psi_0, \psi_1\rangle = |0, 1\rangle$ is

$$R_5 \propto 1 - \cos\phi. \tag{4}$$

From Eqs. (3) and (4), we see that R_{55} will have the oscillation period $\lambda/2$, while R_5 has the period λ. This oscillation period $\lambda/2$ for the two-photon counting rate R_{55} is attributable to the photonic de Broglie wavelength λ/N for the biphoton ($N=2$) state.

3 Results and Discussion

The HOM interference, i.e., coincidence counting rate detecting the photons after the two output ports of BS1 as a function of the optical path-length difference (ΔL_1) between the two input ports, is presented in Fig. 2. The visibility of the HOM interference was 0.97, guaranteeing that the photon pair was almost perfectly traveling together along either arm of the interferometer at $\Delta L_1=0$. Figure 3 shows the measured interference pattern for both one-photon (upper graph) and two-photon (lower graph) de-

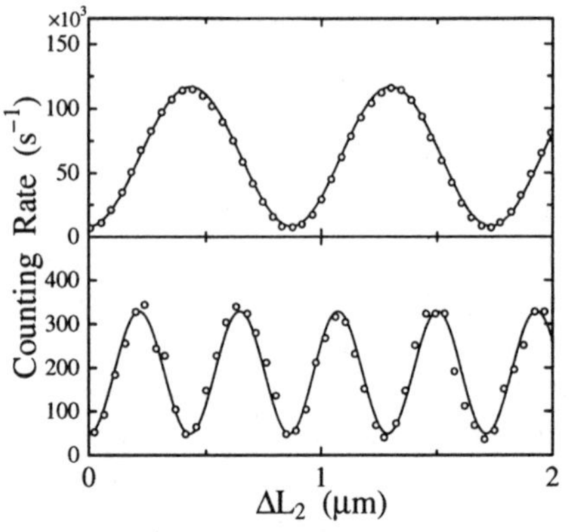

Figure 3. Interference patterns in the one-photon (upper) and two-photon (lower) counting rates at path-length difference around 0 μm.

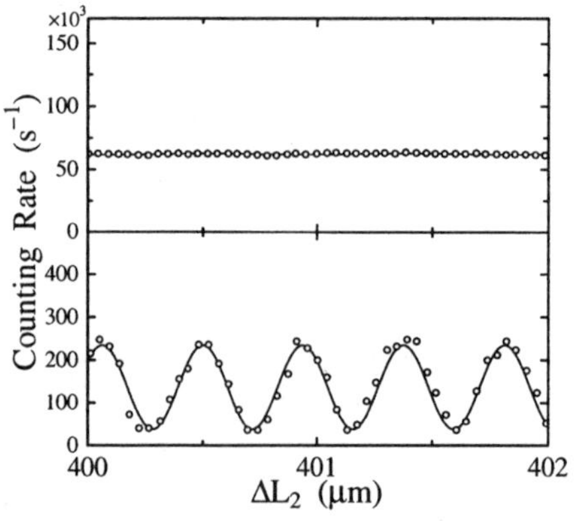

Figure 4. Interference patterns in the one-photon (upper) and two-photon (lower) counting rates at path-length difference around 400 μm.

tection, as a function of path-length difference (ΔL_2) between the two arms of the interferometer, around $\Delta L_2 \sim 0$ μm. Note that one of the input ports of the interferometer was blocked when measuring the one-photon counting rate, otherwise no interference is expected. One can see that the interference patterns for the one- and two-photon counting rates are well reproduced by Eqs. (4) and (3), respectively, and that the interference in the one-photon counting rate has a period of approximately 860 nm, whereas the interference period in the two-photon counting rate is approximately 430 nm. This result clearly indicates that the biphoton state exhibits the interference as a "wave" with half wavelength of the one-photon state. Thus, we have observed the photonic de Broglie wavelength of the biphoton state.

Furthermore, we have also observed the difference of the coherence length between the one- and two-photon counting rates as demonstrated in Fig. 4. Although the oscillational interference in the one-photon counting rate disappear at $\Delta L_2 \sim 400$ μm, the interference of the two-photon counting rate still remains for much larger path-length difference, indicating that biphotons have much longer coherence length than the single photons. Since the spontaneous parametric down-converted photons have considerably wide spectral width, the coherence length of the single-photon counting rate is governed by the spectral bandwidth $\Delta\lambda$ of the interference filters placed in front of the detectors. Thus, the coherence length of the single-photon counting rate becomes very short ($\lambda_c^2/\Delta\lambda \sim 70$ μm). On the other hand, the coherence length of the two-photon counting rate is governed by the spectral width of the sum frequency of signal (ν_s) and idler (ν_i) photons, that is identical to the frequency of pump photons ($2\nu_0$) of the parametric down conversion. Since we used the second harmonic of the single longitudinal mode continuous laser as the pump source, its coherence length is very long ($c/\Delta\nu_0 \sim 400$ cm). As a result, clear interference fringe was observed for the two-photon counting rate even at $\Delta L_2 \sim 400$ μm, whereas almost no fringe was observed for the one-photon counting rate. This is the direct consequence of the frequency correlation:

$$\nu_s + \nu_i = 2\nu_0 \tag{5}$$

between the constituent signal and idler photons of the biphoton. Thus, the fringe interval and coherence length of the two-photon counting rate consistently indicate that the biphoton is associated with the photonic de

Broglie wavelength:

$$\lambda_b = \frac{c}{\nu_s + \nu_i} = \frac{c}{2\nu_0} = \frac{\lambda_0}{2}, \qquad (6)$$

where refractive index dispersion is neglected.

So far, there have been a number of works concerning two-photon interference using parametric down-converted photons and a Mach-Zehnder or Michelson interferometer.[7,8,9,10] However, the previous experiments did not intend to observe the photonic de Broglie wave. Most of these experiments[7,8,10] detected two photons at both output ports of the interferometer. In our experiment, by detecting the two-photon counting rate at one of the output ports, we directly showed that the observed biphoton interference manifests the concept of photonic de Broglie wavelength. Finally, we note the relationship between our experiment and the non-local nature of the correlated two photons, i.e., biphotons, generated by parametric down-conversion or atomic cascade fluorescence. As previously proposed[11] and demonstrated,[12,13] two-photon quantum interference occurs for biphotons even using two spatially separated interferometers. Thus, we understand that the interferometric properties of the biphoton originate from its non-local quantum correlation between the constituent photons, but not from the spatial closeness of the two photons.

In conclusion, we have successfully measured the photonic de Broglie wavelength of the biphotons generated by parametric down-conversion utilizing a Mach-Zehnder interferometer, and showed that the nature of biphoton interference is essentially governed by the frequency correlation between the constituent two photons.

Acknowledgments

This work was supported by a Grant-in-Aid for Scientific Research from the Ministry of Education, Culture, Sports, Science and Technology of Japan.

References

1. J. Jacobson, G. Björk, I. Chuang, and Y. Yamamoto, Phys. Rev. Lett. 74, 4835 (1995).
2. E.J.S. Fonseca, C.H. Monken, and S. Pauda, Phys. Rev. Lett. 82, 2868 (1999); E.J.S. Fonseca, C.H. Monken, S. Pauda, and G.A. Barbosa, Phys. Rev. A 59, 1608 (1999).
3. A.N. Boto, P. Kok, D.S. Abrams, S.L. Braunstein, C.P. Williams, and J.P. Dowing, Phys. Rev. Lett. 85, 2733 (2000).
4. M. D'Angelo, M.V. Chekhova, and Y. Shih, Phys. Rev. Lett. 87, 013602 (2001).
5. C.K. Hong, Z.Y. Ou, and L. Mandel, Phys. Rev. Lett. 59, 2044 (1987).
6. *e.g.* L. Mandel and E Wolf, *Optical Coherence and Quantum Optics* (Campridge University Press, 1995), Sec. 12.12.
7. J.G. Rarity, et al., Phys. Rev. Lett. 65, 1348 (1990).
8. Z.Y. Ou, X.Y. Zou, L.J. Wang, and L. Mandel, Phys. Rev. A 42, 2957 (1990).
9. J. Brendel, E. Mohler, and W. Martienssen, Phys. Rev. Lett. 66, 1142 (1991).
10. Y.H. Shih, et al., Phys. Rev. A 49, 4243 (1994).
11. J.D. Franson, Phys. Rev. Lett. 62, 2205 (1989).
12. Z.Y. Ou, X.Y. Zou, L.J. Wang, and L. Mandel, Phys. Rev. Lett 65, 321 (1990).
13. P.G. Kwiat, A.M. Steinberg, and R.Y. Chiao, Phys. Rev. A 47, R2472 (1993).

EXCITON-PHOTON INTERACTION FOR A HIGH-EXCITON-DENSITY QUANTUM WELL PLACED IN A MICROCAVITY

YU-XI LIU,[(A)] N. IMOTO,[(A,B)] AND Ş. K. ÖZDEMIR[(A,B)]

(A) The Graduate University for Advanced Studies (SOKEN), Hayama, Kanagawa, 240-0193, Japan

(B) CREST Research Team for Interacting Carrier Electronics

GUANG-RI JIN, C. P. SUN

Institute of Theoretical Physics, The Chinese Academy of Sciences, P. O. Box 2735, Beijing 100080, China

We investigate the non-resonant interaction between the high-density excitons in a quantum well and a single mode cavity field. An analytical expression of the physical spectrum of the excitons is obtained. We study the spectral properties of the excitons which are initially prepared in the number states by the resonant femtosecond pulse pumping experiment. We numerically analyze the physical spectrum and mainly discuss the detuning effect on it.

1 Introduction

Recently, the optical properties of the low dimensional semiconductor structure (LDSS) have been investigated intensively both in experimental and theoretical works. This field is quite fascinating because of its related subjects, such as semiconductor microcavity (SMC) quantum electrodynamics (QED) [1,2], quantum dot microlaser and turnstile devices[3], quantum computer with quantum dot [4,5] etc.

It is known that the interaction between the light and these semiconductor microstructures, for example quantum well, may occur via exciton. In this paper we will discuss the detuning effect on the physical spectrum of the high density excitons in a quantum well surrounded by a micro-cavity.

2 Theoretical model and analytical solution

We assume that the cavity finesse is extremely high, and the cavity is in a extremely low temperature circumstance. The exciton operators are described approximately as hypothetical bosonic operators. The deviation of the exciton operators from the ideal bosonic model is corrected by introducing an effective non-linear interaction between these hypothetical ideal bosons, and the phase space filling effect is neglected. Then under the rotating-wave approximation, the Hamiltonian for the excitons and a single-mode cavity field can be written as:

$$H = \hbar\omega_1 a^\dagger a + \hbar\omega_2 b^\dagger b + \hbar g(a^\dagger b + b^\dagger a) + H' \tag{1}$$

with $H' = \hbar A b^\dagger b^\dagger b b$ where $b^\dagger(b)$ are creation (annihilation) operators of the excitons with frequency ω_2. They are assumed to obey the bosonic commutation relation $[b, b^\dagger] = 1$. $a^\dagger(a)$ are the creation (annihilation) operators of the cavity field with frequency ω_1. g is coupling constant between exciton and single mode cavity field. A represents an effective interaction constant between the excitons, and is assumed as a positive real number. Using the Schwinger's representation of the angular momentum by two bosons, we can construct the angular momentum operators

$$J_x = \frac{1}{2}(a^\dagger b + b^\dagger a), J_y = \frac{1}{2i}(a^\dagger b - b^\dagger a),$$
$$J_z = \frac{1}{2}(a^+ a - b^+ b), J^2 = \frac{\hat{N}}{2}(\frac{\hat{N}}{2} + 1) \tag{2}$$

from the ladder operators $a^\dagger, a$ of the cavity field and the exciton operator $b^\dagger, b$, where the total number operator $\hat{N} = a^+a + b^+b$ is the constant of motion. Then the Hamiltonian $H_0 = H - H'$ can be simplified into

$$\begin{aligned} H_0 &= \hbar\Omega\hat{N} + \hbar G(sin\theta J_x + cos\theta J_y) \\ &= \hbar\hat{N} + \hbar G e^{-i\theta J_y} J_z e^{i\theta J_y} \end{aligned} \quad (3)$$

in terms of an $SO(3)$ rotation $e^{i\frac{\pi}{2}J_y}$ of $\hbar\Omega\hat{N} + \hbar G(sin\theta J_x + cos\theta J_y)$ with $G = \sqrt{\Delta^2 + 4g^2}$, $tg\theta = \frac{2g}{\Delta}$, $\Omega = \frac{1}{2}(\omega_1 + \omega_2)$, and $\Delta = \frac{1}{2}(\omega_1 - \omega_2)$. For any fixed total particle number $\mathcal{N}$, the common eigen-states of J^2 and J_z are

$$|jm\rangle = \frac{(a^\dagger)^{j+m}(b^\dagger)^{j-m}}{\sqrt{(j+m)!(j-m)!}}|0\rangle \quad (4)$$

with the eigenvalues $j = \frac{\mathcal{N}}{2}$, and $m = -\frac{\mathcal{N}}{2}, \ldots, \frac{\mathcal{N}}{2}$. Using above expressions, the eigenfunctions and eigenvalues of the Hamiltonian (3) can be written respectively as

$$|\psi_{jm}^{(0)}\rangle = e^{-i\theta J_y}|jm\rangle, E_{jm}^{(0)} = \hbar(\mathcal{N}\Omega + mG). \quad (5)$$

The eigenvalues and eigenfunction of the whole Hamiltonian are obtained by perturbation theory as

$$E_{jm} = E_{jm}^{(0)} + \hbar A\langle jm|e^{i\theta J_y} b^\dagger b^\dagger bb e^{-i\theta J_y}|jm\rangle, \quad (6)$$

and

$$|\psi_{jk}\rangle = |\psi_{jk}^{(0)}\rangle + \hbar A \sum_{n\neq k} \frac{\langle\psi_{jn}^{(0)}|b^\dagger b^\dagger bb|\psi_{jk}^{(0)}\rangle}{E_{jk}^{(0)} - E_{jn}^{(0)}}|\psi_{jn}^{(0)}\rangle. \quad (7)$$

Then the time evolution operators of the system can be easily written as

$$U(t) = e^{-\frac{i}{\hbar}Ht} = \sum_{j=0}^{\infty}\sum_{m=-j}^{j} e^{-it\frac{E_{jm}}{\hbar}}|\psi_{jm}\rangle\langle\psi_{jm}|. \quad (8)$$

The time-dependent wave function can be obtained as $|\psi(t)\rangle = U(t)|\psi(0)\rangle$.

3 Radiation spectrum of exciton

Under the assumption of ideal cavity and extremely low temperature the only broadening mechanism comes from the detecting spectrometer for which the physical spectrum can be defined as [6]

$$\begin{aligned} S(\omega) = 2\gamma \int_0^t \mathrm{d}t_1 \int_0^t \mathrm{d}t_2 \Big\{ & e^{-(\gamma - i\omega)(t-t_2)} \\ & \times e^{-(\gamma+i\omega)(t-t_1)} G(t_1, t_2) \Big\} \end{aligned} \quad (9)$$

where γ is the half-bandwidth of spectrometer, t is time length of the excitation in the cavity, the dipole correlation function of the excitons $G(t_1, t_2) = \langle\psi(0)|b^\dagger(t_2)b(t_1)|\psi(0)\rangle$ with initial state $|\psi(0)\rangle$ of the system. Taking into account the fact that $b(t) = U^\dagger(t)bU(t)$, we can give the stationary physical spectrum which is similar to reference [7] as

$$\begin{aligned} S(\omega) = \sum_{j,l,k,m} & \frac{2\gamma}{\gamma^2 + (\omega - \omega_{jl,km})^2} \\ & \times |\langle\psi(0)|\psi_{jl}|^2 |\langle\psi_{jl}|b^\dagger|\psi_{km}\rangle|^2 \end{aligned} \quad (10)$$

with $\omega_{jl,km} = (E_{jl} - E_{km})/\hbar$. The selection ruler for Eq.(10) is $j = k + \frac{1}{2}$ and $l = m - \frac{1}{2}$. The difference of the eigenvalues $E_{jl} - E_{km}$ determine the position of the spectral component and $|\langle\psi(0)|\psi_{jl}|^2|\langle\psi_{jl}|b^\dagger|\psi_{km}\rangle|^2$ determine the intensity of the spectral lines. When the system is initially in the bare exciton states such as in a single exciton state, the physical spectrum from $\mathcal{N} = 1$ to $\mathcal{N} = 0$ can be obtained as

$$S(\omega) = \frac{2\gamma(sin\frac{\theta}{2})^4}{\gamma^2 + (\omega - \Omega - \frac{G}{2})} + \frac{2\gamma(cos\frac{\theta}{2})^4}{\gamma^2 + (\omega - \Omega + \frac{G}{2})} \quad (11)$$

with double peaks which locate in $\Omega + \frac{G}{2}$ and $\Omega - \frac{G}{2}$. From eq.(11), we find when the cavity field interacts resonantly with the excitons, the two peaks have the same height which is exactly equal to that of the two-level atomic system. But in the non-resonant case, the detuning makes them having different height and enlarges the distance between them. With the increase of the detuning quantity Δ, the height of the peak locating in $\Omega + \frac{G}{2}$ is reduced gradually, but the height of the peak at $\Omega - \frac{G}{2}$ is increased. When the detuning quantity Δ and the coupling constant g satisfy the condition $\Delta \gg 2g$, then

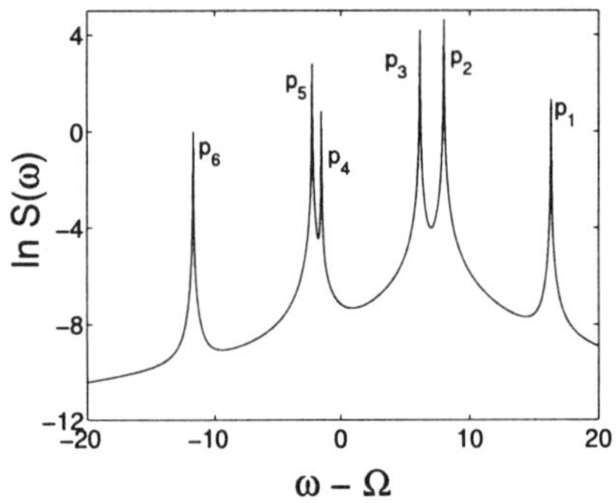

Figure 1. $S(\omega)$ is plotted as a function of the frequency $\omega - \Omega$ for a set of parameters $g = 5$ meV, $A/g = 0.6$, $\gamma = 0.01$ meV $\Delta = 2$ meV.

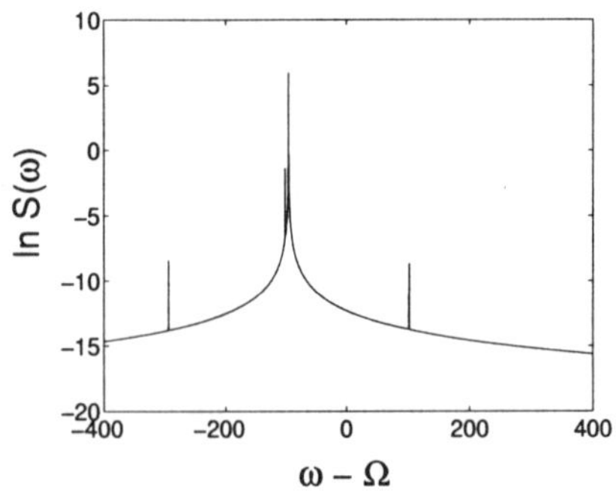

Figure 2. $S(\omega)$ is plotted as a function of the frequency $\omega - \Omega$ for a set of parameters $g = 5$ meV, $A/g = 0.6$, $\gamma = 0.01$ meV, $\Delta = 200$ meV.

$\theta \approx 0$, the physical spectrum has approximatively one peak.

When the exciton number becomes high, the emission spectrum of the system becomes complexity. The physical spectra are plotted in Fig. 1 when the system has initially two excitons. The positions of the six peaks are labelled as p_1, $p_2 .... p_6$ in the spectrum of Fig. 1. Numerical results show that with the increase in the detuning, the frequency difference between p_2 and p_3 becomes so small that they can not be resolved. Peaks p_1 and p_6, respectively, shift to more positive and negative sides of the spectrum, and the height of p_1 becomes so small that we cannot observe this peak. The positions of p_4 and p_5 are interchanged with increasing Δ, and after this interchange their relative positions are kept the same. We find four peaks in Fig. 2. If the detuning increases more and more, the numerical results show that the height of peaks p_2, p_3 and p_6 gradually becomes small. But the height of p_5 becomes the highest one, the height of p_4 is lower than that of p_5, both of them is so close that we almost cannot resolve them. In this case only one main peak is kept.

4 Conclusion

The physical spectrum of the non-resonant interaction between the high-density excitons in a quantum well and a single mode cavity field is investigated. The model and the discussions presented in this paper are valid only when the excitons and cavity field have zero linewidth. Comparing the resonant interaction between the cavity field and the excitons [8], we find when the system initially has one exciton, the non-resonant interaction will change both the heights of peaks and the frequency difference between them, and if the detuning becomes larger, then only one peak can be found. But in the resonant case, we always can find two peaks with the same height. When the system initially has two excitons, the physical spectrum becomes complex, with the increase of the detuning the positions of some peaks are interchanged, the heights of some peaks are reduced.

Acknowledgments

The authors thank M. Koashi, A. Miranowicz and T. Yamamoto for helpful and simulating discussions. Yu-xi Liu is supported by Japan Society for the Promotion of Science (JSPS). C. P. Sun is supported by NSF of China. This work also is supported by Grant-in-Aid for Scientific Research (B) (Grant No. 12440111) by Japan Society for the Promotion of Science.

References

1. Y. Yamamoto, S. Machida, and G. Bjork, Phys. Rev. A **44**, 657(1991).
2. C. Weibuch, M. Nishioka, A. Ishikawa, and Y. Arakawa, Phys. Rev. Lett. **69**, 3314(1992).
3. J. Kim, O. Benson, H. Kan, Y. Yamamoto, Nature **397**, 500(1999).
4. D. Loss, M. Scherwin, A. Small, Phys. Rev. Lett. **83**, 4204(1999).
5. G. Chen, N. H. Bonadeo, D. G. Steel, D. Gammon, D. S. Katzer, D. Park, and L. J. Sham, Science **289**, 1906 (2000); M. Bayer, P. Hawrylak, K. Hinzer, S. Fafard, M. Korkusinski, Z. R. Wasilewski, O. Stern, and A. Forchel, Science **291**, 451 (2001).
6. J. H. Eberly, and K. Wodkiewice, J. Opt. Soc. Am. **67**, 1252(1977).
7. Yu-xi Liu, N. Imoto, and Ş. K. Özdemir, Guang-ri Jin, C. P. Sun, Phys. Rev. A **65**, 023805(2002).
8. Yu-xi Liu, C. P. Sun, and S. X. Yu, Phys. Rev. A **63**, 023802(2001).

QUBIT-STATE GENERATION USING PROJECTION SYNTHESIS

ŞAHİN KAYA ÖZDEMİR,[A] ADAM MIRANOWICZ,[A,B] MASATO KOASHI,[A]
AND NOBUYUKI IMOTO[A,C]

(A) CREST Research Team for Interacting Carrier Electronics, Graduate University for Advanced Studies (SOKEN), Hayama, Kanagawa 240-0193, Japan

(B) Institute of Physics, Adam Mickiewicz University, 61-614 Poznań, Poland

(C) NTT Basic Research Laboratories, 3-1 Morinosato Wakamiya, Atsugi, Kanagawa 243-0198, Japan

We discuss the preparation of qubit states of the form $C_0|0\rangle + C_1|1\rangle$ using *quantum scissors device*, which exploits projection synthesis and prepares the desired state by truncating a coherent light. A feasible experimental scheme is proposed and optimized to obtain the desired qubit state with the highest attainable fidelity by tuning the intensity of the input coherent light.

1 Introduction

Quantum engineering of light offers promising ways to improve manipulation and transmission of information. Quantum states containing a definite number of photons and/or their superposition play a key role in quantum optics and quantum information processing. Such states show the quantum nature of light and are necessary for a theoretical study of a wide range of optical phenomena. Besides, they can be used for more applied phenomena such as teleportation, quantum dense coding and quantum cryptography all of which have first been realized in the quantum optical domain. Superposition of vacuum and single-photon states in a certain optical mode can be used to implement a qubit, which is the basic ingredient in quantum computation and information processing. Among a large variety of proposals put forward for the generation of qubit states, the scheme of Pegg, Phillips and Barnett[1] of the so-called the *quantum scissors device* (QSD) is very promising. In this scheme, a weak intensity coherent state is truncated up to its single-photon state generating a superposition of vacuum and single-photon states, which has the same relative phase and amplitude as those of the input coherent state. Recently, we have proposed a feasible experimental scheme to realize the state truncation with QSD and shown that QSD can be implemented with high fidelity and efficiency by using current level of photon detection and single-photon-state generation[2]. In this, study, we will discuss the preparation of arbitrary superposition states using this scheme, and study their fidelity to the desired ones.

2 Experimental Scheme

The proposed scheme is depicted in Fig. 1. The scheme consists of a parametric down conversion crystal (PDCC) for a type-I phase matching to emit photon pairs of the same polarization in two modes. The output of the three conventional photon counting detectors, which can only discriminate between the presence and absence of the photons but not discriminate between the presence of n and $n+1$ photons, are used to realize conditional measurement. D_1 is used as a gating detector, where recording a "click" ensures the presence of photons at one of the input ports of the BS1, the other port of which is left at vacuum. One of the output ports of BS1 is used as an input to BS2 whose second input port is fed by a weak intensity coherent state. Both output modes of BS2 are detected by the photon counters. Depending on the observed result, different condi-

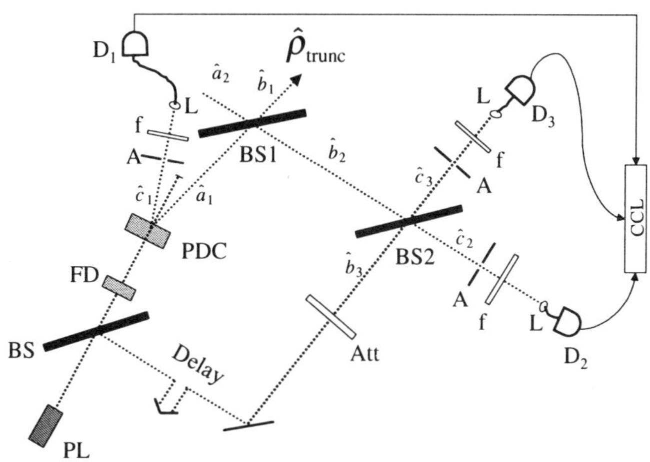

Figure 1. Schematic diagram of the QSD scheme. Key: PL – pulsed laser; FD – frequency doubler; PDC – parametric down conversion crystal; Att – strong attenuator; A – aperture; f – narrow band filter; L – lens; CCL – coincidence counter and logic; $\hat{\rho}_{\text{trunc}}$ – prepared output state; BS1 and BS2 are beam splitters; and D_1, D_2, and D_3 are photon counters.

tional output states $\hat{\rho}_{\text{trunc}}$ are obtained at the output port of BS1. The event of detecting one "click" at D_2 and no "clicks" at D_3 corresponds to the preparation of the qubit $C_0|0\rangle + C_1|1\rangle$. The relative phase between the vacuum and one photon components is equal to the phase of the input coherent state. This coherent light is obtained by weakening the output light of the pulsed laser before it is frequency doubled. The frequency doubled portion of the light from the laser source is used as a pump for the PDCC. The input to the QSD scheme can be written as $\hat{\rho}_{\text{in}} = \hat{\rho}_{(a_1,c_1)} \otimes |0\rangle_{a_2\ a_2}\langle 0| \otimes |\alpha\rangle_{b_3\ b_3}\langle\alpha|$ where the output of the PDC is a mixed state given by

$$\hat{\rho}_{(a_1,c_1)} = (1-\gamma^2)\Big[|00\rangle\langle 00| + \gamma^2|11\rangle\langle 11| + \gamma^4|22\rangle\langle 22| + \cdots\Big]_{(a_1,c_1)} \quad (1)$$

with γ^2 being the single-photon-pair generation rate, and $|\alpha\rangle_{b_3}$ is a weak-intensity coherent state. Then the state just before photodetection is given as $\hat{\rho}_{\text{out}} = \hat{R}_2\hat{R}_1(\hat{\rho}_{(a_1,c_1)} \otimes |0\rangle_{a_2a_2}\langle 0| \otimes |\alpha\rangle_{a_3a_3}\langle\alpha|)\hat{R}_1^\dagger\hat{R}_2^\dagger$ with $\hat{R}_1$ and $\hat{R}_2$ being the forward action of the BS1 and BS2, respectively. Then the conditional output state upon detection becomes

$$\hat{\rho}_{\text{trunc}} = \frac{\text{Tr}_{(c_1,c_2,c_3)}(\Pi_1^{c_1}\Pi_1^{c_2}\Pi_0^{c_3}\hat{\rho}_{\text{out}})}{\text{Tr}_{(b_1,c_1,c_2,c_3)}(\Pi_1^{c_1}\Pi_1^{c_2}\Pi_0^{c_3}\hat{\rho}_{\text{out}})}\,, \quad (2)$$

$$\Pi_0 = \sum_{m=0}^{\infty} e^{-\nu}(1-\eta)^m|m\rangle\langle m|$$
$$\Pi_1 = 1 - \Pi_0\,. \quad (3)$$

with $\Pi_1^{c_1}$, $\Pi_1^{c_2}$ and $\Pi_0^{c_3}$ being the elements of the positive-operator-valued measure (POVM) for the detectors D_1, D_2 and D_3, respectively; 0 and 1 correspond to the number of clicks recorded at the detectors. Mean dark count and efficiency of the detector are represented by ν and η, respectively.

This scheme has been shown to work with high fidelity and efficiency to truncate a coherent state up to its single-photon number state. For the numerical simulations in this study, we assumed that the pump had a repetition frequency of 100 MHz, $O(\gamma^2) = 10^{-4}$, $\nu = 100$ and terms higher than γ^6 are neglected. In Fig. 2, we depict the effects of in-

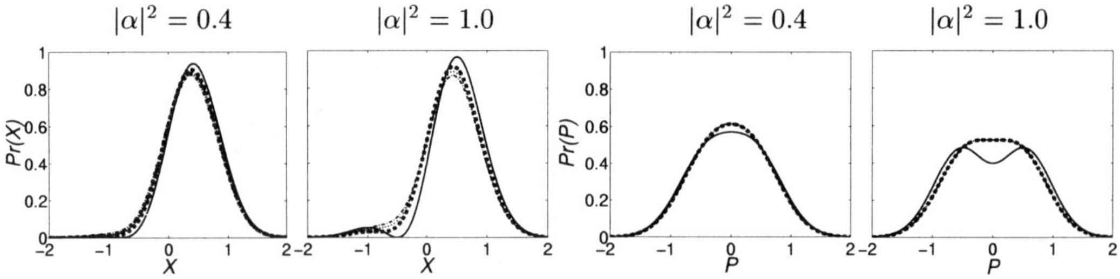

Figure 2. Effect of detector efficiency on the marginal distributions of the output states generated by truncating coherent states of different intensities. Solid curve is for the perfect QSD, and dash-dotted, dotted and dashed curves are for $\eta = 0.5, 0.7$ and 1, respectively.

tensity of the coherent light $|\alpha|^2$ and η on the quasi-distributions of the states prepared by the experimental scheme. We observe that for $|\alpha|^2 < 1.0$, detector losses and source imperfections do not have a significant effect on quasi-distributions. Non-classical structure of the states can be observed even for very low η, however with increasing $|\alpha|^2$, the η effect becomes stronger. When the coherent state is a strong light, the quasi-distributions of the states prepared by the proposed scheme differ strongly from those of the ideal case suggesting that the truncation process is more efficient and closer to the ideal case for weak intensity coherent lights.

3 Preparation of arbitrary optical qubit states by tuning the intensity of input coherent light

In this scheme, the ratio C_1/C_0 of the coefficients of the superposition state $C_0|0\rangle + C_1|1\rangle$ can be adjusted either by using tunable beam splitters or by tuning of the intensity of the input coherent state. The first method can be realized by replacing the beam splitters BS1 and BS2 by Mach-Zehnder interferometers as was done by Paris[3]. This will result in a more complicated set-up and will introduce the problem of controlling the stability and balance of the interferometers. The latter method can be realized without introducing additional components except a controllable tuning of the intensity of the coherent light. In that case, we choose BS1 and BS2 as 50 : 50 beam splitters because they give the highest probability and fidelity for state truncation using the ideal QSD scheme. The highest fidelity in generating arbitrary qubit states from a coherent state can be achieved if $arg(\alpha) = arg(C_1)$, which is assumed to be the case in this study.

Figure 3 shows the results of numerical simulations. It is seen that balanced superposition of vacuum and single-photon states can be prepared with high fidelity even for η as low as 0.1 if the intensity of the coherent light is kept as $|\alpha|^2 \leq 1$. For $\eta = 0.5$, the optimum value for $|\alpha|^2$ is found as 0.72 to prepare a balanced superposition state with the highest fidelity value of 0.89. Preparation rate of such a state is found as 4533 s^{-1} with the simulation parameters given in the previous section. Increasing η from 0.5 to 0.7 will increase the generation rate to 8524 s^{-1} when the optimum value of $|\alpha|^2 = 1.06$ is used. Fidelity is changed only by 0.11%.

It is understood that for a given η, one can always find an optimum value for $|\alpha|^2$ which will result in the highest fidelity for the desired $|C_1/C_0|$, however the fidelity value may not be close to one. The fidelity of the qubit state preparation depends heavily on the relative weights of the vacuum and single-photon states in the superposition, as seen in Fig. 3(b). If the vacuum is heavily weighted

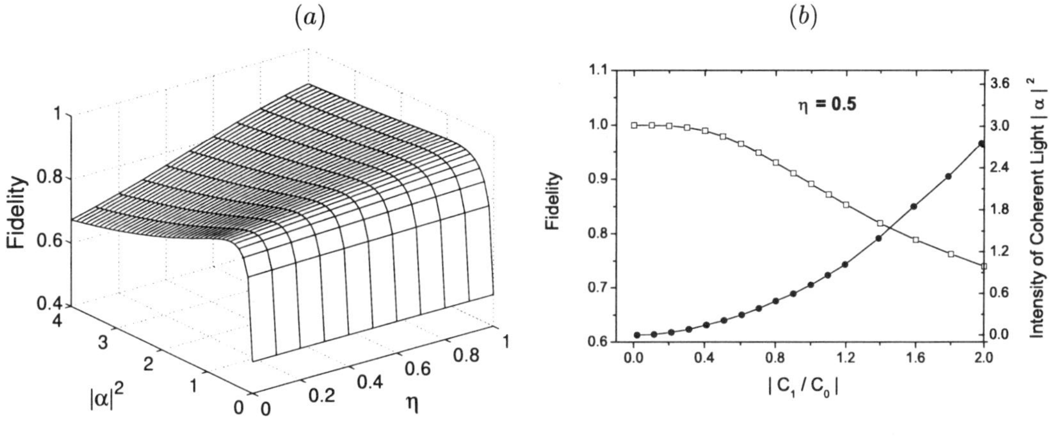

Figure 3. (a) Effect of detector efficiency η and intensity of the coherent light $|\alpha|^2$ on the fidelity of preparing balanced superposition $|C_1|/|C_0| = 1$ of vacuum and single-photon states; (b) Optimized intensity of the coherent light (marked by squares) to prepare arbitrary qubit states with the maximum attainable fidelity (circles) at $\eta = 0.5$.

in the desired qubit state, then fidelity values higher than 0.9 can be obtained for η as low as 0.1. For $\eta = 0.5$ ($\eta = 0.7$), fidelity of preparing a qubit state of $|C_1/C_0| = 0.5$ is 0.979 (0.983) with the optimum $|\alpha|^2 = 0.21$ ($|\alpha|^2 = 0.23$). However, when the weights of vacuum and single photon are reversed, $|C_0/C_1| = 0.5$, the fidelities drop to 0.740 and 0.764, respectively, for $\eta = 0.5$ and $\eta = 0.7$.

4 Conclusion

We have proposed a scheme for generating arbitrary qubit states. The scheme is based on QSD and tuning of input coherent light intensity enables the preparation of the desired qubit. We have presented the effects of the input coherent light intensity and the realistic photon counting process on the marginal distributions of the generated output state. It is observed that fidelity of the generated states strongly depends on the weights of the vacuum and single-photon states in the qubit. Superposition states, for which vacuum component is much stronger than the one-photon component, can be generated with high fidelity using this scheme.

Acknowledgments

This work was supported by a Grant-in-Aid for Encouragement of Young Scientists (Grant No. 12740243) and a Grant-in-Aid for Scientific Research (B) (Grant No. 12440111) by Japan Society for the Promotion of Science.

References

1. D. T. Pegg, L. S. Phillips and S. M. Barnett, *Phys. Rev.* Lett. **81**, 1604 (1998).
2. Ş. K. Özdemir, A. Miranowicz, M. Koashi, and N. Imoto, *Phys. Rev.* A **64**, 063818 (2001); quant-ph/0107048.
3. M. G. A. Paris, *Phys. Rev.* A **62**, 033813 (2000).

QUANTIZED VORTICES IN A BOSE-EINSTEIN CONDENSATE

V. BRETIN, F. CHEVY, K.W. MADISON, P. ROSENBUCH, AND J. DALIBARD
Laboratoire Kastler Brossel, 24 rue Lhomond, 75005 Paris, France*

When a superfluid medium is forced into rotation, Quantum Mechanics imposes strong constraints on its velocity field. The superfluid is set into motion through the nucleation of lines of singularity or *vortex lines*, where the density vanishes and around which the circulation of the velocity is quantized. We report the observation of such vortices in a gaseous Bose-Einstein condensate of ^{87}Rb atoms, which is stirred using a rotating laser beam. We characterize the phase and the angular momentum of a single vortex, and we discuss the maximal number of vortices which can be nucleated in this system. Finally we address the problem of the critical rotation frequency at which the first vortex appears.

1 Introduction

Superfluidity, originally discovered and studied in the context of superconductors and later in the system of superfluid liquid Helium, is a hallmark property of interacting quantum fluids and encompasses a whole class of fundamental phenomena[1,2]. With the achievement of Bose-Einstein condensation in atomic gases[3], it became possible to study these phenomena in an extremely dilute quantum fluid, thus helping to bridge the gap between theoretical studies, only tractable in dilute systems, and experiments.

A striking consequence of superfluidity is the response of a quantum fluid to a rotating perturbation. In contrast to a normal fluid, which at thermal equilibrium rotates like a solid body with the perturbation, the thermodynamically stable state of a superfluid involves no circulation, unless the frequency of the perturbation is larger than some critical frequency, analogous to the critical velocity[1,4]. When the superfluid does circulate, it can only do so by forming vortices in which the condensate density vanishes and for which the velocity field flow evaluated around a closed contour is quantized:

$$\oint \mathbf{v} \cdot d\mathbf{r} = n\frac{h}{M} \tag{1}$$

where n is an integer and M the mass of a particle of the fluid.

In this paper, we report the observation of vortices in a stirred gaseous Bose-Einstein condensate (BEC) of atomic rubidium. We present a direct observation of the phase of a rotating condensate and we show that the angular momentum per atom jumps from 0 to $\hbar$ at the threshold for the nucleation of the first vortex. We also study the situation where several vortices are nucleated in the condensate, and we show that the velocity field obtained in this case is close to the one corresponding to a rigid body rotation. Finally we discuss the mechanism for vortex nucleation in our setup which is observed to occur via a *dynamical* rather than *thermodynamical* instability of the condensate in a rotating potential.

2 Experimental setup

For the preparation of the condensate[5] we start with 10^9 ^{87}Rb atoms in a cylindrically symmetric Ioffe–Pritchard magnetic trap at a temperature $\sim 200\ \mu$K. The oscillation frequency of the atoms along the longitudinal axis of the trap (horizontal in our experiment) is $\omega_z/2\pi \sim 10$ Hz. For the results presented here, the transverse frequencies $\omega_\perp/2\pi$ have been varied between 100 and 200 Hz by adjusting the bias field at the center of the trap.

The standard experimental sequence consists in three steps: (i) condensation *via* evaporative cooling, (ii) vortex nucleation, and (iii) characterization using absorption imaging after a time-of-flight (TOF) expan-

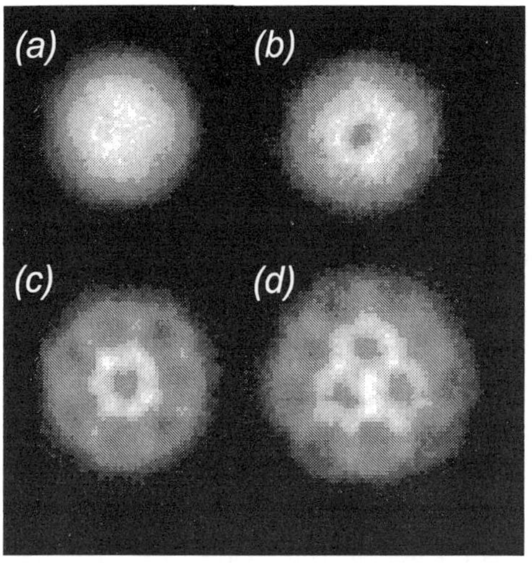

Figure 1. Density profile of a stirred condensate after time-of-flight. (a) Rotation frequency below the critical frequency Ω_c. (b) Rotation frequency just above Ω_c. (c) and (d) Rotation frequency notably above Ω_c. On fig. (d), the triangular structure of the vortex array is clearly visible (Abrikosov lattice). The size of an image is $\sim 300\ \mu$m.

sion of 25 ms. The probe laser for the imaging propagates along the $z-$axis, and the image gives the transverse xy distribution of atomic positions after the expansion. From each image we extract the temperature of the cloud, the size of the condensate in the xy plane, and the number of vortices which have been nucleated.

We evaporatively cool the atoms with a radio-frequency sweep. The condensation threshold is reached at 550 nK. We continue the evaporative cooling to a temperature below 80 nK at which point approximately 3×10^5 atoms are left in the condensate. This number is evaluated from the size of the condensate after expansion.

After the end of the cooling phase we switch on the stirring laser beam, which is parallel with the long axis (z) of the condensate. The position of the stirring beam in the transverse plane ($z = 0$) is controlled by two acousto-optic modulators, and the temporal variation of the stirring laser intensity and position is chosen to create a dipole potential well approximated by the harmonic terms $M\omega_\perp^2(\epsilon_X X^2 + \epsilon_Y Y^2)/2$ with $\epsilon_X = 0.05$ and $\epsilon_Y = 0.15$. The X, Y basis rotates at constant angular frequency Ω with respect to the fixed x, y basis, so that the time dependent stirring potential δU (proportional to $\epsilon_X - \epsilon_Y$) is in the laboratory frame:

$$\delta U \propto (x^2 - y^2)\cos(2\Omega t) + 2xy\sin(2\Omega t)\ . \quad (2)$$

The stirring phase lasts up to 900 ms which is well beyond the typical vortex nucleation time found experimentally to be about 450 ms. This transposition to a gaseous BEC of the famous rotating bucket experiment[2] has also been implemented successfully by other groups using either a laser stirrer[6] or a magnetic one[7,8]. Another method for nucleating a single vortex uses a combination of a laser and a microwave field to print the desired velocity field onto the atomic wave function[9].

Fig. 1 shows typical time-of-flight images as a function of the stirring frequency. For a stirring frequency Ω lower than a threshold value Ω_c, the condensate is not affected by the stirring laser (Fig. 1a). Just above this critical frequency (within 1 or 2 Hz), a density dip appears at the center of the cloud, with a reduction of the optical thickness at this location which reaches 50% (Fig. 1b). When we stir the condensate at a frequency notably higher than Ω_c, more vortices are nucleated (Fig. 1c,d).

3 The single vortex state

3.1 Phase pattern

In order to measure the phase pattern of the single vortex state shown in Fig. 1b, we use an interferometric method[10,11,12,13]. We divide the wave function of the condensate into two components separated in position and momentum. In the location where these components overlap, a matter-wave interference fringe pattern is produced which reveals the phase difference between the two components[14]. This homodyne detection of the phase pattern of a vortex state is complementary to the heterodyne detection[9] performed in Boulder in 1999.

To realize the amplitude splitting interferometer, we use a succession of two radio frequency (rf) pulses, which resonantly drive transitions between different internal states of each rubidium atom. These pulses coherently split and recombine the condensate cloud. The atomic rubidium condensate is initially polarized in the $F = m = 2$ Zeeman substate. The first rf pulse couples this magnetic spin state to the four other Zeeman substates of the $F = 2$ manifold. Among these four, only the $m = +1$ is also magnetically trapped. However, due to gravity, the center of its confining potential is vertically displaced with respect to the center of the $m = +2$ trap by an amount $\Delta x_0 = g/\omega_\perp^2 \approx 6\ \mu$m ($\omega_\perp/2\pi = 192$ Hz in this experiment). After this first pulse, the atoms are allowed to evolve in the harmonic trapping potential. During this time, those atoms driven into the other Zeeman substates fall or are ejected out of the trap. What remains are the two trapped clouds now with slightly different momenta and positions as determined by their respective phase space trajectories. When the second rf pulse is applied, each trapped state generates a copy in the other trapped Zeeman sublevel making a total of four wave packets present in the trap (two in $m = +1$ and two in $m = +2$). We then perform the TOF expansion and we record the atomic interference pattern between the overlapping wave packets of a given internal Zeeman spin state.

We have first measured the interference pattern in the absence of any vortices. As expected, this pattern consists in straight fringes, and a typical result for the $m = +2$ channel is shown on Fig. 2a. When we stir the atomic cloud at a frequency of 130 Hz (close to the critical frequency for these experimental parameters), a well centered, single vortex is created in the condensate. The result from an interference experiment performed in this case is shown on Fig. 2b. This pattern, which is very different from the one obtained with no vortex, clearly shows the telltale "H" shape characteristic of the 2π phase winding due to the presence of a vortex. We repeated this experiment many times for the same initial condition, and have found that the interference patterns produced have a similar shape but with a relative phase between the two components which fluctuates from one shot to another so that the bright and dark regions in Fig. 2b can be exchanged. Note that this technique can be used as a very sensitive tool to diagnose the presence of vortices generated by the fast motion of a macroscopic object in a condensate. This was demonstrated very recently at MIT, in an experiment where the interference pattern between a condensate with vortices and a condensate at rest has been recorded[15].

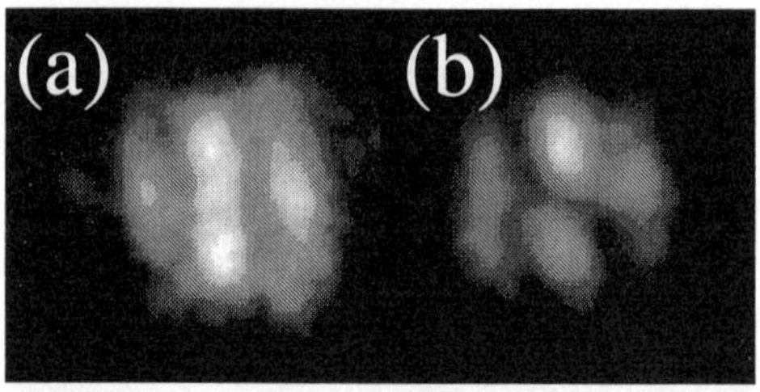

Figure 2. Interference pattern resulting from an homodyne detection of the phase of condensate. (a) Fringe pattern obtained when no vortex is present. (b) Phase pattern obtained with a single vortex. The fringe dislocation reveals the 2π phase winding around the vortex line.

3.2 *Measurement of angular momentum*

In order to measure the angular momentum of the condensate along the symmetry axis z of the trap, we study the two transverse quadrupole modes[16,17,18,19] carrying angular momenta of $m = \pm 2$ along the z axis. In the absence of vortices, the frequencies $\omega_\pm/2\pi$ of these two modes are equal as a consequence of the reflection symmetry about the xy plane.

By contrast, for $L_z \neq 0$, this degeneracy is lifted by an amount[16]:

$$\omega_+ - \omega_- = 2\,L_z/(M r_\perp^2)\ , \qquad (3)$$

where $r_\perp^2$ stands for the average value of $x^2 + y^2$ for the condensate. Consequently the measurements of $\omega_+ - \omega_-$ and of the transverse size of the condensate provide the angular momentum of the gas[20].

In order to measure the frequency difference $\omega_+ - \omega_-$, we excite the transverse quadrupolar oscillation using the dipole potential created by the stirring laser now on a fixed basis $(X, Y = x, y)$ and with a 10-times larger intensity. This potential acts on the atoms for a 0.3 ms duration at the end of the vortex nucleation phase. The transverse quadrupolar mode excited is a linear superposition of the $m = \pm 2$ modes. The lift of degeneracy between the frequencies of these two modes causes a precession of the eigenaxes of the quadrupole mode at an angular frequency given by $\dot{\theta} = (\omega_+ - \omega_-)/4$. Therefore the measurement of $\dot{\theta}$ together with the size of the condensate gives access to L_z.

To determine $\dot{\theta}$ we let the atomic cloud oscillate freely in the magnetic trap for an adjustable period τ (between 0 and 8 ms) after the quadrupole excitation. We then perform the TOF + absorption imaging sequence. A typical result is shown in fig. 3. For this measurement the measured number of atoms was 3.7 10^5. The transverse frequency $\omega_\perp/2\pi$ equals 171 Hz, and the threshold frequency $\Omega_c/2\pi$ for nucleating a vortex is 115 Hz. The sequence of pictures in fig. 3 corresponds to $\tau = 1, 3$ and 5 ms for which the ellipticity in the xy plane is maximum. They have been obtained after stirring the condensate at a frequency $\Omega/2\pi = 120$ Hz for which a single vortex is systematically nucleated.

As expected, the axes of the quadrupole oscillation precess. The precession rate $\dot{\theta}$, obtained from the analysis of sequences of images taken every 0.5 ms, is 5.9 degrees per millisecond. The fit of the image of the condensate after expansion gives an average radius equal to 103 (± 6) μm from which we deduce the value before expansion $r_\perp = 2.0$ μm. Equation 3 then yields $L_z/\hbar = 1.2$ (± 0.1). The average angular momentum per particle is thus of the order of $\hbar$ when a single vortex is visible in the condensate.

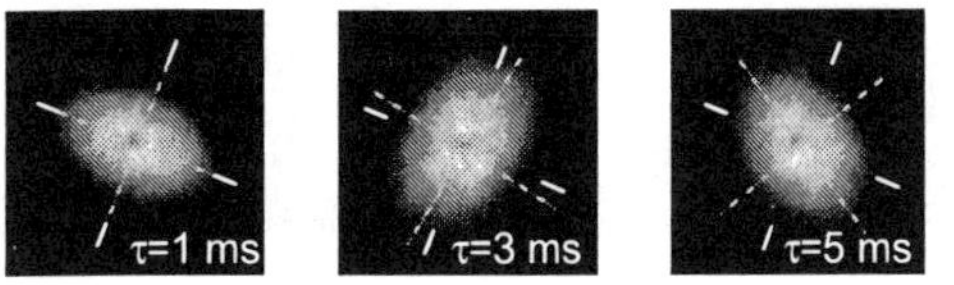

Figure 3. Transverse oscillations of a condensate with a single vortex, with $N = 3.7\ 10^5$ atoms, $\omega_\perp/2\pi =$ 171 Hz and $\Omega/2\pi = 120$ Hz. The fixed axes indicate the excitation basis and the rotating ones indicate the condensate axes. We deduce from these images that the angular momentum per particle is $\sim \hbar$.

4 The many-vortex state

We now turn to the discussion of the vortex pattern obtained for larger stirring frequencies. The regularity of the vortex lattices obtained in this case (see figs. 1cd) is a consequence of the balance between the repulsive interaction between two vortex lines and the restoring force that acts on a vortex line centering it on the condensate.

In the large vortex number limit, the density of vortices can be deduced from the *correspondence principle.* In this limit[21], the coarse grain average (on a scale larger than the distance between two vortices) of the quantum velocity field should be the same as that for classical, rigid-body rotation $\mathbf{v} = \mathbf{\Omega} \times \mathbf{r}$.

In order to recover this linear variation of the velocity field with the distance from the rotation axis, the surface density of vortices ρ_v must be uniform. The circulation of the velocity field $\oint \mathbf{v} \cdot d\mathbf{r}$ on a circle of radius R centered on the condensate is then $\mathcal{N}(R)\ h/M$, where $\mathcal{N}(R) = \rho_v \pi R^2$ is the number of vortices contained in the circle (see Eq. 1). This is equivalent to the rigid body

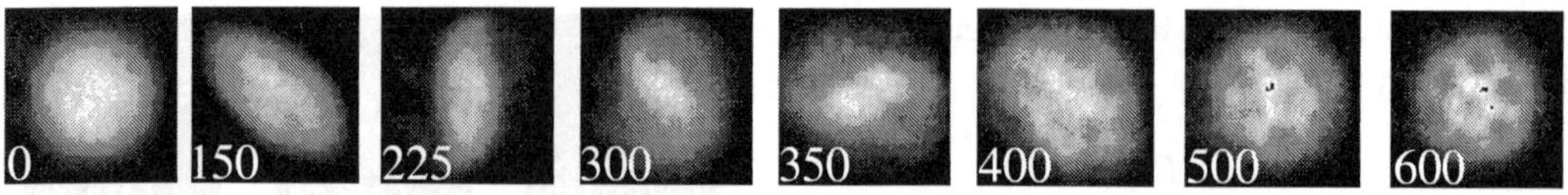

Figure 4. Nucleation of vortices in a stirred BEC. The time after stirring is indicated in ms. These pictures show first the transverse profile of the elliptic state generated by the laser stirrer, and then the nucleation and ordering of the resulting vortex lattice.

circulation $2\pi R^2\Omega$ if

$$\rho_v = \frac{2M\Omega}{h} \; . \tag{4}$$

Consider for instance the vortex lattice shown in fig. 1d, which contains $\mathcal{N} = 12$ visible vortices. Before the time-of-flight expansion, the condensate has a radius $R = 5\ \mu$m and the stirring frequency is $\Omega/2\pi = 77$ Hz ($\omega_\perp/2\pi = 103$ Hz in this case). This yields the ratio between the average velocity $\bar{v}$ on the circle and the velocity v_r corresponding rigid body rotation:

$$\frac{\bar{v}}{v_r} = \frac{\mathcal{N}\hbar}{M\Omega R^2} \simeq 0.7 \; . \tag{5}$$

This shows that the coarse grain average velocity field of the condensate shown in fig. 1d is indeed close that of a rigid body rotating at the stirring frequency Ω.

Using much larger condensates (5 10^7 sodium atoms) and therefore much larger radii, the MIT group has recently succeeded in nucleating large arrays of vortices (up to 150), also in good agreement with the number $\mathcal{N}$ expected for a rigid body rotation[6].

5 The critical rotation frequency

For very weak stirring intensities, corresponding to $\epsilon_X, \epsilon_Y \ll 1$, we observe that we can nucleate vortices only in a small domain of stirring frequencies Ω. This region is centered on the value $\omega_{QP}/2$, where $\omega_{QP} = \omega_\perp\sqrt{2}$ is the quadrupole oscillation frequency of the condensate. This coincidence indicates that the mechanism for vortex nucleation in our experiment involves the resonant excitation of the rotating quadrupole mode at frequency ω_{QP} by the time dependent perturbation at frequency 2Ω created by the stirring laser (see equation (2)).

More precisely we observe that during the first stage of the stirring phase, the condensate deforms in the xy plane and that it acquires a very strong ellipticity (see the first three pictures in fig. 4). This elliptical rotating state[22] corresponds to an irrotational flow of the condensate with a phase pattern proportional to XY. We observe that this rotating state possesses an intrinsic dynamical instability[22,23,24] and that this instability leads to the transformation of the elliptic state into a state of one or more vortices (last four pictures of fig. 4). This scenario is confirmed by a numerical simulation of the Gross-Pitaevskii equation[25,26]. Note that the ordering of the vortices leading to the regular lattice shown in the last picture of fig. 4 is found numerically only if one adds a dissipative term to the Gross-Pitaevskii equation[26].

The fact that vortex nucleation seems to occur only via a dynamic instability explains why the frequency range over which vortices are generated is notably smaller than that expected from thermodynamics[27,28]. When a vortex state is thermodynamically allowed but no dynamical instability is present, the time scale for vortex nucleation is probably too long to be observed in this system. When a stirring potential is used for which the non-harmonic terms are significantly large, the frequency domain of vortex nucleation is larger due to the possibility of exciting higher order, rotating multipole oscillations which can also disintegrate into vortex states[29,30].

For this reason, when many of the higher order terms contribute significantly to the stirring potential, one might expect to recover a nucleation domain which approaches the thermodynamic prediction.

6 Conclusion

We have presented in this paper a study of states of a gaseous Bose-Einstein condensate with one or several vortices. We have characterized the phase pattern and the angular momentum of a single vortex state, and we have related the maximum number of vortices which can be observed in this experiment to the rigid body rotation velocity field. We have also shown that the nucleation of vortices in this system occurs via the excitation of a rotating quadrupolar state of the condensate. A natural extension of this study concerns the dynamics of the vortices themselves, such as the lifetime of these objects[31,5,8] and their possible oscillations (Thomson modes[1]). We plan to address this subject in a future work.

References

[*] Unité de Recherche de l'Ecole normale supérieure et de l'Université Pierre et Marie Curie, associée au CNRS.

1. E.M. Lifshitz and L. P. Pitaevskii, *Statistical Physics, Part 2*, chap. III (Butterworth-Heinemann, 1980).
2. R.J. Donnelly, *Quantized Vortices in Helium II*, (Cambridge, 1991).
3. For a review see e.g. *Bose-Einstein Condensation in Atomic Gases*, Proceedings of the International School of Physics "Enrico Fermi", Course CXL, eds M. Inguscio, S. Stringari, and C.E. Wieman (IOS Press, Amsterdam, 1999).
4. A. Fetter and A. Svidzinsky, Jour. Phys. Cond. Matt. **13**, R135 (2001).
5. K.W. Madison *et al.*, Phys. Rev. Lett. **84**, 806 (2000).
6. J. R. Abo-Shaeer *et al.*, Science **292**, 467 (2001).
7. E. Hodby *et al.*, cond-mat/0106262.
8. P. Haljan *et al.*, cond-mat/0106362.
9. M. R. Matthews *et al.*, Phys. Rev. Lett. **83**, 2498 (1999).
10. E. L. Bolda and D.F. Walls, Phys. Rev. Lett. **81**, 5477 (1998).
11. J. Tempere and J. T. Devreese, Solid State Commun. **108**, 993 (1998).
12. Y. Castin and R. Dum, Eur. Phys. J. D **7**, 399 (1999).
13. L. Dobrek *et al.*, Phys. Rev. A, **60**, R3381 (1999).
14. F. Chevy *et al.*, cond-mat/0104545.
15. S. Inouye *et al.*, cond-mat/0104444.
16. F. Zambelli and S. Stringari, Phys. Rev. Lett. **81**, 1754 (1998).
17. R. Dodd *et al.*, Phys. Rev. A **56**, 587 (1997).
18. S. Sinha, Phys. Rev. A **55**, 4325 (1997).
19. A. Svidzinsky and A. Fetter, Phys. Rev. A. **58**, 3168 (1998).
20. F. Chevy *et al.*, Phys. Rev. Lett. **85**, 2223 (2000).
21. R.P. Feynman, in "Progress in Low Temperature Physics", vol. 1, ed. by C. J. Gorter (North-Holland, Amsterdam, 1955)
22. A. Recati, *et al.*, Phys. Rev. Lett. **86**, 377 (2001).
23. Y. Castin and S. Sinha, cond-mat/0101292.
24. K.W. Madison *et al.*, Phys. Rev. Lett. **86**, 4443 (2001)
25. D. Feder *et al.*, Phys. Rev. Lett. **86**, 564 (2001).
26. M. Tsubota *et al.*, cond-mat/0104523.
27. G. Baym and C.J. Pethick, Phys. Rev. Lett. **76**, 6 (1996).
28. E. Lundh *et al.*, Phys. Rev. A **55**, 2126 (1997).
29. F. Chevy *et al.*, cond-mat/0104218
30. C. Raman *et al.*, cond-mat/0106235.
31. P. Fedichev and G. Shlyapnikov, Phys. Rev. A **60**, R1779 (1999).

VORTEX EXCITATIONS IN A BOSE-EINSTEIN CONDENSATE

S. INOUYE*, J.R. ABO-SHAEER, A.P. CHIKKATUR, A. GÖRLITZ, S. GUPTA, T.L. GUSTAVSON, A.E. LEANHARDT, C. RAMAN, T. ROSENBAND, J.M. VOGELS, K. XU, D.E. PRITCHARD, AND W. KETTERLE

Department of Physics, MIT-Harvard Center for Ultracold Atoms and Research Laboratory of Electronics, MIT, Cambridge, MA 02139, USA

We experimentally studied vortex excitations in a dilute gas Bose-Einstein condensate. First, the phase profile of a vortex excitation was observed using an interferometric technique. Vortices were created by moving a laser beam through a condensate, and they were observed as dislocations in the interference fringes formed by the stirred condensate and a second unperturbed condensate. Secondly, we have observed the formation of highly ordered vortex lattices in a rotating Bose-condensed gas. These triangular lattices contained over 100 vortices with lifetimes of several seconds. Individual vortices persisted up to 40 seconds. The gaseous Bose-Einstein condensates can be a model system for the study of vortex matter.

Quantized vortices play a key role in superfluidity and superconductivity. Magnetic fields can penetrate type-II superconductors only as quantized flux lines. Vorticity can enter rotating superfluids only in the form of discrete line defects with quantized circulation. These phenomena are direct consequences of the existence of a macroscopic wavefunction, the phase of which must change by integer multiples of 2π around magnetic flux or vortex lines.

Recently, vortices in a Bose-Einstein condensate have been realized experimentally [1,2,3]. This gives a unique opportunity to study this quantum mechanical phenomenon, since the low density of the gas allows direct comparison with first principle theories. Here, we studied both the static and the dynamical properties of vortex excitations in condensates. The phase profile of a vortex was observed using an interferometric technique [4], and the formation of highly ordered vortex lattices in a rotating condensate was also observed [5].

1 The phase profile of vortex excitations

Bose-Einstein condensate is characterized by a macroscopic wavefunction $\psi(\vec{r}) = \sqrt{\rho(\vec{r})}\exp(i\phi(\vec{r}))$, which satisfies a non-linear Schrödinger equation. The density $\rho(\vec{r})$ and the velocity field $\vec{v}_s(\vec{r})$ in the hydrodynamic equations can now be replaced by the square of the wavefunction ($\rho(\vec{r}) = |\psi(\vec{r})|^2$) and the gradient of the *phase* of the wavefunction

$$\vec{v}_s(\vec{r}) = \frac{\hbar}{m}\nabla\phi(\vec{r}), \qquad (1)$$

where m is the mass of the particle.

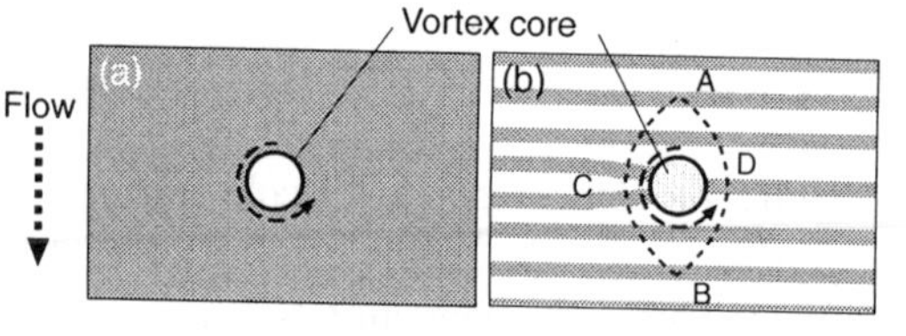

Figure 1. Density (a) and phase (b) profile of a moving condensate with singly-charged ($n = 1$) vortex. The density profile shows the vortex core, whereas the phase pattern features a fork-like dislocation at the position of the vortex. Interference between two initially separated, freely expanding condensates produces exactly the same pattern as shown in (b), if one of the condensate contains a vortex.

The line integral of Eq. (1) around a closed path gives the quantization of circulation:

$$\int \vec{v}(\vec{r}) \cdot d\vec{r} = \frac{\hbar}{m}\left(\phi(\vec{r}_f) - \phi(\vec{r}_i)\right). \qquad (2)$$

If the path is singly connected, there is no circulation. If the path is multiply connected (like around a vortex core) the circulation can take values nh/m (integer multiples of h/m), since the phase is only defined modulo 2π. As a result, the phase accumulated between two points A and B can be different depending on the path (Fig. 1). The integer quantum number n is called the charge of the vortex. When the phase gradient is integrated along a path to the left of the vortex (path ACB), the accumulated phase differs by $2n\pi$ from the path to the right (ADB).

This phase difference can be visualized with interferometric techniques. When two condensates with relative velocity v overlap, the total density shows straight interference fringes with a periodicity h/mv. If one of the condensates contains a vortex of charge n, there are n more fringes on one side of the singularity than on the other side (Fig. 1b). The change in the fringe spacing reflects the velocity field of the vortex. An observation of this fork-like dislocation in the interference fringes is a clear signature of a vortex [6,7].

Our setup for the interferometric observation of vortices is essentially a combination of two experiments conducted in our lab in the past [8,9]. Briefly, laser cooled sodium atoms were loaded into a double-well potential and further cooled by rf-induced evaporation below the BEC transition temperature. The double-well potential was created by adding a potential hill at the center of a cigar-shaped magnetic trap. For this, blue-detuned far off-resonant laser light (532 nm) was focused to form an elliptical $75\,\mu$m $\times$ $12\,\mu$m (FWHM) light sheet and was aligned to the center of the magnetic trap with the long axis of the sheet perpendicular to the long axis of the condensate. The condensates produced in each well were typically $20\,\mu$m in diameter and $100\,\mu$m in length. The height of the optical potential was ~ 3 kHz, which was slightly larger than the chemical potential of the condensate. A more intense light sheet would have increased the distance between the condensates, thus reduced the fringe spacing [8].

After two condensates each containing $\sim 1 \times 10^6$ atoms in the $F = 1, m_F = -1$ state were formed in the double-well potential, we swept a second blue-detuned laser beam through one of the condensates using an acousto-optical deflector (Fig. 2). The focal size of the sweeping laser beam ($12\,\mu$m $\times$ $12\,\mu$m, FWHM) was close to the width of the condensate. The alignment of this beam was therefore done using an expanded condensate in a weaker trap where the beam profile created a circular "hole" in the condensate density distribution.

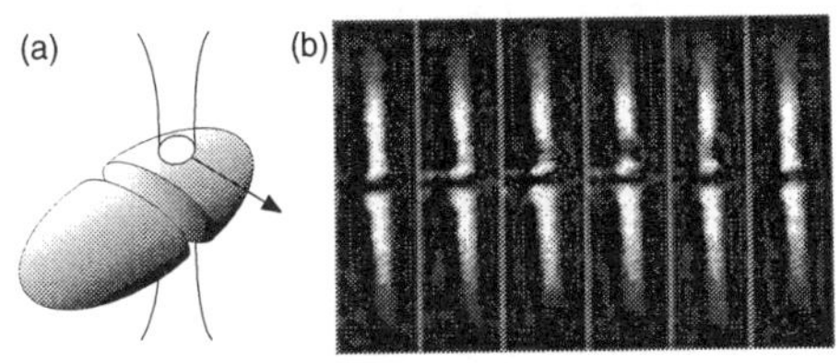

Figure 2. Schematic (a) and phase-contrast images (b) of the condensates used for the experiment. A blue-detuned laser beam (not shown in the figure) was focused into a light sheet separating the two condensates in the magnetic trap. Another tightly focused laser beam was swept through one of the condensates (the upper one in image (b)) to excite vortices. The intensity of each laser beam was a factor of four higher than in the experiments to enhance the depleted regions in the images. The images in (b) have a field of view of $100\,\mu$m $\times$ $380\,\mu$m. Each image was taken with a stationary stirrer, which was advanced from left to right in steps of $5\,\mu$m.

After sweeping the beam once across the "sample" condensate, the magnetic and optical fields were switched off and the two condensates expanded and overlapped during 41 ms time-of-flight. The atoms were then optically pumped into the $F = 2$ hyperfine ground state for $80\,\mu s$ and subsequently probed for $20\,\mu$s by absorption imaging tuned to the $F = 2$ to $F' = 3$ cycling transition.

Images of interfering condensates show a qualita-
tive difference between stirred (Fig. 3(b-d))

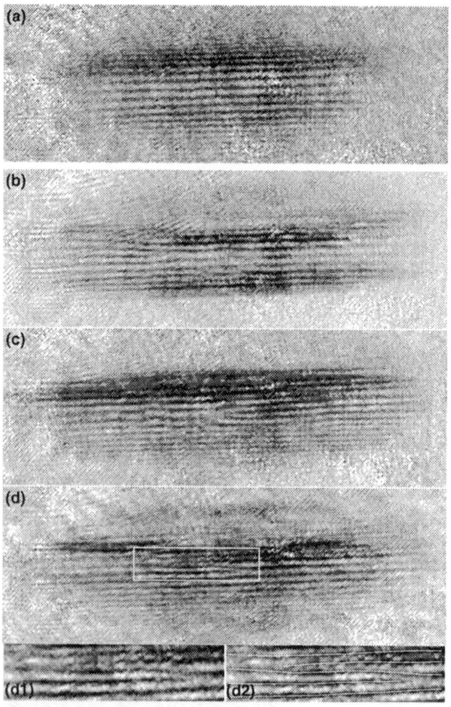

Figure 3. Observation of the phase singularities of vortices created by sweeping a laser beam through a condensate. Without the sweep, straight fringes of $\sim 20\,\mu$m spacings were observed (a), while after the sweep, fork-like dislocations appeared (b-d). The speed of the sweep was $1.1\,\mu$m/ms, corresponding to a Mach number of ~ 0.18. The field of view of each image is 1.1 mm x 0.38 mm. Fig. (d) shows a pair of dislocations with opposite circulation characteristic of a vortex pair. At the bottom, magnified images of the fork-like structures are shown (d1) with lines to guide the eye (d2). The orientation of the condensates is the same as in Fig. 2(b).

and unperturbed states (Fig. 3(a)). Fork-like structures in the fringes were often observed for stirred condensates, whereas unperturbed condensates always showed straight fringes. The charge of the vortices can be determined from the fork-like pattern. In Fig. 3(b), vortices were excited in the condensate on top, and the higher number of fringes on the left side indicates higher relative velocity on this side, corresponding to counterclockwise flow. Fig. 3(c) shows a vortex of opposite charge. The double fork observed in Fig. 3(d) represents the phase pattern of a vortex pair. Multiply charged vortices, which are unstable against the break-up into singly charged vortices, were not observed.

Interferometric techniques have previously been applied either to simple geometries such as trapped or freely expanding condensates [8,10,11], or to read out a phase imprinted by rf- or optical fields [1,12,13]. Here we employed the interferometric technique to visualize turbulent flow. This technique is suited for the study of complicated superfluid flows, e.g., when multiple vortices with opposite charges are present.

2 Vortex lattices

A condensate can be subjected to a rotating perturbation by revolving laser beams around it. This technique was used to study surface waves in a trapped BEC [14], and subsequently for the creation of vortices [2]. In 1997, we tried unsuccessfully to detect quantized circulation as a "centrifugal hole" in ballistic expansion of the gas [6,15]. Theoretical calculations [16,17,18] and ultimately the pioneering experimental work [2] showed that vortices can indeed be detected through ballistic expansion, which magnifies the spatial structure of the trapped condensate.

Experiments were performed in cylindrical traps with widely varying aspect ratios. Most of the results and all of the images were obtained in a weak trap, with radial and axial frequencies of $\nu_r = 84\,$Hz and $\nu_z = 20\,$Hz (aspect ratio 4.2), respectively. In this weak trap inelastic losses were suppressed, resulting in larger condensates of typically 5×10^7 atoms. Such clouds had a chemical potential (μ) of 310 nK (determined from time-of-flight imaging), a peak density of $4.3 \times 10^{14}\,\mathrm{cm}^{-3}$, a Thomas-Fermi radius along the radial direction (R) of $29\,\mu$m, and a healing length (ζ) of about $0.2\,\mu$m.

Vortex lattices were produced by rotating the condensate around its long axis with the optical dipole force exerted by blue-detuned laser beams at a wavelength of 532 nm. A two-axis acousto-optic deflector generated a pattern of two laser beams rotating symmetrically around the condensate at variable drive frequency ν [14]. The two beams were separated by one Gaussian beam waist (w = 25 mm). The laser power of 0.7 mW in each beam corresponded to an optical dipole potential of 115 nK. This yielded a strong, anharmonic deformation of the condensate. After the condensate was produced, the stirring beam power was ramped up over 20 ms, held constant for a variable stirring time, and then ramped down to zero over 20 ms. The condensate equilibrated in the magnetic trap for a variable hold time (typically 500 ms). The trap was then suddenly switched off, and the gas expanded for 35 ms to radial and axial sizes of $l_r > 1000\,\mu$m and $l_z > 600\,\mu$m, respectively. We probed the vortex cores using resonant absorption imaging. To avoid blurring of the images due to bending of the cores near the edges of the condensate, we pumped a thin, 50 to 100 μm slice of atoms in the center of the cloud from the $F = 1$ to the $F = 2$ hyperfine state [8]. This section was then imaged along the axis of rotation with a probe pulse resonant with the cycling $F = 2 \rightarrow 3$ transition. The duration of the pump and probe pulses was chosen to be sufficiently short (50 and 5 μs, respectively) to avoid blurring due to the recoil-induced motion and free fall of the condensate. We observed highly ordered triangular lattices of variable vortex density containing up to 130 vortices (Fig. 4). A striking feature is the extreme regularity of these lattices, free of any major distortions, even near the boundary. Such "Abrikosov" lattices were first predicted for quantized magnetic flux lines in type-II superconductors [19]. Tkachenko showed that their lowest energy structure should be triangular for an infinite system [20]. For a trapped condensate with maximum vortex density, we infer that the distance between the vortices was $\sim 5\,\mu$m. The radial size of the condensate in the time-of-flight images was over 10% larger when it was filled with the maximum number of vortices, probably due to centrifugal forces.

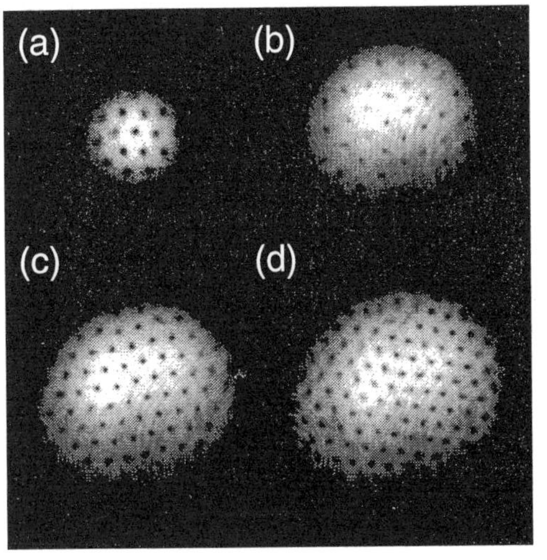

Figure 4. Observation of vortex lattices. The examples shown contain approximately (a) 16, (b) 32, (c) 80, and (d) 130 vortices. The vortices have "crystallized" in a triangular pattern. The diameter of the cloud in (D) was 1 mm after ballistic expansion, which represents a magnification of 20. Slight asymmetries in the density distribution were due to absorption of the optical pumping light.

The vortex lattice had lifetimes of several seconds. The observed stability of vortex arrays in such large condensates is surprising because in previous work the lifetime of vortices markedly decreased with the number of condensed atoms [2]. Theoretical calculations predict a lifetime inversely proportional to the number of vortices [21]. Assuming a temperature $k_BT > \mu$, where k_B is the Boltzmann constant, the predicted decay time of > 100 ms is much shorter than observed. After 10 s, the number of vortices had diminished considerably. In several cases we observed a single vortex near the condensate center after 30 to 40 s. This dwell time is much longer than that observed for elongated clouds (1 to 2 s) [2] and for nearly spherical condensates (15 s) [22]. We estimate that during its lifetime, the super-fluid flow field near the central vortex core had completed more than 500,000 revolutions and the lattice itself had rotated ~ 100 times.

Our experiments show that vortex formation and self-assembly into a regular lattice is a robust feature of rotating BECs. Gaseous condensates may serve as a model system to study the dynamics of vortex matter. Of particular interest are collective modes of the lattice. In liquid helium, transverse oscillations in a vortex lattice (Tkachenko oscillations) have already been investigated [23,24]. Further studies may address the nucleation, ordering, and decay of lattices, in particular to delineate the role of the thermal component [21], and possible phase transition associated with melting and crystallization.

Acknowledgments

We thank J.R. Anglin, D. Feder, R. Onofrio, and L. Levitov for useful discussions. This work was supported by NSF, the ONR, ARO, NASA, and the David and Lucile Packard Foundation. A.E.L. and A.P.C. acknowledge additional support by fellowships from NSF and JSEP, respectively.

* Present address: JILA, University of Colorado, Boulder, CO 80309-0440, USA

References

1. M.R. Matthews *et al.*, Phys. Rev. Lett. **83**, 2498 (1999).
2. K.W.Madison, F.Chevy, W. Wohlleben, and J. Dalibard, Phys. Rev. Lett. **84**, 806 (2000).
3. B.P. Anderson *et al.*, Phys. Rev. Lett. **86**, 2926 (2001).
4. S. Inouye *et al.*, Phys. Rev. Lett. **87**, 080402 (2001).
5. J.R. Abo-Shaeer, C. Raman, J.M. Vogels, and W. Ketterle, Science **292**, 476 (2001).
6. M.R.Andrews, Ph.D. Thesis, MIT (1998).
7. E.L. Bolda and D.F. Walls, Phys. Rev. Lett. **81**, 5477 (1998); J. Tempere and J.T. Devreese, Solid State Commun. **108**, 993 (1998).
8. M.R. Andrews *et al.*, Science **275**, 637 (1997).
9. C. Raman *et al.*, Phys. Rev. Lett. **83** 2502 (1999).
10. J.E. Simsarian *et al.*, Phys. Rev. Lett. **85**, 2040 (2000).
11. I. Bloch, T. W. Hänsch, and T. Esslinger, Nature **403**, 166 (2000).
12. D.S. Hall, M.R. Matthews, C.E. Wieman, and E.A. Cornell, Phys. Rev. Lett. **81**, 1543 (1998).
13. J. Denschlag *et al.*, Science **287**, 97 (2000).
14. R. Onofrio *et al.*, Phys. Rev. Lett. **84**, 810 (2000).
15. Reviewed in W. Ketterle, D. S. Durfee, D. M. Stamper-Kurn, in Bose-Einstein Condensation in Atomic Gases, Proceedings of the International School of Physics Enrico Fermi, Course CXL, M. Inguscio, S. Stringari, C. Wieman, Eds. (International Organisa- tions Services B.V, Amsterdam, 1999), pp. 67-176.
16. E. Lundh, C. J. Pethick, H. Smith, Phys. Rev. A **58**, 4816 (1998).
17. F. Dalfovo, M. Modugno, Phys. Rev. A **61**, 023605 (2000).
18. Y. Castin, R. Dum, Eur. Phys. J. D **7**, 399 (1999).
19. A. A. Abrikosov, J. Exp. Theor. Phys. **5**, 1174 (1957) [Zh. Eksp. Teor. Fiz. **32**, 1442 (1957)].
20. V. K. Tkachenko, J. Exp. Theor. Phys. 22, 1282 (1966) [Zh. Eksp. Teor. Fiz. 49, 1875 (1965)].
21. P.O. Fedichev and A.E. Muryshev, e-print cond-mat/0004264.
22. P. C. Haljan, B. P. Anderson, I. Coddington, and E. A. Cornell, Phys. Rev.

Lett. **86**, 2922 (2001).
23. V. K. Tkachenko, J. Exp. Theor. Phys. 23, 1049 (1966) [Zh. Eksp. Teor. Fiz. 50, 1573 (1966)].
24. S. J. Tsakadze, Fiz. Nizk. Temp. 4, 148 (1978) [Sov. J. Low Temp. Phys. 4, 72 (1978)].

COLLAPSING DYNAMICS OF TRAPPED BOSE-EINSTEIN CONDENSATES WITH ATTRACTIVE INTERACTIONS

HIROKI SAITO AND MASAHITO UEDA

Department of Physics, Tokyo Institute of Technology, Tokyo 152-8551, Japan
E-mail: hsaito@stat.phys.titech.ac.jp

The dynamics of collapsing and exploding trapped Bose-Einstein condensates triggered by switching the sign of the interaction from repulsive to attractive is studied by numerically integrating the Gross-Pitaevskii equation with atomic loss. Our simulations reproduce some experimental results, and predict new phenomena.

1 Introduction

The static and dynamical properties of Bose-Einstein condensate (BEC) of trapped atomic vapor crucially depend on the sign of the interatomic interaction. When the interaction is attractive, BEC in a spatially uniform 3D system is unstable to collapse into a denser phase. In a spatially confined system, on the other hand, the quantum pressure arising from Heisenberg's uncertainty principle counterbalances the attractive force, allowing metastable BEC to be formed if the number of BEC atoms is below a certain critical number (typically ~ 1000) above which BEC collapses[1]. The BEC with attractive interactions, therefore, has been restricted to a small and uncontrollable number of atoms[2]

In recent experiments performed at JILA[3,4], the interaction was controlled in both strength and sign using the technique of the Feshbach resonance. Using this technique, almost pure condensates with large number of atoms can be obtained, since they are prepared in the repulsive regime, and then the interaction is switched to attractive. This technique enabled us to study the collapse of BEC with large number of atoms, which exhibits explosive atomic ejection from collapsing BEC. This phenomenon is called "Bosenova", and its theoretical explanation is controversial[5,6,7].

In this paper, we study the collapsing dynamics of BEC in such situation by numerically integrating the Gross-Pitaevskii (GP) equation with atomic loss. We show that our simulations reproduce the "Bosenova" and some accompanying phenomena observed in the experiments[4], suggesting that our method is at least qualitatively valid. We find that the collapse is not a single and collective one but a sequence of localized implosions that occur rapidly and intermittently. We also predict pattern formation in the atomic density caused by the switch of the sign of the interaction.

2 The Gross-Pitaevskii equation with atomic loss

We consider a system of Bose-condensed atoms with mass m and s-wave scattering length a, confined in a trap potential $V_{\rm trap}$. We assume that the mean-field or GP approximation is applicable, and employ the GP equation with atomic loss processes as[5]

$$i\hbar\frac{\partial}{\partial t}\psi = -\frac{\hbar^2}{2m}\nabla^2\psi + V_{\rm trap}(\mathbf{r})\psi + \frac{4\pi\hbar^2 a}{m}|\psi|^2\psi - \frac{i\hbar}{2}K_3|\psi|^4\psi, \quad (1)$$

where K_3 is the loss-rate coefficient of three-body recombination (including the Bose-statistical factor 1/6). The last term of the right-hand side of Eq. (1) describes atomic loss in proportion to cube of the density. We assume that the high energy (~ 1 mK) atoms and molecules produced by three-body recombination escape from the trap without affecting the condensate. The values of K_3 far from the Feshbach resonance are measured for several atomic species, which agree with the theoretical predictions within a factor of ten. However, just near the Feshbach resonance, behavior of K_3 is not well understood, and no precise values are available[8].

3 Numerical analysis of collapsing dynamics

We performed numerical integration of the GP equation (1) using the method in Ref. [1]. Initially we prepared the ground-state wave function for the s-wave scattering length $a_{\rm init} \geq 0$, and at $t = 0$ the interaction is suddenly switched to attractive $a_{\rm collapse} < 0$. For simplicity, we used the spherically symmetric trap $V_{\rm trap} = m\omega^2 r^2/2$ with $\omega = 12.8 \times 2\pi$ Hz, which is the geometric mean frequency of the cigar-shape trap in the experiments[3,4].

3.1 Intermittent implosion and atomic decay

Figure 1 shows time evolution of the peak height $|\psi(r = 0)|$ of the wave function, the total number of atoms in the trap $N_{\rm total}$, and the number of BEC atoms left at the center of the trap $N_{\rm center} \equiv \int_0^R 4\pi r^2 |\psi|^2 dr$, where we adopt $R = 6(\hbar/m\omega)^{1/2}$. The initial number of BEC atoms (^{85}Rb) is $N_0 = 6000$, and $a_{\rm init} = 0$ is switched to $a_{\rm collapse} = -30a_0$ at $t = 0$, where a_0 is the Bohr radius. The three-body recombination rate is assumed to be $K_3 = 10^{-28}$ cm^6/s.

The striking feature in Fig. 1 is that the sequence of rapid implosions occurs over ~ 10 ms intermittently[6]. The implosion occurs very rapidly ($\lesssim 0.1$ ms) within a very localized region ($\lesssim 0.1$ μm). This behavior is explained as follows: When the condensate shrinks towards the central region and the peak density becomes $4\pi\hbar^2 a|\psi|^2/m \sim \hbar K_3|\psi|^4/2$, i.e., $|\psi|^2 \sim 8\pi\hbar a/(mK_3)$, the collisional loss rate becomes comparable to the accumulation rate of atoms at the center. Since the kinetic and interaction energies are proportional to the atomic density and its square, respectively, the total energy increases upon the loss of atoms. The atoms near the center of BEC thus acquire outward momentum, and are ejected as atomic burst. After the implosion and explosion, inward flow outside the region of the implosion replenishes the peak density, inducing the subsequent implosion.

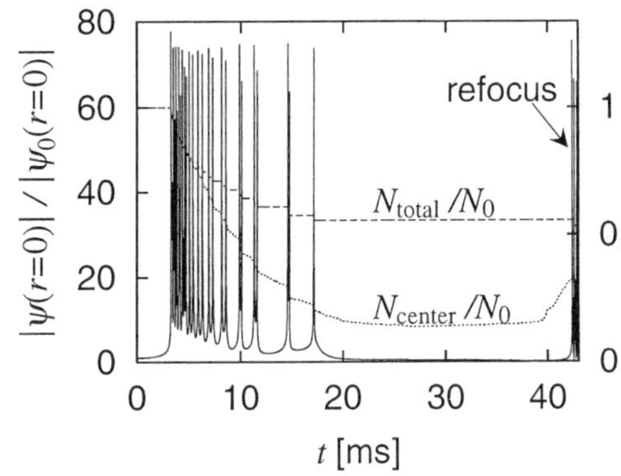

Figure 1. Time evolution of the peak height of the wave function normalized by the initial one $|\psi(r = 0)|/|\psi_0(r = 0)|$ (referring the left axis), the fraction of the number of atoms in the trap $N_{\rm total}/N_0$, and the fraction around the center of the trap $N_{\rm center}/N_0$ (referring the right axis), where the initial number of ^{85}Rb BEC atoms is $N_0 = 6000$. At $t = 0$ the s-wave scattering length is changed from 0 to $-30a_0$, where a_0 is the Bohr radius.

The number of atoms in the central BEC $N_{\rm center}$ remains constant for some time and suddenly decays, since the implosions suddenly begin. This duration in which $N_{\rm center}$ is constant decreases with increasing $|a_{\rm collapse}|$, and our numerical results quantitatively agree with those of the experiment[6]. In each implosion, several tens of atoms are lost from the trap due to three-body recombination, and then $N_{\rm total}$ decreases in a step-like manner. The number of BEC atoms remaining at the center of the trap $N_{\rm center}$ gradually decays in addition to the step-like decrease because of the ejection of the atomic burst, where the number of burst atoms is given by $N_{\rm total} - N_{\rm center}$. The behavior of

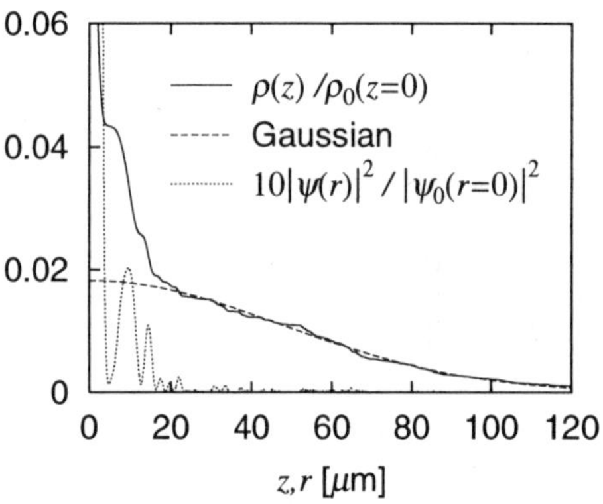

Figure 2. The one-dimensional density distribution $\rho(z)/\rho_0(z=0)$ (solid curve) and the radial density $10|\psi(r)|^2/|\psi_0(r=0)|^2$ (dotted curve) at $t = 30$ ms in the same condition as Fig. 1, where $\rho(z) \equiv \int |\psi|^2 dxdy$, and $\rho_0(z=0)$ and $|\psi_0(r=0)|^2$ are the initial ones at the center of the trap. The dashed curve shows the Gaussian fitted to the tail ($z \geq 20$ μm) of the solid curve.

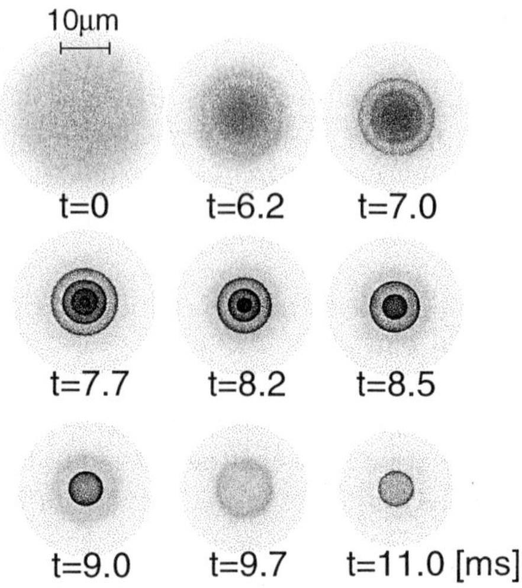

Figure 3. Time evolution of the column density of ^{85}Rb BEC, where the initial number of atoms is $N_0 = 5 \times 10^4$, and at $t = 0$ the s-wave scattering length is switched from $a_{\rm init} = 400a_0$ to $a_{\rm collapse} = -310a_0$, where a_0 is the Bohr radius. The three-body recombination rate is $K_3 = 8 \times 10^{-26}\text{cm}^6/\text{s}$.

$N_{\rm center}$ shown in Fig. 1 well describes the experimental observation[4], suggesting that the intermittent implosion occurs in the experiments.

3.2 Atomic burst

In Fig. 1, implosions revive due to the refocus of the burst, which occurs at the center of the trap in the spherically symmetric trap. Since the cigar-shape trap is used in the experiments[4], on the other hand, the axial and radial refocuses occur alternately. When the radial refocus occurs, the burst cloud becomes like a stripe along the trap axis, and this image was taken in the experiments[4]. In order to compare this burst distribution with our result, we show the one-dimensional density distribution $\rho(z) \equiv \int |\psi|^2 dxdy$ in Fig. 2, where the data was taken after the sequence of implosions ($t = 30$ ms in Fig. 1). The curve of $\rho(z)$ is very similar to that of the experiment[4], in that there is a distinction between the burst and remnant atoms, and the burst part is well fitted by the Gaussian (dashed curve in Fig. 2). This indicates that the burst energy can be described by "temperature", since the thermal distribution is given by $\exp[-m\omega^2 z^2/(2k_B T)]$. In the case of Fig. 2, the temperature of the burst atoms is estimated to be $T \simeq 160$ nK. There are shoulders in the remnant part of $\rho(z)$, which originate from the fringe in the radial density as shown in the dotted curve in Fig. 2.

3.3 Pattern formation

We next consider the case of larger N_0, $a_{\rm init}$, and $|a_{\rm collapse}|$ than the current experimental parameters. Figure 3 shows time evolution of column density of ^{85}Rb BEC, where $N_0 = 5 \times 10^4$, and the s-wave scattering length is switched from $a_{\rm init} = 400a_0$ to $a_{\rm collapse} = -310a_0$. The three-body recombination rate is $K_3 = 8 \times 10^{-26}$ cm^6/s at this $a_{\rm collapse}$[8]. When the interaction is switched to attractive, the condensate expanded by the repulsive interaction begins to shrink, and ripples in the atomic density arise. These ripples become pronounced, and grow to the concentric spherical shells at $t = 7.7$ ms in Fig. 3. The ripples are caused by the interference between the atoms going inward and the inside atoms, and the growth of the ripples is attributed to the self-focusing effect due to the attractive interaction. The shells go inward and collapse one by one at the center of the trap. After the all shells collapse, another shell is formed from the remaining BEC at $t = 11$ ms in Fig. 3. We note that such shell structure should be distinguished from the density oscillation as shown in Fig. 2 (dotted curve). The oscillation in Fig. 2 arises as a result of violent implosion and explosion, while the shell-structure formation occurs in the process of shrinking.

In the case of an axi-symmetric trap, the pattern of the atomic density is sensitive to the asymmetry of the trap[6]. For a cigar-shape trap, compression of BEC by change in the sign of the interaction is larger in the radial direction than in the axial direction. Hence, the ripples arise in the radial direction, which results in a cylindrical shell structure. For a pancake-shape trap, the ripples arise in the axial direction, leading to a layered structure. Once the implosion occurs, various patterns such as rings and clusters can be formed in axi-symmetric systems. In our preliminary simulations for a cigar-shape trap, atomic clusters are formed along the trap axis, which suggests the occurrence of the multiple implosions. This might be related with the multiple jets observed in the experiment[4], which are manifestations of local spikes in the atomic density formed along the trap axis.

The resolution limit of the imaging system in the experiment[4] is 7 μm FWHM, and then somewhat inadequate for observing the patterns. Since expansion of BEC before imaging will blur out the patterns, improvement of the *in situ* imaging method or enlargement of patterns by use of larger size of BEC is required to observe the pattern formation.

4 Conclusions

We studied the collapsing and exploding dynamics of BEC by numerically integrating the GP equation with atomic loss. We showed that the GP equation describes the experimental results[4] well, and found two new phenomena: intermittent implosion and pattern formation. We believe that the intermittent implosion is the origin of the experimental observations, such as exponential decay of BEC atoms and burst production.

Acknowledgments

This work was supported by a Grant-in-Aid for Scientific Research (Grant No. 11216204) by the Ministry of Education, Science, Sports, and Culture of Japan, and by the Toray Science Foundation.

References

1. P. A. Ruprecht, M. J. Holland, K. Burnett, and M. Edwards, Phys. Rev. A **51**, 4704 (1995).
2. J. M. Gerton, D. Strekalov, I. Prodan, and R. G. Hulet, Nature **408**, 692 (2000).
3. S. L. Cornish, N. R. Claussen, J. L. Roberts, E. A. Cornell, and C. E. Wieman, Phys. Rev. Lett. **85**, 1795 (2000).
4. E. A. Donley, N. R. Claussen, S. L. Cornish, J. L. Roberts, E. A. Cornell, and C. E. Wieman, Nature **412**, 295 (2001).
5. Yu. Kagan, A. E. Muryshev, and G. V. Shlyapnikov, Phys. Rev. Lett. **81**, 933 (1998).
6. H. Saito and M. Ueda, Phys. Rev. Lett. **86**, 1406 (2001); Phys. Rev. A **63**, 043601 (2001); cond-mat/0107248.
7. R. A. Duine and H. T. C. Stoof, Phys. Rev. Lett. **86**, 2204 (2001).
8. J. L. Roberts, N. R. Claussen, S. L. Cornish, and C. E. Wieman, Phys. Rev. Lett. **85**, 728 (2000).

CRITERION FOR BOSE-EINSTEIN CONDENSATION IN A TRAP

HIROSHI EZAWA
Department of Physics, Gakushuin University, Mejiro, Toshima-ku, Tokyo 171-8588, Japan
E-mail: hiroshi.ezawa@gakushuin.ac.jp

KEIJI WATANABE
Department of Physics, Meisei University, Hino, Tokyo 191-8506, Japan
E-mail: watanabk@phys.meisei-u.ac.jp

AND

KOICHI NAKAMURA
Division of Natural Science, Meiji University, Izumi Campus, Eifuku, Suginami-ku, Tokyo 168-8555, Japan
E-mail: knakam@isc.meiji.ac.jp

With modified Bogoliubov replacement $a_0^\# \to \sqrt{N_0} + a_0^\#$, we study the fluctuation in number of the trapped condensate particles around the average $N_0 \gg 1$, taking interaction among them into account. A criterion is obtained for the stability of the fluctuation, thus also for the Bose-Einstein condensation. In the case of weak interaction, the condensation takes place if the interaction is in a sense repulsive, and probably the condensation does not take place if attractive.

1 Introduction

It is well-known that the *free* Bose gas condenses in momentum space at sufficiently low temperature and 'high' (actually low) density such that $\rho\lambda^3 > 2.162$, where $\lambda = \sqrt{2\pi\hbar^2/mk_BT}$. Then, what if the *interactions* among particles are present?

It is customary to treat the Bose-Einstein condensate by replacing its annihilation/creation operators by a c-number, $\sqrt{N_0}$, with N_0 being its average occupation number (Bogoliubov approximation[1]) violating the canonical commutation relations. Under what conditions is this approximation valid?

We address ourselves to these two questions in the case of the Bose particles in a trap. In 1965, one of the authors (H.E.[2, 3]) proposed to use $\sqrt{N_0} + a_0^\#$ for the annihilation/creation operators of the condensate mode with operators satisfying $[a_0, a_0^\dagger] = 1$, and examine under what conditions the fluctuations of these variables $a_0^\#$ remain small in comparison with $\sqrt{N_0}$. But, in the infinite volume case, the Goldstone theorem prohibits the energy gap at zero momentum, making the fluctuation of $a_0^\#$ infinitely large, so that we were confronted with a difficult task of estimating the finite volume effects. We now consider the same problem in the case where the Bose system is trapped in a finite volume.

For the Bose particles in a trap and at sufficiently low temperature and 'high' density, we shall show that, roughly speaking, if the interaction $V(\boldsymbol{x})$ among particles is repulsive (in a sense to be specified in the final section), then $a_0^\#$ are effectively of the order of $N_0^{1/4}$, which is much smaller than $N_0^{1/2}$ when $N_0 \gg 1$, justifying the Bogoliubov approximation. It then follows that the Bose-Einstein condensation takes place. If on the contrary the interaction is attractive, the effective Hamiltonian we construct for $a_0^{\#}$ has no lowest energy eigenstate, implying that the the Bogoliubov approximation invalid. It seems in this connection that the failure of the Bogoliubov approximation implies the failure of the Bose-Einstein condensation, since if the condensation would take place then seemingly the Bogoliubov approximation should work. If this is the case, then the criterion for the validity of the Bogoliubov approximation is at the same time the criterion for the Bose-Einstein condensation. Some further discussions will be given in the final section.

2 Hamiltonian

In terms of the Bose field operator $\phi_A(\boldsymbol{x})$, the Hamiltonian of the system is given by

$$\begin{aligned}\mathcal{H} &= \int d\boldsymbol{x}\, \phi_A^\dagger(\boldsymbol{x})\Big\{ -\frac{\hbar^2}{2M}\Delta + v(\boldsymbol{x}) - \mu\Big\}\phi_A(\boldsymbol{x}) \\ &+\frac{1}{2}\int d\boldsymbol{x}\, d\boldsymbol{x}' \phi_A^\dagger(\boldsymbol{x})\phi_A^\dagger(\boldsymbol{x}')V(\boldsymbol{x}-\boldsymbol{x}') \\ &\times\phi_A(\boldsymbol{x}')\phi_A(\boldsymbol{x}), \end{aligned} \tag{1}$$

where $v(\boldsymbol{x})$ is the trap potential and $V(\boldsymbol{x}) = V(-\boldsymbol{x})$ the interaction, the former varying much more slowly than the latter and μ is the chemical potential.

Suppose tentatively that the system has $N_0 \gg 1$ condensate particles in average in the lowest single particle mode $u_0(\boldsymbol{x})$ to suggest the use of $\sqrt{N_0}+a_0^{\#}$ for the mode. Then, the field operator takes the form,

$$\phi_A(\boldsymbol{x}) = \sqrt{N_0}u_0(\boldsymbol{x}) + \phi(\boldsymbol{x}). \tag{2}$$

where

$$\phi(\boldsymbol{x}) = \sum_n a_n u_n(\boldsymbol{x}). \tag{3}$$

with a complete set of orthonormal system $\{u_n(\boldsymbol{x})\}$.

If we put (2) in (1), then terms linear in $\phi(\boldsymbol{x})$ arise, which can be eliminated by introducing the Hartree potential $v_{\rm H}$ as defined below. Namely, we rewrite (1) as

$$\begin{aligned}\mathcal{H} = &\int d\boldsymbol{x}\, \phi_{\rm A}^\dagger(\boldsymbol{x})\{-\frac{\hbar^2}{2M}\Delta + v(\boldsymbol{x}) + v_{\rm H}(\boldsymbol{x}) \\ &-\mu\}\phi_{\rm A}(\boldsymbol{x}) + \frac{1}{2}\int d\boldsymbol{x}\, d\boldsymbol{x}'\phi_{\rm A}^\dagger(\boldsymbol{x})\phi_{\rm A}^\dagger(\boldsymbol{x}') \\ &\times V(\boldsymbol{x}-\boldsymbol{x}')\phi_{\rm A}(\boldsymbol{x})\phi_{\rm A}(\boldsymbol{x}') \\ &-\int d\boldsymbol{x}\, \phi_{\rm A}^\dagger(\boldsymbol{x})v_{\rm H}(\boldsymbol{x})\phi_{\rm A}(\boldsymbol{x}) \end{aligned}\tag{4}$$

by adding and subtracting the term with the Hartree potential,

$$v_{\rm H}(\boldsymbol{x}) := N_0\int V(\boldsymbol{x}-\boldsymbol{x}')u_0^2(\boldsymbol{x}')d\boldsymbol{x}'. \tag{5}$$

For u_n's ($n \geq 0$), we take the solutions to

$$\begin{aligned}&\Big\{ -\frac{\hbar^2}{2M}\Delta + v(\boldsymbol{x}) \\ &+N_0\int d\boldsymbol{x}'V(\boldsymbol{x}-\boldsymbol{x}')u_0^2(\boldsymbol{x}')\Big\}u_n(\boldsymbol{x}) \\ &= \varepsilon_n u_n(\boldsymbol{x}). \end{aligned}\tag{6}$$

We note that (6) is different from the Gross-Pitaevski equation[4] in that it gives a complete orthonormal set $\Big\{u_n(\boldsymbol{x}) \;\Big|\; n = 0,\cdots\Big\}$. After putting (2) in (4), the Hamiltonian are arranged in powers of $\lambda = 1/\sqrt{N_0}$:

$$\mathcal{H} = \lambda^{-1}\mathcal{H}^{(-1)} + \mathcal{H}^{(0)} + \lambda\mathcal{H}^{(1)} + \lambda^2\mathcal{H}^{(2)}, \tag{7}$$

and the chemical potential is expanded accordingly,

$$\mu = \mu^{(0)} + \lambda\mu^{(1)} + \cdots. \tag{8}$$

We have

$$\mathcal{H}^{(-1)} = (\epsilon_0 - \mu^{(0)})(a_0^\dagger + a_0), \tag{9}$$

$$\begin{aligned}\mathcal{H}^{(0)} &= \sum_n(\epsilon_n - \mu^{(0)})a_n^\dagger a_n \\ &+\frac{1}{2}\sum_{mn}{}' J_{mn}(a_m^\dagger + a_m)(a_n^\dagger + a_n) \\ &+\sum_n{}' J_{n0}(a_n^\dagger + a_n)(a_0^\dagger + a_0) \\ &+\frac{1}{2}J_{00}(a_0^\dagger + a_0)^2 - \mu^{(1)}(a_0^\dagger + a_0), \end{aligned}\tag{10}$$

and the terms $\mathcal{H}^{(1)}$ and $\mathcal{H}^{(2)}$, cubic and quartic respectively, in $a_0^{\#}$ and $a_k^{\#}$, where the prime on the summation symbol $\sum_n{}'$, say, means excluding $n = 0$-term from the sum, and

$$J_{mn} = N_0 \int d\boldsymbol{x}\, d\boldsymbol{x}' u_m(\boldsymbol{x}) u_n(\boldsymbol{x}') \times V(\boldsymbol{x} - \boldsymbol{x}') u_0(\boldsymbol{x}') u_0(\boldsymbol{x}). \quad (11)$$

Since u_n's are normalized, it is of the order of (volume of the system)$^{-1/2}$, so that

$$J_{mn} = O\left(\frac{N_0}{\text{volume}}\right) = O(\text{number density of the condensate}),$$

which we assume to be of $O(1)$ with respect to $N_0 \to \infty$.

3 Diagonalizing the Non-Condensate Part of $\mathcal{H}^{(0)}$

In the following, we shall construct an effective Hamiltonian for the condensate fluctuation by a perturbation theory. At each order j of the perturbation calculation, the chemical potential $\mu^{(j)}$ is determined such that the terms linear in $a_0^\dagger + a_0$ ($= b_0^\dagger + b_0$ in terms of the operators (15) to be introduced later) are removed. So, in the lowest order, (9) gives

$$\mu^{(0)} = \epsilon_0, \quad (12)$$

so that

$$\mathcal{H}^{(-1)} = 0. \quad (13)$$

We diagonalize $\mathcal{H}^{(0)}$ with respect to the non-condensate ($k \neq 0$) modes to

$$\mathcal{H}^{(0)} = \sum_k{}' \omega_k b_k^\dagger b_k + \frac{1}{2} J_0 (b_0^\dagger + b_0)^2 - \mu^{(1)} (b_0^\dagger + b_0) \quad (14)$$

by a transformation,

$$a_n = \sum_k{}' (c_{kn} b_k - s_{kn} b_k^\dagger) - \alpha_n (b_0^\dagger + b_0)$$
$$a_0 = b_0 - \sum_k{}' \beta_k (b_k^\dagger - b_k), \quad (15)$$

where

$$c_{kn} = \frac{1}{2}\left(\sqrt{\frac{\omega_k}{\epsilon_n}} + \sqrt{\frac{\epsilon_n}{\omega_k}}\right) T_{kn}$$
$$s_{kn} = \frac{1}{2}\left(\sqrt{\frac{\omega_k}{\epsilon_n}} - \sqrt{\frac{\epsilon_n}{\omega_k}}\right) T_{kn}$$
$$\alpha_n = \frac{1}{2}\sqrt{\epsilon_n} \sum_k{}' \frac{\lambda_k}{\omega_k^2} T_{kn} \quad (16)$$
$$\beta_k = \frac{1}{2} \frac{\lambda_k}{\omega_k^{3/2}},$$

with

$$\lambda_k = 2 \sum_n{}' T_{kn} \sqrt{\epsilon_n} J_{n0} \quad (17)$$

and an orthogonal matrix T_{km} satisfying

$$\sum_{mn}{}' T_{km} (\epsilon_n^2 \delta_{mn} + 2\sqrt{\epsilon_m} J_{mn} \sqrt{\epsilon_n}) T_{ln} = \omega_k^2 \delta_{kl}. \quad (18)$$

The third term in (14), linear in $b_0^\dagger + b_0$ is removed by putting

$$\mu^{(1)} = 0. \quad (19)$$

Then, we rewrite the Hamiltonian (7) in terms of $b_k^\#$'s, and regroup the terms in the following way:

$$\mathcal{H} = \mathcal{H}_{\rm B}^{(0)}(b_k^\#) + \mathcal{H}_{\rm C}^{(0)}(b_0^\#) + \lambda \mathcal{H}^{(1)}(b_k^\#, b_0^\#) + \lambda^2 \mathcal{H}^{(2)}(b_k^\#, b_0^\#) \quad (20)$$

where $b_k^\#$ stands for those with $k \neq 0$, and $\mathcal{H}_{\rm B}^{(0)}$ and $\mathcal{H}_{\rm C}^{(0)}$ are given by the first term and the second term in (14), respectively.

Let the eigenstates of $\mathcal{H}_B^{(0)}$ be $|n\rangle$.

4 Effective Hamiltonian for the Condensate Fluctuation

To construct the effective Hamiltonian for the condensate fluctuation, we follow Ezawa and Luban[3].

Let the Hilbert space of the condensate ($k = 0$) be $\mathsf{H}_{\rm C}$ and the Hilbert space of the non-condensate ($k \neq 0$) be $\mathsf{H}_{\rm B}$, then the Hilbert space of the total system is given by $\mathsf{H}_{\rm B} \otimes \mathsf{H}_{\rm C}$.

We solve the eigenvalue problem in $\mathsf{H}_{\rm B}$,

$$\mathcal{H}\psi_n = \psi_n \Lambda_n, \quad (21)$$

by perturbation theory as

$$\psi_n = (1 + \lambda\mathcal{U}_n^{(1)} + \lambda^2\mathcal{U}^{(2)} + \cdots)|n\rangle,$$
$$\Lambda_n = \Lambda_n^{(0)} + \lambda\Lambda_n^{(1)} + \lambda^2\Lambda_n^{(2)} + \cdots, \qquad (22)$$

where $\mathcal{U}_n^{(j)}$'s are operators on $\mathsf{H}_\mathrm{B}\otimes\mathsf{H}_\mathrm{C}$ involving both $b_0^\#$ and $b_k^\#$, and ψ_n a state vector in H_B and operator on H_C involving $b_0^\#$.

Λ_n, an operator on H_C, is the effective Hamiltonian for the condensate fluctuation for the case where the unperturbed state of the non-condensate is $|n\rangle$. If $|\chi_{nl}\rangle$ is the eigenstate of Λ_n with eigenvalue E_{nl}, then $\psi_n|\chi_{nl}\rangle$ is the eiegenstate of the total Hamiltonian. In fact, by (21),

$$\mathcal{H}\psi_n|\chi_n\rangle = \psi_n\Lambda_n|\chi_{nl}\rangle = E_{nl}\psi_n|\chi_{nl}\rangle. \quad (23)$$

The reaction of the condensate fluctuation on the state of the non-condensate is included in $\psi_n|\chi_{nl}\rangle$ since ψ_n is a state vector in H_B and at the same time an operator on H_C.

Note that ψ_n and Λ_n are not unique, as

$$\psi_n' = \psi_n\mathcal{V}, \qquad \Lambda_n' = \mathcal{V}^\dagger\Lambda\mathcal{V} \qquad (24)$$

can also satisfy the equation, where $\mathcal{V}$ is any unitary operator on H_C. It can be shown that, using this freedom, we can require that

$$\langle n|\mathcal{U}^{(1)}|n\rangle = 0, \qquad (25)$$

and

$$\langle n|\mathcal{U}^{(2)}|n\rangle = \langle n|\mathcal{U}^{(2)\dagger}|n\rangle. \qquad (26)$$

Then, we find

$$\Lambda_n^{(0)} = \epsilon_n + \mathcal{H}_C^{(0)}$$
$$\Lambda_n^{(1)} = \langle n|\mathcal{H}^{(1)}|n\rangle \qquad (27)$$
$$\Lambda_n^{(2)} = \frac{1}{2}\sum_m{}' \left(\langle n|\mathcal{H}^{(1)}|m\rangle\langle m|\mathcal{U}_n^{(1)}|n\rangle + \text{h.c.}\right) + \langle n|\mathcal{H}^{(2)}|n\rangle$$

where $\langle m|\mathcal{U}^{(1)}|n\rangle$ is determined from

$$(\epsilon_n - \epsilon_m)\langle m|\mathcal{U}^{(1)}|n\rangle = \langle m|\mathcal{H}^{(1)}|n\rangle + [\mathcal{H}_C^{(0)}, \langle m|\mathcal{U}^{(1)}|n\rangle]. \quad (28)$$

When $|n\rangle$ is the ground state $|0\rangle$ of $\mathcal{H}_\mathrm{B}^{(0)}$, we find

$$\Lambda_0 = A\left\{\xi^2 + \frac{1}{\nu}(\xi\eta^2 + \eta^2\xi) + \frac{\eta^4}{\nu^2}\right\} + \frac{B_2}{\nu^2}\xi^2 + \frac{B_3}{\nu}\xi^3 + \frac{B_4}{\nu^2}\xi^4 + \frac{C}{\nu^2}\eta\xi^2\eta + O\left(\frac{1}{\nu^3}\right), \qquad (29)$$

where

$$\xi = \frac{1}{\sqrt{2}}(b_0^\dagger + b_0), \quad \eta = \frac{i}{\sqrt{2}}(b_0^\dagger - b_0), \quad (30)$$

which satisfy

$$[\eta, \xi] = -i, \qquad (31)$$

and

$$A = J_{00} - 2\sum_n{}' J_{n0}\alpha_n \qquad (32)$$

with J_{mn} and α_n being given in (11) and (16), respectively, and $B_2, \cdots, C$ are also c-numbers of $O(1)$ with respect to

$$\nu = 2\sqrt{2N_0}, \qquad (33)$$

and are related to the matrix elements like J_{mn} of the interaction potential. Miraculously, Λ has no η^2 term.

By a unitary transformation U, and change of variable by

$$\eta = \frac{\nu}{\rho}\sinh\frac{\rho x}{\sqrt{\nu}}, \qquad (34)$$

we have

$$U^\dagger\Lambda U = \frac{A'}{\nu}\left[-\frac{d^2}{dx^2} + \nu V_c(\sqrt{\nu}x)\right], \qquad (35)$$

where

$$\nu V_c(\sqrt{\nu}x) = \frac{\rho^2}{4\nu\cosh^2(\rho x/\sqrt{\nu})} + \frac{\nu^3}{\rho^2}\frac{\sinh^6(\rho x/\sqrt{\nu})}{\cosh^2(\rho x/\sqrt{\nu})} + \frac{\rho^2}{4\nu}, \qquad (36)$$

$$A' = A + O\left(\frac{1}{\nu^2}\right), \qquad (37)$$

and

$$\rho^2 = \frac{\sigma + 3\kappa_3}{A} \qquad (38)$$

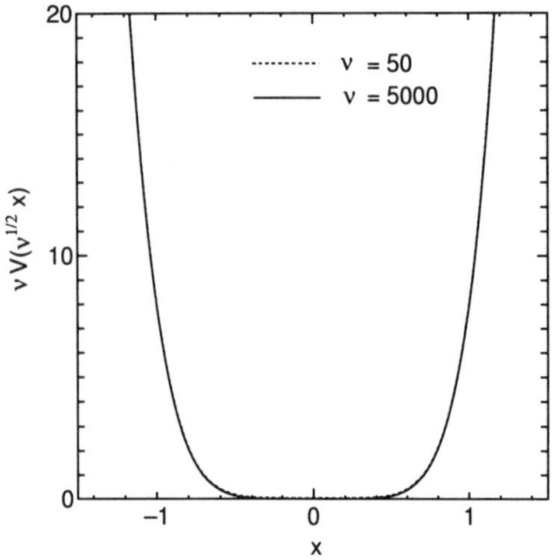

Figure 1. Effective potential

with

$$\begin{aligned}\sigma &= 2J_{00} - 8J^{\alpha} + 8J'^{\alpha\alpha} - 16{\sum_k}' \frac{A_{3k}}{\omega_k} \\ \kappa_3 &= 2J_{00} - 12J^{\alpha} + 16J^{\alpha\alpha} \\ &\quad +8J'^{\alpha\alpha} - 16K^{\alpha\alpha\alpha}\end{aligned} \tag{39}$$

Here, $J^{\alpha\alpha}, J'^{\alpha\alpha}, K^{\alpha\alpha\alpha}$, and A_{3k} are of $O(1)$ with respect to (33), and $O(g^3)$ or even higher if the strength of the interaction is of order $g \to 0$.

For $\nu \gg 1$, we have

$$\nu V_c(\sqrt{\nu}x) = \rho^2 x^6 + \text{const.} \tag{40}$$

and the potential is quite like a square well potential (Fig.1, drawn for $\rho^2 = 8$; see (51) below).

Therefore, the effective Hamiltonian for the condensate fluctuation has a discrete spectrum if

$$\rho^2 > 0. \tag{41}$$

If further

$$A' > 0, \tag{42}$$

the spectrum is lower bounded, implying that the condensate is stable.

These two conditions, (41) and (42), comprise the criteria for Bose-Einstein condensation. In fact, when these two conditions are met, the extension of the condensate wave function in x space is of $O(1)$, so that

$$\eta = O(\sqrt{\nu}) = O(N_0^{1/4}) \tag{43}$$

by (34), and then $\xi = O(N_0^{-1/4})$. Thus,

$$b_0^{\#} \quad \text{is effectively} \quad O(N_0^{1/4}), \tag{44}$$

which is much smaller than $N_0^{1/2}$ if $N_0 \gg 1$. Since the excitation energies of the condensate fluctuation are small, it is important to notice for finite, but low temperature that the potential $V_c(x)$ is like a square well potential if $\rho^2 > 0$, $A' > 0$ and therefore that the extensions of the wave functions are almost constant in any excited states which could mix with the ground state at non-zero temperature.

5 Conclusion

One remark was in order before we had stated the criteria of the Bose-Einstein condensation in the previous section, because, while we have started in Introduction by saying that the effective smallness of $a_0^{\#}$ in comparison with $N_0^{1/2}$ would give the criteria for the condensation, we have ended up with the criteria for smallness of $b_0^{\#}$.

If we rewrite the field operator (2) in terms of $b_0^{\#}$ and $b_k^{\#}$'s

$$\begin{aligned}&\phi_A(\boldsymbol{x}) \\ &= \sqrt{N_0}u_0(\boldsymbol{x}) + \{u_0(\boldsymbol{x}) - {\sum_n}' \alpha_n u_n(\boldsymbol{x})\}b_0 \\ &\quad -\{{\sum_n}' \alpha_n u_n(\boldsymbol{x})\}b_0^{\dagger} \\ &\quad +{\sum_k}'\{{\sum_n}' c_{kn}u_n(\boldsymbol{x}) + \beta_k u_0(\boldsymbol{x})\}b_k \\ &\quad -{\sum_k}'\{{\sum_n}' s_{kn}u_n(\boldsymbol{x}) + \beta_k u_0(\boldsymbol{x})\}b_k^{\dagger}\end{aligned} \tag{45}$$

The (average) total number of particles in the ground state of the system $\psi_0|0\rangle|\chi_{00}\rangle$ is given roughly by the expectation value,

$$N_{\rm tot} = \langle\chi_{00}|\langle 0| \int \phi_{\rm A}^{\dagger}(\boldsymbol{x})\phi_{\rm A}(\boldsymbol{x})d\,\boldsymbol{x}\,|0\rangle|\chi_{00}\rangle. \tag{46}$$

Then, the number of condensate particles is given by the expectation value at $|\chi_{00}\rangle$ of

$$\begin{aligned}&\int [\sqrt{N_0}u_0(\boldsymbol{x}) + \{u_0(\boldsymbol{x}) - w(\boldsymbol{x})\}b_0^{\dagger} - w(\boldsymbol{x})b_0]\\ &\times[\sqrt{N_0}u_0(\boldsymbol{x}) + \{u_0(\boldsymbol{x}) - w(\boldsymbol{x})\}b_0 - w(\boldsymbol{x})b_0^{\dagger}]\\ &\times d\boldsymbol{x}\\ &= N_0 + \sqrt{N_0}(b_0^{\dagger} + b_0) + (1 + 2{\sum_n}' \alpha_n^2)b_0^{\dagger}b_0\\ &\quad -({\sum_n}' \alpha_n^2)(b_0^{\dagger}b_0^{\dagger} + b_0b_0),\end{aligned} \tag{47}$$

of which N_0 dominates if the criteria we obtained are met, where

$$w(\boldsymbol{x}) = {\sum_n}' \alpha_n u_n(\boldsymbol{x}).$$

Thus, our criteria serves to guarantee the Bose-Einstein condensation.

Of the two criteria, (41) and (42), for the Bose-Einstein condensation at low temperature and 'high' density, (42) is rather simple:

$$A' \sim A = J_{00} - 2{\sum_n}' J_{n0}\alpha_n > 0, \tag{48}$$

but the other one, (41), involves a complex quantity (38).

So, let us see what these conditions mean in the weak coupling limit where the interaction is replaced by $gV(\boldsymbol{x})$ with $g \to 0$.

Since $J_{mn}, J'_{mn} = O(g)$, we have

$$\alpha_n = O(g), \tag{49}$$

so that to the first order in g we have

$$A = J_{00}, \quad \kappa_3 = 2J_{00}, \quad \sigma = 2J_{00}. \tag{50}$$

Therefore, in the weak-coupling limit, one of the criteria (41) become trivial,

$$\rho^2 = \frac{\sigma + 3\kappa_3}{A} + O(g^2) \sim 8 > 0, \tag{51}$$

and the other (42) turns out to give

$$A' \sim A \sim J_{00} + O(g^2) > 0, \tag{52}$$

which, if written out explicitly by (11), means

$$J_{00} = \int V(\boldsymbol{x} - \boldsymbol{x}')u_0^2(x)u_0^2(\boldsymbol{x}')d\,\boldsymbol{x}\,d\boldsymbol{x}' > 0. \tag{53}$$

Thus, in the case where the Bose particles interact weakly, the system at sufficiently low temperature and high density undergoes the Bose-Einstein condensation if the interaction is repulsive in the sense of (53), and probably does not otherwise.

Appendix

Coefficients of Λ_0

We list here the coefficients of Λ_0 to $O(g^2)$, where g is the strength of the interaction gV.

$$A' = A + \frac{\kappa_2}{\nu^2}, \quad A = J_{00} - 2J^{\alpha},$$

$$\rho^2 = 8 - 48\frac{J^{\alpha}}{J_{00}},$$

where

$$\kappa_2 = 2(27J^{\alpha} - 5\tau), \qquad \tau = {\sum_k}' J_{kk}^{cs},$$

$$J^{\alpha} = {\sum_n}' \alpha_n J_{n0}, \qquad J_{kk}^{cs} = {\sum_{m,n}}' c_{km}c_{kn}J_{mn}$$

References

1. N.N. Bogoliubov, J. Phys. **11**, 23 (1947).
2. H. Ezawa. *J. Math. Phys.* **6**, 380 (1965). Eq. (6.15b) should be replaced by $\omega_0 = 0$. Discussions are given in ref. 3.
3. H. Ezawa and M. Luban, *J. Math. Phys.* **8**, 1285 (1967).
4. L.P. Pitaevski, *Zh. Exp. Theo. Fys.* **40**, 646 (1961) [*Sov. Phys.* **13**, 451 (1961)]; E.P. Gross, *NCA* **20**, 454 (1961); *J. Math. Phys.* **4**, 195 (1963).

GIANT QUANTUM REFLECTION OF NEON ATOMS FROM SILICON SURFACE

FUJIO SHIMIZU

Institute for Laser Science, University of Electro-Communications, Chofu-shi, Tokyo 182-8585, Japan

JUN-ICHI FUJITA

NEC Fundamental Research Laboratories, Tsukuba-shi, Ibaraki 305-8501, Japan

Cohent reflection of a cold atomic beam with macroscopic dimension from a solid surface caused by attractive van der Waals potential has been studied experimentally. The reflectivity of an ultra-cold metastable neon atomic beam from a silicon surface was measured as a function of normal incident velocity. The reflectivity was shown to improve drastically when the flat surface was changed to a grating structure with narrow ridges. As an example of practical applications the atomic beam was manipulated by a reflective hologram encoded on the silicon surface.

1 Introduction

Diffraction and interference have been one of the most important physical phenomena to verify the wave nature of matter. Interference of even a complex particle such as an atom or a molecule was experimentally confirmed shortly after the establishment of quantum mechanics[1]. Owing to recent technical advances in laser cooling, the interference of atoms is used not only for a demonstration of the rule of modern physics but also for practical applications such as an extremely sensitive metrological instrument[2]. A wave has another basic property, ie reflection at a sharp boundary of refractive index or local wave vector, which has been almost completely neglected in the field of matter waves. This is a sharp contrast to the optical wave for which the reflection at a boundary of refractive index has been the most important mechanism to construct stable and highly accurate optical instruments widely available today.

The reason is obvious. It is not possible to produce a large potential step within the de Broglie wavelength of the matter wave in free space. However, it has been realized for some time that the attractive van der Waals potential near a solid surface can function as a step to cause reflection of neutral atoms. This reflection has been called quantum reflection and was verified subsequently from the reduction of sticking coefficient of light atoms on superfluid liquid helium surface[3,4], and from a sharp increase on reflectivity of helium atoms from liquid helium surface as the kinetic energy of the colliding atom approaches zero[5,6]. We have recently measured the reflectivity of a laser-cooled neon beam from a flat silicon surface and confirmed the quantum reflection on a solid surface[7]. The reflectivity from a flat surface, however, is usually extremely small. We invented subsequently a technique to increase the reflectivity drastically.

In the next section we describe characteristics of quantum reflection due to van der Waals potential. In the third section we discuss a technique to improve reflectivity and describe our experimental result. Then, we show the reflective type atom hologram as a first demonstration of a complex coherent atom optical element that uses a solid surface.

2 Quantum Reflection

A wave feels potential variation sufficiently steep when the variation of its wave vector within the distance of its wavelength is larger than the wavevector itself.

$$\phi = \frac{1}{k^2}|\frac{\partial k}{\partial z}| > 1, \quad (1)$$

where $k = \sqrt{k_0^2 - 2mU/\hbar^2}$ is the local wavevector, U is the potential, z the distance from the solid surface, k_0 is the wave vector in the potential free region, and k_0 is assumed to be positive over entire z in consideration. For a power low potential $U = -C_n/r^n$ with $n > 2$, such as van der Waals potential, ϕ takes maximum ϕ_{max} at a finite distance $z = z_{max}$.

$$\phi_{max} = \frac{1}{z_{max}k_0}\frac{(n+1)(n-2)^{1/2}}{3^{3/2}n^{1/2}}, \quad (2)$$

$$z_{max} = \mathcal{N}'\left(\frac{C_n m}{\hbar^2}\right)^{\frac{1}{n-2}}, \quad (3)$$

This means that an atomic wave is partially reflected at around z_{max} on the negative slope of van der Waals potential. Its reflectivity is a function of ϕ_{max} which diverges as $k_0 \to 0$ and, therefore, approaches unity as the atomic velocity goes towards zero. When the velocity is increased the reflectivity drops off rapidly. For most atoms the maximum velocity to obtain a measurable reflectivity is extremely low. Since atoms on earth are constantly accelerated by gravity it is not practical to provide such low velocity atoms.

To overcome this problem we measured the reflectivity by shooting a well collimated atomic beam at a very small angle to the surface. Metastable neon atoms in the $1s_5$ state were cooled and trapped in a magneto-optical trap with a large magnetic field gradient. A high gradient provided an intense small cloud of neon atoms. A focused laser beam at 598nm was sent into the cloud of the neon atoms and optically transfer to the $1s_3(J = 0)$ metastable state. The $1s_3$ atom does not interact with trapping laser or magnetic field and is accelerated downwards by gravity. A silicon plate was placed 44 cm below the trap and also immediately below a 0.1 mmϕ pinhole. With this atomic beam we could measure the reflectivity at a minimum angle of 0.5 mRad and the normal incident velocity of as slow as 1.5 mm/s. The atoms were detected by a micro-channel plate detector (MCP) that was placed 112 cm below the atomic source. The coherently reflected atoms formed a diffraction limited spot on the MCP and were easily distinguished from diffusive scattering. Figure 1 shows the reflectivity as a function of normal incident velocity of neon atoms. The van der Waals potential at close distance from the surface is $-C_3/z^3$. At a larger distance it changes to $-C_4/z^4$ due to the retardation of interacting photons. We fitted the data in Fig 1 by using a model potential

$$U_{int} = -\frac{C_4}{(r + \lambda/2\pi)r^3}, \quad (4)$$

We numerically integrated one dimensional Schroedinger equation with the above potential and obtained the values of C_4 and λ. The range within σ confidence was $6.7 - 8.4 \times 10^{-56}$ Jm^4 for C_4 and $0 - 4.7$ μm for λ. For a conductor C_4 can be calculated from the polarizability of the atom using the Casimir's formula. This C_4 give $\alpha = 2.0 - 2.5 \times 10^{-39}$ Fm^2, which is approximately in agreement with the value 2.8×10^{-39} Fm^2 determined from DC Stark shift.

3 Giant Quantum Reflection

The quantum reflection provides means to make a reflector for an atomic wave with macroscopic dimension. However, the reflectivity from a flat surface is too small for a practical applications. We describe in this section a technique to improve reflectivity by modifying the surface[8]. As it is understood from Eqs. (2) and (3) ϕ, therefore, the reflectivity increases as the potential constant

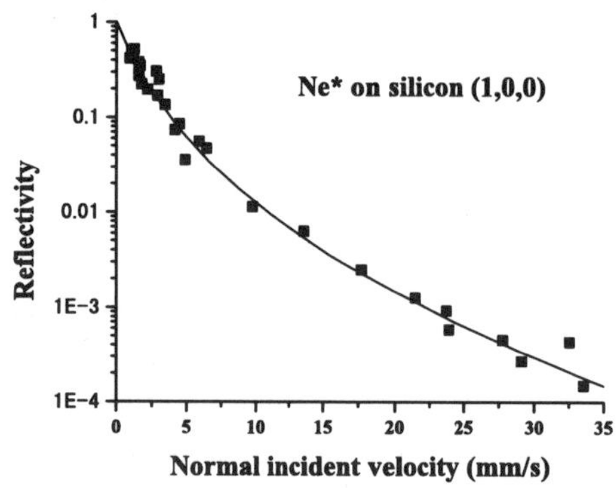

Figure 1. Reflectivity vs normal incident velocity of Ne atoms from silicon surface. The solid line is the calculated reflectivity for the Casimir van der Waals potential.

C_n decreases. Since C_n is approximately proportional to the density of the solid near surface, a higher reflectivity is expected if the effective density near the surface is reduced. This peculiar behavior results that the atom is reflected not at the distance where the potential gradient is maximum. For the power law potential with $n > 2$ the potential gets steeper as the atom approaches the surface. However, the fractional change of the wave vector ϕ takes a maximum at a finite distance because of the positive kinetic energy of the atom. When C_n is large, the distance of ϕ_{max} is also large. Therefore, the atom is reflected at a distance where ϕ is small, and the reflectivity is small. When C_n is small, the peak approaches the surface, and ϕ_{max} become large. As a result the reflectivity increases. The simple estimate from Eqs. (2) and (3) shows that the wave vector k with the same reflectivity varies as C_3^{-1} for $n = 3$ and $C_4^{-1/2}$ for $n = 4$. Figure 1 indicates that the reflectivity is approximately 10% at the incident angle of 1mRad. We can expect the same reflectivity at the incident angle of 0.1 Rad, if the effective density is reduced to 1%.

As a simple method to reduce the effective density we etched the silicon surface to form a narrow-ridge grating structure. Figure 2 shows the structure. The original silicon surface had the (0,0,1) face. The grating structure was formed by using differential etching technique. The ridge was formed along the (1,1,0) direction. The wall of the ridge was of the $(\pm1, \mp1, 1)$ face. The top of the ridge was the original face and was kept flat. We made various samples with the periodicity of the ridge ranging 10 to 100 μm, the width of the top of the ridge 30 nm to 10 μm and the height of the ridge was 1 and 5 μm. In the reflectivity measurement the atomic beam was sent at a small angle to the surface and perpendicular to the ridge of the grating. Figure 3 is an example of the reflectivity measurement. The general characteristics are similar to that of the flat surface. The reflectivity approaches unity as the velocity goes to zero, and decreases monotonically as the velocity is increased. The slope, however, was much smaller than that for a flat surface. For the sample with a large width-of-the-top to periodicity ratio ($\sim 1:100$) the slope was nearly an order of magnitude smaller. As a result the reflectivity was more than two orders of magnitude larger.

4 Reflection Type Hologram for Atoms

We show in this section an example of a coherent reflective atom-optical element that uses enhanced quantum reflection described in the previous section. A binary amplitude hologram was encoded on a silicon surface and an atomic beam reflected from the surface draw a pattern on the screen. We used the difference of the reflectivity between the flat and grating surfaces to encode the binary pattern. Figure 4 shows a part of the hologram. The pattern was composed by a 50×100 μm rectangular cell. The cell that represent absorbing area was a flat surface. The reflecting cell was composed of eleven

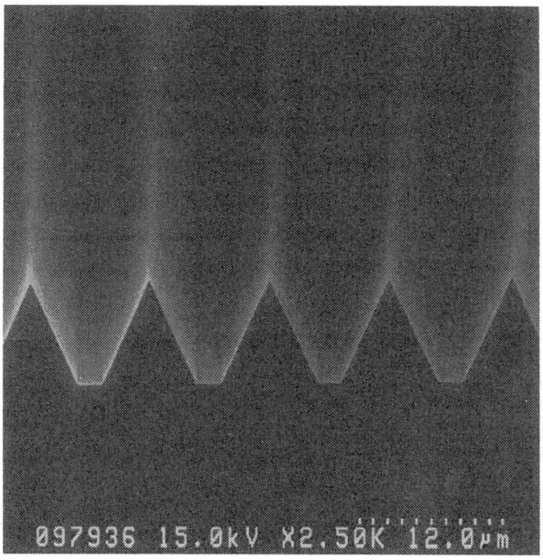

Figure 2. Scannig microscope photograph of the grating structure. The spacing of the ridges was 10 μm.

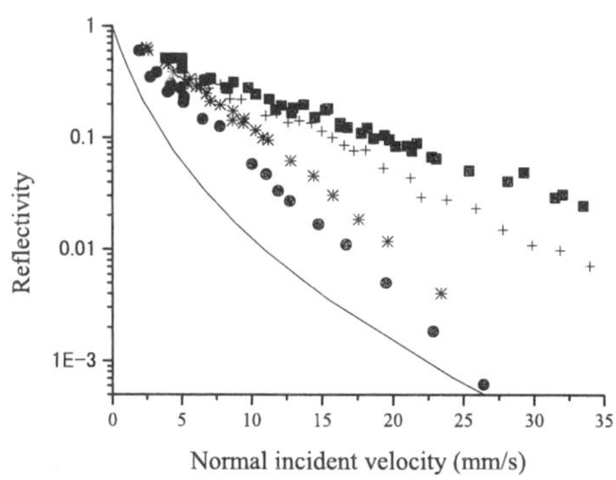

Figure 3. Reflectivity vs normal incident velocity of Ne atoms from the grating structure for various spacings and the top of the ridge. The spacing and the width of the ridge were, circles: 100 μm and 11 μm, stars: 100 μm and 1 μm, crosses: 30 μm and 100 nm, squares: 10 μm and 30 nm, respectively.

ridges with the periodicity of 10 μm parallel to the 50 μm side. Since the hologram was used at glazing angle, the hologram pattern cannot be calculated by simple Fourier transform of the object pattern. We expressed the black and white object by an aggregation of white spots. Then, the complex amplitude $A(i,j)$ of each cell was calculated by summing the contributions from all bright spots

$$A(i,j) = \sum_k e^{i\Phi_{rand}(k)} e^{-i\Phi_k(i,j)} D(x_i - X_k), \tag{5}$$

where (i,j) represents the cell numbers of the hologram perpendicular (x) and parallel (y) to the direction of the atomic beam, x_i is the coordinates of the cell i, X_k is the coordinate of the kth bright spot on the image plane. $\Phi_{rand}(k)$ is a random phase factor assigned for the white spot k, $\Phi_k(i,j)$ is the accumulated phase of the atomic path from the source to the kth spot on the image plane through the cell (i,j). $D(x_i - X_k)$ is the amplitude of the diffraction pattern from a cell of the hologram along the x axis. This allowed us to write a pattern with an arbitrary size along the x axis. The position of the reflecting cell was determined by setting a threshold on A as

$$\mathcal{R}\left(A(i,j)\right) > a_{th}, \tag{6}$$

where the threshold a_{th} was typically 1/10 of the peak amplitude. Figure 5 shows the atomic pattern constructed from the hologram shown in Fig. 4. The hologram was 10 cm long along the y axis and approximately 5 mm along x axis. The atomic beam hit the hologram with an incident angle of 2.5 mRad. The object was the series of capital letters "SURFACE" that are drawn by lines of white spots separated by approximately 100 μm. The horizontal length of the atomic image on the MCP was approximately 8 mm. Since the diffraction from a single cell of the hologram spread the atom along the x direction only for 100 μm, the image was formed, the pattern along the x direction was close to

Figure 4. A part of the hologram pattern. Actual size of a cell was 100 μm for vertical (y) direction and 50 μm for holozontal (x) direction.

Figure 5. Reconstructed atomic pattern. The upper thick black line was the pattern specularly reflected from the surface. (Zeroth order image.) The lower line was made by atoms that did not hit the surface. Effective accumulation time was approximately 2 hours.

the projection image from the hologram. The pattern along the Y direction was made from interference from all reflecting cells along the y direction.

5 Conclusions

We have shown for the first time coherent reflection of an atomic wave from a solid surface that is caused by an attractive interaction of van der Waals force. We invented a technique that improves reflectivity more than two orders of magnitude by forming a grating structure with narrow ridges on the surface. As an example of practical application, we demonstrated holographic manipulation of atoms by a hologram that was encoded on a silicon surface.

The quantum reflection from a $-C_3/z^3$ potential scales with the inverse of the potential constant C_3 and the square inverse of the atomic mass m. The present result on neon atom should be improved more than one order of magnitude if helium or hydrogen atoms are used, and a practical reflector that functions at much larger angle is possible. The quantum reflection is also an useful tool to study characteristics of solid surfaces. The distance of reflection can be adjusted from micrometers to nanometers by changing the normal incident velocity of the colliding atom. This changes the depth from the surface that affects the reflection many orders of magnitude, and therefore variety of physical processes can be studied. Present result on the giant quantum reflection shows the sensitivity of the reflection itself on the structure of the surface.

Acknowledgments

This work was partly supported by the Grants in Aid for Scientific Research (Contract Number 11216202) from the Ministry of Education, Science, Sports and Culture.

References

1. Estermann, I., and Stern, O. (1930) *Z. Phys.* **61**, 95.

2. See for example, *Atom Interferometry* Ed. P. R. Berman, (Academic Press, San Diego, 1997).
3. J. M. Doyle *et al.*, Phys. Rev. Lett. **67**, 603 (1991).
4. I. A. Yu *et al.* Phys. Rev. Lett. **71**, 1589 (1993).
5. V. U. Nayak, D. O. Edwards, and N. Masuhara, Phys. Rev. Lett. **50**, 990 (1983).
6. J. J. Berkhout *et al.*, Phys. Rev. Lett. **63**, 1689 (1989).
7. F. Shimizu, Phys. Rev. Lett. **86**, 987 (2001).
8. F. Shimizu and J. Fujita, submitted for publication.

A DC STARK FREE ATOM INTERFEROMETER HAVING TWO ARMS WITH DIFFERENT ZEEMAN SUBLEVELS FOR MEASUREMENT OF THE AHARONOV-CASHER EFFECT

SHINYA YANAGIMACHI, MASAKI KAJIRO AND ATSUO MORINAGA

Faculty of Science and Technology, Science University of Tokyo, 2641 Yamazaki, Noda-shi, Chiba 278-8510, Japan

We have observed the Aharonov-Casher effect using a dc Stark-free atom interferometer whose wavepackets are in the different magnetic sublevels. A novel atom interferometer using a thermal Ca atomic beam is comprised of two separated light fields, each of which consists of two laser beams with differnt polarizations and frequencies, and yields Ramsey fringes as a function of detuning of the Zeeman shift frequency. Quantitative dependence of the phase shift on the electric field agree with theoretical prediction of AC effect within 5%.

1 Introduction

In 1984, Aharonov and Casher predicted that a neutral particle with a magnetic moment μ traveling a closed path around a line charge experiences a phase shift given by

$$\Delta\psi_{AC} = \frac{1}{\hbar c^2}\oint \boldsymbol{\mu} \times \mathbf{E} \cdot d\mathbf{r}, \qquad (1)$$

where E is the electric field due to the line charge [1]. This Aharonov-Cahser (AC) effect is analogous to the Aharonov-Bohm effect [2], in which an electro-magnetic vector potential can shift the phase of the de Broglie wave of charged particles even if classical forces are absent.

At the first, the AC effect was demonstrated using a neutron interferometer with an experimental accuracy of 16 % [3], and the measured value agree with the theoretical value within nearly twice as much as standard deviation. In contrast to neutrons, the use of atoms have two advantages to measure the AC effect with regard to reduction of the uncertainty. First, the magnetic moment is about two thousand times larger than those of neutrons. Secondly, atomic beams with high flux are easy to obtaine. In a Rb atomic system, A. Gorlits et.al. measured the AC phase shift using the ground state of $F = 3$ with $\Delta m = 2$ spin coherence [4]. Their value was in good agreement with predictions within 1.4%. The AC effect of atoms was also measured by K. Zeiske et. al. using a Ca atom interferometer [5], where the frequency shift of an optical Ramsey resonance in a homogeneous electric field is equivalent to the phase shift of the AC phase and dc Stark phase which is ten thousands times larger than the AC phase. As a result, the accuracy enough to verify the AC shift was obtained at an integration time of as much as 8000 s per a data. Furthermore, these experiments required the information of atom's velocity in order to estimate the phase shift.

The present authors have already developed symmetrical Ramsey-Bordé atom interferometers composed of three or four copropagating traveling laser beams and have achieved a visibility of 0.25 using a Ca atomic beam [6,7].These symmetrical atom interferometers work like white-light interferometers, and offer us to measure the atomic phase without atom's velocity. In principle, the phase shift of these interferometer does not depend on laser frequency. Therefore, the symmetrical Ramsey-Bordé interferometer promise to detect a small phase shift, such as the AC effect which needs a long integration time to measure. However, in order to measure the AC effect accurately, two electrodes with oppsite electric fields must be applied to two arms in the symmetrical interferoemter to eliminate dc Stark effect. In our simple experiment it was revealed that an imperfect symmetry between the two eletrodes introduces a residual dc Stark phase shift that washes out fringes as a result of dispersion.

To eliminate dc Stark shift perfectly, the atom interferometer with $m = +1$ and $m = -1$ magnetic states in its arms is required. In this study, we report on development of a novel atom interferometer composed of two seperated light fields, each of which has two different σ^+ and σ^- polarizations. The atomic wavepackets are split into $m = +1$ and $m = -1$ states at the first interaction, and combined into $m = +1$ or $m = -1$ state at the second interaction. This interferometer eliminates the dc Stark phase shift perfectly under the single electric field, since two atomic wavepackets move the same path. Then the AC effect can be measured without dc stark phase.

2 The dc Stark free Ca atom interferometer

Figure 1 shows the configuration of our new interferometer. We consider the Ca atom as a two-level state atom. The ground state is the $^1S_0$ state and the excited state is the $^3P_1$ state with a life time of 0.57 ms. The atomic beam in the ground state moves in the x direction and interacts with the two laser beams, which propagate along the z direction, through a homogeneous magnetic field also applied in the z direction. Each laser beam has two components, one of which is circularly σ^+ polarized and shiftted in frequency $+\Delta\nu$ from the carrier frequency of the laser, ν_L, another is circularly σ^- polarized and shifted in frequency $-\Delta\nu$ from ν_L. The frequency shift $\Delta\nu$ is close to the Zeeman frequency shift $\Delta\nu_B = \mu_B B/h$. In the first excitation zone, an atom interacts with two components of the laser beam, and a part of wavepacket is excited to the $^3P_1$, $m = \pm1$

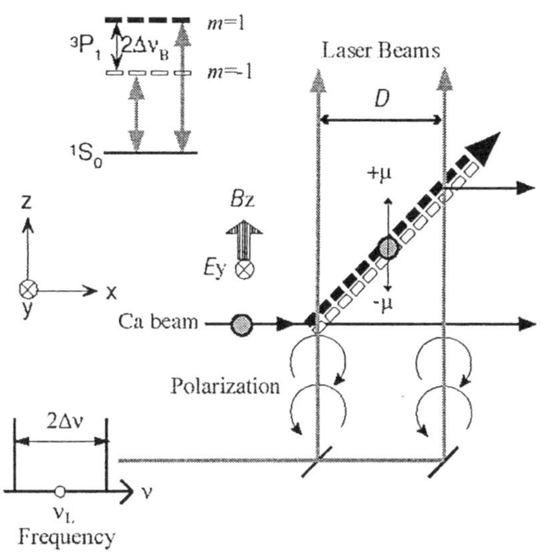

Figure 1. Configuration of the dc Stark free atom interferometer, and level diagram of calcium. Solid line, ground state ; broken black lined, excited m=+1 state ; broken white line, excited m=-1 state

states with recoil velocity. After the atom moves the same path for beam spacement D in the homogeneous magnetic field, the atom interacts with the second laser beam with the same components as the first laser beam. In the second excitation zone, the wavepacket in the $^3P_1$, $m = \pm1$ states, as is an orthogonal state each other, is combined by the second laser beam to the $m = +1$ or $m = -1$ state. Recently we have developed one of such an atom interferometer using a pair of two opposite circularly polarized laser beams which are tuned to each resonance frequency [8]. The phase difference of this interferometer can be calculated based on the phase shift arising from the interaction of the atom with light and the phase shift due to the free evolution of atoms between interactions. The total phase shift $\Delta\psi$ is given by

$$\Delta\psi = 2\pi\left(\Delta\nu - \Delta\nu_B\right)\frac{2D}{v_x} \qquad (2)$$

where v_x is the initial velocity of atoms in the x direction. Figure 2 shows a typical interference Ramsey fringes obtained from singal of the fluorescence after two interactions

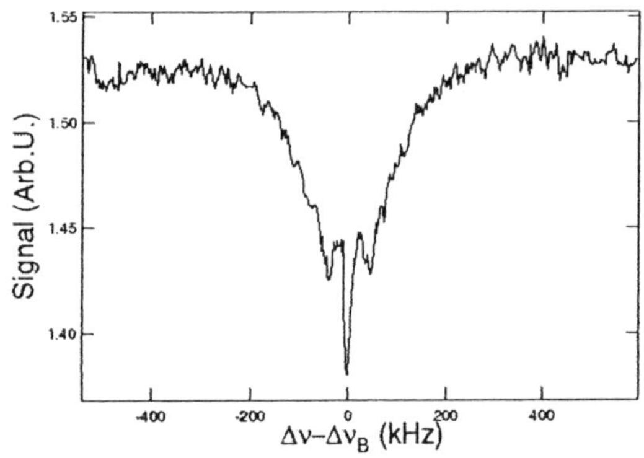

Figure 2. Ramsey fringe as a function of detuning of the Zeeman frequency; Beam spacing D=8.3mm; Signals are detected at an integration time of 20ms every 2kHz.

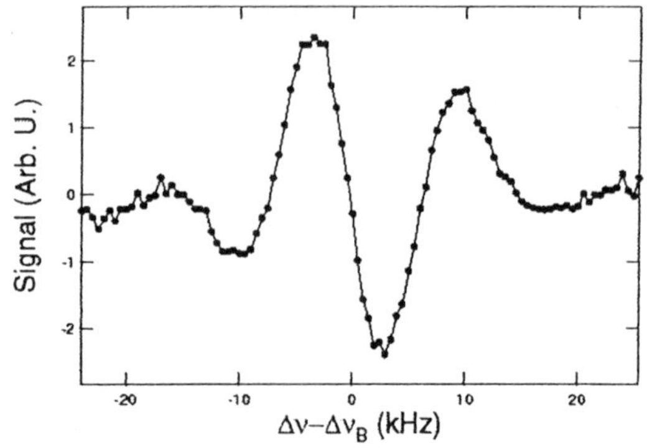

Figure 3. Derivative Ramsey fringes as a function of the detuning of the Zeeman frequency shift at a beam spacing D of 24.9mm; Signals are detected at an integration time of 1s every 500Hz.

with a beam spacement D of 8.3mm. Around the resonance of $\Delta\nu - \Delta\nu_B$, a saturation dip appears with a depth of about 15% and a FWHM of about 400kHz, which corresponds to the transit width. Near the center dip, the Ramsey fringes with a period of about 50kHz appears. If there is a nondispersive phase shift, the pattern of the Ramseys fringes will change, and the phase shift results in the signal change at the center of the fringe.

3 Measurement of the Aharonov-Cahser effect

As noted before, according to the original AC configuration, two arms in the same state enclose the line charge. In our case two arms with the different magnetic moment are placed through a uniform electric field applied in the y direction, as shown in Fig 1. In such an electric field, the atom in each arm acquires the phase shift $\Delta\psi\,(m = \pm 1)$ given by

$$\Delta\psi\,(m = \pm 1) = -\left(\frac{\Delta\alpha E^2 D}{2\hbar v_x} \pm \frac{\mu E D}{\hbar c^2}\right), \tag{3}$$

where the first term represent the dc Stark shift, and $\Delta\alpha$ is the difference of polarizability between atomic state of two arms. The phase difference of the interferometer in the electric field is given by $\Delta\psi\,(m = +1) - \Delta\psi\,(m = -1)$, that is the reason why we can cancel the dc Stark shift completely. It should be noted that even if the inhomogeous electric field is applied, it can be canceled out, becuase two wavepackts travel the same trajectory. Figure 3 shows a typical derivative signal of the interference Ramsey fringes obtained from the fluorescence after two interactions with a beam spacement D of 24.9mm. Around the resonance of $\Delta\nu - \Delta\nu_B$, the Ramsey fringes with a period of about 15kHz appears. The intensity of the fluorescence of the excited atoms was recoreded at an integration time of 1s. Figure 4 shows an example of the phase shift at an applied electric field of +10 kV/cm and -10kV/cm for measurement time of 30s.

At first we obtained the maximum and minimum signal in the fringes, the difference of which is equivalent to the phase of π. This experimental result provides us the AC phase shift of 150±6 mrad at an electric field of 20kV/cm. Figure 5 shows the measured AC phase shift vs the applied electric field. The line is obtained by a least-squares fit to the data. In our preliminary experiment, the value of the slope has been determined to be 7.5 mrad/(kV/cm) without the information

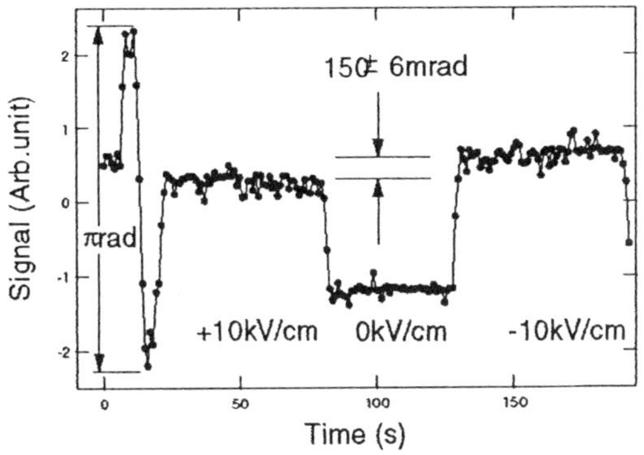

Figure 4. Typical experiment result of determination of the AC phase shift at an electric field of 20kV/cm

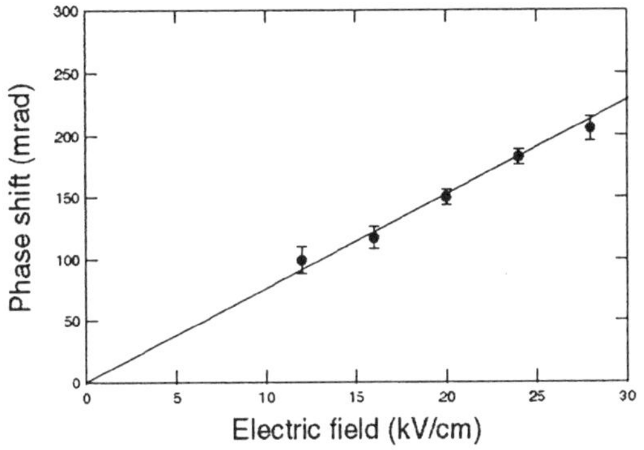

Figure 5. AC phase shift vs electric field together with a fitting line

of the velocity of atoms. The ratio

$$\frac{\Delta\psi_{\mathrm{exp}}}{\Delta\psi_{theory}} = 1.03 \pm 0.05 \tag{4}$$

verifies the absolute magnitude of the AC phase shift with an uncertainty of 5%.

4 Summary

We have measured the Aharonov-Chaser phase shift in a calcium atomic interferometer with an uncertainty of 5%. The experimental result will be improved by the coming more precise experiment. We also have demonstrated the Ramsey-Bordé atom interferometer having two arms with different Zeeman sublevels, and that allows us to measure the AC effect free from the dc Stark shift.

References

1. Y. Aharonov and A. Casher, Phys. Rev. Lett. 53 319 (1984)
2. Y. Aharonov, and D. Bohm, Phys. Rev. 115, 485 (1959)
3. A. Cimmino, G.I. Opat, A. G. Klein, H. Kaiser, S. A. Werner, M. Arif, and R. Clothier, Phys. Rev. Lett. 63, 380 (1989)
4. A. Gorlits, B. Schuh, and A. Weis, Phys. Rev. A 51, R4305 (1995)
5. K. Zeiske, F. Riehle, G. Zinner, and J. Helmcke, Appl. Phys. B 60, 205 (1995)
6. A. Morinaga, and Y. Ohuchi, Phys. Rev. A 51, R1746 (1995)
7. S. Yanagimachi, Y. Omi, and A. Morinaga, Phys Rev. A 57, 621 (1998)
8. S. Yanagimachi, K. Mizobuchi and A Morinaga, Phys Rev. A 64R (To be published)

ATOMIC MULTIPLE-BEAM INTERFEROMETER PHASE-SHIFTED BY THE MULTIPLE MAGNETIC PULSE FIELDS

KENJI SHINOHARA, TAKATOSHI AOKI AND ATSUO MORINAGA

Department of Physics, Faculty of Science and Technology, Science University of Tokyo, 2641 Yamazaki, Noda-shi, Chiba 278-8510, Japan

The interactions of a number of co-propagating stimulated Raman pulses with trapped atoms can make the Ramsey type multiple atom interferometer, as the effective wave number vector k_{eff} of Raman beams is nearly zero. The interference fringes were obtained by shifting the atomic wave phase using the perturbation of weak magnetic pulse fields applied between the adjacent Raman pulses.

1 Introduction

Recently, precise measurements using the atomic interferometer have been studied prosperously[1]. The de Broglie wave length of an atom is shorter than the light wave, so that a higher sensitivity can be obtained. The Ramsey-Bordé atom interferometer can be found using the interaction of a thermal atomic beam with two separated laser fields. Furthermore, the development of an atomic multiple beam interferometer such as a Fabry-Perot interferometer in the field of light optics, is expected so that the sensitivity of the phase measurements will be remarkably improved. However, the spatial and apparatus limitation limit the number of fields.

On the other hand, recent technologies of laser cooling and trapping atoms have enabled fabrication of the time domain atom interferometer. The atoms that were trapped at a point in space are excited by pulsed laser beams separated in time. Hinderthür et al. succeeded to observe clear sharp Airy peaks with a finesse of 140 in the optical Ramsey geometry[2].

We have developed the atomic multiple beam interferometer comprised of co-propagating stimulated Raman pulses using laser cooled and trapped sodium atoms, and demonstrated interference fringes with a finesse of 180 with 200 successive Raman pulses, as shown in Fig.1[3].

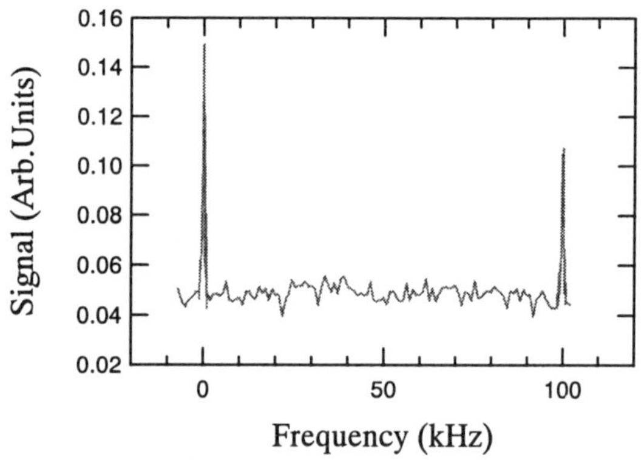

Figure 1. Typical atomic multiple interference signals obtained from 200 laser pulses. Pulse width and pulse separation are $3\mu s$ and $7\mu s$, respectively.

However, in the above experiments the phase is different between two arms of the interferometer results and the interactions of atoms with the laser beams. Now, we study the multiple atomic beam interference fringes versus a real atomic phase shift by applying a weak magnetic field. To our knowledge, there is no report on time domain atom interferometer phase-shift by the magnetic pulse fields. To do this experiment, we applied the homogeneous magnetic field to atoms in order to select the Zeeman sublevel and then applied weak magnetic pulses to change the phase of the atoms.

In this paper, we describe the principle of this interferometer and introduce the preliminary experimental results.

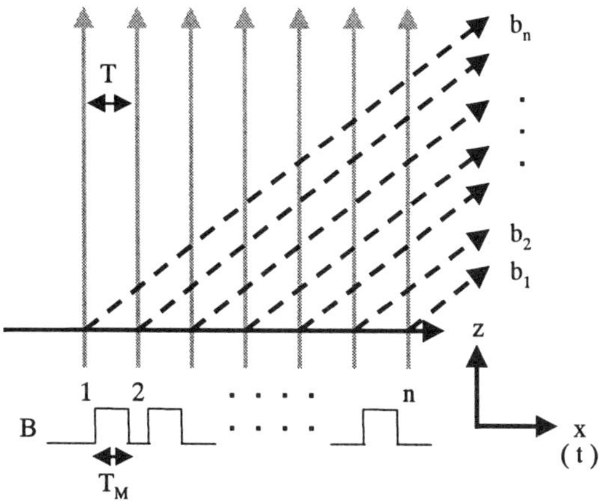

Figure 2. Interaction geometry of Ramsey atomic interferometer comprised of n laser pulses and n-1 weak magnetic pulse fields. The solid line and dotted line show ground and excited states, respectively.

2 Principle

Figure 2 shows the interaction geometry of the Ramsey atomic interferometer comprised of n laser pulses and n-1 weak magnetic pulse fields between the adjacent laser pulses. When the trapped atoms in the ground state, which are moving in the x direction, interact with light pulses propagating in the z direction, the trajectory of the atom splits into two paths and the excited atomic state gets a recoil velocity in the z direction. We assume that the excitation probability is very weak. In this case, the atomic trajectory in the excited state splits into n paths, as shown in Fig. 2.

We consider that a weak magnetic field B is applied to atoms during the time of T_M. Then the phase of atomic wave shifts by a quantity given by

$$\begin{aligned}\psi &= \exp\left[-ig_F\mu_B M_F B T_M/\hbar\right] \\ &= \exp\left[-i\phi_M\right], \qquad (1)\end{aligned}$$

where μ_B is Bohr magneton, g_F is Landé's g factor, m_F is the magnetic quantum number,

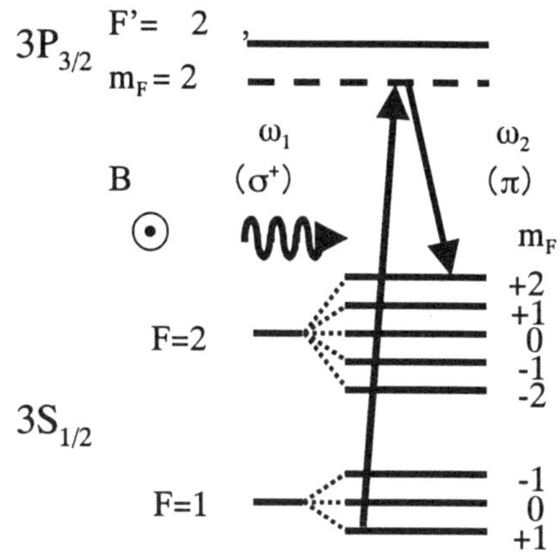

Figure 3. Energy diagram of sodium atom with Raman transition and each polarization of two photons.

and h is Planck's constant. From Eq. (1), if the atom interferometer is composed of the excited and ground states with different m_F or g_F, it makes the phase different between the wavefunctions of the excited and ground states. Therefore, each path of the excited state has a different phase because the time of the applied magnetic pulse fields in the excited state are different. As we use the co-propagating stimulated Raman pulse, the effective wave number vector $\mathbf{k}_{eff} = \mathbf{k}_1\text{-}\mathbf{k}_2$ is nearly zero. Therefore those paths overlap after the interactions with the n laser pulses and they generate the interference fringes.

In order to avoid the shift of resonance frequency, depending on the magnetic field, we apply a magnetic pulse field when the laser pulses absent.

As a result, the probability of the exicited atoms after interaction with n laser pulses and n-1 weak magnetic pulse fields under weak laser power, bb* is given by

$$bb^* \simeq \alpha^{2n-2}\beta^2 \left[n + \sum_{m=1}^{n-1} 2m\cos\left[(n-m)\,\phi_M\right]\right], (2)$$

where β is the probability amplitude that de-

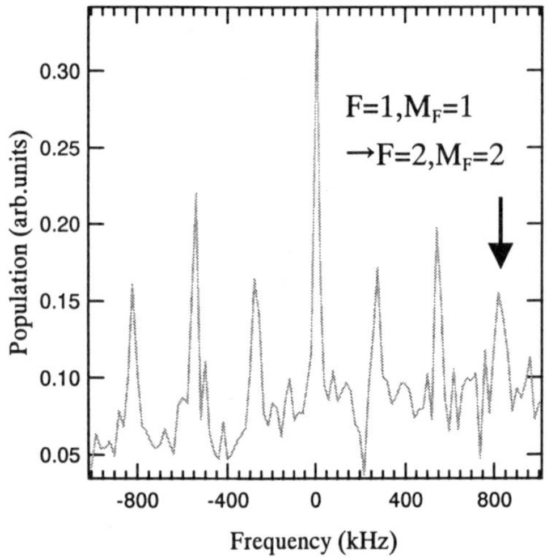

Figure 4. Zeeman splitting of sodium hyperfine structure.

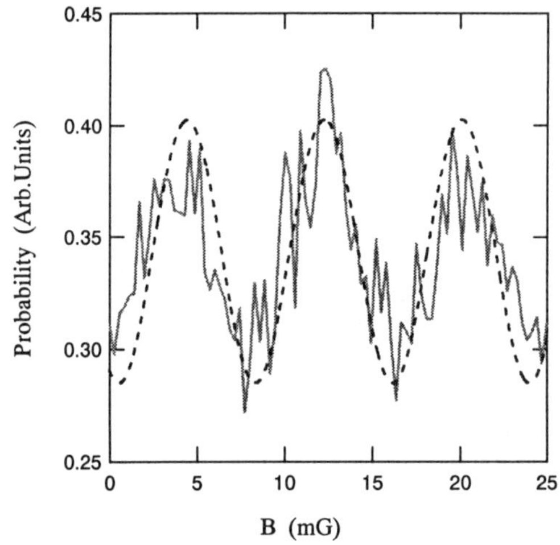

Figure 5. Interference fringes with two pulses of beams due to the dynamical magnetic phase shift, the solid line is the experimental result and the dotted line is the theoretical calculation .

flects the atomic trajectory, while α is the probability amplitude that does not change the trajectory. It is found that the probability consists of many cosine functions with periods of 2π (the fundamental period), $\pi, \ldots,$ and to $2\pi/(n\text{-}1)$. Therefore we obtain the interference fringes like an Airy function[4].

3 Experiment and result

Figure 3 shows the partial energy diagram of a sodium atom under a weak magnetic field. The states we used as the ground and excited states in the interferometer were $3S_{1/2}$,F=1, m_F=1 state and $3S_{1/2}$,F=2, m_F=2 state. The phase difference between both states under the magnetic field B with a duration of T_M was $1.5\mu_B BT_M/\hbar$. Both states were coupled coherently by a stimulated Raman transition. In order to select these states we used a half-wave plate which rotated the azimuth of a linear polarization by 45° from the quantitized axis, so that laser beams acted as σ^+ and π polarization of stimulated Raman beams, as shown in Fig. 3. The two Raman frequency is generated by electro optical modulater.

The homogeneous magnetic field applied the trapped atomic cloud which was made from the Helmholtz coil. Figure 4 shows the Zeeman splitting of the stimulated Raman transition of the F=1 to F=2 under a homogeneous magnetic field of 0.4 G. At first, we tuned the frequency different between two laser beams to the resonance frequency of the Zeeman splitting within 1 kHz.

The sodium atomic ensemble in the $3S_{1/2}$, F=2 state with a density of 5×10^{10} atom/cm^3 was created by the conventional magneto-optical trap and cooled down to 240 μK by a polarization gradient cooling. After the sodium atoms were trapped, the quadrupole magnetic field and optical molasses beams were switched off. Then sodium atoms were optically pumped to the F=1 state by applied optical pumping pulse. After this initialization, the Raman laser pulse and magnetic pulse field applied to the atoms sequentially.

Figure 5 shows the interference fringes as a function of the magnetic field strength when two Raman pulses and one magnetic field pulse were applied to the atom. The

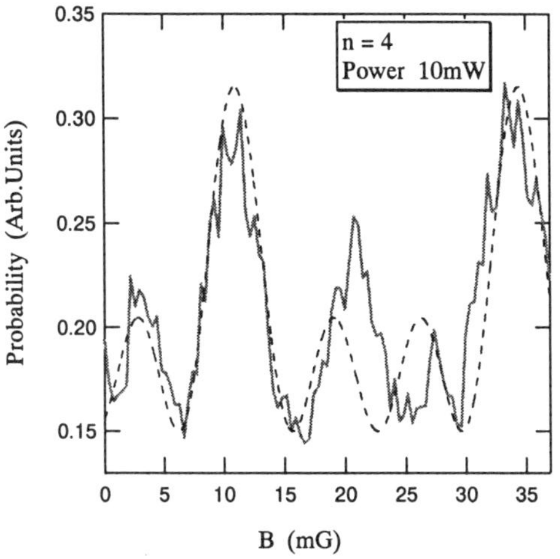

Figure 6. Multiples Interference fringes with four pulses of laser beams, the solid line is the experimental result and the dotted line is the theoretical calculation.

Raman beam power was about 35 mW and pulse width was 20 μs. The strength of magnetic pulse field with a pulse width of 60 μs was varied over 25 mG. The interference fringes with a period of 7.8 mG were clearly observed, according to the theoretical sinusoidal curve.

On the other hand, Figure 6 shows the interference fringes when the four Raman pulses and three magnetic field pulses were applied to the atoms. The Raman beam power was 10 mW and pulse width was 5 μs. The magnetic field pulse width was 20 μs and the strength was varied over 35 mG. The interference fringes with a period of 23.4 mG were observed. The pattern was consistent with the theoretical curve for $\beta = 0.48$. These results show that we succeeded in shifting the phase of atomic wave by applying the magnetic pulse field between the adjacent laser beams for the sodium atomic ensemble.

4 Summary

We could change the phase of the time domain atom interferometer by applying weak magnetic pulse fields and demonstrating the interference fringes of the atomic multiple-beam. In the future we shall remove the influence of the inhomogeneous field, increase the number of magnetic pulse fields and demonstrate the high-finesse multiple-beam interference signal which will be highly sensitive to the magnetic field.

References

1. *Atom interferometry*, edited by P. Berman (Academic, San Diego, 1997).
2. H. Hinderthür, F. Ruschewitz, H. J.-Lohe, K. Sengstock and W. Ertmer Phys. Rev. A **59**, 2216 (1999).
3. T. Aoki, K. Shinohara, and A. Morinaga Phys. Rev. A **63**, 063611 (2001).
4. A. Morinaga and T. Aoki, in *Quantum Coherence and Decoherence ISQM-Tokyo'98.* Ed.by Y. A. Ono et al. (Elsevier1999) p.115.

EFFECTS OF DECOHERENCE ON ENTANGLED ATOMIC WAVE FUNCTIONS IN MICROCAVITIES

A. S. MAJUMDAR

S. N. Bose National Centre for Basic Sciences, Block JD, Sector III, Calcutta 700098, India
E-mail: archan@boson.bose.res.in

N. NAYAK

Department of Physics, Texas A&M University, College Station, Texas 77843-4242, USA
E-mail: nayak@atlantic.tamu.edu

We consider an experimental set-up where two-level atoms are streamed through a microcavity in such a manner that at most one atom is present inside the cavity at any instant of time. The interaction of a single atom with the cavity photons leaves an imprint on the steady-state cavity density operator. The wave function of the next atom that passes through the cavity gets entangled with the cavity photons and subsequent secondary correlations develop between two or more atoms in this way. After leaving the cavity the atoms pass through an electromagnetic field that is tuned to give a $\pi/2$ pulse to the atoms with varied phase for different atoms. The atoms are then detected in either of their upper or lower states. The secondary correlations between two or more subsequent atoms can be exploited to formulate Bell-type inequalities for their detection probabilities. We investigate the effects of decoherence on atomic entanglement brought about by both atomic decay and cavity dissipation through interaction with their respective reservoirs. We show by using realistic models for the micromaser as well as the microlaser that effects of decoherence on the Bell sum can be experimentally monitored and observed in a controlled fashion.

1 Introduction

Micromaser dynamics is a natural candidate for studying varied aspects and consequences of entanglement of atomic wave functions with cavity photons[1]. In this article we consider the violation of a Bell-type inequality using a realistic model for the micromaser[2]. Our formalism also enables us to similarly analyse the microlaser within the same framework but with different choices of the parameters. Our purpose is to study the effect of both cavity dissipation and atomic decay towards diminishing atom-photon entanglement in an actual experiment[3].

The generation of correlations between the wave functions of two or more atoms is achieved in the following way. A two-level atom initially in its upper excited state $|e>$ traverses a high-Q single mode cavity. The cavity is in a steady state and tuned to a single mode resonant with the transition $|e> \rightarrow |g>$. The emerging single-atom wavefunction is a superposition of the upper $|e>$ and lower $|g>$ state, and it leaves an imprint on the photonic wavefunction in the cavity. After leaving the cavity, the atom passes through an electromagnetic field which gives it a $\pi/2$ pulse the phase of which can be varied for different atoms. The atom then reaches the detector, placed at a sufficient distance, capable of detecting the atom only in the upper or lower state. This process is repeated for a similar second atom. Though there is no spatial overlap between the two atoms, the entanglement of their wavefunctions with the cavity photons can be used to formulate a Bell-type inequality on the detection probabilities for the two atoms[1]:

$$B < 0 \tag{1}$$

with the Bell sum B given by

$$B \equiv |E^\lambda(\phi_1, \phi_2) - E^\lambda(\phi_1, \phi_3)| + sign(E_0)[E^\lambda(\phi_2, \phi_3) - E_0] \tag{2}$$

where λ is the hidden variable and ϕ's are

the phases of the $\pi/2$ pulse. $E^{\lambda}(\phi_1, \phi_2)$ is the ensemble average for double-click events in a local realist theory with $E_0 = E^{\lambda}(\phi_1 = \phi_2)$. The quantum mechanical analog of E is given by

$$E(\phi_1, \phi_2) = P_{ee}(\phi_1, \phi_2) + P_{gg}(\phi_1, \phi_2) - P_{eg}(\phi_1, \phi_2) - P_{ge}(\phi_1, \phi_2) \quad (3)$$

where $P_{eg}(\phi_1, \phi_2)$ stands for the probability that the first atom is found to be in state $|e>$ after traversing the $\pi/2$ pulse with phase ϕ_1, and the second atom is found to be in state $|g>$ with the phase of the $\pi/2$ pulse being ϕ_2 for its case.

The combination of atomic and photonic statistics plays a key role in the setting up of entanglement. The initial state of the cavity is built up by the passage of a large number of atoms, but only one at a time, through it. The values of the cavity pump parameter and the atom-photon interaction time determine the resultant photonic wavefunction which in turn controls the nature of entanglement between two successive experimental atoms detected in their upper or lower states by the detector. Our aim is to control the dissipation due to the interaction of the pumping atoms with their reservoir, as well as the loss of cavity photons, and study their effect on the statistics of detected atoms.

2 Dissipation and decoherence in a real micromaser and a microlaser

Our purpose first is to calculate the Bell sum using a realistic model for the micromaser and see how it behaves as a function of the cavity parameters and decay rates. For doing so we consider the standard atom-cavity field interactions with their reservoirs and solve the equation

$$\dot{\rho} = \dot{\rho}|_{atom-reservoir} \quad (4)$$
$$+ \dot{\rho}|_{field-reservoir} + \dot{\rho}|_{atom-field} \quad (5)$$

where the first two terms on the r.h.s are given respectively by

$$\dot{\rho}|_{atom-reservoir} = -\gamma(1+\bar{n}_{th})(s^+s^-\rho - 2s^-\rho s^+ + \rho s^+s^-) -\gamma\bar{n}_{th}(s^-s^+\rho - 2s^+\rho s^- + \rho s^-s^+) \quad (6)$$

and

$$\dot{\rho}|_{field-reservoir} = -\kappa(1+\bar{n}_{th})(a^\dagger a\rho - 2a\rho a^\dagger + \rho a^\dagger a) -\kappa\bar{n}_{th}(aa^\dagger\rho - 2a^\dagger\rho a + \rho aa^\dagger) \quad (7)$$

where γ and κ are the atomic and cavity decay constants, respectively, and $\bar{n}_{th}$ is the average number of thermal photons. The atom-field interaction is governed by the standard Jaynes-Cummings Hamiltonian.

The dynamics for an intial steady-state cavity density operator has been solved[2]. This solution can be used to calculate the probabilities in the Bell sum treating cavity dissipation and atomic decay in a realistic fashion, in the following way. The experimental atoms on which we plan to test the Bell's inequality (BI), encounter this steady state radiation field $\rho_f^{(ss)}$. The probability of detection of the first experimental atom in the upper state $|e>$ and the lower state $|g>$ can be written respectively as

$$P_e = \mathrm{Tr}._f\mathcal{P}_e$$
$$P_g = \mathrm{Tr}._f\mathcal{P}_g \quad (8)$$

with

$$\mathcal{P}_e = \frac{1}{2}[\mathcal{A}\rho_f^{(ss)}\mathcal{A}^\dagger + \mathcal{D}\rho_f^{(ss)}\mathcal{D}^\dagger - \{e^{-i\phi_1}\mathcal{A}\rho_f^{(ss)}\mathcal{D}^\dagger + e^{i\phi_1}\mathcal{D}\rho_f^{(ss)}\mathcal{A}^\dagger\}]$$
$$\mathcal{P}_g = \frac{1}{2}[\mathcal{A}\rho_f^{(ss)}\mathcal{A}^\dagger + \mathcal{D}\rho_f^{(ss)}\mathcal{D}^\dagger + \{e^{-i\phi_1}\mathcal{A}\rho_f^{(ss)}\mathcal{D}^\dagger + e^{i\phi_1}\mathcal{D}\rho_f^{(ss)}\mathcal{A}^\dagger\}] \quad (9)$$

where trace is taken over the cavity field and the operators $\mathcal{A}$ and $\mathcal{D}$ are given by

$$\mathcal{A} = \cos(gt\sqrt{a^\dagger a+1})$$
$$\mathcal{D} = -ia^\dagger\frac{\sin(gt\sqrt{a^\dagger a+1})}{\sqrt{a^\dagger a+1}} \quad (10)$$

After the passage of the first atom through the cavity and its detection in, for example, the state $|e>$, the second atom encounters the cavity field with density operator $\rho_f^{(2)}$ given by

$$\rho_f^{(2)} = \mathcal{A}\rho_f^{(ss)}\mathcal{A}^\dagger + \mathcal{D}\rho_f^{(ss)}\mathcal{D}^\dagger$$
$$-\{e^{-i\phi_1}\mathcal{A}\rho_f^{(ss)}\mathcal{D}^\dagger + e^{i\phi_1}\mathcal{D}\rho_f^{(ss)}\mathcal{A}^\dagger\} \quad (11)$$

(since $\mathrm{Tr}._f\mathcal{P}_e = 1/2$). The phase of the $\pi/2$ pulse is set to ϕ_2 for the second atom. $\mathcal{P}_e$ for the second atom is given by

$$\mathcal{P}_e^{(2)} = \frac{1}{2}[\mathcal{A}\rho_f^{(2)}\mathcal{A}^\dagger + \mathcal{D}\rho_f^{(2)}\mathcal{D}^\dagger$$
$$-\{e^{-i\phi_1}\mathcal{A}\rho_f^{(2)}\mathcal{D}^\dagger + e^{i\phi_1}\mathcal{D}\rho_f^{(2)}\mathcal{A}^\dagger\}] \quad (12)$$

The conditional probability $P_{ee}(\phi_1, \phi_2)$ is thus given by

$$P_{ee}(\phi_1, \phi_2) = \mathrm{Tr}._f\mathcal{P}_e^{(2)} \quad (13)$$

P_{gg} is obtained similarly, and using the relations $E(\phi_1, \phi_2) = 2P(\phi_1, \phi_2) - 1$ and $P(\phi_1, \phi_2) = P_{ee}(\phi_1, \phi_2) + P_{gg}(\phi_1, \phi_2)$ one obtains the values of E in the Bell sum (1).

We are able to take into account dissipation even during the short atom-field interaction time. This is important since decoherence effects are inherent in the build up of the cavity field to its steady state as it encounters a large number of atoms over the time required for the steady state to be reached. The resultant steady-state photon statistics in our model is clearly different and more realistic compared to the micromaser model used by Loffler *et al*[1] where cavity decay and atomic loss is neglected during the atom-field interaction times.

3 Results and discussion

We compute the Bell Sum numerically with the values of the phases set to $\phi_1 = 0$, $\phi_2 = \pi/3$, and $\phi_3 = 2\pi/3$. The steady-state dynamics for microcavities considered here is suitable for describing both the micromaser as well as the microlaser. However, their distinctive features are manifested in the choice of parameters which we use to study the violation of BI in both separately. While considering micromaser dynamics, one can safely set atomic decay to zero. This is because the Rydberg levels involved in the micromaser have a lifetime of about $0.2s$, whereas the atomic flight time through the cavity (atom-field interaction time) is typically $35\mu s$. However, atomic decay is an important factor in the microlaser where atomic levels at optical frequencies are involved. Although decay is unimportant for the dynamics of a single atom interacting with the field up to a certain interaction time, its accumulated effect for a large number of atoms is crucial for the evolution of the microlaser field to a steady state. Furthermore, the interaction time of the atom with the $\pi/2$ pulse (in this case it is $gt = \pi/2$) is far less compared to even individual atom-field interaction times of interest in microlaser dynamics. For this reason the effect of atomic decay can be neglected during interaction of the atom with the $\pi/2$ pulse. Our results for the micromaser and the microlaser are presented separately in Figures 1 and 2 respectively.

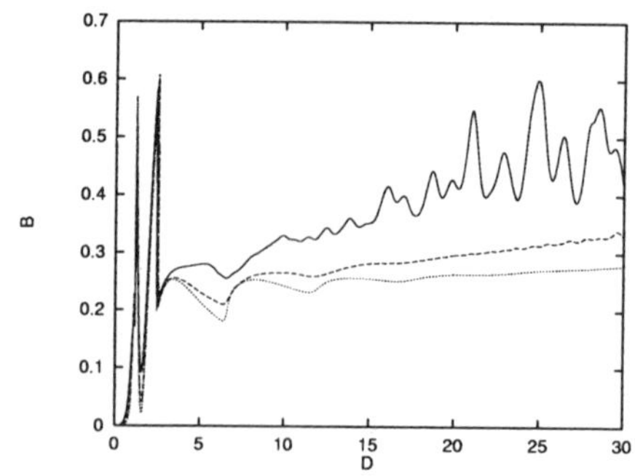

Figure 1. Violation of Bell's inequality in a micromaser. The pump rate N, the number of individual atoms that pass through the cavity in a photon lifetime, = 20 (full), 50 (broken), and 100 (dotted).

In figures 1 and 2 the Bell sum B is plotted versus the pump parameter $D = \phi\sqrt{N}$ where the Rabi angle $\phi = g\tau$. B decreases with the increase of pump rate N for a large range of interaction times τ. Dissipa-

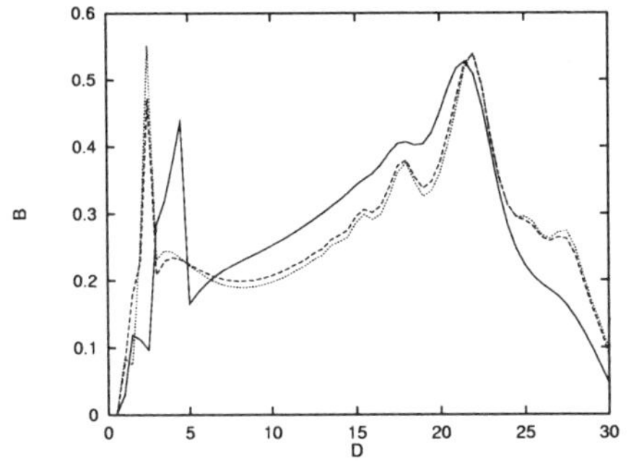

Figure 2. Violation of Bell's inequality in a microlaser. We take $N = 100$, $\gamma/g = 0.1$ and $\bar{n}_{th} = 0$. The cavity leakage rate is $\kappa/g = 0.01$ (full), 0.001 (broken) and 0.0001 (dotted).

tive effects creep into the dynamics through two parameters N and τ^2. For shorter values of single atom interaction times we find that the correlations build up sharply with τ, and the peak value of B signifying maximum violation of BI is larger for higher values of N, (as a magnification of Figure 1 reveals[4]). The maximum violation (the value of B at its second peak) is 0.5812 for $N = 20$ (full line), 0.6079 for $N = 50$ (broken) and 0.6241 for $N = 100$ (dotted) This feature is a curious example of multiparticle induced nonlocality[5].For short interaction times, naturally the effects of decoherence are too small to affect the correlations.

To conclude, our study serves to highlight the fact that systems such as the micromaser[3] and the microlaser[6] can be effectively used to study fundamental problems of quantum theory like entanglement and decoherence.

References

1. M. Loffler, B.-G. Englert and H. Walther, *Appl. Phys.* B **63**, 511 (1996).
2. N. Nayak, *Opt. Commun.* **118**, 114 (1995).
3. G. Rampe, F. Schmidt-Kaler and H. Walther, *Phys. Rev. Lett.* **64**, 2783 (1990).
4. A. S. Majumdar and N. Nayak, *Phys. Rev.* A**64**, 013821 (2001).
5. N. D. Mermin, *Phys. Rev. Lett.* **65**, 1838 (1990); S. M. Roy and V. Singh, *Phys. Rev. Lett.* **67**, 2761 (1991); D. Home and A. S. Majumdar, *Phys. Rev.* A**52**, 4959 (1995).
6. K. An, J. J. Childs, R. R. Dasari and M. S. Feld, *Phys. Rev. Lett.* **73**, 3375 (1994).

Quantum Theory of Continuous-Wave Atom Laser by non Born-Markov Approach

Hiroshi IWASAWA and Kuniaki MATSUO

Hiroshima Kokusai Gakuin University,6-20-1 Nakano,Aki-ku,Hiroshima,739-0321,Japan

The Heisenberg picture approach of the continuous-wave atom laser with an output coupler was investigated theoretically. The mechanism of the output coupler will be considered without using the so-called Born-Markov Approximation. The quantum theoretical treatment of the continuous-wave atom laser in this paper will demonstrate that it is a self-sustained oscillator.

1. Introduction

Nowadays the realization of a Bose-Einstein condensation (BEC) becomes a standard technique in the laboratory. First, a pulsed atom laser was realized[1], and recently the experiments for the continuous-wave (CW) atom laser have been reported[2].

Since 1995, several theoretical studies on the atom lasers have been reported on the basis of the Born-Markov Approximation (BMA) which have been used frequently in the field of the quantum optics [3]. However, it has been pointed out that the assumption of the BMA is inadequate for the real parameters of atom lasers[4].

In this report, we study a CW atom laser with an output coupler with the two case; the non BMA and the Born approximation with the non-Markov process.

2. Model of a CW Atom Laser

The two-body elastic collision model with the evaporative cooling by Holland et al. [3], is modified by taking into account of an output coupler as shown in Fig. 1. This is an open system model , which contains three reservoirs. In this scheme, the BMA is assumed for the reservoirs 1 and 2. However, the BMA is not adapted for the reservoir 0 which is the external space.

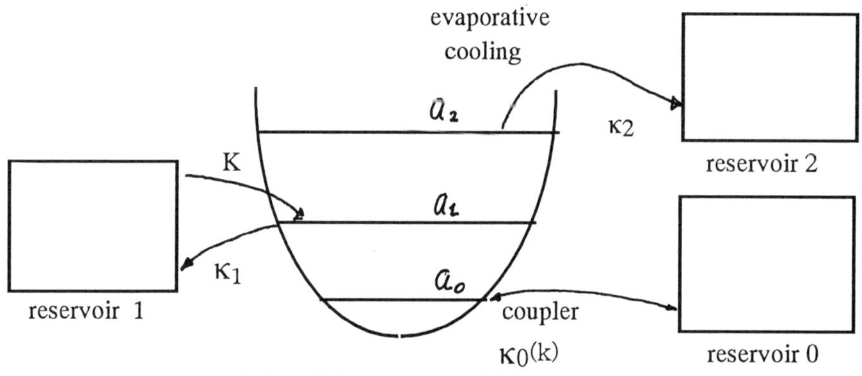

Fig.1 Model of a CW atom laser with an output coupler

The total Hamiltonian of the atom laser for this three level model can be written in the Heisenberg picture(HP) as follows.

$$H_{tot} = H_{tr} + H_{int} + H_{ext} + \Sigma H_j + \Sigma H_{aj}, \quad j = 1, 2, \qquad (1)$$

$$H_{tr} = \Sigma \hbar\omega_i a_i^\dagger(t)a_i(t) + V(t),$$

$$V(t)= i\hbar\Sigma\ g_{ijkl}\ a_i^\dagger(t)a_j^\dagger(t)\ a_k(t)a_l(t)\ ,$$

$$i \leqq j\ ,\ k \leqq l, \quad (2)$$

$$H_{ext} = \int dk\hbar\omega_k b_k^\dagger(t)b_k(t), \quad (3)$$

$$H_{int} = -i\hbar\Gamma^{1/2}\int dk\{\kappa_0(k)b_k(t)a_0^\dagger(t) - \kappa_0^*(k)b_k^\dagger(t)a_0(t)\}, \quad (4)$$

where H_{tr} is the Hamiltonian of the harmonic trap. This consists of the oscillation energy at frequency ω_i and the interaction Hamiltonian V(t) which describes the redistribution of the atoms by elastic collisions. The operator a_i is the second quantized annihilation operator for the mode i, and the coefficients g represent the total transfer rates between levels by collisions.

Using the rotating wave approximation we can reduce the interaction term V(t) to the energy conservation terms. Some terms of them can be ignored, because the level 2 decays very fast by the effect of the evaporative cooling .

The term H_{ext} is the Hamiltonian for the out-put reservoir and H_{int} represents the interaction between the laser mode a_0 and the output free space field b_k.

The coupling constant, $\kappa(k) = \Gamma^{1/2}\kappa_0(k)$, depends on k and its dispersion relation is expressed by $\omega_k = \hbar k^2/2M$ (M is the mass of the atoms) which differs from that for the light. Hence, we treat this coupling to the external free fields without the BMA, and assume that the external field is empty at the initial time, since initially there is no matter. The term H_j is the Hamiltonian of the reservoir j (j = 1, 2), and H_{aj} is the interaction energy between the reservoir j and the atoms in the trap level j, in which the BMA is assumed.

3. Fundamental Equations in HP

3.1 Quantum Mechanical Langevin Equations

First, eliminating the operators of the reservoirs 1 and 2 from the Heisenberg equations for each operator, we obtain the coupled quantum mechanical Langevin equations of the annihilation operators a_2,a_1,a_0 and b_k for the higher, pumping, and lasing modes, and the external field, respectively. In Fig.1, κ_2 and κ_1 are the decay rates of the higher mode a_2 and of the pumping mode a_1, respectively.

In order to build up a condensate in the ground state, it is necessary for the rates to obey the inequality

$$\kappa_2 >> \kappa_1 >> \kappa_0. \quad (6)$$

Under the condition (6), we derive the equation of the lasing mode $a_0(t)$ by eliminating a_2 and a_1 adiabatically as follows:

Step1: let $a_i(t) = \hat{a}_i(t)e^{-i\omega_i t}$, i =0,1, 2 and eliminate $\hat{a}_2$ adiabatically from both equations of $\hat{a}_1$ and $\hat{a}_0$.

Step 2: Eliminate the pumping mode $\hat{a}_1$ adiabatically from the equation of $\hat{a}_0$.

Having performed these steps, we finally obtain the following Langevin equation of the lasing mode $\hat{a}_0$,

$$da_0(t)/dt = -\gamma_0'(t)+ (\alpha+i\Omega_1)\hat{a}_0(t) - (\beta+ i\Omega_0)n_0(t)\ \hat{a}_0(t) - \xi(t)+F_{0t}(t), \quad (7)$$

$$\gamma_0'(t) = \gamma_0(t) - \kappa_0,$$

$$\gamma_0(t) = \Gamma\int f(t-t')\hat{a}_0(t')dt, \quad (8)$$

$$\hat{f}(t) = \int dk|\kappa_0(k)|^2 e^{-i\delta_k t}, \quad (9)$$

$$\xi(t) = \int dk\kappa_0(k)e^{-i\delta_k(t-t_0)}\,\hat{b}_k(t_0)\ , \quad (10)$$

$$\delta_k = \omega_k - \omega_0, \quad (11)$$

where

$$\alpha = \Lambda\,(K/\kappa_1)2 - \kappa_0,$$

$$\beta = 4\Lambda\,(K/\kappa_1)(\kappa_0/\kappa_1),$$

$$\Omega_0 = 2g_{0000}\ ,\quad \Omega_1 = (K/\kappa_1)g_{1010},$$

$$\Lambda = |g_{0211}|^2/\kappa_2.$$

Here, α is the effective gain coefficient , β

is the self saturation coefficient of the gain, Λ is the transition rate from the pumping mode to the laser mode, and Ω_0, Ω_1 present the self energy shift, respectively , which affect the spectral property of the atom laser. In the coefficients α and β, K is the pumping rate from the reservoir 1 to the level 1 in the trap , and it is introduced phenomenologically in the equation of the pumping mode a_1.

Equation (7) is the fundamental equation of the lasing mode, which contains time convolution term $\gamma_0(t)$ due to non Born-Markovian coupling to the external field. The terms $\gamma_0(t)$ and $\xi(t)$ correspond to the damping factor and the fluctuation operator, respectively, which appear in the usual laser theory.

On the other hand, the equation of the external field $\hat{b}_k(t)$ is solved directly and the solution is given by

$$\hat{b}_k(t)=\hat{b}_k(t_0)e^{-i\delta_k t} +\Gamma^{1/2}\kappa_0^*(k) \int dt'\hat{a}_0(t')e^{i\delta_k(t-t')}. \qquad (12)$$

The Heisenberg equation of the expectation value of $a_0(t)$ is obtained from Eq. (7),

$$d\langle a_0(t)\rangle/dt = -\langle\gamma'_0(t)\rangle- [\alpha- i (\omega_0+\Omega_1) - (i \Omega_0 + \beta n_0)] \langle a_0(t)\rangle- \Gamma \langle\xi(t)\rangle . \qquad (13)$$

When the BMA is valid, we obtain

$$d\langle a_0(t)\rangle/dt= [(\alpha -i\omega_0)- \beta n_0]\langle a_0(t)\rangle, \qquad (14)$$

where Ω_0 and Ω_1 are ignored for simplicity when we are interested only in the occupation number of the ground state.

The Heisenberg equations (13) and (14) of the lasing mode show

1)Equation (13) contains the effects of non Born-Markovian output coupler.

2) Equations (13) and (14) show that a CW atom laser behaves as a self-sustained oscillator, since they have a third-order non-linear saturation term. Similar discussions from another approach have been reported independently by Scully[5]).

3) It is worth noting that Eq.(14) is closely analogous to the single mode laser equation.

3.2 Solution

For simplicity, we ignore the fluctuation force operator F_{0t} from the Markovian reservoirs 1 and 2, and the nonlinear saturation term. In order to solve the Langevin equation, Laplace transformation $\pounds[g(s)]=\int g(t)e^{-st}ds$ and inverse Laplace transformation, $\pounds^-[\quad]$ are applied to the final Eq. (7). The solutions for the laser mode and output mode are given by

$$\hat{a}_0(t)=\hat{a}(o)\pounds^-[\{\hat{a}_0(0)-\pounds[\xi(s)]\} \{s+(i\Omega_1-\alpha)+\Gamma\pounds[\hat{f}(s)]\}^{-1}], \qquad (15)$$

$$b_k(t) = b_k(t_o)\, e^{-i\omega_k t} -\Gamma^{1/2}\, k_0^*(k)a_0(0)\, e^{i\omega_0 t}\, h(t) - \Gamma\kappa_0^*(k)\, e^{-i\omega_0 t}\int h(t-t')\xi(t')dt' , \qquad (16)$$

$$h(t) = \pounds^- [(s+\delta_k)^{-1}\{s+(i\Omega_1 - \alpha) +\Gamma\pounds[f(s)]\}^{-1}](t). \qquad (17)$$

The expectation value of the output intensity $\langle n_k(t)\rangle$ is obtained from Eq. (16) as

$$\langle n_k(t)\rangle=\langle b_k^\dagger(t)b_k(t)\rangle = \Gamma|k_0(k)|^2\langle a_0^\dagger(0)a_0(0)\rangle |h(t)|^2 , \qquad (18)$$

where initially external reservoir is empty i.e., $\langle b_k(0)\rangle = 0$.

For a pulsed atom laser, we can ignore the pumping mechanism, so we put $\alpha = 0$ and $\beta = 0$. In addition we ignore Ω_1 for simplicity. Then the solution is given by

$$\langle a_0(t)\rangle = \langle a_0(0)\rangle\, e^{-i\omega_0 t}\, \pounds^{-1}[s + \pounds [f(s)]]^{-1}(t). \qquad (19)$$

There is no further restrictions except $\langle n_k(0)\rangle = 0$, or $\langle b_k(0)\rangle = 0$.

In this stage, we do not assume the BMA.

4. Projection Operator Approach by the Heisenberg Picture (HP)

In this section, We assume Born approximation with the non-Markovian process. We treat the simplest model of the atom laser without the pumping by using the time convolution decomposition method with the projection operator in the HP, which have been recently reported[6].

The Hamiltonian of this system is given by

$$H = H_0 + H_1 , \qquad (20)$$

$$H_0 = \hbar\omega_0 a_0^\dagger a_0 + \int dk \hbar\omega_k b_k^\dagger b_k ,$$

$$H_1 = - i\hbar\Gamma^{1/2}\int dk\{\kappa_0(k) b_k a^\dagger - \kappa_0^*(k) b_k^\dagger a_0\}.$$

The starting point is the Liouville-von Neumann equation for an operator A(t) of the atom laser system, for example, A(t) = $a_0(t)$ or $b_k(t)$. The Liouville-von Neumann equation is expressed as

$$dA(t)/dt = iL(t)A(t) = i(L_0 + L_1)A(t),$$

$$L_i = \hbar^{-1} [H_i, A(t)] \quad i = 0,1 . \qquad (21)$$

We transform L_1 and A(t) into the interaction picture as follows:

$$\hat{L}_1 = e^{iL_0(t-t_0)} L_1(t) e^{-iL_0(t-t_0)} ,$$

$$\hat{A}(t) = \hat{U}_-(t, t_0) A,$$

$$\hat{U}_-(t, t_0) = T_-[\exp\{\int dt' i\hat{L}_1(t')\}], \qquad (22)$$

where the symbol T_- indicates an increasing time ordering from the left to the right. Starting with the time evolution equation

$$d\hat{U}_-(t,t_0)/dt = \hat{U}_-(t,t_0) i\hat{L}_1(t), \qquad (23)$$

we obtain the fundamental equation of A(t) in the time convolution decomposition.

When $A = a_0$, we obtain as follows.

$$d\hat{a}_0(t)/dt = \hat{U}_-(t, t) Pi\hat{L}_1(t) a_0 + \Sigma\int dt_1 \int dt_2 \cdots \int dt_{n-1} \hat{U}_-(t_{n-1},t_0) Pi\hat{L}_1(t_{n-1}) Qi\hat{L}_1(t_{n-2}) \cdots Q\, i\hat{L}_1(t)\, a_0 + \hat{J}_-(t), \qquad (24)$$

where P represent the projection operator in HP, and Q=1-P and P^2=P. The operation of P takes the role of an appropriate averaging operation on a system and on a reservoir.

Here, each term of Eq. (24) is expressed as follows.

$$Pi\hat{L}_1(t) a_0 = 0 ,$$

$$P\, i\hat{L}_1(t_1) Q\, i\hat{L}_1(t) a_0 ,$$

$$= - \Gamma\int dk\, |\kappa_0(k)|^2 \exp[-i\delta_k(t - t_1)] a_0 ,$$

$$Pi\hat{L}_1(t_n) Q \cdots\cdots i\hat{L}_1(t_1) Qi\hat{L}_1(t)\, a_0 = 0 , \quad n>2$$

and

$$J_{-1} = Qi\hat{L}_1(t) a_0 = - \Gamma^{1/2} \int dk\, \kappa^*(k) \exp[i\delta_k(t - t_1)] b_k,$$

$$J_{-2} = \int dt_1 Qi\hat{L}_1(t_1) Qi\hat{L}_1(t) a_0, = (1 - P)\Gamma\int dk\, |k_0(k)|^2 a_0 ,$$

$$J_{-n} = 0, \quad \text{for } n > 3.$$

Finally, we obtain

$$d\langle a_0(t)\rangle/dt = - \Gamma\int dt_1 \int dk |\kappa(k)|^2 e^{-i\delta_k(t - t_1)} \langle a_0(t_1)\rangle. \qquad (25)$$

Solution: Initial condition is same as that for Eq. (15)

$$\langle a_0(t)\rangle = \langle a_0(0)\rangle e^{-i\omega_0 t} \times \pounds^{-1}[s + \pounds [f(s)]]^{-1}(t). \qquad (26)$$

6. Concluding Remark

1) We obtained the quantum mechanical Langevin equation of the lasing mode for the CW atom laser with the output coupling without the Born-Markov approximation.

2) We demonstrated that CW atom laser is a self-sustained oscillator using the Heisenberg equation .

3) The simplest case, pulsed atom laser, we chose the projection operator approach by the HP under the Born Approximation with non-Markovian process. It is noted that the solution (26) is accord with that of Eq. (19) and Ref. 3 when the reservoir 0 is initially empty.

4) The detail investigations of the section 3,4 and the property of the coherence will be published elsewhere.

Acknowledgment: The one of us (H.I.) thanks to Dr.T.Uchiyama for fruitful discussions on the projection operator approach in the Heisenberg picture.

References:

[1]M.Q.Mewes, et.al.Phys.Rev.Lett.**78** (1997)582; M.R. Andrews, et al., Science **275** (1997) 637.

[2] E.W. Hagley et.al., Science **275** (1997) 637; L.Bloch,T.W.Hansch, T. Esslinger, Phys. Rev. Lett. **82** (1999)3008.

[3]M. Holland et al, Phys.Rev. **A54**(1996) R1758, A.M .Guzman, M. Moore and P. Meystre, Phys. Rev. **A53** (1996)977.

[4]J.J.Hope, Phys. Rev. **A 55**(1997) R2531., G.M.Moy, J.J.Hope and C.M.Savage, **A59** (1999) 667.

[5]M.O.Scully, Phys. Rev. Lett. **82** (1999) 3927.

[6]T.Uchiyama and F.Shibata, Phys. Rev.**E 60** (1999)2636.

ENVIRONMENTAL EFFECTS ON QUANTUM REVERSAL OF MESOSCOPIC SPINS

R. GIRAUD, I. CHIORESCU, W. WERNSDORFER, AND B. BARBARA
Laboratoire de Magnétisme Louis Néel CNRS, BP166, 38042 Grenoble, Cedex-09, France

A. G. M. JANSEN
Laboratoire des Champs Magnétiques Intenses, BP166, 38042 Grenoble, Cedex-09, France

A. CANESCHI
Department of Chemistry, University of Florence, 50144, Italy

A. MUELLER
Fakultat fur Chemie, Universitat Bielefeld, D-33501, Bielefeld, Germany

A.M. TKACHUK
S.I. Vavilov State Optical Institute, 199034 St Petersburg, Russia

We describe what we learnt these last years on quantum reversal of large magnetic moments, using mainly conventional SQUID or micro-SQUID magnetometry. Beside the case of ferromagnetic nanoparticles with $10^3 - 10^5$ atoms (e.g. Co, Ni, Fe, Ferrites), most fruitful systems appeared to be ensembles of magnetic molecules. These molecules, generally arranged in single crystals, carry relatively small magnetic moments ($S = 10$ in Mn_{12}-ac and Fe_8). They are sufficiently apart from each other not to be coupled by exchange interactions. The ground multiplet is split over an energy barrier of tens of kelvin (≈ 67 K for Mn_{12}) by a strong local crystal field, leading to an Ising-type ground-state. Only weak inter-molecular dipolar interactions are present, as well as intra-molecular interactions, such as hyperfine interactions. Quantum properties of molecule spins are crucially dependent on their magnetic environment of electronic and nuclear spins (the spin bath). Energy fluctuations of the spin bath of about 0.1 K are important, especially at very low temperatures. In particular, they are much larger than the ground-state tunnel splitting of large-spin molecules in low applied fields, of about 10^{-8} K or even less (such a low value is due to the presence of large energy barriers). Theoretical predictions are experimentally checked for tunneling effects in the presence of non-equilibrated or equilibrated spin-energy distribution. It is also shown that the phonon-bath plays no role in low field, except when the temperature approaches the cross-over temperature to the thermal activation regime. In fact, spin–phonon transitions can play a role only if the tunnel splitting is not too small in comparison with k_BT. This is the case both for large-spin molecules in a large magnetic field (e.g. Mn_{12}-ac in a few tesla) and for low-spin molecules, as shown with the study of the molecule V_{15} (Hilbert space dimension as large as 2^{15} and spin 1/2). We also give our latest results on the extension of these studies beyond molecular magnetism. Single-ion slow quantum relaxation is observed in rare-earth Ho^{3+} ions highly diluted in an insulating matrix $LiYF_4$. This relaxation is due to the coherent tunneling of individual Ho^{3+} spins strongly coupled to their nuclear spins, leading to electro-nuclear entangled states at avoided level crossings. In fact tunneling of the spin system is induced by the hyperfine coupling. Together with the important role of the "spin bath", the roles of cross–spin and spin–phonon relaxations are also considered. All these results confirm the emergence of a new field of research: "mesoscopic magnetism".

1 Introduction

Quantum tunneling of the magnetization (QTM[1]) is one of the most striking features observed in nanomagnets with a large uniaxial anisotropy[2]. It is associated with low-frequency quantum fluctuations in the magnetic system, leading to staircaselike hysteresis loops at very low temperatures. This phenomenon is now evidenced both in molecular magnets (Mn_{12}, Fe_8, Mn_4,...) and in highly diluted rare-earth ions in a nonmagnetic single crystal. Furthermore, the dynamics of the magnetization reversal can also be limited by couplings to the phonon bath as shown in molecular magnets (V_{15} and other low-spin molecules) or in diluted rare-earth ions.

The paper is divided in three parts. We begin with the low temperature study of the high-spin molecule Mn_{12}-ac, showing the crossover from thermally assisted quantum tunneling to ground-state tunneling, when decreasing the temperature. In the second part, we present some hysteresis loops measured in the low-spin molecule V_{15}, underlying the relevance of spin–phonon transitions for large tunneling gaps. Finally, single-ion slow quantum relaxation in $LiY_{0.998}Ho_{0.002}F_4$ points out an intermediate case with two disctinct regimes showing either QTM or the role of spin–phonon and spin–spin transitions, both resulting in specific hysteresis loops at low temperatures. All the measurements were performed on single crystals.

2 Low temperature study of the high-spin molecule Mn_{12}-ac

The system Mn_{12}-ac is constituted of magnetic molecules with 12 Mn ions (spins 3/2 and 2), strongly coupled by super-exchange interactions. The resulting spin $S = 10$ is defined over $\approx 1\text{nm}^3$, with 10^3 atoms of different species (Mn, O, C, H). The Hilbert space dimension is huge: 10^8. However, in the mesoscopic approach where the spin $S = 10$ is assumed to be rigid, a Hilbert space dimension reduced to $2S + 1 = 21$, which is sufficient to account for most observations. Due to a large energy barrier between $m' = -10$ and $m = 10$ ($-DS^2 - BS^4 \approx$ 67 K, for second and fourth-order anisotropy from EPR[3]), the zero-field tunnel splitting of

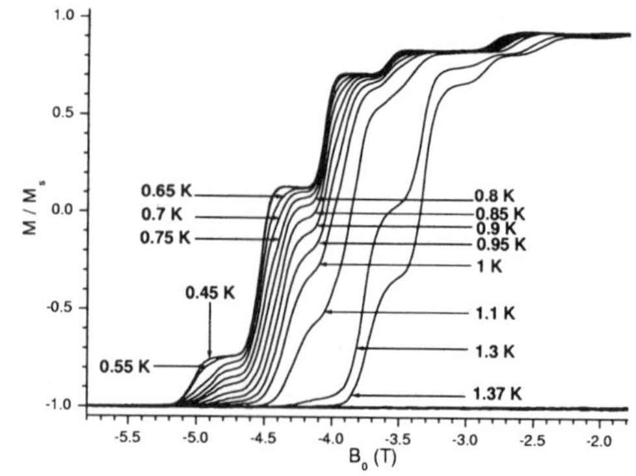

Figure 1. Hysteresis loops of Mn_{12}-ac, extracted from torque measurements with an applied field along the c–axis (sweeping rate r = 11 mT/s). Below a temperature depending on the involved resonance, the curves are temperature-independent, indicating tunneling from the ground-state. Transition widths result from the distribution of dipolar fields.

the ground state is extremely small (about 10^{-10} K). Quantum relaxation is thus slow and must be influenced by environmental degrees of freedom. Above 1.5 K, a single crystal of Mn_{12}-ac shows staircaselike hysteresis loops when the magnetic field is applied along the c–easy axis of magnetization[4] and references therein (for oriented powder experiments see[5]). Tunneling rates deduced from relaxation and ac-susceptibility experiments show maxima precisely at the fields where hysteresis loops make steps: $H_n \approx 0.44n$, with $n = 0, 1, 2, \ldots$[4]. These measurements give strong evidence for thermally assisted tunneling in Mn_{12}-ac and allow to study the role of spin–spin and spin–phonon transitions in the tunneling mechanism and the passage to the classical regime[6,7].

The results presented here, in the low temperature regime, determine the nature of the transition between thermally assisted tunneling and ground state tunneling and show some interesting effects of the spin and phonon baths on quantum relaxation. For that our measurements are extended in high fields and subkelvin temperatures. The magnetization of a small single crystal of Mn_{12}-ac was obtained from magnetic torque ex-

periments performed at the High Magnetic Field Laboratory in Grenoble (LCMI) below 1.4 K. The way to extract the longitudinal or transverse magnetization (parallel or perpendicular to the easiest $c-$axis) is analytical and unambiguous[8]. Figure1 shows several hysteresis loops obtained in a longitudinal field (the positive field part is rigorously symmetrical). The fields H_n are larger than those observed above 1.5 K. Moreover, they become independent of temperature below a crossover temperature $T_c(n)$[9]. These features suggest that tunneling takes place from the ground-state $m' = -10$ to $m = 10 - n$[10,11]. In our case $n = 7$, 8, 9 and 10. In Fig. 2, we visualize more clearly the resonances $n = 7$, 8 and 9, by plotting the field derivative of the hysteresis loops dM/dB_0 vs. B_0, measured at three different temperatures. Each resonance consists of two resolved peaks of widths ≈ 0.15 T, a value close to the width of the distribution of dipolar fields E_D.
Contrary to the high temperature case, where spin-phonon transitions at energy $\approx \Delta \approx E_D$, give homogeneous line broadenings[7], here resonance linewidths result from inhomogeneous distributions of dipolar fields (with $\Delta << E_D$). Regarding peaks position, a comparison between measured and calculated H_n makes their interpretation obvious. Calculated H_n are given by level crossings in Fig. 3b). Intercepts of the ground-state $m' = -S = -10$ with levels $m = 3$, $m = 2$, $m = 1$ and $m = 0$ give the H_n indicated by straight lines in Fig. 3a ($-10 \to 3$, $-10 \to 2$, $-10 \to 1$ and $-10 \to 0$; respectively noted as (7-0), (8-0), (9-0), (10-0) in Fig. 3). They coincide very well with the first peaks observed at low temperatures. Similar intercepts of $m' = -9$ with $m = 2$, $m = 1$, $m = 0$ and $m = -1$ are indicated by $-9 \to 2$, $-9 \to 1$, $-9 \to 0$ and $-9 \to -1$; these are noted as (7-1), (8-1), (9-1), (10-1) respectively, in Fig. 3. The agreement with the second peak is also very good. The overall agreement shows unambiguously that each

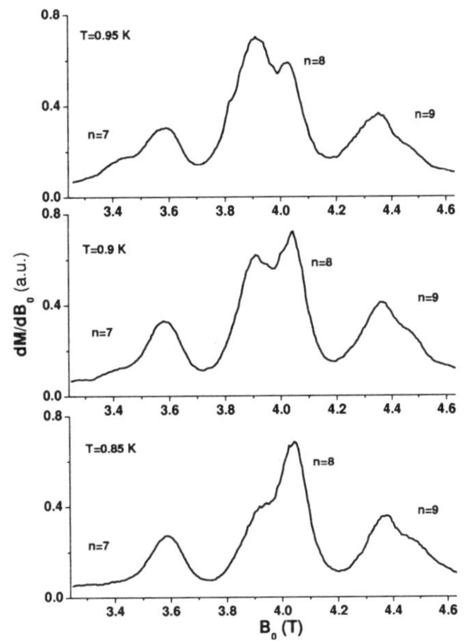

Figure 2. Derivative dM/dB_0 of the hysteresis loops of Fig.1 at different temperatures. Each measured resonance, with n=7, 8 and 9, is split in two: tunneling from the ground-state $m' = -10$ or from the first excited state $m' = -9$. For instance, below $T = 0.85$ K, the resonance $n = 7$ only shows one peak (the higher field one) associated with ground-state tunneling. In the same trend, above $T = 1$ K, the resonance $n = 9$ only shows one peak (the lower field one) associated with thermally assisted quantum tunneling.

resonance, corresponding to a given value of n, is in fact a "group of resonances". In each group, only two resonances are well resolved. The first one is associated with tunneling from the ground-state $m' = -10$ while the second is associated with tunneling from the first excited level $m' = -9$. The third resonance (tunneling from $m' = -8$) is also seen in the temperature dependence of H_n for $n = 7$, 8 and 9. The other ones, $m' = -7$, -6, -5... get closer and closer (Fig. 3b) and cannot be resolved experimentally. This is the reason why H_n appears to be dependent on temperature in the high temperature side of Fig. 3a. Note that the H_n can also be calculated analytically from the equality condition $E(m) = E(m' = -n - m)$, with $E(m) = -Dm^2 - Bm^4 - g\mu_B m H_z$ (the corre-

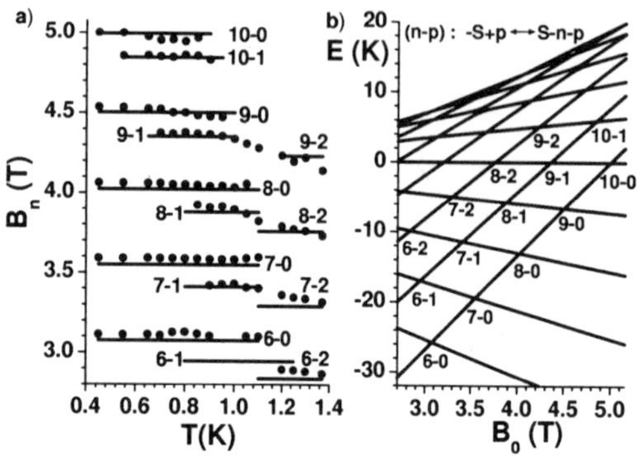

Figure 3. **a**: Position of peak maxima vs. temperature. The lines indicate fields at which level crossings (-n-m, m) are calculated. **b**: Energy levels spectrum calculated by exact diagonalization in a longitudinal field, for a spin $S = 10$ in a crystal field of tetragonal symmetry, with the Hamiltonian: $H = -DS_z^2 - BS_z^4 - C(S_+^4 + S_-^4) - g\mu_0\mu_B S_z H_z$. We used the parameters $D = 0.56$ K, $B = 1.1$ mK, $C = 3 \cdot 10^{-5}$ K obtained in EPR. The diagonal anisotropy term of fourth order shifts the resonances belonging to the same n and with different m to the left (lower fields) when m decreases.

sponding Hamiltonian is given in the caption of Fig. 3b [3]). The result for $H_n(m)$ is given by

$$H_n(m) = \frac{nD}{g\mu_B}[1 + \frac{B}{D}((m+n)^2 + m^2)] \quad (1)$$

which shows that shifts in H_n for the same n, are important only if m is large, i.e. at low temperature. At a higher temperature, resonances $H_n(m)$ with smaller m are close to each other and superposed. This is why they can be fitted to a single Lorentzian[7,12].

Regarding peak intensities, the area of deconvoluted peaks in each resonance group (n fixed) determine the fraction of tunneling events taking place from the ground state (N_{10}) or from the first excited state (N_9). Since the tunneling rate is proportional to $\Delta^2/(2m-n)$ [13], we obtain

$$\ln \frac{N_{10}}{N_9} \approx \frac{E_{10-9}}{kT} + 2\ln\frac{\Delta_{10}}{\Delta_9} \quad (2)$$

where E_{10-9} is calculated through the exact diagonalization of the Hamiltonian (see Fig. 3b). Systematic temperature dependences of tunnel fractions, analysed with the above expression, show that all the observed resonances have comparable tunneling gaps. We consider here only the case of the resonance $n = 8$ where quantum relaxation is dominated by the transverse fourth order crystal field term C (n being even, transverse magnetic fields have a negligible influence). Figure2 shows that the areas N_9 and N_{10} become equal at ≈ 0.93 K. This temperature is at the crossover between tunneling from $m' = -10$ and $m' = -9$. It is given by $kT_c(n = 8) \approx E_{10-9}/2\ln(\Delta_9/\Delta_{10})$. The calculated value for E_{10-9} is approximately equal to 7.3 K for $n = 8$, which gives $\ln(\Delta_9/\Delta_{10}) \approx 3.88$ corresponding to a transverse field of ≈ 0.31 T (which is associated with a misalignment of $\approx 4.5°$). It is worth noting that this cannot be attributed to our experimental misalignment, which is less than three degrees, so that the discepancies with the results obtained from the Hamiltonian used in the calculations come from a lack of accuracy in the knowledge of this Hamiltonian. Actually, the S_4 point symmetry of Mn_{12}-ac leads to the full Hamiltonian: $H = B_2^0O_2^0 + B_3^2O_3^2 + B_4^0O_4^0 + B_4^4O_4^4 + B_5^2O_5^2 + B_6^0O_6^0 + B_6^4O_6^4 + ...$ where O_l^m are the Steven's equivalent operators. Interestingly, the additional odd terms have no diagonal contribution, which is consistent with the good agreement we obtain regarding the resonance position. This also shows that diagonal terms of order ≥ 6 are negligible. However, Cornia *et al.*[15] pointed out very recently the existence of local distortions, breaking the S_4 symmetry. Including a second-order transverse anisotropy term in our calcuation gives tunneling gaps of about 10^{-5} K for odd values of n (e.g. $-10 \to 3$, $-10 \to 1$, $-9 \to 2$, $-9 \to 0$), as observed experimentally.

All these results give evidence that the transition from ground-state tunneling to thermally activation is not abupt, as in standard models and specific applications to

Mn_{12}-ac[14], but is of a new nature. Furthermore, if the applied field is perpendicular to the easiest axis of magnetization, the barrier remains symmetrical and we have shown [11] that ground-state to ground-state tunneling takes place. In addition, magnetic relaxation experiments both in longitudinal and transverse fields, show that tunneling takes place while the spin system is not at equilibrium with the bath at short time scales $M \propto \sqrt{\Gamma_{sq}t}$, and recovers equilibrium after a certain delay $M \propto exp - (\Gamma_{exp}t)$ [11,13]. The important effect of phonon recovery in the tunneling-induced relaxation was also discussed.

3 Hystresis loop measurements in the low spin molecule V_{15}

In this section we discuss the molecular complex $K_6[V^{IV}_{15}As_6O_{42}(H_2O)]\cdot 8H_2O$ (so-called V_{15}) [16]. Despite the absence of an energy barrier, magnetic hysteresis is observed over a timescale of several seconds in this spin 1/2 molecular complex. Such hysteresis loops can be understood on the basis of a dissipative two-level model, in which fluctuations and tunnel splittings are of the same energy, under a phonon bottleneck regime. Thus, spin-phonon couplings lead to long relaxation times and to a particular "butterfly" hysteresis loop. The crystal is made of molecules with fifteen V^{IV} ions of spin $S = 1/2$, placed in a quasi-spherical layered structure formed of a triangle, sandwiched by two hexagons. The unit-cell contains two V_{15} clusters, and it is large enough so that dipolar interactions between different molecules are negligible (a few mK). All intra-molecular exchange interactions are antiferromagnetic, resulting in a $S = 1/2$ total spin for each molecule. Such a small spin has zero energy barrier and a relatively large splitting in zero applied field ($\sim 10^{-2}$ K). Although the spin entanglement results in 2^{15} eigenstates per molecule, the magnetization curves can be interpreted in terms of a dissipative two-level model[17,18,19], as shown in Fig. 4.

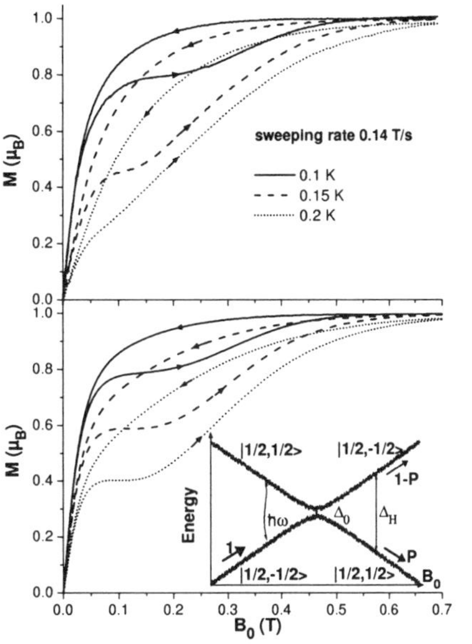

Figure 4. Measured (*top*) and calculated (*bottom*) hysteresis loops for three temperatures and for a given field sweeping rate 0.14 T/s. The plateau is more pronounced at low T. The inset is a schematic representation of a two-level system $S_Z = \pm 1/2$ with repulsion due to non-diagonal matrix elements. In a swept field the switching probability P is given by the Landau-Zener formulae. The two-levels are broadened by the hyperfine fields and the absorption or the emission of phonons can switch the polarization state of spins.

The measurements were performed with the micro-SQUID technique [21]. Only the positive parts of hysteresis loops are given (the other ones being rigorously symmetrical). When the field increases, coming from the negative saturation, the magnetization curve passes through the origin of the coordinates, reaches a plateau and then approaches saturation. This leads to a winged hysteresis loop characterized by the absence of irreversibility near zero field, as shown in Fig. 4. Nevertheless, the initial susceptibilities being larger the faster sweeping field (see Fig. 5), the magnetization is out of equilibrium also near zero field where it appears to be reversible.

The wings depend sensitively on temperature T and field sweeping rate r. In Fig.4a, where three hysteresis loops are presented at three different temperatures for a given sweeping rate, the plateau is higher and more pronounced at low temperature. The same tendency is observed at a given temperature and faster sweeping rates (Fig.5a). When compared to its equilibrium limit (dotted curve in Fig.5), each magnetization curve shows a striking feature: the plateau intersects the equilibrium curve and the magnetization becomes smaller than that at equilibrium. Equilibrium is then reached in higher fields near saturation.

Contrary to the case of a 2-level system described by the Landau-Zener model[18] in the non-adiabatic regime, the plateau of Fig.5 increases if the sweeping rate is increased. Taking the typical value $r = 0.1$ T/s and the zero-field splitting $\Delta_0 \cong 0.05\text{K}^{22,24,25,23}$, one gets a ground-state switching probability P very close to unity: in the absence of dissipation the spin 1/2 must adiabatically follow the field changes. Extremely large sweeping rates ($\approx 10^9$ T/s) would be needed to get into the quantum non-adiabatic regime $P < 1$. The mark of the V_{15} system is that the dissipative spin-phonon coupling is acting also near the zero applied field because $\hbar\omega \approx \Delta_0$ is of the order of the bath temperature, which is not the case for large spin molecules where $\Delta_0 << k_B T$. The spin temperature T_S is such that $n_1/n_2 = \exp(\Delta_H/k_B T_S)$, where $\Delta_H = \sqrt{\Delta_0^2 + (2\mu_B B_0)^2}$ is the two-level field-dependent separation, and $n_{1,2}(n_{1,2eq})$ is the non equilibrium (equilibrium) level occupation numbers. In the magnetization curves at 0.1 K (Fig.5a), the spin temperature T_S is significantly lower than the bath temperature T ($n_1 > n_{1eq}$, $T_S < T$) at fields between -0.3 T (when the magnetization curve departs from the equilibrium one) and 0.15 T (the field at which the magnetization curve intersects the equilibrium one).

After this intersect, T_S is larger than

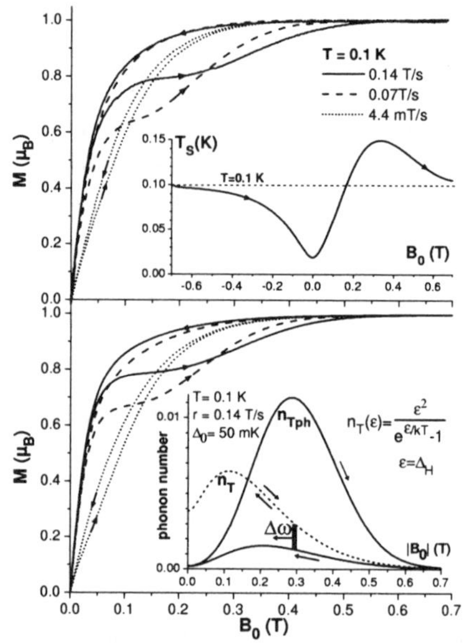

Figure 5. Measured (*top*) and calculated (*bottom*) hysteresis loops for three field sweeping rates at $T = 0.1$ K. The observed plateau is more pronounced at high sweeping rate. The equilibrium curve can be approximated by the median of the two branches of the low sweeping rate hysteresis loop (dotted curve). In the *top* inset is plotted the spin and phonon temperature $T_S = T_{ph}$ for $T = 0.1$ K and $r = 0.14$ T/s, when the field is swept from negative values. T_S decreases until zero-field and then increases linearly within the plateau region. Then it overpasses the bath temperature to finally reach the equilibrium. In the *bottom* inset the calculated number of phonons with $\hbar\omega = \Delta_H$ is plotted vs. the sweeping field modulus (note the arrows) at equilibrium ($T_{ph} = T_S = T$, dashed line) and out-of-equilibrium ($n_{T_{ph}} = n_{T=T_S}$, $r = 0.14$ T/s, black line). The difference between the two curves (thick segment $\Delta\omega$) suggests the moving hole in the phonon distribution, while their intersection gives the plateau intercept of the equilibrium magnetization curve.

the bath temperature ($n_1 < n_{1eq}$, $T_S > T$), and at sufficiently high fields (about 0.5 T) it reaches the equilibrium value ($n_1 = n_{1eq}$, $T_S = T$).

In a direct process, the spins at the temperature T_S should relax to the phonons temperature within a timescale τ_1, the phonons being at the bath temperature. However, even with a silver sample holder, it is not possible to maintain the phonon temperature equal to the temperature of the bath. This is because in V_{15} below 0.5 K, the heat capacity of the phonons C_{ph} is very much smaller than that of the spins C_S, so that the energy exchanged between spins and phonons will very rapidly adjust the phonons temperature T_{ph} to the spin one T_S. Furtheremore, the energy is transfered from the spins only to those phonon modes with $\hbar\omega = \Delta_H$ (within the resonance line width). The number of such lattice modes being much smaller than the number of spins, energy transfer between the phonons and the sample holder must be very difficult. This is a phenomenon known as the phonon bottleneck [26,27].

To conclude with this part, the V_{15} molecular complex constitutes an example of a dissipative two-levels system of mesoscopic size. The total spin 1/2 being formed of a large number of interacting spins, its splitting results from the structure of the molecule. Due to the multi-spins character, the ground-state is formed of two doublets. Intra-molecular Dzyaloshinsky-Moriya interactions, allowed by the symmetry of the molecule, should split each doublet in two singlets of the same energy at zero field. This scenario, explaining our first observation of a zero-field splitting with a half-integer spin, is not in contradiction with the Kramer's theorem, due to the fact that the degeneracy of the non-splitted ground-state is larger than $2S+1$, where S is the ground-state spin. Furthermore, the observed splitting is rather large (a fraction of Kelvin), due to the abscence of barrier (whereas in large-spin molecules the presence of large energy barriers lowers the splittings by orders of magnitude). Spin-phonon transitions within the tunneling gap are thus important, and the V_{15} system shows spin rotation under a strong coupling to the phonon bath contrary to large-spin molecules where resonant phonon transitions are irrelevant, unless between states at different energies [28] or in the presence of a transverse field large enough to create a tunnel splitting of the order of the temperature energy scale [29].

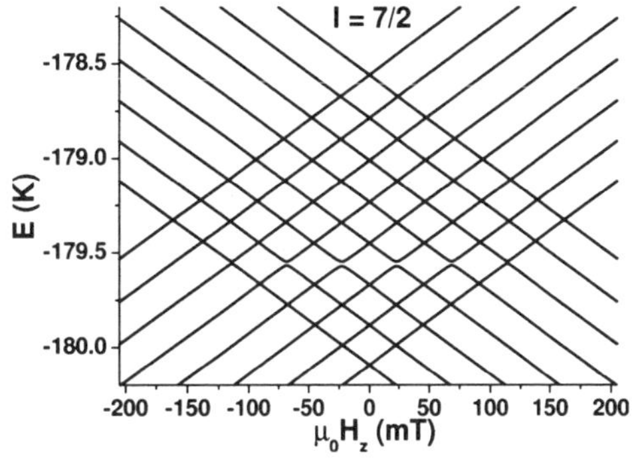

Figure 6. Splitting of the electronic ground-state doublet by the hyperfine interaction ($A_J/k_B \approx$ 38.6 mK; see below). The level crossings occur for resonant values of the longitudinal field H_n ($-7 \leq n \leq 7$). Some are avoided level crossings and hyperfine levels repulsion is then induced by the electronic level repulsion in the excited states.

4 Single ion slow quantum relaxation in $LiY_{0.998}Ho_{0.002}F_4$

Staircaselike hysteresis loops of the magnetization can also be observed in a $LiY_{0.998}Ho_{0.002}F_4$ single crystal, at subkelvin temperatures and low field sweep rates[30]. This behavior results from quantum dynamics at avoided level crossings of the energy spectrum of single Ho^{3+} ions in the presence of hyperfine interactions.

The crystal has a tetragonal scheelite structure with a S_4 point symmetry group at Ho^{3+} (for $LiHoF_4$, unit cell parameters are $a = b = 5.175$ Å and $c = 10.74$ Å [31]). Because of a very strong spin–orbit coupling, each magnetic ion of ^{165}Ho is characterized by its $J = 8$ ground-state manifold ($g_J = 5/4$), split through the crystal field effects. These give rise to a large uniaxial magnetic anisotropy, leading to an Ising-type system with an energy barrier hindering the magnetic moment reversal. The isolated magnetic moments are weakly coupled by dipolar interactions ($\mu_0 H_{\mathrm{dip}} \sim$ few mT) so that this very diluted insulator exhibit a nearly single ion quantum behavior. At very low temperatures, the system should be equivalent to a two-level system, but this picture is strongly modified when hyperfine interaction with the rare-earth nuclear spin is taken into account ($A_J \neq 0$), leading to the following Hamiltonian:

$$H = B_2^0 O_2^0 + B_4^0 O_4^0 + B_4^4 O_4^4 + B_6^0 O_6^0 + B_6^4 O_6^4$$
$$+ A_J \vec{J} \cdot \vec{I} - g_J \mu_0 \mu_B \vec{J} \cdot \vec{H} \qquad (3)$$

where O_l^m are the Steven's equivalent operators. This results in a more complex diagram in the electronic ground-state, showing several level crossings for resonant values H_n ($-7 \leq n \leq 7$) in Fig. 6. The transverse hyperfine contribution $\frac{1}{2}A_J(J_+I_- + J_-I_+)$ induces some avoided level crossings between $|\psi_1^-, I_{z1}>$ and $|\psi_1^+, I_{z2}>$, with $\Delta I = |I_{z2} - I_{z1}|$, only when $\Delta I/2$ is an odd integer so that the two electronic low-lying states $|\psi_1^\pm>$ are coupled through nondegenerated excited electronic levels. Note that the degeneracy of the other level crossings is also removed by internal fields fluctuations.

Magnetic measurements were made at $0.04 < T < 1$ K and for $\mu_0 H < 2$ T, with a micro-SQUID magnetometer [21] allowing field sweep rates up to 1 T/s. The crystal is first saturated in a large positive field applied along the $c-$axis $\mu_0 H_{\mathrm{sat}} \approx 0.3$ T, and then the field H_z is swept between $\pm H_{\mathrm{sat}}$. At slow field sweep rates, an *isothermal* process occurs, leading to staircaselike hysteresis loops at $T \leq 200$ mK. These well-defined steps

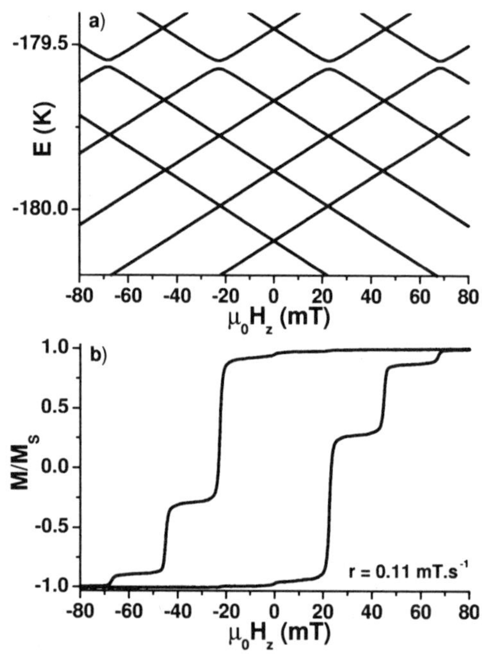

Figure 7. **a**: Zeeman diagram of the split electronic ground-state doublet by the hyperfine interaction (low-energy part). **b**: Hysteresis loops at $T \approx 40$ mK and for $r = 0.11$ mT/s showing quantum tunneling of the magnetization. The resonant values of the longitudinal field H_n coincide with the level crossings shown above.

come from quantum relaxation at avoided level crossings. At $T = 40$ mK, the lowest energy level is mainly populated and magnetization steps are observed for $-1 \leq n \leq 3$, as shown in Fig. 7b. The QTM in zero field is mainly associated with the dynamics of the lowest avoided level crossing (the first excited crossing is at $\Delta E/k_{\mathrm{B}} = g_{\mathrm{eff}}\mu_{\mathrm{B}}\mu_0 H_1/k_{\mathrm{B}} \approx 205$ mK, assuming $g_{\mathrm{eff}} \approx 13.3$[32]). The amplitude of the next step, the resonance $n = 1$ at $\mu_0 H_1 = 23$ mT, is much larger, suggesting a larger tunnel splitting Δ. Indeed, Fig.7a shows that the hyperfine induced tunnel splitting of the third excited avoided level crossing is large enough to render the barrier transparent ($\Delta \approx 25$ mK). The relaxation time is thus simply given by thermal activation $\tau = \tau_0 \exp(2\Delta E/k_{\mathrm{B}}T)$, with a long τ_0 because spin–lattice relaxation time T_1 can be hours at very low temperatures and/or as

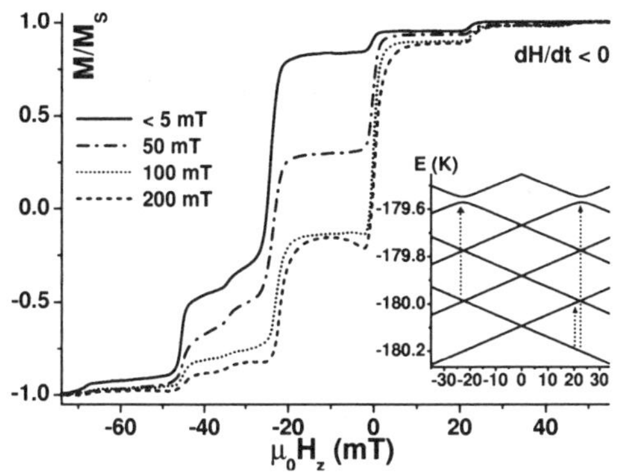

Figure 8. Hysteresis loops in a constant transverse field at $T = 50$ mK and for $r = 0.55$ mT/s. A transverse field enhances the quantum fluctuations in zero longitudinal applied field leading to a larger magnetization step. Inset: details of the Zeeman diagram around zero field. Thermally activated tunneling shows two possible channels over the first and the third, more efficient, excited avoided level crossings.

a result of internal fields fluctuations. It is worthy to note that the width of the resonant transitions is about $\mu_0 \Delta H = 2 - 3$ mT, which is expected from dipolar broadening. Similarly to molecular magnets, quasi-static fields due to dipolar interactions lead to a distribution of internal fields whereas field fluctuations, essentially of F^- nuclear spins, give homogeneous level broadening.

The quantum relaxation is strongly enhanced by a constant transverse field, as a result of an increase of the tunnel splittings (see Fig. 8). This allows the study of the relative magnitude of tunnel splittings. In zero longitudinal field, the small tunnel splittings rapidly increase and hysteresis vanishes. A saturation of the magnetization at $M \approx 0$ is observed in transverse fields larger than 100 mT, when the barrier is nearly transparent, and the small "overshot" with an oscillation in M may be due to spin–phonon transitions. As expected for a large tunnel splitting, sensitivity to a small transverse field is very weak for the resonance $n = -1$.

At faster sweep rates, nonequilibrated spin–phonon and spin–spin transitions, mediated by weak dipolar interactions, lead to magnetization oscillations and additional steps. A hysteresis loop measured at $T = 50$ mK for a much faster field sweep rate ($r = 0.3$ T/s) is shown in Fig.9a. A succession of equally spaced large and weak magnetization steps occur at fields H_n, with $-14 \leq 2n \leq 14$. The larger ones, with integer n, are associated with several equally spaced level crossings and the smaller steps, with half integer n, fall just in between when the levels are equally spaced (see Fig.6). The dm/dH is used to determine the H_n values plotted in Fig.9a inset. From the slope, we accurately obtain $\mu_0 H_n = n \times 23$ mT. The electronic ground doublet is thus split by hyperfine interaction in eight doublets over an energy range of about 1.44 K. We deduce $A_J/k_B \approx 38.62$ mK, to be compared to $A_J/k_B \approx 40.95$ mK[32]. The observed hysteresis loops depend sensitively on sample thermalization, showing that the spin–phonon system is not at equilibrium with the cryostat, leading to a phonon bottleneck[26,27]. At a fast field sweep rate $r = 0.3$ T/s, the system enters such a regime at $T \approx 1$ K (moderate sample thermalization) showing hysteresis without any magnetization steps down to $T \approx 600$ mK. When the field is swept back and forth, a stationary regime occurs and hysteresis loops become nearly temperature-independent below a temperature $T_c(r)$ depending on sample thermalization ($T_c \approx 200$ mK for $r = 0.3$ T/s). Below $T \approx 600$ mK, a nearly *adiabatic* process occurs, due to a much longer spin–lattice relaxation time T_1. The spin system becomes more and more isolated from the phonon bath, and energy exchange between electronic and nuclear spins is only possible at fields H_n. Equilibrium within the spin system is due to either quantum fluctuations at avoided level crossings (integer n) or to spin–phonon transitions and/or cross-spin relaxation, allowed by weak dipolar interactions, when energy levels are almost equally spaced (integer and half integer n)[33]. Spin–

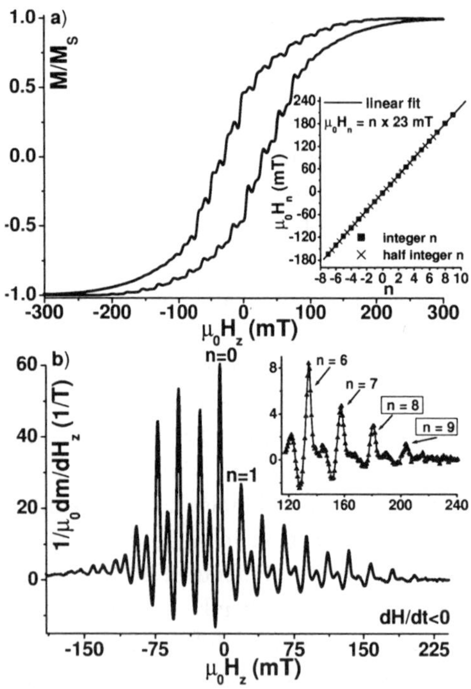

Figure 9. **a**: Hysteresis loops at T = 50 mK and for r = 0.3 T/s. Several magnetization steps are observed for resonant values of the applied field $\mu_0 H_n \approx n \times 23$ mT [see inset; H_n values are deduced from Fig. 9b)]. **b**: Derivative of the loop shown in **a** for a decreasing field. The two additional measured steps shown in the inset, for $n = 8$ and $n = 9$, are associated with cross-spin relaxation only.

spin interactions allow two additional steps for $n = 8$ and $n = 9$, at fields with equally spaced levels but no level crossing [Fig.9b inset]. A small transverse applied field only increases the zero-field magnetization step, showing the weak effect of enhanced quantum fluctuations on hysteresis loops in this regime. Other resonances and small magnetization steps, dominated by cross-spin relaxation, are not affected by a small transverse field, if small enough ($\mu_0 H_T \leq 80$ mT). If the field sweep is suddenly stopped, the spin–phonon system exchanges energy with the cryostat and the magnetization relaxes toward the equilibrium curve.

We thus have shown that diluted rare-earth ions in a nonmagnetic insulating single crystal are very suitable to study the possible entanglement of nuclear and electronic moments, when tunneling occurs. Very diluted Holmium doped $LiYF_4$ is indeed a model system to study tunneling of an electronic moment strongly coupled to its nuclear spin. Similarly to the case of high-spin molecules, the quantum rotation of weakly coupled magnetic moments of Ho^{3+} ions can be driven, and even monitored in this case, by hyperfine couplings at very low temperatures. In a constant transverse field, the magnetization steps, associated with incoherent tunneling at the avoided level crossings, increase very rapidly, as a consequence of the increase of the tunnel splittings. At faster field sweep rates, additional magnetization steps are observed and attributed to cross-spin relaxation and spin–phonon transitions in a phonon bottleneck regime, showing the relevance of couplings to the phonon bath in the fast sweep rate regime.

Conclusion

We have shown that the new field of "mesoscopic magnetism", which studies the tunneling of large magnetic moments in the presence of phonons and spins, can now be studied in very versatile systems such as molecular complexes or highly diluted rare-earth ions. The problem of quantum dynamics of a two-level system coupled to an environment (boson or fermion bath), which is at the core of mesoscopic physics [17], can thus be investigated in details. In particular, after the first studies on the large-spin molecules Mn_{12}-ac [4,5,34] and Fe_8 [35], the role of the spin bath on the tunnel mechanism was shown [7,13,36,37]. In molecules as well as in diluted rare-earth ions, quasistatic fields due to dipolar interactions between molecules lead to a distribution of internal fields, and field fluctuations, essentially of nuclear spins, give homogeneous level broadening allowing the restoration of tunneling in a finite energy window, at low temperature (this broadening being much larger than the phonon one, it is more relevant to induce tunneling). This mechanism is efficient unless all nuclear spins of the molecule are frozen, which occurs only below the mK scale. In low-spin molecules, large tunneling gaps favor spin–phonon transitions. Although the hyperfine induced level broadening is the same as in large-spin molecules, the phonon bath becomes as important as the spin bath [20]. Both regimes, that is the decay out of a field-induced metastable state (QTM) or the decay of a field-induced out-of-equilibrium state mediated by spin–spin and spin–phonon transitions, were observed in a 0.2% Holmium doped $LiYF_4$ single crystal. In all these cases, the role of field fluctuations was clearly evidenced.

Acknowledgments

We thank T. Goto, S. Maegawa, S. Miyashita, N. Prokof'ev, P. Stamp, I. Tupitsyn and A. Zvezdin, for fruitful on-going collaboration and discussions.

References

1. *Quantum Tunneling of Magnetization – QTM '94*, NATO ASI, edited by L. Gunther & B. Barbara, Series E **301**, (Kluwer Publishing, 1995).
2. B. Barbara and L. Gunther, Physics World **12**, 35 (1999).
3. A. L. Barra, D. Gatteschi, R. Sessoli, Phys. Rev. B **56**, 8192 (1997).
4. L. Thomas, F. Lionti, R. Ballou, R. Sessoli, D. Gatteschi and B. Barbara, Nature **383**, 145 (1996).
5. J. R. Friedman, M. P. Sarachik, J. Tejada, R. Ziolo: Phys. Rev. Lett. **76**, 3830 (1996); J. M. Hernandez, X. X. Zhang, F. Luis, J. Bartolome, J. Tejada, R. Ziolo, Europhys. Lett. **35**, 301 (1996).
6. L. Thomas, A. Caneschi, B. Barbara, Phys. Rev. Lett. **83** 2398 (1999); L. Thomas and B. Barbara, J. of Low Temp. Phys. **113**, 1055 (1998).
7. B. Barbara *et al.*, J. Magn. Magn. Mat. **200**, 167 (1999).
8. I. Chiorescu, Ph.D. thesis, Université Joseph Fourier, Grenoble (France), 2000.
9. J. A. A. J. Perenboom *et al.*, Phys. Rev. B **58**, 333 (1998); A. D. Kent *et al.*, EuroPhys. Lett. **49**, 521 (2000).
10. L. Bokacheva, A. D. Kent, M.A. Walters, Phys. Rev. Lett. **85**, 4803 (2000).
11. I. Chiorescu, R. Giraud, A. Caneschi, A. G. M. Jansen, B. Barbara, Phys. Rev. Lett. **85**, 4807 (2000).
12. B. Barbara, L. Thomas, F. Lionti, A. Sulpice, A. Caneschi, J. Magn. Magn. Mat. **177-181**, 1324 (1998).
13. N. V. Prokof'ev and P. C. E. Stamp, Phys. Rev. Lett. **80**, 5794 (1998); J. of Low Temp. Phys. **113**, 1147 (1998); Rep. Prog. Phys. **63**, 669 (2000).
14. E. M. Chudnovsky and D. A. Garanin; Phys. Rev. Lett. **79**, 4469 (1997); Phys. Rev. B **56**, 11102 (1997). In these papers the authors analyse the passage from ground-state to thermally activated tunneling in terms of a "phase transition". We disagree with this point of view.
15. A. Cornia *et al.* cond-mat/0112112.
16. D. Gatteschi, L. Pardi, A. L. Barra, A. Müller, J. Döring, Nature **354**, 465 (1991).
17. A. J. Leggett, S. Chakravarty, A. T. Dorsey, M. P. A. Fisher, A. Garg, W. Zwerger, Rev. Mod. Phys. **59**, 1 (1987).
18. C. Zener, Proc. R. Soc. London, Ser. **A137**, 696 (1932); L.D. Landau, Phys. Z. Sowjetunion **2**, 46 (1932).
19. H. De Raedt *et al.*, Phys. Rev. B **56**, 11761 (1997); S. Miyashita, J. Phys. Soc. Jpn. **64**, 3207-3214 (1995); Y. Kayanuma, H. Nakayama, Phys. Rev. B **57**, 13099 (1998); E. Shimshoni, A. Stern, Phys. Rev. B **47**, 9523 (1993).
20. I. Chiorescu *et al.*, Phys. Rev. Lett. **84**,

3454 (2000).

21. W. Wernsdorfer *et al.*, Phys. Rev. Lett. **79**, 4014 (1997).
22. For an isolated Kramers spin 1/2, $\Delta_0 = 0$. However, the total spin 1/2 of V_{15} comes from 15 coupled spins and different intra-molecular couplings such as hyperfine ($A\vec{I} \cdot \vec{S}, I = 7/2, A \approx 10$ mK [23]) and Dzyaloshinsky-Moriya interactions [24,25] could generate a splitting.
23. G. C. Carter, L. H. Bennett, D. J. Kahan, Metallic Shifts in NMR, Progr. in Mat. Sc **20**, part I, 364, Pergamon Press Ltd., (1977).
24. M. I. Katsnelson, V. V. Dobrovitski, B. N. Harmon, Phys. Rev. B **59**, 6919 (1999); cond-mat/9906375.
25. B. Barbara, L. Thomas, F. Lionti, A. Sulpice, A. Caneschi, J. Magn. Magn. Mat. **177-181**, 1324 (1998).
26. J. H. Van Vleck, Phys. Rev. **59**, 724 (1941); K. W. H. Stevens, Rep. Prog. Phys. **30**, 189 (1967).
27. A. Abragam and B. Bleaney, *Electronic Paramagnetic Resonance of Transition Ions* (Clarendon Press, Oxford, 1970).
28. F. Hartmann-Boutron, P. Politi and J. Villain, Int. J. Mod. Phys. B **10**, 2577 (1996); M. Leuenberger, D. Loss, Euro-Phys. Lett. **46**, 5, 692 (1999) and cond-mat/9907154.
29. G. Belessa, N. Vernier, B. Barbara, D. Gatteschi, Phys. Rev. Lett. **83**, 2, 416 (1999).
30. R. Giraud *et al.*, Phys. Rev. Lett. **87**, 57203 (2001).
31. E.R. Thoma *et al.*, J. Phys. Chem. **65**, 1906 (1961).
32. J. Magariño *et al.*, Phys. Rev. B **13**, 2805 (1976).
33. J.C. Verstelle and D.A. Curtis, *Paramagnetic Relaxation*, in *Handbuch Der Physik, Magnetismus* (Springer-Verlag, Berlin, 1968).
34. B. Barbara *et al.*, J. Magn. Magn. Mater. **140-144**, 1825 (1995); A. Novak and R. Sessoli, in [1] p. 171.
35. C. Sangregorio *et al.*, Phys. Rev. Lett. **78**, 4645 (1997); W. Wernsdorfer and R. Sessoli, Science **284**, 133 (1999).
36. W. Wernsdorfer *et al.*, Phys. Rev. Lett. **84**, 2965 (2000).
37. I.Tupitsyn and B.Barbara, in *Magnetoscience-From Molecules to Materials*, edited by J.S. Miller and M. Drillon (Wiley VCH, Weinheim, 2000); cond-mat/0002180.

RESISTANCE OF GEOMETRICALLY CONFINED MAGNETIC DOMAIN WALL

T. ONO, A. KOGUSU, T. FUKUDA, D. FUSE AND S. NASU
Graduate School of Engineering Science, Osaka University, Toyonaka 560-8651
E-mail: ono@mp.es.osaka-u.ac.jp

K. MIYAKE, K. SHIGETO, K. MIBU AND T. SHINJO
Institute for Chemical Research, Kyoto University, Uji 611-0011

Small contact structure between two NiFe wires was fabricated by an electron beam lithography and a lift-off method and the magnetoresistance was measured. The magnetization switching process was artificially controlled by engineering the sample geometry, and a single domain wall was trapped in the small contact area. The contribution of the domain wall to the resistance was negative, which can be attributed to the anisotropic magnetoresistance. At low temperatures, the resistance increases as the temperature decreases, and the resistance with the domain wall shows almost the same temperature dependence as that without the domain wall.

1 Introduction

Magnetism and transport on mesoscopic materials have attracted considerable attention due to developments in lithography and nano-technology. How do the quantum coherence and decoherence emerge in nano-magnets? One of the interests is the effect of the magnetic domain wall on the electric transport phenomena in ferromagnetic wires.

Theories predict a large positive contribution of the domain wall to the electrical resistance if the domain wall is thin enough compared to the Fermi wave length of the conduction electrons[1-3]. Since the Fermi wave length of metal is very small, this condition requires a domain wall with several atomic length. Such a thin domain wall can be realized only for materials with huge magnetocrystalline anisotropy or very small exchange stiffness constant. However, this situation is changed, if the magnets have a very small constriction. As theoretically proposed by Bruno[4], the domain wall is geometrically confined in such a small constriction, and the width of the domain wall becomes as small as the size of the constriction. Garcia *et al.* experimentally observed large magnetoresistance up to 300 % at room temperature for point contacts between two macroscopic Ni wires[5]. They insist that their results can be attributed to the formation of the domain wall where the spin rotation finishes within several atoms[3]. The switching of quantum conductance by the external magnetic filed of Ni point contacts reported by Ono *et al.* seems also related to a trapped domain wall at the point contacts[6, 7]. However, there is no clear evidence for the existence of the domain wall at the contacts in above two experiments. To elucidate the relation between the wall resistance and the wall width, experiments on a well-defined domain wall are needed.

On the other hand, in diffusive transport region, it has been pointed out from the theoretical viewpoints that the domain wall suppresses the quantum interference effects, such as weak localization and electron-electron interaction effects[8, 9]. The confined domain wall is a good candidate to investigate these theoretical predictions, since the phase coherence of electron should be comparable to the domain wall width at low temperatures.

In this contribution, we report the geometrical confinement of a domain wall in a nano-contact between two NiFe wires, which was fabricated by an electron beam lithography and a lift-off method. The results of the magnetoresistance measurements can be interpreted as the anisotropic magnetoresistance (AMR) effect caused by the confined domain wall. At low temperatures, the resistance increases as the temperature decreases, and the resistance with the domain wall shows almost the same temperature dependence as that without the domain wall.

2 Experiment

A schematic illustration of the sample structure is shown in Fig. 1. The width of the wires connected with each other is different. Furthermore, the wire with wider width is connected to a pad (large area) at an end[10], and the wire with narrower width has a sharp-pointed shape at an end[11]. These special shapes are introduced for the purpose of the control of magnetization switching process[12]. The expected switching process is as follows. When the magnetic field is applied parallel to the wire axis, a single domain wall is injected from the pad and the magnetization reversal takes place in the wider wire. The domain wall is trapped at the small contact between two wires until the magnetic field increases up to a critical value. The sharp end prevents a domain wall from nucleating at this end. When the field exceeds the critical field, the domain wall is depinned from the contact area and all the switching process is completed.

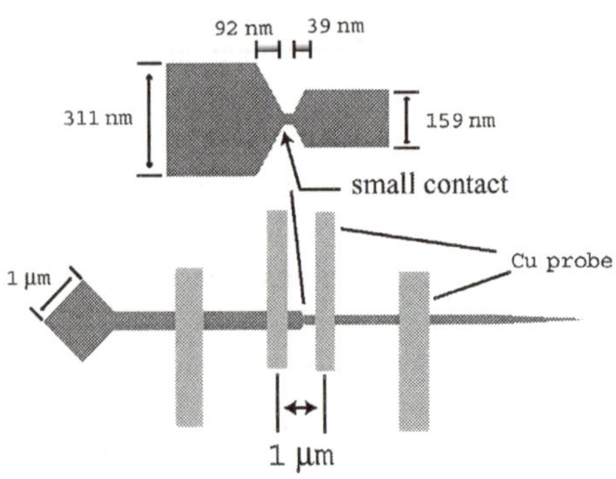

Fig. 1 Typical sample shape for resistance measurement. The size parameters were estimated from the SEM image.

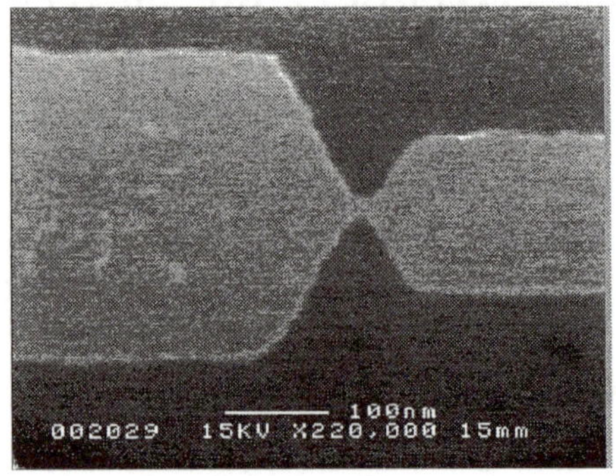

Fig. 2 SEM image of a small contact.

The samples were prepared on thermally oxidized Si substrates using an electron beam lithography and lift-off process. Many samples with different designed distance between two wires were patterned on one substrate. After deposition and a lift-off, some samples incidentally connected with a very small contact can be obtained. For example, the size of the contact shown in Fig. 2 is 15 × 20 nm^2. The thickness of the prepared samples for the magnetoresistance measurements is 7 nm. The samples have four current-voltage probes made of a nonmagnetic material, Cu, in order to prevent the influence from magnetic probes. The distance between voltage probes is 1 µm.

3 Results and Discussion

Figure 3(a) shows a typical

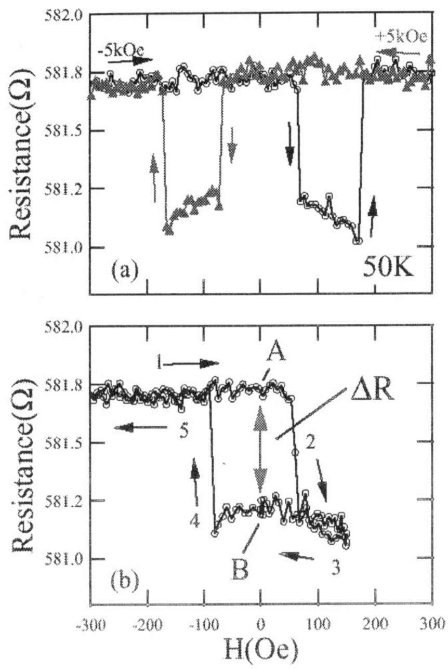

Fig. 3 Hysteresis loop of magnetoresistance at 50 K.
(a) Full hysteresis from -5 to 5 kOe.
(b) Minor curve: (A) and (B) correspond to the state without and with a domain wall, respectively.

magnetoresistance curve measured at 50 K. The contact size of the sample is 22 × 34 nm^2. Prior to the measurement, a magnetic field of -5 kOe was applied parallel to the wire axis in order to align the magnetization in one direction and then the field was swept to the counter direction. Abrupt negative jump was observed at +65 Oe and the reistance gradually decreases until the field reaches +172 Oe, where the resistance suddenly changed to almost the same value as that before the negative jump at +65 Oe. Similarly two jumps were observed for the measurement from +5 to –5 kOe. The negative jump at +65 Oe can be interpreted as the injection and the trapping of the domain wall.

Schematic illustrations of the inferred domain structures from the

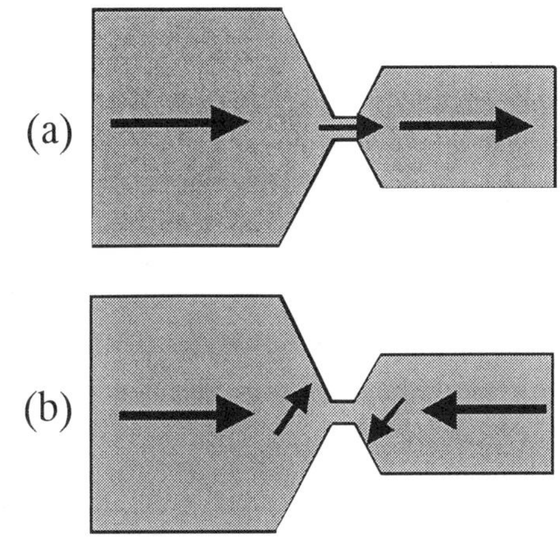

Fig. 4 Schematic illustratoins of the inferred magnetic structures around a small contact from the magnetoresistance measurements. (a) and (b) correspond to the state (A) and (B) in Fig. 3(b), respectively.

magnetoresistance measurements are shown in Fig. 4. Figs. 4(a) and 4(b) correspond to the state (A) and (B) in a minor loop shown in Fig. 3(b), respectively. Here, the magnetic field was decreased from +150 Oe before the positive jump as shown by arrows in the figure. In the state (B) in Fig. 3(b), the magnetic moments around the contact turn transversely to the wire axis in the sample plane, and a 180 degree domain wall is confined in the small contact.

The gradual change in resistance between +65 and +150 Oe is almost reversible, indicating the reversible rotation of the magnetic moments in the domain wall. The resistance at zero field returning from +150 Oe is smaller than that at zero field from -5 kOe. This difference in resistance between the point (A) and the point (B) both at zero field is due to the existence of the domain wall. The domain wall decreases the resistance. The value of the domain wall resistance, ΔR = -0.513 Ω, can be understood on the basis of the AMR, since the trapped domain wall has the magnetic moments tilted transversely to the current direction, resulting in the decrease in resistance due to the AMR effect. For rough estimation, we assume that the resistance change originates from magnetization rotation of only the narrow part around the small contact. The resistance of the narrow part is estimated to be 185 Ω. The ratio of the ΔR to 185 Ω corresponds to -0.3 %. The AMR ratio of -1.1 % was obtained for the same sample from a magnetoresistance measurement, where a magnetic field was applied perpendicular to the wire axis in the sample plane. Therefore, the change in resistance less than 1.1 % can be explained only by the AMR effect.

The resistance of the sample decreases with a decrease of temperature down to 20 K and then starts to increase. Figure 5 shows the temperature dependence of the resistance at low temperatures. The circles and crosses indicate the results with and without the domain wall, respectively. The observed log$T$ dependence of the increase in resistance is consistent with the theory of two-dimensional weak localization or of two-dimensional electron-electron interaction (EEI) effect. The increase in resistance at low temperatures may be due to the EEI effect[13], which was also observed in Ni wires with 20 nm in width[14]. The resistance with the domain wall shows almost the same temperature dependence as that without the domain wall.

4 Conclusion

In conclusion, we succeeded in confining a domain wall in a small contact fabricated by an electron beam lithography and a lift-off method. The resistance with the domain wall confined in the small contact was lower than that without the domain wall. This negative contribution of the domain wall to the measured resistance can be understood on the basis of the AMR effect. The wall in the investigated sample is not thin enough to obtain large positive magnetoresistance effect. The resistance of the sample decreases as the temperature decreases. The effect of the domain wall on the temperature dependence of the resistance could not be observed.

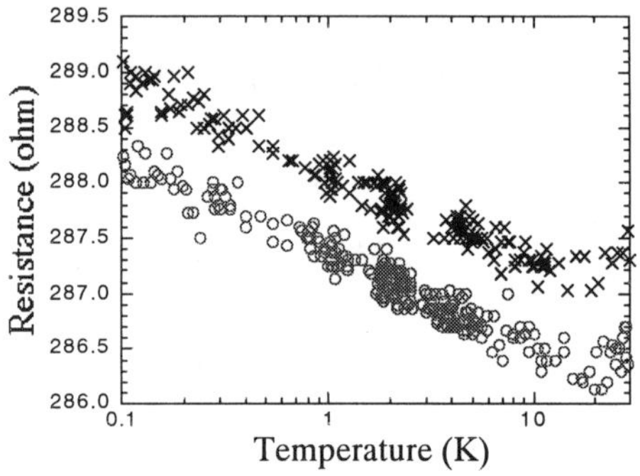

Fig. 5 Temperature dependence of resistance. The circles and crosses indicate the results with and without the domain wall, respectively.

Acknowledgments

The authors would like express thanks to Dr. Y. Suzuki for his valuable discussions. This work is partly supported by Grants-in-Aid for Research for the Future from Japan Society for the Promotion of Science, and for COE Research (10CE2004) from the Ministry of Education, Science, Sports and Culture.

References

1. Cabrera and L. M. Falicov, Phys. Status Solidi B **61**, 539 (1974).
2. van Hoof *et al.*, Phys. Rev. B **59**, 138 (1999).
3. Tatara *et al.*, Phys. Rev. Lett., **83**, 2030 (1999).
4. P. Bruno, Phys. Rev. Lett., **83**, 2425

(1999).

5. N. Garcia, M. Munoz, and Y. W. Zhao, Phys. Rev. Lett. **82**, 2923 (1999).
6. T. Ono, Y. Ooka, H. Miyajima and Y. Otani, Appl. Phys. Lett. **75,** 1622 (1999).
7. H. Imamura, N. Kobayashi, S. Takahashi and S. Maekawa, Phys. Rev. Lett., **84**, 1003 (2000).
8. G. Tatara and H. Fukuyama, Phys. Rev. Lett., **78**, 3773 (1997).
9. Y. Lyanada-Geller, I. L. Aleiner and P. M. Goldbart, Phys. Rev. Lett. **81**, 3215 (1998).
10. K. Shigeto, T. Shinjo and T. Ono, Appl. Phys. Lett. **75**, 2815 (1999).
11. T. Schrefl, J. Fidler, K. J. Kirk and J. N. Chapman, J. Magn. Magn. Mater. **175**, 193 (1997).
12. Y. Yokoyama, Y. Suzuki, S. Yuasa, K. Ando, K. Shigeto, T. Shinjo, P. Gogol, J. Miltat, A. Thiaville, T. Ono and T. Kawagoe, J. Appl. Phys. **87**, 5618 (2000).
13. B.L. Altshuler and A.G. Aronov, in *Electron-Electron Interactions in Disordered Systems*, edited by M. Pollak and A.L. Efros (North-Holland, Amsterdam, 1985), p. 1.
14. T. Ono, Y. Ooka, S. Kasai, H. Miyajima, K. Mibu and T. Shinjo, J. Magn. Magn. Mater., in press.

BLOCH OSCILLATION OF A MAGNETIC DOMAIN WALL

JUNYA SHIBATA

Department of Physics, Osaka University, Toyonaka, Osaka 560-0043, Japan
E-mail: shibata@acty.phys.sci.osaka-u.ac.jp

KOMAJIRO NIIZEKI

Department of Physics, Tohoku University, Sendai 980-8578, Japan
E-mail: niizeki@cmpt.phys.tohoku.ac.jp

SHIN TAKAGI

Fuji Tokoha University, Fuji 417-0801, Japan
E-mail: takagi@fuji-tokoha-u.ac.jp

We show that the quantum dynamics of a domain wall in a quasi-one dimensional mesoscopic ferromagnet is equivalent to that of a Bloch particle within a one-band basis. Collective degrees of freedom of the domain wall, namely the center position and the chirality, correspond to position coordinate and periodic momentum operators, respectively, which are mutually canonically conjugate. Due to the periodicity of the momentum of the Bloch particle, it is shown that position coordinate of the domain wall is quantized in unit of $a/2S$, where a is the lattice constant of the spin chain and S is the magnitude of spin. Various dispersion relations of energy-band for the domain wall are derived from transverse anisotropy or external magnetic field perpendicular to the easy axis. As an expected phenomenon, under a uniform magnetic field, the domain wall oscillates along the chain. This corresponds to the Bloch oscillation. Using the spin-coherent state path integral on the basis of stationary action approximation, we show how this oscillation manifests itself in the transition probability.

1 Introduction

Recent nanostructure technology enables us to study mesoscopic magnetic domain wall, which has stimulated our interest in macroscopic quantum phenomena (MQP) [1]. In existing literature for MQP, the domain wall has been treated as a usual quantum mechanical particle, which is described by the relevant collective degrees of freedom, namely, the center position and the chirality [2]. While, a Bloch particle (electron), (we mean by a Bloch particle the one whose state is confined to the space of a single band), is a purely quantum-mechanical particle because its position is quantized in unit of the lattice spacing, while the canonically conjugate variable to the position is the quasi-momentum which has a periodicity [7]. Recently, Braun and Loss [4] proposed an analogy between a domain wall and the Bloch particle. They assumed a periodic pinning potential in the magnetic system for the domain wall to obtain an energy band. While, Kyriakidis and Loss [5] showed that magnetic solitons in anisotropic spin-1/2 chains exhibit Bloch oscillations [6] under an external magnetic field. In fact, we have found that a magnetic domain wall in a spin chain formed of spins with an arbitrary magnitude S inherently has the property of a Bloch particle even in the absence of periodic pinning potential. In this proceedings, we demonstrate the equivalence between the quantum dynamics of a domain wall and that of a Bloch particle, and show that the magnetic domain wall is found to exhibit the Bloch oscillation under a uniform external magnetic field. This oscillation is quantized into equally spaced energy levels, yielding the so-called Wannier-Stark ladder [7].

2 Quantum dynamics of a domain wall

We replace a quasi-one-dimensional ferromagnet by a chain of spins, each with magnitude S. We take the x axis in parallel to the spin chain and assume that the easy and the hard axes of the ferromagnet are parallel to the z and x axes, respectively. Accordingly, we adopt the following Hamiltonian which is justified in the quasi-continuous regime:

$$\hat{H} = \int_{-L/2}^{L/2} \frac{dx}{a} \left\{ Ja^2 \partial_x \hat{\boldsymbol{S}}(x) \cdot \partial_x \hat{\boldsymbol{S}}(x) - \frac{K_z}{2} \hat{S}_z^2(x) + \frac{K_x}{2} \hat{S}_x^2(x) \right\} + \hat{H}_{\text{ext}}, \quad (1)$$

where L is the length of the ferromagnet, a is the lattice constant of spin chain, J is the exchange coupling constant, while K_z and K_x are longitudinal and transverse anisotropy constants, respectively. These parameters are all assumed to be positive. The second term of the right-hand side of (1) is the Hamiltonian representing the effect of a uniform external magnetic field:

$$\hat{H}_{\text{ext}} = -g\mu_{\text{B}} \boldsymbol{B} \cdot \int_{-L/2}^{L/2} \frac{dx}{a} \boldsymbol{S}(x), \quad (2)$$

where g is the g-factor and μ_{B} is the Bohr magneton. In the absence of the transverse anisotropy and the magnetic field, a domain wall is defined as a classical stationary-solution for the Hamiltonian. It has two parameters, Q and ϕ, which specify the center position and the chirality, respectively (Fig.1); the chirality is the azimuthal angle around the z axis of each spins in the region of the domain wall projected onto the x-y plane. The width of the domain wall is given by $\lambda = \sqrt{\frac{Ja^2 S}{K_z(S-\frac{1}{2})}}$. The energy of the domain wall is given by $E_{\text{DW}} := 2N_{\text{DW}} K_z S(S-1/2)$, where $N_{\text{DW}} := \lambda/a$ stands for the number of spins in the domain wall. Note that $N_{\text{DW}} \gg 1$ in the quasi-continuous regime, where $\lambda \gg a$.

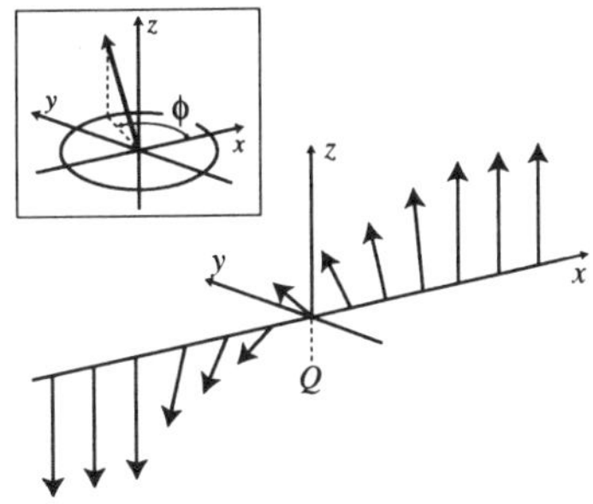

Figure 1. A domain wall in a quasi-one-dimensional ferromagnet; Q and ϕ represent the center and the chirality of the domain wall, respectively.

For the quantum-mechanical treatment of the domain wall, we construct the *domain wall state* [8] by use of the spin coherent state [9] at each site as

$$|z\rangle = \bigotimes_x |\xi^{\text{s}}(x; z)\rangle, \quad (3)$$

$$\xi^{\text{s}}(x; z) = \exp\left(-\frac{x}{\lambda} + z\right), \quad (4)$$

where $z := Q/\lambda + i\phi$, $\xi^{\text{s}}(x; z)$ represents the domain wall configuration related to the one-soliton solution as $\xi^{\text{s}}(x; z) = e^{i\phi} \tan\theta(x - Q); \theta(x - Q) = \arctan e^{-(x-Q)/\lambda}$.

In Refs. 7 and 9, the quantum dynamics of the domain wall is discussed for the Hamiltonian (1) without the magnetic field, by using the method of collective degrees of freedom in spin coherent state path integral. The main result is that the spin variables are transformed into collective variables and environmental variables, where the collective variables are the center position and the chirality of the domain wall. Alternatively, the complex variable $z = Q/\lambda + i\phi$ can be used as the collective variable. According to the result in Refs. 7 and 9, the transition amplitude between the two domain wall states, $|z_{\text{I}}\rangle$ and $|z_{\text{F}}\rangle$, is written in the discrete-time path-integral formalism as

$$\langle z_{\text{F}}| e^{-i\hat{H}T/\hbar} |z_{\text{I}}\rangle \simeq \lim_{N\to\infty} \int \prod_{n=1}^{N-1} N_{\text{DW}} S \times \frac{dz_n dz_n^*}{2\pi i} \exp\left(\frac{i}{\hbar} S[z^*, z]\right), \quad (5)$$

$$
\begin{aligned}
&\frac{i}{\hbar}\mathcal{S}[z^*,z] = \\
&-\frac{N_{\rm DW}S}{2}\sum_{n=1}^{N-1}(|z_n|^2+|z_{n-1}|^2+2z_n^*z_{n-1}) \\
&-\frac{iN_{\rm DW}S}{\hbar}\sum_{n=1}^{N}\epsilon\Big\{\hbar\Omega\cosh\tilde{\Delta}z_n-\alpha\cosh\frac{\tilde{\Delta}z_n}{2} \\
&-i\beta\sinh\frac{\tilde{\Delta}z_n}{2}+\hbar\omega_{\rm B}(z_n^*+z_{n-1})\Big\}, \qquad (6)
\end{aligned}
$$

$$
\Omega \equiv \frac{K_x}{2\hbar}\left(S-\frac{1}{2}\right), \quad \omega_{\rm B}\equiv\frac{g\mu_{\rm B}B_z}{\hbar}, \qquad (7)
$$

$$
\tilde{\Delta}z_n \equiv z_n^* - z_{n-1}, \qquad (8)
$$

where ϵ ($:= T/N$) is the infinitesimal time, z_n ($:= Q_n/\lambda + i\phi_n$) is the value of the variable z at the discrete time $n\epsilon$, and the parameters α and β are proportional to B_x and B_y, respectively. The symbol $\simeq$ in (**??**) means that the environmental degrees of freedom are ignored in the right-hand side, which is justified in the case of weak transverse anisotropy, $K_x \ll K_z$, and the quasi-classical situation, $S \gg 1$. Note that our formula formally coincides with the coherent-state path integral for a conventional particle [11]. However, there is an important difference between the two formulae. When we express the action (6) with the real variables, Q_n and ϕ_n, the chirality variable ϕ_n appears in the form of the difference, $\phi_n - \phi_{n-1}$, which must be interpreted to be $\Delta_n := \phi_n - \phi_{n-1} - 2\pi m$, where m is an integer chosen so that Δ_n belongs to the interval $[-\pi, \pi]$. This modification is derived from the periodicity of the chirality, which is different from the case of the coherent state path integral for a conventional particle.

Furthermore, this formula formally coincides with the coherent state path integral for the Bloch particle within a one-band basis [12]. The equivalence of the two path integral formalisms means that the quantum mechanics of a single domain wall is equivalent to that of the Bloch particle, where the center position and the chirality of the domain wall correspond to the position coordinate and the dimensionless quasi-momentum, which are mutually canonically conjugate. Also, the domain wall state $|z = Q/\lambda + i\phi\rangle$ corresponds to a sort of coherent state for the Bloch particle. The mean value of the position and the quasi-momentum are given by Q and ϕ, respectively, while their spreads by $\sqrt{a\lambda/(2S)}$ and $\sqrt{a/(2S\lambda)}$, respectively.

From this equivalence, the corresponding Hamiltonian for (6) may be reduced to the following form:

$$
\begin{aligned}
&\hat{H}_{\rm cdf}/N_{\rm DW}S = \hbar\Omega\cos 2\hat{\varphi} - \alpha\cos\hat{\varphi} \\
&-\beta\sin\hat{\varphi} + 2\hbar\omega_{\rm B}\hat{X}/\lambda, \qquad (9)
\end{aligned}
$$

where $\hat{X}$ and $\hat{\varphi}$ are the position and the dimensionless quasi-momentum operators, respectively. They should obey the following commutation relation:

$$
[\hat{X}, e^{-i\hat{\varphi}}] = \frac{a}{2S}e^{-i\hat{\varphi}}. \qquad (10)
$$

Hence, the position of the domain wall as a quantum mechanical particle is quantized in unit of $a/2S$ [13]. It is natural that the position is related to the z component of the total angular momentum, $\hat{J}_z := \int_{-L/2}^{L/2}\frac{dx}{a}\hat{S}_z(x)$, which is quantized in unit of 1. The commutation relation (10) is also derived from $[\hat{J}_z, \hat{J}_\pm] = \pm\hat{J}_\pm$. Furthermore, from (9), we find that the dispersion relation of energy-band for the domain wall are derived from the transverse anisotropy and the external magnetic field perpendicular to the easy axis. In the case of $\alpha = \beta = 0$, the Hamiltonian (9) is formally equivalent to the Hamiltonian of the Bloch particle under a uniform electric field along the crystal axis. Thus, the motion of the domain wall exhibits "Bloch oscillations" [3,6]. The frequency of this oscillation is given by $2\omega_{\rm B}$ whose factor 2 is due to the dispersion relation derived from the transverse anisotropy. Also, the amplitude, which is called to be "localization length", is given by $L_{\rm B} := \lambda\Omega/\omega_{\rm B}$.

Let us demonstrate this phenomena through the transition probability on the basis of stationary action approximation [8]. We obtain

$$|\langle z_{\rm F}|e^{-i\hat{H}T/\hbar}|z_{\rm I}\rangle|^2 \simeq \frac{1}{\sqrt{1+C^2(T)}} \times \exp\left[- N_{\rm DW}S\left\{ \frac{1}{\lambda^2}\frac{(Q_{\rm F}-Q(T))^2}{1+C^2(T)} + (\phi_{\rm F}-\phi_{\rm I}+\omega_{\rm B}T)^2\right\}\right], \quad (11)$$

$$Q\ (T) := Q_{\rm I} - L_{\rm B}\sin\omega_{\rm B}T\sin(\phi_{\rm F}+\phi_{\rm I}), \quad (12)$$

$$C\ (T) := \frac{L_{\rm B}}{\lambda}\sin\omega_{\rm B}T\cos(\phi_{\rm F}+\phi_{\rm I}). \quad (13)$$

It is noted that the last term of the exponent in the right-hand side of (11) must be interpreted to be $\phi_{\rm F} - \phi_{\rm I} + \omega_{\rm B}T - 2\pi m$ as previously mentioned. The transition probability is regarded as a function of $Q_{\rm F}$, $\phi_{\rm F}$, $Q_{\rm I}$, $\phi_{\rm I}$, and T. For a given initial condition $(Q_{\rm I}, \phi_{\rm I})$, the wave packet centroid in the extended phase space, $Q_{\rm F}$-$\phi_{\rm F}$, exhibits the Bloch oscillations along the position axis. While, the time-dependence of the spreads of the wave packet in the phase space are given by

$$\delta Q(T) = \sqrt{\frac{a\lambda}{S}(1+C^2(T))}, \delta\phi = \sqrt{\frac{a}{\lambda S}} \quad (14)$$

along the position and the momentum axes, respectively. Thus, the wave packet along the position axis exhibits a breathing behavior similar to the behavior of the squeezed state of the harmonic oscillator. Since $\delta\phi \ll 1$, we can substitute $\phi_{\rm I} - \omega_{\rm B}T$ for $\phi_{\rm F}$. Thus, the transition probability can be regarded as a function of $Q_{\rm F}$ and T. We depict this in Fig.2.

If the Bloch oscillation is treated quantum mechanically, we obtain equally spaced energy levels, which is called a Wannier-Stark ladder [7]. The level spacing is given by $\Delta E = \hbar\omega_{\rm B}$, which is identical to the Zeeman splitting.

References

1. *Quantum Tunneling of Magnetization*, Proceedings of the NATO workshop, Chichilianne, France, 1994, edited by L. Gunther and B. Barbara (Kluwer Academic, Norwell, MA, 1995).
2. S. Takagi and G. Tatara, Phys. Rev. B **54**, 9920 (1996).
3. F. Bloch, Z. Phys. **52**, 555 (1928).
4. H.-B. Braun and D. Loss, Phys. Rev. B**53**, 3237 (1996); H.-B. Braun and D. Loss, J. Appl. Phys. **76**, 6107 (1994); 6177(1994).
5. J. Kyriakidis and D. Loss, Phys. Rev. B**58**, 5568 (1998).
6. C. Zener, Proc. R. Soc. London Ser. A**145**, 523 (1934).
7. G. H. Wannier, Phys. Rev. **117**, 432 (1960).
8. J. Shibata and S. Takagi, Phys. Rev. B**62**, 5719 (2000).
9. J. M. Radcliffe, J. Phys. A4, 313 (1971).
10. J. Shibata and S. Takagi, *XXIII International Colloquium on Group Theoretical Methods in Physics, Dubna, 2000* to be published in Phys. Atom. Nucl. **12** (2001); cond-mat/0105365.
11. L. S. Schulman, *Techniques and Applications of Path Integration* (Wiley, New York, 1981).
12. J. Shibata and K. Niizeki, accepted for publication in J. Math. Phys. ;cond-mat/0105453.
13. F. D. M. Haldane, Phys. Rev. Lett. **57**, 1488 (1986).

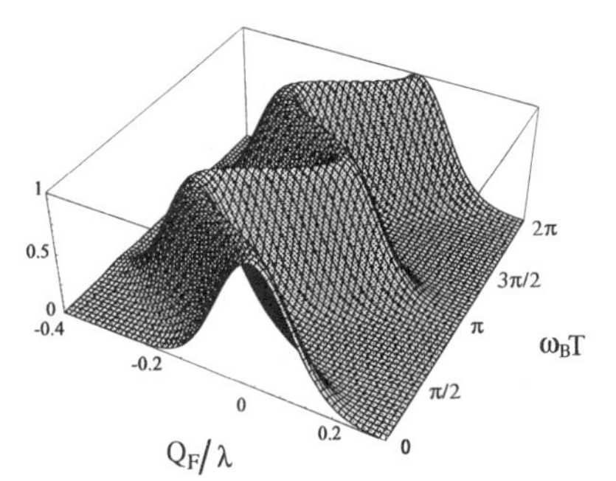

Figure 2. Motion of wave packet as a function of $Q_{\rm F}$ and T. We take $Q_{\rm I} = 0$, $\phi_{\rm I} = \pi/2$, $N_{\rm DW}S = 100$, and $L_{\rm B}/\lambda = 0.2$.

MAGNETIC STRUCTURES AND TRANSPORT PROPERTIES IN FERROMAGNETIC TUNNELING JUNCTION

M. ICHIMURA[1], S. KOKADO[2], T. ONOGI[1], J. HAYAKAWA[2], AND K. ITO[2]
[1]*Advanced Research Lab., Hitachi, Ltd., Hatoyama, Saitama 350-0395, Japan*
[2]*Central Research Lab., Hitachi, Ltd., Kokubunji, Tokyo 185-9601, Japan*

On the basis of the density-functional theory, we calculated the interfacial magnetism of the slab model in ferromagnetic tunneling junction Co/Al-oxide/Co. We found that in the Al/Co interface the interfacial Al layer exhibits a *positive* spin polarization (SP) and that both the sign and value of SP in the Al layer are in agreement with the experimental data. This result suggests that in the junction Co/Al-oxide/Co, the tunneling of s-character electron in Al is favored. Based on a model exhibiting the positive SP at the interfacial Al layer, we calculated the tunneling conductance. It was found that a change from the tunneling magnetoresistance (TMR) scheme to the giant MR (GMR) scheme occurs only for an ultra-thin film insulator.

1 Introduction

Since the discovery of large tunneling magnetoresistance (TMR) effect at room temperature in ferromagnetic tunneling junctions,[1] this research area has been intensively studied aiming at many possible applications including new magnetic sensors and random-access memory elements. Using the density of states (DOS) model,[2] the magnetoresistance (MR) ratio in the TMR at low temperature is expressed by $2P_LP_R/(1+P_LP_R)$, where P_L (P_R) is the spin polarization (SP) for the left (right) side of the ferromagnetic metal (FM) attached to oxide barrier. SP is defined as $P = (D_\uparrow - D_\downarrow)/(D_\uparrow + D_\downarrow)$, where $D_\uparrow(D_\downarrow)$ is the DOS with $\uparrow$ ($\downarrow$) spin at the Fermi energy. When we adopt the DOS model using the SP obtained by the experiments of FM/Al_2O_3/superconductor junctions,[3] the MR ratio in the TMR is explained qualitatively. However, the sign of the observed SP for Co is positive, which contradicts the negative sign for the bulk Co predicted by electronic structure calculations.

In this study, we calculate the interfacial magnetism of the slab model including Al/Co interface by the first-principles electronic structure calculation, and obtain the SPs at the interfacial layers. It is found that the electronic structure of interfacial Al layer is modulated by the hybridization of Al s- and Co d-orbitals, and that the SP at Al layer is positive. This result suggests that the insulating layer plays important roles in interfacial magnetism and tunneling current in real junctions, and it is necessary to consider the positive SP and the tunneling process due to s-character electrons. Previously, the TMR effect in Co/Al-oxide/Co junctions was discussed theoretically based on the tight-binding model where only the d-orbitals of Co contribute to the tunneling electrons. In additin, we discuss the TMR effect qualitatively by applying the Landauer approach to a simple model reflecting the result of the calculated electronic structure.

2 Interfacial Magnetism

We have examined the electronic and magnetic structures of TMR junction, Co/Al-oxide/Co, by the first-principles electronic structure calculation using the full-potential augumented plane waves (FLAPW) method.[4] Since we are interested in the interfacial magnetism of the junction, we adopt the simple slab model shown in Fig.1 which contains the interface with Al and fcc Co(001). In this model, the unit cell length $a(= b)$ is set as 2.51 Å. We construct the supercell structure by adding the vacuum re-

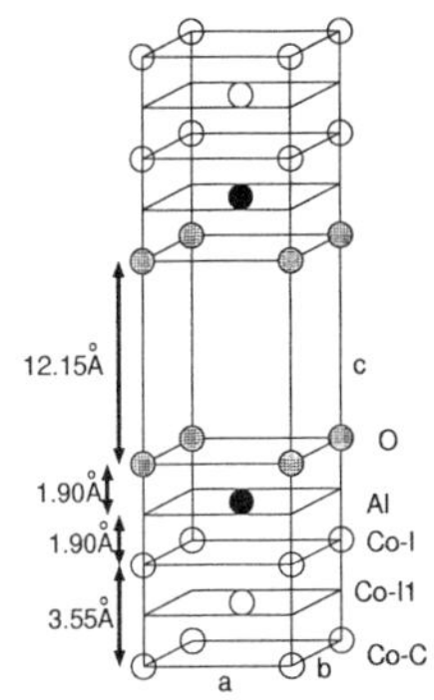

Figure 1. Unit cell of slab model. Open circles represent Co atoms. Closed and shaded circles represent Al and O atoms, respectively.

gion which is three times as long as the fcc Al unit cell (12.15 Å). The vacuum region is sufficiently large, since no essential difference could be observed in the results between three and five times of this unit cell. The resulting lengths of unit cells c is 26.85 Å. No surface/interface reconstruction was considered for the present calculation.

The self-consistent electronic structures have been obtained.[5] The magnetic moment of interfacial Co (Co-I) and Al layers are $1.11\mu_B$ and $-0.08\mu_B$, respectively. The Co-I layer possesses a reduced magnetic moment compared with the fcc bulk Co ($1.70\mu_B$). Although the magnetic moment at the interfacial Al layer is very small in absolute value, this shows the opposite direction to the magnetic moment at the Co-I layer.

We present the calculated result of SP at interfacial layers, here the DOS with each spin is averaged over the energy range of $\pm$ 0.1 eV around the Fermi energy. This averaging is required due to the position of the Fermi energy in the energy interval of about 0.02 eV and the oscillating behavior in DOS. This average range of energy corresponds to seven sampling points in our DOS calculation.

As shown in Fig. 2, the hybridization of Al s- and Co d-orbitals modulates the local DOS at the interface. Due to the hybridization, the SP for the Al layer is positive with a value of +39%, which is in agreement with the experimental value,[3] while the SP for the Co layer is in contrast to -69% for Co layer. At the Al layer, angular momentum is decomposed into +58% in the s-orbital and +23% in the p-orbital. These results suggest the folloing : 1) The insulating layer yields interfacial magnetism and tunneling current in the real junctions in addition to being an insulating barrier. 2) In the junction Co/Al-oxide/Co, the tunneling of s-character electrons in Al is favored over that of d-character electrons in Co.

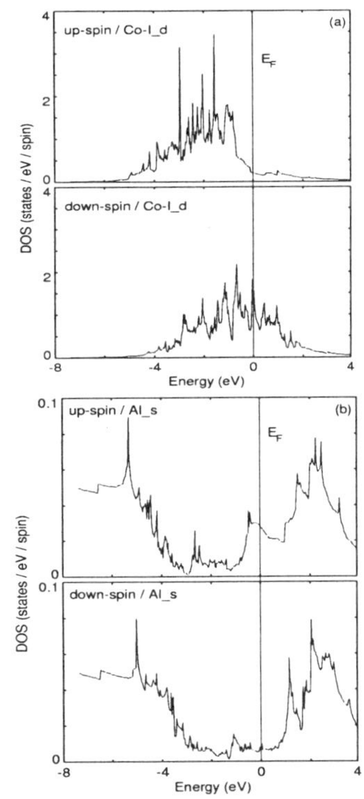

Figure 2. Spin-dependent local density of states for (a) d band in Co-I layer, and (b) s band in Al layer. The Fermi energy is denoted by the vertical lines.

The value of SP at the Al layer is in a quantitative agreement with the experimental result of the Co/Al_2O_3/superconductor junction.[3] We may conclude that the interfacial magnetism is dominated by the coupling of Co and Al even though we take into account the amorphous structure in the Al-oxide barrier.

3 Transport properties

Based on a simple tight-binding model exhibiting the positive SP at the interfacial Al layer, the TMR effect can be qualitatively discussed in the linear response regime using the Landauer approach. In the present model, the Al layer is regarded as the terminating layer of ferromagnetic electrode, and the s-character electrons in the Al layer behaves as the tunneling electrons. The structure of the model is Co/Al/insulator/Al/Co. For simplicity, the Co layers are set to include ten monolayers in a simple cubic structure. An Al monolayer in a square lattice structure is stacked at the hollow sites on a Co(001) surface. The n insulating layers, represented by a single orbital at each site, are stacked on the Al layer. We take into account the Al s-, Co s-, and five degenerate Co d-orbitals. The exchange potential due to the local moment of Co atoms is included in the on-site energy of Co d-orbitals in the present tight-binding model.

With the use of the Landauer formula,[6] the tunneling conductance for spin σ at zero temperature is written as $\Gamma_\sigma = \sum_{\mathbf{k}} \Gamma_\sigma(\mathbf{k}) = \sum_{\mathbf{k}} (4e^2/h)|T_\sigma(\mathbf{k})|^2 \mathrm{Im}\,[G_{L,\sigma}(\mathbf{k})]\,\mathrm{Im}\,[G_{R,\sigma}(\mathbf{k})]$, with $\mathbf{k} \equiv (k_x, k_y)$, where the conservation of $\mathbf{k}$ should be kept in mind. The $G_{j,\sigma}(\mathbf{k})$ (j=L or R) is the Green's function of the interfacial Al layer adjacent to the left or right side Co electrodes, and is given by $G_{j,\sigma}(\mathbf{k}) = \langle \Phi_{\mathrm{Al}}(\mathbf{k})|(E_{\mathrm{F}} - \mathcal{H}_{j,\sigma} + i0^+)^{-1}|\Phi_{\mathrm{Al}}(\mathbf{k})\rangle$, where $|\Phi_{\mathrm{Al}}(\mathbf{k})\rangle$ is the Bloch wave of the Al s-orbital. Here, $\mathcal{H}_{j,\sigma}$ is the tight-binding Hamiltonian of j side electrode. The $|T_\sigma(\mathbf{k})|^2$ represents the transmission coefficient, and can be evaluated in an analytic form using $G_{j,\sigma}(\mathbf{k})$ and $g_{ij}(\mathbf{k})$, where $g_{ij}(\mathbf{k})$ (i, j=1 or n) is the bare propagator between i- and j-th layer in the n insulating layers. The $T_\sigma(\mathbf{k})$ is approximately composed of $g_{1n}(\mathbf{k})$ in a numerator and the self-energy correction in a denominator, and the self-energy correction include the product of $G_{j,\sigma}(\mathbf{k})$ and $g_{ij}(\mathbf{k})$.[7]

We calculated insulator thickness dependence of the TMR ratio, defined by $(\Gamma_\uparrow^{\mathrm{P}} + \Gamma_\downarrow^{\mathrm{P}} - \Gamma_\uparrow^{\mathrm{AP}} - \Gamma_\downarrow^{\mathrm{AP}})/(\Gamma_\uparrow^{\mathrm{P}} + \Gamma_\downarrow^{\mathrm{P}})$. Here, $\Gamma_\sigma^{\mathrm{P}}$ and $\Gamma_\sigma^{\mathrm{AP}}$ denote the conductance for spin σ in parallel (P) and anti-parallel (AP) magnetization configurations, respectively. Figure 3 shows the TMR ratio as a function of insulator thickness with the specific values of bare insulating barrier height and hybridization between Al and insulating layers. In the inset the experimental result of R. Arai *et al.*, (unpublished.) is shown. Figure 3 shows that the TMR ratio decreases rapidly as the insulator becomes thinner ($n < 3$), although it has a nearly constant value for a thick insulator ($n \geq 3$).[7]

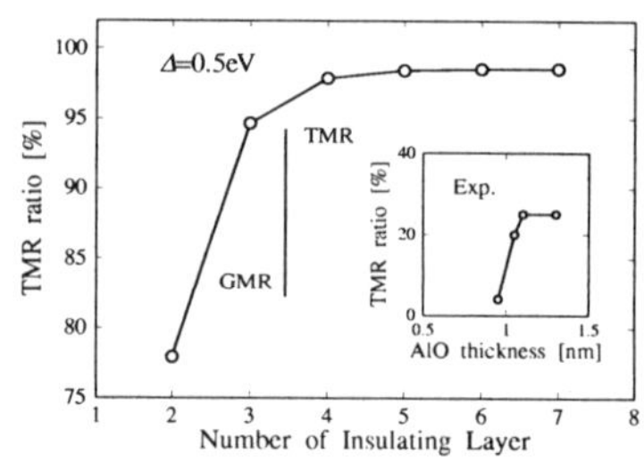

Figure 3. The TMR ratio as a function of insulator thickness. The inset shows the experimental result [R. Arai *et al.*, unpublished.].

In order to study the decreasing TMR ratio, we decomposed $|T_\sigma(\mathbf{k})|^2$ for each $\mathbf{k}$. We focused on the $|T_\sigma(\mathbf{k}^*)|^2$, where $\mathbf{k}^*$ gives a predominant component of $|\Gamma_\sigma(\mathbf{k})|^2$, and found that $|T_\sigma(\mathbf{k}^*)|^2$ does not show exponential decay with respect to n in a thinner insulator ($n < 3$), although $|g_{1n}(\mathbf{k}^*)|^2$ shows exponential decay with n. Such behavior of $|T_\sigma(\mathbf{k}^*)|^2$ is explained by focusing on the self-energy correction in $T_\sigma(\mathbf{k})$ due to the electrodes,[8] which includes multiplication of $g_{1n}(\mathbf{k})$ and $G_{j,\sigma}(\mathbf{k})$. With the parallel magnetization configuration of the two electrodes, the self-energy correction becomes larger, and the deviation from exponential decay of $|T_\sigma(\mathbf{k}^*)|^2$ also becomes larger, compared with the anti-parallel case. This reflects the relatively larger hybridization between the Al monolayer and the electrode, i.e. relatively larger local DOS at the Al monolayer. Then this behavior of $|T_\sigma(\mathbf{k}^*)|^2$ with the parallel configuration reduces the TMR ratio. We can regard the transport through the ultra-thin insulator as metallic conduncion, i.e. the giant MR (GMR) seen in junctions FM/normal metal/FM. Since we have assumed here the tunneling process with conservation of $\mathbf{k}$ and this reduction of TMR is induced mainly by $\mathbf{k}^*$, we suppose that the reduction of TMR effect should be observed in the experiment with high-quality junctions prepared by single crystal.[9]

4 Concluding Remarks

We examined the interfacial magnetism of the slab model including Al/Co interface, on the basis of density-functional theory. We obtained the local magnetic moments and spin polarizations (SPs) at the interfacial layers. The interfacial Al layer exhibits positive SP although the sign of SP at the interfacial Co layer is negative. The value of SP at the Al layer is approximately +40%, and the decomposition of angular momentum is +58% in the s-orbital, and +23% in the p-orbital. The sign and value of SP at the Al layer show a quantitative agreement with the experimental values of the ferromagnet/Al_2O_3/superconductor junction.[3] For the ferromagnetic tunneling junction Co/Al-oxide/Co, our result suggests that the insulating layer yields interfacial magnetism and spin-dependent tunneling current in the real junctions in addition to being an insulating barrier. It is necessary to take into account the positive SP and the tunneling of s-character electrons, for an analysis of transport properties. Based on the tight-binding model reflecting the above requirement, we studied the tunneling conductance qualitatively in the linear response regime using the Landauer formula. In the present model, the Al layer is regarded as the terminating layer of ferromagnetic electrode, and the s-character electrons in Al layer behaves as the tunneling electrons. It was found that a change from the tunneling magnetoresistance (TMR) scheme to the giant MR (GMR) scheme takes place for an ultra-thin film insulator, since the self-energy correction becomes large for the parallel magnetization configuration. This change of scheme shows the qualitative agreement with the experiment.

References

1. T. Miyazaki and N. Tezuka, J. Magn. Magn. Mater. **139**, L231 (1995); J. S. Moodera *et al.*, Phys. Rev. Lett. **74**, 3273 (1995).
2. M. Julliere, Phys. Lett. **54 A**, 225 (1975); S. Maekawa and U. Gäfvert, IEEE Trans. Magn. **18**, 707 (1982).
3. P. M. Tedrow and R. Meservey, Phys. Rev. B **7**, 318 (1973); R. Meservey and P. M. Tedrow, Phys. Rep. **238**, 173 (1994).
4. P. Blaha, K. Schwarz, and J. Luitz, WIEN97, Vienna University of Technology, 1997, improved and updated Unix version of the original copyrighted WIEN code. Original WIEN code was published by P. Blaha, K. Schwarz, P. Sorantin, and S. B. Trickey, Comput. Phys. Commun. **59**, 399 (1990).
5. M. Ichimura *et al.*, Jpn. J. Appl. Phys. **40**, 4528 (2001).
6. J. Mathon, Phys. Rev. B **56**, 11810 (1997).
7. S. Kokado *et al.*, Appl. Phys. Lett. **79**, 3986 (2001).
8. S. Datta, *Electronic Transport in Mesoscopic Systems* (Cambridge University Press, Cambridge, U.K., 1995).
9. S. Yuasa *et al.*, Europhys. Lett. **52**, 344 (2000).

FERROMAGNETIC AND ANTI-FERROMAGNETIC RECONSTRUCTION IN SEMICONDUCTOR QUANTUM DOTS

M. STOPA

ERATO - Tarucha Mesoscopic Correlation Project
NTT Basic Research Laboratories
3-1 Morinosato-Wakamiya, Atsugi-shi
Kanagawa-ken, 243-0198 Japan
e-mail stopa@tarucha.jst.go.jp
phone: 81-(0)46-248-4016. FAX: 81-(0)46-248-4014

We present results of spin density functional calculations for the electronic structure of small ($N \sim 30-50$), GaAs-AlGaAs quantum dots in a transverse magnetic field. We demonstrate that two types of reconstruction of the dot ground state, for N even, can occur when the single particle levels near the Fermi surface cross. In the first case, typically at lower B, the dot reconstructs anti-ferromagnetically, with the total spin remaining zero. In the latter case, a more standard Hund coupling occurs and the dot reconstruction results in the total dot spin increasing to unity.

In recent years, quantum dots have functioned as unique, small laboratories where fundamental quantum mechanical ideas are explored and tested. At the same time nanotechnology, has begun to employ quantum dots for a wide variety of practical applications; both with regard to their transport and optical properties. Future applications, from digital processing electronic devices to, perhaps some day, quantum computers, are also envisioned. Thus, the confluence of new technology and foundations of quantum mechanics is perhaps nowhere more pronounced than it is in the study of quantum dots.

One aspect of quantum dot electronic structure that is currently exciting the nanotechnology community is the behavior of spin. Here too, one of the focii of research has been pure, fundamental physics as exemplified by the Kondo effect in transport through Coulomb blockaded quantum dots. Additionally, however, the potential for future application of quantum dots to magneto-electronics and to spintronics has led to substantial increases in research funding in these areas [1].

Recently we have shown, using spin density functional theory (SDFT), that small (electron number $N \sim 50$) lateral, GaAs-AlGaAs heterostructure-based quantum dots exhibit a rich pattern of ground state spin as a function of a transverse magnetic field B and N [2]. In particular, we have investigated the low B regime ($B \lesssim 1.5\ T$) and found that, for N even, the spin fluctuates in a quasi-periodic fashion between zero and one, the so-called "singlet-triplet" (S-T) transition [3]. These fluctuations occur, as B increases, due to the development of incipient Landau levels which result in level crossings (or close anti-crossings) at the Fermi surface. To the extent that the confinement can be approximated by a circular, parabolic well, the crossings exhibit a complex regularity (in the extreme, symmetric dot case we have the Darwin-Fock spectrum [4]) and, consequently, so do the regions of spin polarization. In addition, in regions where three or more orbitals come close together at the Fermi surface it is also possible, for N odd, to observe a transition from a spin 1/2 ground state to a spin 3/2 ground state. Obviously these occurrences are more rare than for the N even, S-T transitions.

However, for N even, the spontaneous polarization to $S = 1$ is not the only type of exchange-induced reconstruction that can occur. We have also determined that, typically

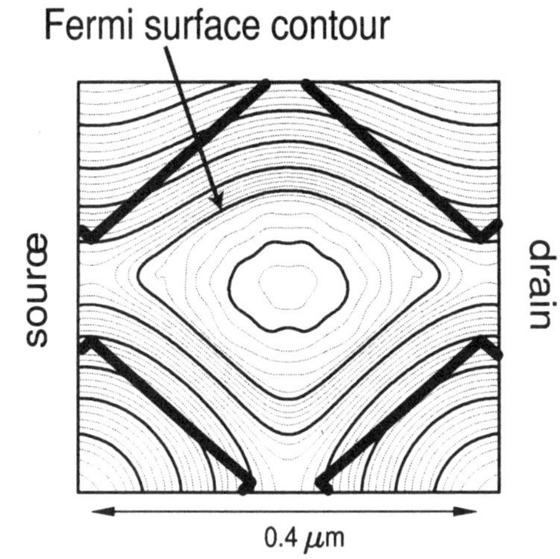

Figure 1. Typical potential contour (effective 2D, i.e. lowest subband) for small dot from reference [1]; $N = 36$, $B = 0\ T$. Potential bottom at dot center about $-1.3\ Ry*$; contours unevenly spaced (in percentiles). Potential at QPC saddle points about $0.5\ Ry*$. Contour of Fermi surface (energy zero) is indicated.

at lower fields, the dot can, in the vicinity of level crossing or, generally, any close level spacing at the Fermi surface (energy E_F), undergo an internal spin polarization. It is the purpose of this paper to illustrate and contrast these two types of spin reconstruction.

In figure 1 we illustrate a typical example of the effective 2D confining potential for electrons in the dot. The growth profile and gate pattern are taken from the experimental device of van der Wiel *et al.* [5]. We have assumed that the device is in the electric quantum limit, i.e. all electrons are confined to the lowest subband in the z (growth) direction. Thus what is plotted in figure 1 is the energy of the lowest subband as a function of position in the $x - y$ plane. The principal feature to note is that the dot is far from circular symmetry. Hence the effective single particle states are not eigenstates of angular momentum and, further, any degeneracy due to circular symmetry is lifted (although there remains an approximate parity symmetry with respect to the vertical center line of

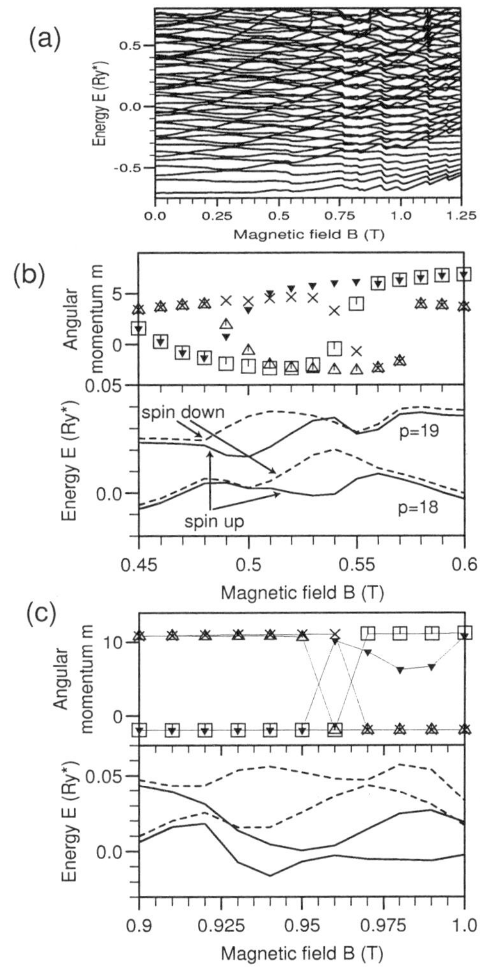

Figure 2. (a) Kohn-Sham energy levels, as a function of B, for spin up electrons (only plotted for clarity). Development of Landau levels and resulting level crossings are clearly seen. Energy zero is Fermi energy of the leads, which, by choice of gate voltages is close to that of the dot. (b) Anti-ferromagnetic reconstruction showing the evolution of the angular momenta and eigenenergies of the highest two occupied ($18th$) and lowest two unoccupied ($19th$) states. In lower panel: solid line spin up, broken line spin down; in upper panel, solid triangle $p = 18 \uparrow$, box $p = 18 \downarrow$, open triangle $p = 19 \uparrow$, cross $p = 19 \downarrow$. Note that at all B the spin remains zero. (c) Ferromagnetic reconstruction where the level crossing results in the total dot spin increasing to $S = 1$. Note that anti-crossings of the separate spin orbitals also occur, but here, the up spins are full and the down spins are empty.

the dot).

In figure 2a we plot the Kohn-Sham levels for the electrons confined in the dot as a function of uniform magnetic field applied transverse to the structure (parallel to z). The Fermi surface of the leads is the energy zero. The dot is constrained in this case to contain $N = 36$ electrons. The gate voltages are set to large enough (negative) values so that the quantum point contacts (QPC) separating the dot from the source and drain leads (and an additional two dimensional electron (2DEG) seen at the bottom of figure 1) are fully depleted and we have a well-defined eigenvalue problem for the electrons in the dot. In the calculation, the charge density in the leads is included through a 2D Thomas-Fermi approximation [6]. The level energies bear some resemblance to the Darwin-Fock spectrum for a pure 2D parabolic potential [2], however there are clear distortions due to the non-circularity as well as fourth order terms in the radial confining potential. Nonetheless, the beginning of the development of Landau levels can clearly be seen. It is therefore clear that B induces a fairly systematic series of level crossings (or close anti-crossings) at E_F. Furthermore, each such crossing will typically involve two states which are of differing angular momentum m. Such states are developing into different Landau levels and, therefore, will also have differing radial distribution. The state which is decreasing in energy with B will be the lower Landau level with higher m and vice-versa for the state increasing in energy with B.

Two examples of level anti-crossings are shown in figures 2b and 2c. In both cases we show the (expectation value of) the angular momentum in the upper panel and the Kohn-Sham energy levels in the lower panel. We show only the $18th$ and $19th$ levels. These are the highest two occupied and the lowest two unoccupied levels (recall $N = 36$). The solid lines are spin up and the dashed lines are spin down. Note that for figure 2b, where the anti-crossing occurs around $B \approx 0.52\ T$, the spin remains zero throughout the plotted region (of the lower two levels, one is always spin up and the other is always spin down; the "core" states of the remaining 34 electrons have net zero spin). This contrasts with the anti-crossing that occurs around $B \approx 0.95\ T$, where, from $B \approx 0.93\ T$ to $0.99\ T$, there are two spin up states filled and two spin down states empty. Thus, in this latter case, there is a net spin $S = 1$ through this interval.

Note, however, that a different type of transition occurs in figure 2b. In particular, the character of the filled up spin state (as revealed by its angular momentum) interchanges with that of the empty up spin state. This occurs near $B \approx 0.49\ T$. At still higher $B \approx 0.55\ T$, the down spin states change character as well. (It should be noted that the sign of B is such that the Zeeman energy of the up spin states is lower than that of the down spin states, although that energy here is only of order $0.005\ Ry^*$ in 2b and $0.01\ Ry^*$ in 2c. This splitting is actually visible at, say, $B = 0.6\ T$ in 2b and $B = 0.9\ T$ in 2c).

It is simple to understand what is occurring in these two cases by referring to figure 3. On the left hand side we begin (top) with two electrons occupying the level with angular momentum $m = m_1$, with presumably $m_1 < m_2$ and the eigenstate corresponding to m_1 located closer to the center of the dot than that of m_2. The direct Coulomb energy penalizes the double occupancy of any particular spatial orbital and so when the levels m_1 and m_2 approach, one of the two electrons invariably redistributes to the outer, higher m state *before the second electron does so.* In the left hand case of figure 3 (corresponding to figure 2b) the spin of the redistributed electron remains anti-parallel whereas on the right hand side, which typically occurs at higher B, the spin flips to become parallel with the spin of the electron left behind. Due to the different spatial distributions of m_1 and m_2 there is still an *internal* polarization which develops even when the total spin remains zero (left side), hence the description "anti-ferromagnetic."

While the results of SDFT calculations are remarkably trustworthy, they are not always fully intuitive. Thus while it is clear *what* is happening in these two cases, it is by no means obvious *why*. In the most elementary approximation, where we focus on only these two electrons at the Fermi surface, we might expect the singlet (i.e. anti-ferromagnetic) reconstruction to dominate at low B due to the reduced kinetic energy of the spatially symmetric two-body state. Further, at sufficiently high B the triplet reconstruction must prevail simply due to Zeeman energy. However this is essentially a hydrogen molecule type of ansatz and does not take into account the type of shell degeneracy effects related to Hund's first rule for atoms. For example, for two electrons in a degenerate shell in an atom, even at zero field, we expect the spins to align. It is obvious that two electrons in higher orbitals of a many-body system is a fundamentally different problem from the two-body problem in a fixed external potential, due to density of states considerations. The Hund's coupling argument, which is generally applicable in atomic systems, would seem to be the more reasonable departing point for understanding the electronic structure at either low or high B.

The question is then why is the singlet state ever lower in energy than the triplet state, which would seem to benefit from the reduced Coulomb energy manifested in exchange (i.e. Hund's coupling) ? It is plausible that the interaction of these two "valence" electrons with the core of other electrons in the dot results in a reduction in energy when there is an overall spin density wave within the dot. In the limiting case if one imagined segmenting all the up spins from the down spins, while keeping the total spin fixed at zero, then the local Hund's coupling would lower the energy of *all* electrons, since all

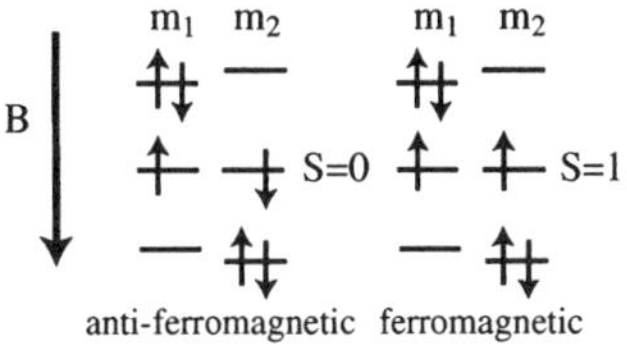

Figure 3. Schematic representation of the two reconstructions shown in figure 2. The "inner" level has angular momentum m_1 and it is assumed that $m_1 < m_2$. In the anti-ferromagnetic case the transition, which results from the direct Coulomb cost of doubly occupying a single orbital, occurs from $S = 0$ to $S = 0$, i.e. the spin of the redistributing electron remains the same. In the ferromagnetic case, Hund's energy and/or Zeeman energy lead to a ground state of total spin $S = 1$.

electrons would be localized in a group of electrons of parallel spin and would therefore have less overlap and lower Coulomb energy. This kind of reconstruction of all the levels in the dot is obviously extreme, but to the extent that there can be a tendency for internal polarization, it is clear that the density of states must be high rather than low. This is because, reasoning from a perturbation theory viewpoint, it is necessary for each eigenstate to have admixtures of higher eigenstates in order for the wavefunctions of the two spin species to diverge. Possibly in the region of higher magnetic field the constraint for the eigenstates to encompass integer number of flux quanta reduces the ability of the core states to polarize and thereby leaves the Hund coupling, triplet state as the best alternative.

While this interpretation is somewhat speculative, it would be interesting to attempt to test the results of this calculation experimentally. The recent observation of a chessboard pattern in the conductivity through quantum dots in the Coulomb blockade regime, which is believed to result from Kondo (anti-ferromagnetic) coupling of the dot electrons to the leads, might provide some insight into the anti-ferromagnetic versus ferromagnetic reconstruction scenarios for the electrons *within* the dot. It is important to note that both of the reconstructions described in this paper are likely to permit a Kondo coupling to the leads for even N. This is because in both cases the electron which is much more strongly coupled to the leads is the single electron in the outer, $m = m_2$ state. In the limit where the tunneling through the inner eigenfunction is completely negligible (due to that eigenfunction's distance from the QPCs), the Kondo effect represents simply the flipping of the spin of the outer electron (or, in a different way of speaking, the screening of the spin of the outer electron by the spins in the leads). If, however, either the singlet or the triplet ground state becomes strongly favored, then the Kondo effect becomes inelastic, since the state to which the system returns after a virtual excitation to $N \pm 1$ electrons must have a different spin (in that outer orbital). Therefore, if the Kondo effect is to proceed through the outer orbital alone, with the inner orbital remaining inert, it is essential that, whichever state, singlet or triplet, is the ground state, the other state must be nearly degenerate in energy. To put this another way, the Kondo temperature will, in this picture, be a sensitive function of the splitting between the ferromagnetic and anti-ferromagnetic states.

As a final note, it is worth pointing out that in the case of a vertical dot we could expect an entirely different transport signature, even if the evolution of the ground state, and the reconstructions which we have discussed, are the same. This is due to the fact that we would expect, generally, the coupling of the inner and outer m levels to be equal (or at least much closer) for vertical tunneling. In this case it is more likely that some type of cooperative Kondo effect would occur due to the additional phase space provided by the proximity of the singlet-triplet degeneracy [7].

In conclusion, we have presented results from spin density functional calculations which have shown that, for N even, as a function of magnetic field, two different kinds of reconstruction within the dot can occur. One of these reconstructions, which occurs more frequently at higher B, is a typical Hund rule type of reconstruction wherein the dot sacrifices kinetic energy when single particle levels approach one another in order to gain exchange energy due to an increase in the total dot spin. In the other case, the dot spin remains equal to zero, but due to the direct Coulomb term and the concomitant redistribution of electrons in the dot, a local spin polarization develops and the coupling of the electrons at the Fermi surface can be regarded as anti-ferromagnetic. We have attempted to understand, from a more fundamental perspective, the physical origin of these different reconstructions, and we have suggested that some features of the singlet-triplet splitting might be observable in the transport properties through the dot in the Coulomb blockade regime.

References

1. X. Hu, R. de Sousa and S. Das Sarma, cond-mat/0108339.
2. M. Stopa, Physica E, **10**, 103 (2001).
3. S. Tarucha, D. G. Austing, Y. Tokura, W. G. van der Wiel and L. P. Kouwenhoven, Phys. Rev. Lett. **84**, 2485 (2000).
4. C. G. Darwin, Proc. Cambridge Philos. Soc. **27**, 86 (1931).
5. W. G. van der Wiel *et al.*, Science **289**, 2105 (2000).
6. M. Stopa, Phys. Rev. B, **54**, 13767 (1996).
7. S. Sasaki *et al.*, Nature **405**, 764 (2000).

A SINGLE-PHOTON DETECTOR IN THE FAR-INFRARED RANGE

O. ASTAFIEV, V. ANTONOV, T. KUTSUWA AND S. KOMIYAMA

Department of Basic Science, University of Tokyo, Building 16-622, 3-8-1 Komaba, Meguro-ku, Tokyo 153-8902, Japan

Japan Japan Science and Technology Corporation (JST), Japan

E-mail: astf@mujin.c.u-tokyo.ac.jp

Single-electron transistor (SET) operation of a quantum dot (QD), fabricated in a $GaAs/Al_xGa_{1-x}As$ heterostructure crystal is demonstrated to serve as an extremely high sensitivity detector of far-infrared (FIR). When the single QD is placed in a high magnetic field, the resonant conductance through the SET switches on (off) upon the excitation of just one electron to a higher Landau level inside the QD, thereby enabling us to detect individual events of FIR-photon absorption ($\lambda = 0.17$ - 0.22 mm). Additionally, we demonstrate that the SET consisting of two parallel QDs operates as a high sensitivity detector of FIR approaching the single-photon counting level in the absence of magnetic fields.

1 Introduction

The single-electron transistor (SET) is extremely sensitive to its electrostatic environment. This gives us a unique opportunity to realize the detection of individual photons by utilizing single-electron photo-excitation mechanisms. Here, we explore (i) cyclotron resonance photo-excitations[1,2] and (ii) Kohn-mode plasma resonance excitations at the characteristic frequency of the parabolic bare confinement potential in the quantum dots (QDs) by measuring transport through QDs operated in the SET regime.

2 Single-photon Detection by Quantum Dots in High Magnetic Fields

The electron energy profile of a relatively large QD in high magnetic fields B is well described by discrete Landau levels (LLs). When the QD is illuminated by far-infrared (FIR) radiation at nearly the cyclotron frequency, ω_c, photons are resonantly absorbed, so that electron-hole pairs are excited on two consecutive Landau levels. This gives rise to an internal charge polarization inside the QD, which, in turn, strongly affects the conductance through the QD in SET regime. Each event of photon absorption can thus be recorded as a dramatic change of the conductance.

A simplified picture of the processes is shown in Fig. 1 (a). Let us choose the magnetic field so that the first Landau level (LL1) is completely filled, while the second Landau level (LL2) is only slightly occupied in the QD. The Landau levels are bent by the electrostatic confining potential, whereby LL1 and LL2 form an "outer-ring" and an "inner-core", respectively, at the Fermi level[3] (here, we neglect spin splitting which is much weaker than the cyclotron energy $\hbar\omega_c$). Conductance through the QD takes place predominantly via the tunneling through the "outer-ring", which has stronger coupling to the external reservoirs. The tunneling is affected by the Coulomb blockade effect, and, when the gate voltage V_g is swept, the conductance has a resonance peak every time when an electrochemical potential of the "outer-ring" lines up with that of the external reservoirs. When an FIR photon is absorbed, an electron is excited into the higher LL2. The excited electron and hole in the lower LL1 rapidly release their excess energy to the lattice and, eventually, relax into the spatially separated "inner-core" and "outer-ring". This makes the QD charge polarized. As a result, the electrochemical potential of the "outer-ring" shifts by

$\Delta\mu \approx -e^2C_2/C_{12}(C_1 + C_2)$, where C_i and C_{ij} are capacitances denoted in Fig. 1 (a) ($C_{12} \gg C_1 + C_2$). As a result, each Coulomb peak is expected to shift towards a position of more negative V_g by $\Delta V_g = \alpha\Delta\mu$, $\alpha/|e| > 1$, as it is shown in Fig. 1 (b). A simple estimation shows that the shift can be as large as 20% of the spacing between the original Coulomb peaks. The shift of the Coulomb peak is maintained until the excited electron-hole pair recombines. This lifetime can be extremely long due to the space separation between the "inner-core" and the "outer-ring". When a second electron-hole pair is excited inside the QD, while the first excited electron-hole pair is still present without recombination, the induced polarization will be doubled, yielding the Coulomb peak at a doubly shifted position $2\Delta V_g$. We report here the observation of the single-photon detection due to the mechanism described in the above in the wavelength range of 0.17 - 0.22 mm.

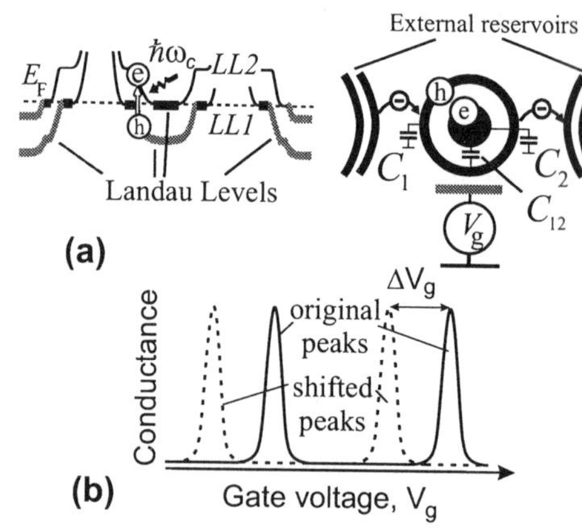

Figure 1. (a)Schematic picture of the QD in a magnetic field. Two edge states, "inner-core" and "outer-ring", are formed from LL2 and LL1, respectively. An electron-hole pair is excited in the QD via absorption of a FIR photon, $\hbar\omega_c$. The electron and the hole relax to the "inner-core" and the "outer-ring" polarizing the QD. C_i denotes the mutual capacitance. (b) conductance of the QD as a function of V_g; conductance has the Coulomb resonance peaks when the electrochemical potential of the reservoirs lines up with that of the QD. The peak pattern shifts to the left by ΔV_g when the electron-hole pair is excited in the QD.

An inset of Fig. 2 (a) shows SEM micrograph of the sample, which is 0.7 μm-size QD fabricated on a GaAs/Al$_x$Ga$_{1-x}$As heterostructure (μ = 80 m^2/Vs and $n_s = 2.4 \times 10^{11}$ cm^{-2} at 4.2 K). The two-dimensional electron gas (2DEG) is confined to form the QD by negatively biasing the metal gates on top of the heterostructure. We estimate ~ 300 electrons are trapped in the QD. The electrical leads to the metal gates of 100 μm length serve as a dipole antenna for the the FIR radiation. The QD sample is placed in the mixing chamber of a dilution refrigerator with a base temperature of 0.05 K. As a source of FIR, we use a high-mobility GaAs/Al$_x$Ga$_{1-x}$As 2DEG Hall bar situated at about 2 K within the same cryostat, which emits relatively narrow cyclotron radiation. The frequency $\omega_c = eB_e/m^*$ (m^* is the effective electron mass) is tunable by scanning the magnetic field B_e for the emitters[4]. To guide the FIR from the emitter to the samples we use an optical scheme similar to that previously described in Ref. 2. The FIR power incident on the effective antenna area ($\sim 100 \times 100\mu$m^2) does not exceed ~ 0.01 fW in this experiment. The SET conductance is measured with an ac-voltage (25 μV and 1 kHz) by standard lock-in technique.

Striking photo-response is observed in the magnetic field range B = 3.4 - 4.2 T, where the LL2 is slightly occupied. In the absence of FIR-illumination, the conductance shows regular Coulomb peaks when V_g is scanned (upper trace in Fig. 2 (a)). The picture drastically changes when the QD is illuminated by weak FIR: the ground-state peaks become unstable, exhibiting switching. At the same time, the conductance spikes appear at positions shifted from the original one by -0.6 mV and -1.2 mV, forming new peaks (lower trace in Fig. 2 (a)). The first shifted peak in Fig. 2 (a) corresponds to the excited state with one electron-hole pair inside the QD, while the second

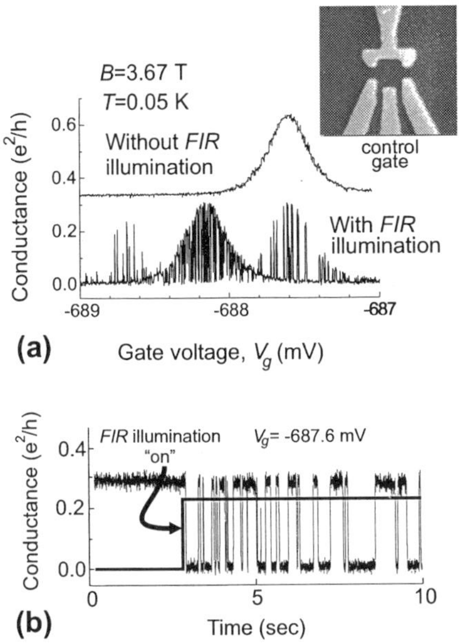

Figure 2. Conductance of the QD as a function of the gate voltage V_g at B = 3.67 T. Initially stable Coulomb resonance peak without FIR (upper trace) becomes unstable under the FIR-radiation, exhibiting switching. At the same time, conductance spikes forming new peaks appear at the positions shifted by -0.6 mV and -1.2 mV from the original peak (lower trace). (b) Conductance time trace measured at the position of the original conductance peak. When the FIR-radiation is turned-on, the conductance exhibits switches, each of which corresponded to an event of the single-photon absorption.

shifted peak to the state with two electron-hole pairs. The observations are consistent with the model discussed above. The rate of switching increases with increasing dissipated emitter power P_{em}. If V_g is fixed at the position of the original Coulomb peak, the random telegraph type switches are observed as it is shown in Fig. 2 (b). The switching behavior is clearly discerned up to T = 0.4 K, although the amplitude of the switches decreases.

On average a lifetime of the excited dot strongly increases as magnetic field increases from 1 ms at B = 3.4 T (the instrumental time constant = 3 ms) up to 20 minutes at B = 4.0 T, above which it rapidly drops to a level below 1 ms. With increasing B, the LL2 depopulates through electron transferring from the "inner-core" to the "outer-ring", reducing the size of the "inner-core". This increases spatial separation between them, so that the probability of the recombination of the excited electron-hole pair decreases.

3 High Sensitivity Detection by Double Quantum Dots

Although single-photon counting in the FIR has been demonstrated by QDs in high magnetic fields, it is highly desirable to realize detection without magnetic fields and to extend the wavelength range. Here, we demonstrate another mechanism of FIR detection by QDs in the absence of magnetic fields with an extremely high sensitivity close to the single-photon detection level.

The mechanism is described by Figs. 3 (a) - (c). The schematic device structure shown on Fig. 3 (a) is reminiscent of a lateral double-QD SET studied earlier by different groups[5,6]. Adjacent to the first QD (D1) that forms an SET, the second QD (D2) is placed and capacitively coupled to D1. D2 is coupled to incident FIR by a planar dipole antenna.

As depicted in Fig. 3 (b), if an electron in D2 gains an excess energy E* through the excitation by FIR, the excited electron escapes either to D1 or to the electron reservoir adjacent to D2 so that the number of electrons, N_2, in D2 decreases by one (ΔN_2 = -1). The electron then rapidly releases its excess energy (via phonon emission or electron-electron interaction) relaxing to the Fermi level, ε_F. The potential barriers, in turn, prevent the "cold" electron from returning to D2, thereby realizing a relatively long lifetime τ_l of the ionized state of D2. Letting C_{12} be an inter-QD capacitance and C_i (i = 1 and 2) capacitances between Di and the environments, the ionization of D2 (ΔN_2 = -1) de-

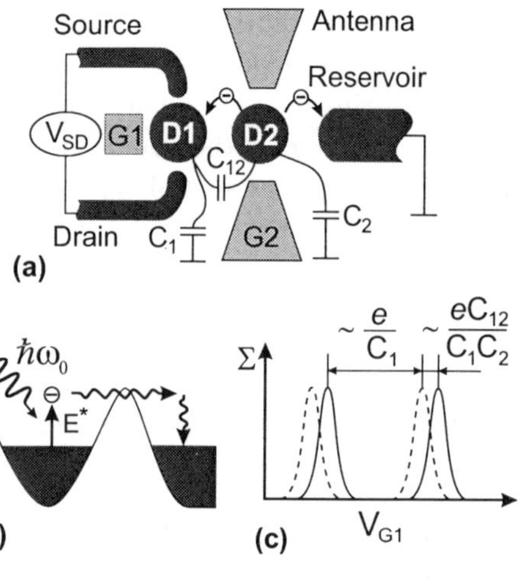

Figure 3. Schematic representations of the FIR photon detection in double QD-geometry. (a) An SET consisting of parallel double QDs. (b) Mechanism of D2 ionization. (c) Conductance peak shift induced by the ionization.

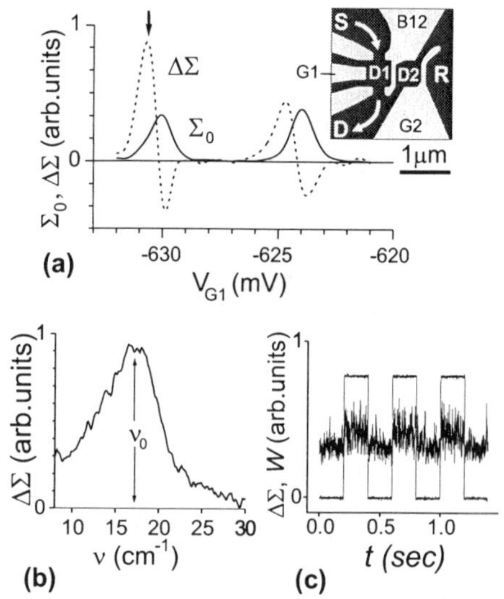

Figure 4. (a) Conductance peaks, Σ_0, without FIR (solid line) and the photoresponse, $\Sigma \equiv \Sigma - \Sigma_0$ (dotted line), as a function of V_{G1}, studied with W = 0.3 fW and V_{G2} fixed at -628 mV. The inset shows schematic view of the device. (b) The excitation spectrum of $\Delta\Sigma$, studied with W = 0.15 fW in the same gate bias condition as that of the V_{G1}-position marked by the arrow in (a). (c) A real-time trace of the conductance at W = 0.15 fW, where the square waveform indicates on-and-off of the FIR.

creases the electrochemical potential of D1, μ_1, by $\Delta\mu_1 \approx -e^2C_{12}/C_1C_2$ (e is the unit charge), where $C_{12} \ll C_1$ and $C_{12} \ll C_2$ hold in the experimental condition. This will result in a shift of the SET conductance peak by $\Delta\mu_1/\varepsilon_{ch} \approx -C_{12}/C_2 = -(3 \sim 15)\%$ in the sweep of V_{G1}, yielding a detectable conductance change as shown in Fig. 3 (c), where $\varepsilon_{ch} = e^2/C_1$ is the charging energy of D1 that determines the period of the Coulomb conductance oscillations.

The inset of Fig. 4 (a) schematically shows the device fabricated on the same wafer as the QD discussed in the previous section. Light areas indicate metal gates deposited on top of the crystal. Negatively biasing the gates depletes the two-dimensional electron gas (2DEG) below the gates and forms D1, D2, the source (S), the drain (D) and the reservoir (R). The lithographic size of each QD is 0.5 × 0.5 μm^2, with about 200 electrons in it. The gate B12 defines the inter-QD potential barrier. The control gate, G2, controls not only the electrochemical potential of D2, μ_2, but also defines the potential barrier between D2 and R. Metal leads for G2 and B12 extend over 200 μm in length, forming a dipole antenna for D2. A coupling efficiency between antenna and the dot is expected to be high because D2 impedance for the present sample is about 30 Ω (the impedance value is of the order of dc resistance of 2DEG bar with 1:1 aspect ratio) being close to estimated antenna impedance, which is of the order of 100 Ω.

We apply weak FIR power, W, incident on the effective antenna area for D2 (about 200 μm diameter), which is roughly estimated to be $W \approx 1$ fW or 3×10^6 photons per second in a band width of 0.6 ± 0.2 mm when electrical input power P_{in} = 1 mW is fed to the n-InSb emitter. The SET conductance is measured with an ac-voltage (25 μV and 1 kHz) while FIR is typically applied in a square waveform at a lower frequency of 7 Hz. We represent the photoresponse by the difference between the dark conductance Σ_0 and the conductance Σ with FIR, $\Delta\Sigma \equiv \Sigma - \Sigma_0$, and study it via a double lock-in technique.

Figure 4 (a) shows a typical photoresponse signal, $\Delta\Sigma$, along with Σ_0 in a sweep of V_{G1}, where V_{B12} = -532 mV, W = 0.3 fW and the effective time constant of the measurements of 1 sec. The curve of $\Delta\Sigma$ versus V_{G1} shows that the FIR causes the conductance peak to shift towards the negative direction of V_{G1} by 3 - 5 %, strongly suggesting the ionization of D2 (ΔN_2 = -1). Though not shown here, we have carefully confirmed that the shape of the $\Delta\Sigma$ versus V_{G1} curve is kept unchanged with increasing W up to 4 fW. The negative peak shift as well as its amplitude, together with the fact that these features are independent of W, definitely indicate that $\Delta\Sigma$ arises from the switch between the ground state (N_2) and the ionized state (N_2 - 1), as we have expected in Figs. 3. On the other hand, the amplitude of $\Delta\Sigma$, linearly increases with increasing W only in a limited weak range of W but is saturated at higher levels. The saturation can be reasonably interpreted as a consequence that the rate of photon absorption at D2 exceeds the inverse lifetime, τ_l^{-1}, at $W \geq 1$ fW.

An excitation spectrum is studied by tuning the wavelength of the cyclotron emission line ($\Delta\nu \approx 1.5$ cm^{-1}) from the emitter over a range 5 cm^{-1} $< \nu <$ 100 cm^{-1}. The radiation intensity is chosen to be in a linear response regime ($W \approx 0.2$ fW). Figure 4 (b) shows the B_{emit}-dependence of $\Delta\Sigma$ at the peak position of the data in Fig. 4 (a), where B_{emit} is converted to the frequency, $\nu = eB_{emit}/(2\pi m^*)$. Distinct resonance is found at $\nu = 17$ cm^{-1} ($h\nu = 2$ meV or $\lambda \approx 0.6$ mm) with a FWHM of $\nu_{FWHM} \approx 3.5$ cm^{-1}. We identify $\nu = 17$ cm^{-1} as the Kohn-mode plasma resonance, because the value agrees with the calculation of the bare confinement potential for D2 as well as with the extrapolation of the (plasma-shifted) cyclotron resonance studied in our previous experiments on QDs in high magnetic fields (see Eq. (13) in Ref. 2). We suppose that the initially excited collective motion of electrons is very rapidly transferred to a single-electron excitation (within a lifetime of $1/2\pi\Delta\nu_{FWHM} \approx$ 2.2 ps), so that the excited electron with E* = $h\nu$ = 2 meV escapes from D2.

The detector sensitivity is extremely high, approaching the single-photon detection level. This is additionally confirmed by the study of a real-time trace of Σ. As shown in Fig. 4 (c), photoresponse arises as irregular conductance spikes. Here, the data are taken with a time constant of 3 ms at W = 0.15 fW in the same gate bias condition as that for the marked peak position in Fig. 4 (a). The density of the conductance spikes increases with W. The positive spikes are ascribed to the switches, $N_2 \rightarrow N_2$ - 1, although individual events of photon absorption cannot be clearly discerned because the excited state relaxation time (0.1 - 1 ms) is shorter than the time constant of the measurements. Taking into account the switching rate we estimate the quantum efficiency for the present detector to be of the order of 1%, which, probably, can be improved by optimizing antenna and dot geometries. We find that the conductance spikes do not completely vanish in the dark condition, probably because the sample is not perfectly shielded against 4.2 K blackbody radiation in the present work.

4 Conclusion

We have studied the effect of individual photon absorption by the QDs in wavelength range of 0.17 - 0.22 mm. Each event of FIR photon absorption at the cyclotron frequency is seen as a dramatic change of the conductance due to internal charge polarization induced by the excitation of the inter Landau level electron-hole pair in the QD. We also have demonstrated ultra-high sensitivity detection of FIR by using an SET consisting of parallel double QDs in the absence of magnetic fields.

5 Acknowledgments

The work is supported by Core Research Evolutional Science and Technology (CREST) of Japan Science and Technology Corporation (JST).

References

1. 1. S. Komiyama, O. Astafiev, V. Antonov, T. Kutsuwa, H. Hirai, Nature **403**, 405 (2000).
2. O. Astafiev, V. Antonov, T. Kutsuwa, and S. Komiyama, Phys. Rev. **B 62**, 16731 (2000).
3. N. C. van der Vaart, L. P. Kowenhowen, M. P. de Rueter van Stereninck, Y. V. Nazarov, C. J. P. M. Harmans, C. T. Foxon, Phys. Rev. **B 55**, 9746 (1997).
4. Y. Kawano, Y. Hisanaga, and S. Komiyama, Phys. Rev. **B 59**, 12537 (1999).
5. F. Hofmann, T. Heinzel, D. A. Wharam, J. P. Kotthaus, G. Böhm, W. Klein, G. Tränkle, and G. Weimann, Phys. Rev. **B 51**, 13872 (1995).
6. Y. Takahashi, A. Fujiwara, K. Yamazaki, H. Namatsu, K. Kurihara, and K. Muraoc, Jpn. J. Appl. Phys. **38**, 2457 (1999).

CORRELATION EFFECTS ON JOSEPHSON CURRENT THROUGH ONE-DIMENSIONAL JOSEPHSON JUNCTION ARRAYS

TAKEO KATO

Department of Applied Physics, Osaka City University, Sumiyoshi-ku, Osaka 558-8585, Japan

E-mail: kato@a-phys.eng.osaka-cu.ac.jp

Josephson junctions hybridized with an array of superconducting islands are studied. Assuming large intra-island Coulomb interaction, a Josephson critical current are calculated as functions of voltages of two leads based on a hard-core Boson model, which corresponds to the quantum XY model with boundary fields. It is shown that the boundary Josephson coupling affects the number of resonant peaks and phase-current relation.

1 Introduction

Josephson junction networks have attracted interest for many years as interacting Boson systems showing zero-temperature quantum transitions. [1] Although transport properties have been studied experimentally for granular superconductors and patterned Josephson junction arrays, it is just recently that systematic control of parameters has been realized. [2] This experimental progress is suggestive of direct control of quantum states in correlated systems. For accurate measurement of quantum states, we can refer recent experiments of resonant tunneling of a Cooper-pair through one superconducting island. [3,4,5]

In this paper, we study the effect of Coulomb interaction between Cooper pairs on Josephson currents through one-dimensional Josephson junction arrays. The interaction effect on the bulk properties of such systems have been studied in many literatures. [6,7,8] In order to clarify novel features of correlated quantum systems, however, it is also necessary to consider a *finite* number of islands and to survey the system-size dependences. In this paper, Josephson junctions hybridized to a finite number of islands are focused on. We show that the boundary Josephson couplings to superconducting leads are important in this system.

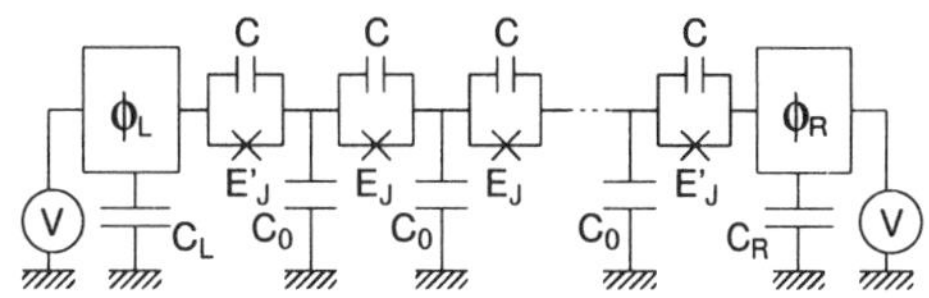

Figure 1. A circuit of a Josephson junction array considered in this paper.

2 Model

We consider a Josephson junction circuit shown in Fig. 1. Each island couples to its neighboring islands with a capacitance C and a Josephson energy $E_{\rm J}$. The islands at two edges couple to large superconducting leads with a capacitance C and a Josephson energy $E'_{\rm J}$. All islands and two leads are coupled to the ground with capacitances denoted with C_0, C_L and C_R, respectively. We assume $C_{\rm L}, C_{\rm R} \gg C_0$, and the electrostatic potentials V are applied to both leads. Dissipation due to quasiparticle tunneling is neglected throughout this paper.

First, the effective charging energy is derived. In this paper, we restrict ourselves to the case $C \ll C_0$, in which only the intra-island interaction $U = (2e)^2/2C_0$ is dominant. The effective charging energy is obtained as

$$H_{\rm C} = \sum_{i=1}^{L} U(n_i - n^*)^2. \tag{1}$$

Here, the charge on the i-th island is denoted with $Q_i = -2en_i$, and the offset charge $Q^* = -2en^*$ is controlled by the gate voltages as $n^* = -C_0V/(2e)$. We further assume $U \gg E_{\rm J}$. Then, the Josephson supercurrent is enhanced only when the offset charge is set near the frustration point $n^* = m + 1/2$. In this condition, each island can take two possible charge states, $n_i = m$ and $n_i = m + 1$. Below, without loss of generality, we set $m = 0$, and restrict the values of n_i to 0 and 1. In this approximation, the distance from the frustration point defined by $n^* = m+1/2+\Delta n^*$ plays a role of a chemical potential as $\mu = 2U\Delta n^*$. Thus, the Hamiltonian of the charging energy is obtained as

$$H_{\rm C} = -\mu \sum_{i=1}^{L} n_i. \tag{2}$$

Next, the Josephson energy is considered. We assume that the low-impedance current source applies a current to superconducting leads. In this situation, the superconducting phase difference between leads, $\phi = \phi_{\rm L} - \phi_{\rm R}$ can be treated as a classical variables. Then, the Josephson energy is expressed as

$$H_{\rm J} = -t \sum_{i=1}^{L-1} \left(b_i^\dagger b_{i+1} + b_{i+1}^\dagger b_i \right)$$
$$-t' \left(b_1 + b_1^\dagger \right) - t' \left(b_L e^{-{\rm i}\phi} + b_L^\dagger e^{{\rm i}\phi} \right), \tag{3}$$

where $t = E_{\rm J}/2$, $t' = E'_{\rm J}/2$, and b_is ($b_i^\dagger$s) are annihilation (creation) operators of Cooper pairs satisfying $b_i^2 = 0$.

We study the Josephson supercurrent based on the model Hamiltonian $H = H_{\rm C} + H_{\rm J}$. Before showing the results, we note that this model Hamiltonian is equivalent to a quantum spin-1/2 model [9]

$$H = -2t \sum_{i=1}^{L-1} (S_x^i S_x^{i+1} + S_y^i S_y^{i+1})$$
$$+\mu \sum_{i=1}^{L} \left(S_z^i + \frac{1}{2} \right) - \mathbf{H}_1 \cdot \mathbf{S}_1 - \mathbf{H}_L \cdot \mathbf{S}_L \tag{4}$$

where $\mathbf{H}_1 = (t', 0, 0)$ and $\mathbf{H}_N = (t' \cos\phi, t' \sin\phi)$ correspond to external magnetic fields applied to the boundary spins along the xy-plane. The effect of boundary fields in the 'z-direction' has been solved for the XY-model, [10,11] and for the XXZ-model by the Bethe-Ansatz [12]. However, it seems that the method used there is not applicable to the spin system with boundary fields along the xy-plane because the total S_z is not conserved.

The spin model (4) can be mapped by the Winger-Jordan transformation [10]

$$S_+^j = S_x^j + {\rm i}S_y^j = c_j^\dagger \exp\left({\rm i}\pi \sum_{l=1}^{j-1} c_l^\dagger c_l \right), \tag{5}$$

to a Fermion model

$$H = -t \sum_{i=1}^{L-1} (c_i^\dagger c_{i+1} + h.c.) - \mu N$$
$$-t' \left(c_1 + c_1^\dagger \right) - t' \left(c_L e^{-{\rm i}\pi N - {\rm i}\phi} + h.c. \right) \tag{6}$$

where $N = \sum_{j=1}^{L} c_j^\dagger c_j$ is a particle number operator. In the absence of the boundary Josephson coupling ($t' = 0$), the model is reduced to a free Fermion model, and one-body energy levels are obtained as

$$\varepsilon_n = -2t \cos\left(\frac{n\pi}{L+1} \right). \tag{7}$$

In the presence of the boundary Josephson coupling, however, the extra factor $e^{-{\rm i}\pi N}$ seems to play an important role. The exact treatment of this factor is not known at present.

3 Maximum Current

In this section, we calculate the zero-temperature Josephson current numerically by

$$I(\phi) = \frac{\partial E_0}{\partial \phi}, \tag{8}$$

where E_0 is a ground state energy. The maximum current $I_0 = \max[I(\phi)]$ is shown as a function of μ for an $L = 6$ array in Fig. 2. For small t'/t, the maximum current increases when the chemical potential agrees with discrete energy levels ε_n of the array. Hence, six resonant peaks can be seen for the $L = 6$ array. As t'/t increases, these peaks are shifted towards the center $\mu = 0$, and beyond a finite t'/t, two central peaks marked by square dots in Fig. 2 disappear. To study details of this behavior, we show the peak positions in Fig. 3 for $L = 2, 4, 6$ as a function of t'/t. For any L, the central peaks disappear at $t' \sim 0.7t$. Thus, the resonant tunneling peak is sensitive to the couplings at boundaries.

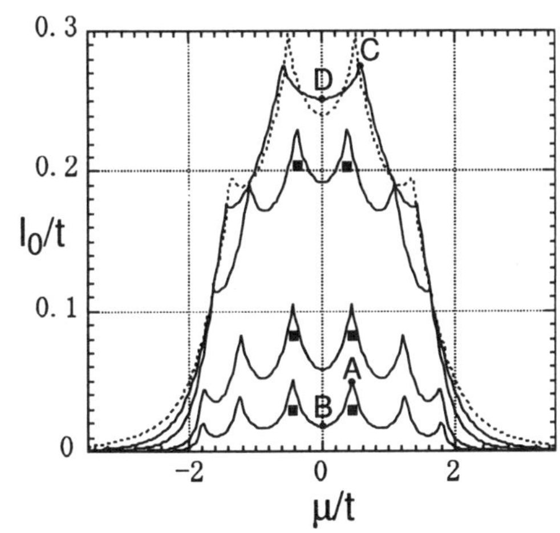

Figure 2. Maximum Josephson currents I_0 versus a chemical potential μ for an $L = 6$ array at zero temperature. The boundary Josephson coupling to leads is changed as $t' = 0.1t, 0.2t, 0.5t, t$ from the bottom to the top curve. The two peaks marked by square do's disappear at $t' \sim 0.7t$. The dotted curve denotes a $L = 4$ array with $t' = t/2$, which corresponds to a $L = 6$ with $t' = \infty$.

Generally, the number of resonant peaks of an array with L islands changes from L to $L - 2$ as t' increases. This effect can be understood as follows. For $t' \gg 1$, the islands located at edges couple to the leads strongly, and are described as

$$|\psi_i\rangle = \frac{1}{\sqrt{2}}\left(|0\rangle + e^{i\phi_i}|1\rangle\right), \tag{9}$$

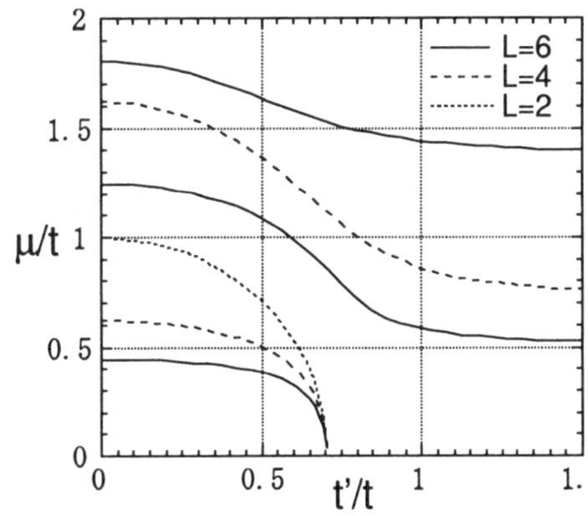

Figure 3. Resonant peak positions for $L = 2, 4, 6$ as a function of t'/t in the range of $\mu > 0$.

where $i = 1$ or L, and phases are given by $\phi_1 = 0$ and $\phi_L = \phi$, respectively. Then, they play a role of new leads, and the system is described by an effective Hamiltonian $\tilde{H} = \langle\psi_1|\langle\psi_L|H|\psi_1\rangle|\psi_L\rangle$ with $L - 2$ islands. It can be checked that the new coupling constant t' is given by $t/2$. For comparison with the $L = 6$ array, the maximum current for $L = 4$ with $t' = t/2$ is shown in Fig. 2 by a dashed curve. It can be seen that the solid curves for $L = 6$ approach this dashed curve as t'/t increases.

4 Phase-Current Relation

The boundary effect due to the coupling t' appears more clearly in the phase-dependence of the Josephson current. In Fig. 4, the Josephson current is shown as a function of the phase difference ϕ. The lower pair of curves shows the $t' = 0.1t$ case, where the chemical potential is taken at on- and off-resonance point denoted with A and B in Fig. 2, respectively. The Josephson current has a maximum for the on-resonant point A near $\phi = \pi$, while for off-resonant point B near $\phi = \pi/2$. This result is the same as the resonant tunneling effect in one-island systems ($L = 1$). On the other hand, the upper pair of curves shows the $t' = t$ case, where the chemical potential is taken at on- and off-resonance point denoted with C and D in Fig. 2, respectively. For both points, the Josephson current has a maximum near $\phi = \pi$. Further, the Josephson current is almost proportional to ϕ for both C and D. Thus, the Josephson couplings at boundaries also affect the phase dependence of the supercurrent.

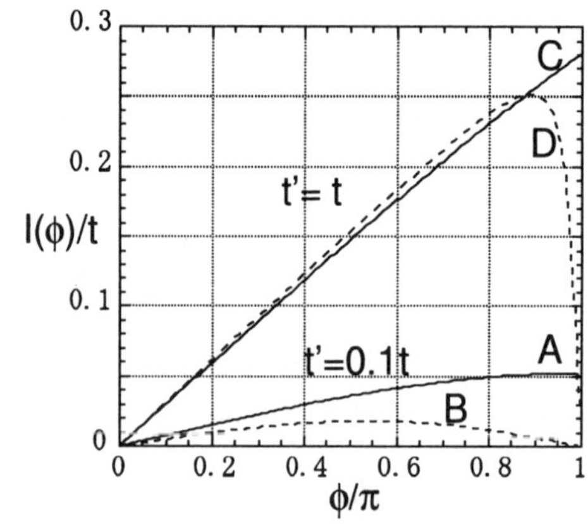

Figure 4. Phase dependence of the Josephson current. The upper and lower pair of curves denote $t' = t$ and $t' = 0.1t$ cases, respectively. Solid(dashed) curves correspond to the on-resonant (off-resonant) points shown by A and C (B and D) in Fig. 2.

5 Summary

Josephson junction arrays coupled to large superconducting leads have been studied. Assuming large on-site Coulomb interactions, the system has been mapped to a hard-core Boson model, which is equivalent to the XY model with boundary fields. It has been found that the number of resonant peaks changes from L to $L-2$ for the array with L islands as the Josephson coupling t' at boundaries increases. It has also been shown that the large boundary coupling changes the phase-dependence of the Josephson current crucially. We expect that these results can be observed in experiments when experimental parameters are properly tuned.

In this paper, only the simplest case is discussed. I expect that Josephson junction networks can be used to study strongly-correlated systems such as spin systems. For example, by introducing nearest neighbor interactions, the Josephson junction network can be related to various types of the XXZ-spin systems. [13] It would be a challenging problem to study what properties of these spin systems can be observed by the Josephson current. These remaining problems are discussed in future.

References

1. R. Fazio, and H. van der Zant, *preprint* (cond-mat/0011152).
2. E. Chow, P. Delsing, and D. B. Haviland, *Phys. Rev. Lett.* **81**, 204 (1998).
3. P. Joyez, P. Lafarge, A. Filipe, D. Esteve, and M. H. Devoret, Phys. Rev. Lett. 72 (1994) 2458.
4. W. J. Elion, M. Matters, U. Geigenmüller, and J. E. Mooij, Nature 371 (1994) 594.
5. M. Matters, W. J. Elion, and J. E. Mooij, Phys. Rev. Lett. 75 (1995) 721.
6. C. Bruder, R. Fazio, and G. Schön, Phys. Rev. B **47**, 342 (1996).
7. A. A. Odintsov, Phys. Rev. **54**, 1228 (1996).
8. L. I. Glazman and A. I. Larkin, Phys. Rev. Lett. **79**, 3736 (1997).
9. T. Matsubara and H. Matsuda, Prog. Theor. Phys. **16**, 569 (1956).
10. E. Lieb, T. Schultz, and D. Mattis, Ann. Phys. **16**, 407 (1961).
11. H. Asakawa and M. Suzuki, J. Phys. A **28**, 4679 (1995).
12. H. Asakawa and M. Suzuki, J. Phys. A **29**, 225 (1995).
13. Y. Kanai, T. Kato, K. Kusakabe, and K. Okunishi, in preparation.

SUPERCONDUCTOR-INSULATOR TRANSITION IN ONE- AND TWO-DIMENSIONAL ARRAYS OF DISSIPATIVE SMALL JOSEPHSON JUNCTIONS

TAKAHIDE YAMAGUCHI, HISAO MIYAZAKI, AKINOBU KANDA, YOUITI OOTUKA

Institute of Physics, University of Tsukuba, 1-1-1 Tennodai, Tsukuba 305-8571, Japan

CREST, Japan Science and Technology Corp., 4-1-8 Honcho, Kawaguchi, 332-0012, Japan

E-mail: yamaguchi@lt.px.tsukuba.ac.jp

We studied one- (1D) and two-dimensional (2D) arrays of small Josephson junctions in which each junction was shunted by a normal metal resistor. For the arrays with the charging energy E_C comparable to the Josephson coupling energy E_J, we find a crossover from insulating to superconducting behavior as the shunt resistance R_S decreases. The critical value of R_S is close to dR_Q (d: dimensionality, $R_Q \equiv h/4e^2 = 6.45$ kΩ), which is consistent with theories of the dissipation-driven phase transition. We discuss the difference between the phase diagrams in the $E_J/E_C - R_Q/R_S$ plane for 1D and 2D arrays.

1 Introduction

Small Josephson junction arrays are model systems showing quantum phase transitions. [1–5] It has been suggested that the competition between the Josephson coupling, the charging effect, and the dissipation leads to a superconductor-insulator (SI) transition at zero temperature. When the charging energy exceeds the Josephson coupling energy, quantum fluctuations destroy the global order of the superconducting phases, and the array undergoes a superconductor-insulator transition.[1–3] In this scenario, the dissipation plays an important role since the quantum fluctuation of the phases strongly depends on the strength of the dissipation. According to theories,[4,5] the dissipation larger than a critical value suppresses the fluctuation, leading to the superconducting state even for arrays with large charging energy. The theories also predict that the quantum phase transition in Josephson junction arrays depends strongly on the dimensionality of the system.

In this report, we present the experimental results on the low-temperature transport properties of both 1D and 2D Josephson junction arrays. To control the junction parameters (E_J and $E_C \equiv e^2/2C$, C: junction capacitance) and the dissipation independently, we fabricated arrays in which each junction (Al-AlO_x-Al)is shunted by a normal metal resistor (Cr). In Josephson junction systems, the inverse of the shunt resistance R_S represents the magnitude of the dissipation; the ratio R_Q/R_S is used as a measure of the dissipation.

2 Experimental results and discussion

The samples were fabricated using the electron beam lithography with a bilayer resist.[6] We attached a Si substrate to a bi-axis-rotation stage in a vacuum chamber and evaporated Cr, Au (for good electric contacts between Cr and Al) and Al successively from different angles, without exposing the substrate to air in the process. We fabricated a set of arrays on a substrate to obtain arrays with nominally the same junction parameters and different shunt resistances. The shunt resistances were varied by changing their lengths, and were $1 - 8$ μm long, 0.15 μm wide and $5 - 8$ nm thick. We fabricated four sets of 2D arrays and three sets of 1D arrays. The junction parameters were varied from one substrate to another. The 2D arrays were 48-junctions long and 40-

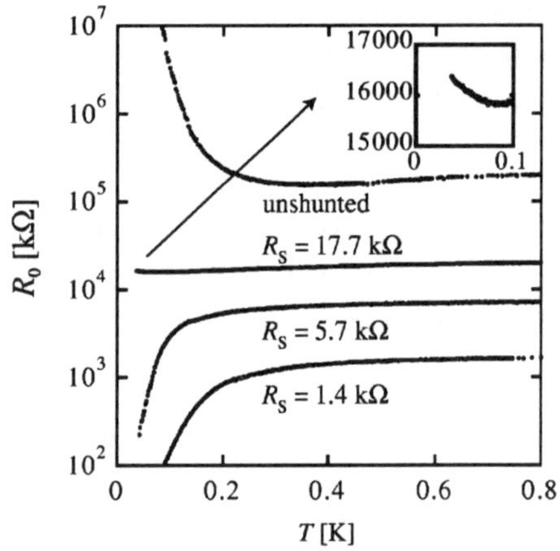

Figure 1. The zero-bias resistances of 2D arrays plotted as a function of temperature. The arrays have nominally the same junction parameters: the junction normal state resistance $R_J = 76.2$ kΩ, $E_J/k_B = 0.14$ K and $E_C/k_B = 1.1$ K.

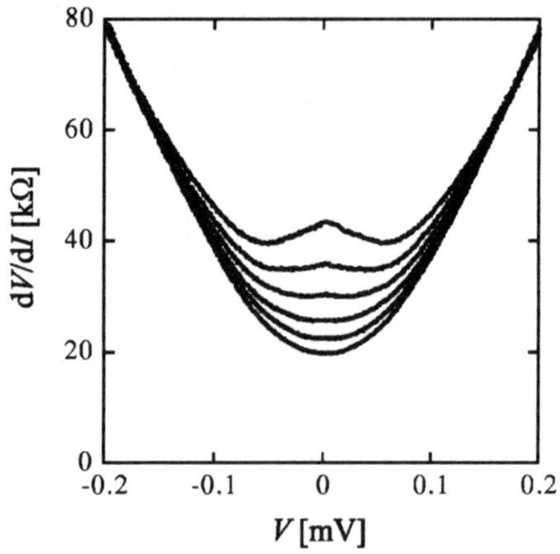

Figure 2. The differential resistance vs. voltage at different magnetic fields B for a 1D shunted array with $R_S = 15$ kΩ and $E_J^0/E_C = 4.0$. From bottom to top, $f(= BS/\Phi_0) = 0.27 - 0.32$, step 0.01.

junctions wide, while the 1D arrays were 47-junctions long. We performed a four-terminal measurement. The voltage probes of 2D arrays were 40-junctions apart in the middle of the array and those of 1D arrays were attached to the ends of the arrays. In the 1D arrays, the neighboring island electrodes are connected by two junctions in parallel. Such SQUID geometry enabled us to tune the effective Josephson coupling between adjacent islands by applying an external perpendicular magnetic field B; $E_J = E_J^0|\cos(\pi BS/\Phi_0)|$ where $\Phi_0 \equiv h/2e = 20.7$ gauss μm^2. The area S of the SQUID loop was 2.4 μm^2.

Figure 1 shows the temperature dependence of the zero-bias resistance for a series of the 2D arrays.[7] They are fabricated on the same substrates and have nominally the same junction parameters but have different shunt resistances R_S. As the temperature is lowered, the resistances of the unshunted array and the array with $R_S = 17.7$ kΩ increase, while those of the arrays with smaller R_S decrease, indicating an existence of superconductor-insulator transition at $T \to 0$.[a] The IV curves of these arrays also exhibit the crossover from the voltage gap of the Coulomb blockade to the Josephson-current-like structure as the shunt resistance decreases.

In the 1D arrays we also observe changes of the transport properties depending on the shunt resistance.[8] Furthermore, the IV curves of some arrays including both shunted and unshunted ones show the crossover from the Josephson-like behavior to the Coulomb blockade as the Josephson coupling energy is suppressed by the external perpendicular magnetic field. Figure 2 shows an example of how the differential resistance versus voltage changes as the magnetic field is varied.

We performed measurements for 13 1D arrays and 17 2D arrays and obtained the phase diagrams in the $E_J/E_C - R_Q/R_S$ plane at $T \to 0$ as shown in Fig. 3. We determined the phase diagram, judging from whether the IV curve shows the Coulomb blockade or the Josephson-current-like structure at the lowest temperature. The vertical lines for the 1D arrays in Fig. 3(b) correspond to the ranges of the effective Josephson coupling energy tuned

[a] We use the term "insulating" in the sense that the Josephson channel of current is closed. In the *shunted* arrays, the resistance remains finite at $T = 0$ even for the "insulating" case.

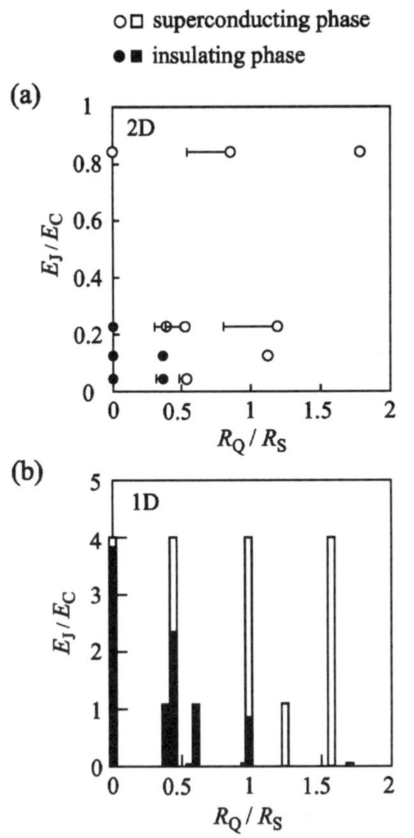

Figure 3. Phase diagram of (a) 2D and (b) 1D arrays. The open and filled symbols show the superconducting and insulating phase respectively. A change from an open bar to a filled one means an occurring of the SI transition due to the magnetic field.

by the magnetic field.

Several theoretical studies have been performed on the phase diagram. The results of the junction-capacitance model, where only the junction capacitance C is included in the capacitance matrix $\{C_{ij}\}$, are summarized as follows. We think the junction-capacitance model is applicable to our experiment at least as a first order approximation, because in real systems fabricated by electron beam lithography the C dominates the other matrix elements ($C/C_0 \sim 10^2 - 10^3$). In d-dimensional square arrays with $E_J << E_C$, the insulating state is located in $R_Q/R_S<1/d$ and the superconducting state is in $R_Q/R_S>1/d$.[5] In 2D arrays, the critical value of R_Q/R_S decreases as the ratio E_J/E_C increases, and becomes zero at $E_J/E_C \sim 1$, which means that 2D unshunted arrays show an SI transition depending on the E_J/E_C.[1-3] On the other hand, the 1D arrays are insulating when $R_Q/R_S<1$ and superconducting when $R_Q/R_S>1$, irrespective of the ratio E_J/E_C.[5,9] That is, the SI transition depending on the E_J/E_C does not occur in 1D arrays.

The experimental results in Fig. 3 indicate that the critical value of R_Q/R_S for small E_J/E_C is close to $1/d$ in the d-dimensional arrays. This is in good agreement with the results of the theories.[5] Furthermore, as shown in Fig. 3(a), the 2D arrays with $R_Q/R_S = 0$ undergoes the SI transition depending on the E_J/E_C. The critical value $(E_J/E_C)_c$ is between 0.23 and 0.84, which is consistent with the result ($(E_J/E_C)_c = 0.55 - 0.67$) of an earlier experiment on unshunted arrays.[10] The $(E_J/E_C)_c$ predicted by the junction-capacitance model (with no dissipation) vary from each other but are values somewhat smaller than 1.[1-3]

On the other hand, the results of the 1D arrays for large E_J/E_C differ from the expectations. Figure 3(b) shows that the arrays with small R_Q/R_S and large E_J/E_C are superconducting. This is contrary to the theoretical prediction. The experiments of 1D *unshunted* arrays, performed by Haviland *et al.*,[11] also show that the transition takes place at $E_J/E_C = 3.0$. A possible origin of this discrepancy is the existence of self-capacitance C_0 in the real system. Analyzing a model where only the C_0 is taken into account, Bradley and Doniach[12] reached a conclusion that the SI transition should occur at a finite value of E_J/E_{C0}, where $E_{C0} \equiv e^2/2C_0$. More recently, Choi *et al.* [9] considered both C and C_0, and found that the SI transition should take place at $E_J/E_C \sim (32/\pi^2)C/C_0$ for $C_0 \ll C$. However, the theory does not explain the experimental results because the observed critical value of E_J/E_C is much smaller than $(32/\pi^2)C/C_0$ which is $10^2 - 10^3$ in the experiments of ours and Haviland *et al.*

Thus, the obtained phase diagram for 1D arrays is qualitatively similar to that of the 2D arrays. However, the critical value of E_J/E_C for $R_Q/R_S = 0$ for 1D arrays is one-order of magnitude larger than that for the 2D arrays, indicating the fluctuation is more important in lower dimensions.

3 Conclusions

We have determined the phase diagram of 1D and 2D arrays of resistively shunted small Josephson junctions. The insulating phase appears when both E_J/E_C and R_Q/R_S are small. The critical values of R_Q/R_S for small E_J/E_C are close to $1/d$, which is in good agreement with the theories of the dissipation-driven SI transition. We also find that the phase diagram of the 1D arrays has a larger region of the insulating phase than that of 2D, indicating the importance of the quantum fluctuation in lower dimensions. There is, however, an unsolved discrepancy between the result of the experiment and theories: the 1D arrays with $R_Q/R_S<1$ and relatively large E_J/E_C show no Coulomb blockade behavior, which does not agree with the theoretical prediction.

Acknowledgments

We thank R. Yagi and S. Kobayashi for collaborations in the early stages of the experiments presented here, and T. Kato for helpful discussions. One of the authors (Y.T.) acknowledges the support of a research fellowship from the Japan Society for the Promotion of Science for Young Scientists. This work has been partially supported by Grants-in-Aid for Scientific Research from the Ministry of Education, Science, Sports and Culture, Japan.

References

1. R. Fazio and G. Schön, Phys. Rev. B **43**, 5307 (1991).
2. C. Rojas and J. V. Jose, Phys. Rev. B **54** 12361 (1996).
3. B.J. Kim *et al.*, Phys. Rev. B **56** 395 (1997).
4. S. Chakravarty *et al* Phys. Rev. B **37**, 3283 (1988); S. Chakravarty *et al.* Phys. Rev. B **35**, 7256 (1987); A. Kampf and G. Schön, Phys. Rev. B **36**, 3651 (1987).
5. M. P. A. Fisher, Phys. Rev. B **36**, 1917 (1987).
6. R. Yagi and S. Kobayashi, J. Phys. Soc. Jpn **66**, 3360 (1997); R. Yagi, S. Kobayashi, Y. Ootuka, J. Phys. Soc. Jpn. **66**, 3722 (1997).
7. Y. Takahide *et al.*, Phys. Rev. Lett. **85**, 1974 (2000); T. Yamaguchi *et al.*, Physica C **352**, 181 (2001).
8. H. Miyazaki *et al.*, in preparation.
9. M.-S. Choi *et al.*, Phys. Rev. B **57**, R716 (1998).
10. L. J. Geerligs *et al.*, Phys. Rev. Lett. **63** 326 (1989); H. S. J. van der Zant *et al.*, Phys. Rev. B **54**, 10081 (1996).
11. D.B. Haviland *et al.*, Phys. Rev. Lett. **81**, 204 (1998); D.B. Haviland *et al.*, Physica C **352**, 55 (2001).
12. R.M. Bradley and S. Doniach, Phys. Rev. B **30**, 1138 (1984).

QUANTIZED CONDUCTANCE OF GOLD NANOWIRE STUDIED BY UHV-ELECTRON MICROSCOPE WITH STM

K. TAKAYANAGI

Tokyo Institute of Technology, Physics Department

2-12-1 Oh-okayama, Meguro-ku, Tokyo 152-8551, Japan

E-mail: takayang@surface.phys.titech.ac.jp

Development of an UHV electron microscope and results on gold QPC are summarized. Quantum point contact(QPC) of gold was made by an STM installed in the UHV electron microsope to study conductance quantization of gold QPCs in relation with their structure simultaneously. The conductance changed in steps in the unit of $2e^2/h$ in case that gold nanowires were formed at the QPC, where e is the electron charge and h, Planck constant. Long gold nanowires were found to have chiral structure with gold atomic rows which coil around the axis of the wire. A single strand of gold atoms synthesized was found to be stable even when their atomic bond distance reaches to 0.4nm.

1. Introduction

Quantum point contact of metals attracts much interest because of the conductance quantization even at room temperature. Many experiments were done using STM geometry[1], mechanically controlled breaking junction[2] and switching relays[3]. In STM geometry, metal nanowires are made at the contact of the STM tip and the substrate, and the conductance change is measured while the tip is retracted. Conductance changes in steps give an indication of quantization because of appearance of the step degitized in a unit of $2e^2/h$, where e is the electron charge and h, Planck constant. Conductance histograms are examined from thousands of experiments, for metal QPCs of Au[4-6], Ag[6,7], Cu[6-8], Pt[6,7], Na[8], A[19], Pb[10], Bi[11], Hg, Sn[12], Fe, Co, and Ni[6,7]. It is important issue to elucidate which structure formed at QPCs gives conductance quantization. We, therefore, developed an UHV electron microscope with a miniaturized STM at the specimen stage of the electron microscope to make simultaneous observation of the structure and conductance possible.

2. Development of UHV electron microscope with STM

UHV electron microscope is specially designed to build a miniaturized STM at the specimen position. The specimen and the STM (tip and piezo drives) are combined at the specimen stage of the electron microscope. The gold sample was made by deposition onto a copper wire, and the gold tip was mechanically sharpened. The gold tip was dipped into the gold substrate several times until regular shape and crystallographic orientation of the contact was generated at the contact. The conductive current of the QPC was measured by the bias voltage of 13mV for most of our experiments. The structure of gold QPC was observed by transmission electron microscopy at atomic resolution, being recorded on a video tape at every 33msec, while the tip was retracted.

3. Conductance quantization of gold QPC

Gold QPC had different structures depending on the crystallographic orientation of the contacting two electrodes[13]. When the both electrodes have the same orientation as [110], [111], or [100], one to one correspondence between the atomic structure and the conductance of QPC was analyzed in detail. As reported previously, QPCs formed between [110] electrodes showed results similar to the one in Fig.1[14]. Transmission electron microscope images shown in Fig.1(a) reveal that a gold nanowire has uniform diameter, and changes in steps. For every step of the diameter change, conductance changes in steps. The step heights of the conductance are integral multiple of the $2e^2/h$, which indicates conductance quantizaiton of the gold QPC. This result indicates that each conduction channel has the transmission probability of unity; T_j=1 in the Landauer formula[15], $G=(2e^2/h)\,\Sigma\, T_j$.

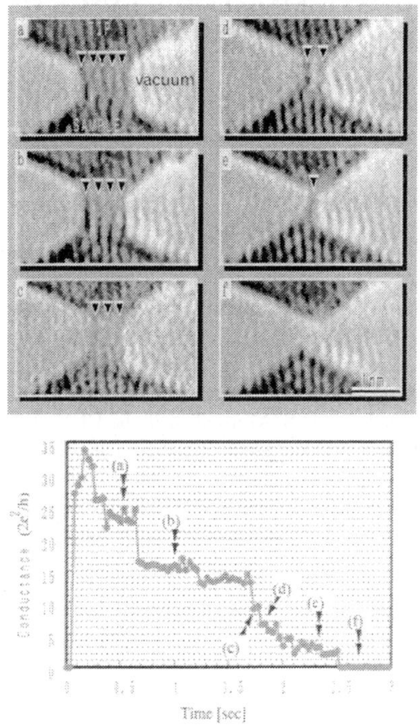

Fig.1 Transmission electron microscope images of a gold QPC(upper panel), and the condactance change(lower panel). The electrodes have the [110] crystallographic orientation. The dark lines in the electrode are the (002) lattice fringes with 0.2nm spacing. The dark lines within the gold nanowire that bridges two electrodes indicate lattice sheet, since the number of atoms in the vertical direction to the plana of the sheet (the direction of the imaging electron beam) is not known without careful profile analysis of the line contrast. Dark lines of the nanowire disappear one-by-one, while conductance reduces in steps; (a) – (f) in the both panels are recorded at the same time in this experiment.

The gold nanowire in Fig.1(f) show a single dark line, while its conductance is $3(2e^2/h)$. The conductance of this wire goes in the next step to $2(2e^2/h)$, and further to zero. Provided that the nanwire in (f) had have three atomic chains connecting the both electrodes, the conductance could change one-by-one in steps as the atomic chain breaks one-by-one. We obtained the electron microscope images of a double chain and a single chain, which gave the conductance of $2(2e^2/h)$ and $(2e^2/h)$, respectively. This proves that the QPC of a single chain of gold atoms has the conductance of $(2e^2/h)$.

4. Synthesis of chiral nanowire of magic number seven

The gold nanowires formed at QPCs are breaks during retraction of the tip, and thin nanowires like in Fig.1 could not be elongated over 10nm. To synthesize longer gold nanowire, we used electron beam thinning technology. A thin gold film (3nm thick) was made by Pashley method, and irradiated by very intense electron beam (100A/cm^2, and 200keV) until the film has many holes. One of the bridges, which remained between the holes, was thinned further with a reduced electron beam intensity to turn into a regular naonowire[16]. When the regular nanowires became thin (less than 2nm), the nanowires presented specific electron microscope images which have never been reported or seen before[17]. The images show a few to several dark lines running along the wire axis and dark dots representing gold atoms. However, no periodic structure could be recognized in the images. From histogram for the apparent widths, W, of the nanowires, the widths are noticed to be peaked at $W(n)$=0.08 n (nm), where n is an integer and $W(n)$ is the distance of the dark lines representing the diameter of the nanowire.

The model structures of gold nanowires are shown in Fig.2: They have helical multi-shell (HMS) structure[17] similar to that of the multi-wall carbon nanotubes (MWNT)[19]. While the honeycomb sheet rolls into the carbon nanotube, a close packed sheet rolls into the HMS. As schematically shown in Fig.2, the *n-n'-n"*HMS nanowires are composed of *n*-atomic rows in outer shell, *n'*-atomic rows in the middle, and *n"*-atomic rows in the center[1]: 7-1, 11-4, 12-5, 13-6, 14-7-1, 15-8-1 HMS

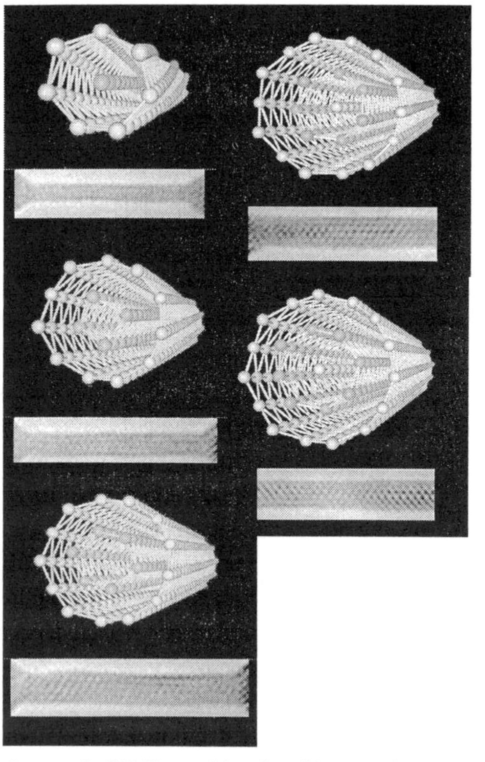

Fig.2 *n-n'-n"*HMS model of gold nanowires. the *n-n'-n"*HMS nanowires are composed of *n*-atomic rows in outer shell, *n'*-atomic rows in the middle, and *n"*-atomic rows in the center[1]: 7-1, 11-4, 12-5, 13-6, 14-7-1, 15-8-1 HMS structures are shown except 12-5HMS.

structures were verified. The diameter D(n) of the n-atomic rows shell is given by $D(n)=\sqrt{3}d/\sin(\pi/n)/\cos H$, where helical angle H is given by $\sin H=n\sqrt{3}/2L$. Here, d =0.288nm (the nearest neighbor distance of the gold atoms in an atomic row), and L is the pitch of the helical nanowires. Since $D(n)$ changes linearly with n (~0.80n), the apparent width of the HMS nanowire changes with the step of W(n)=0.08nm, as was observed experimentally. The observed n-n'-n''HMS nanowires fullfil the relation $D(n)-D(n')=2d$. This relation means that the inner n'-atomic rows tube can be smoothly slide in (out) the outer n-atomic row tube. This relation, $0.08(n-n')=2d$, is satisfied only in the case $n-n'=7$. Thus, gold HMS structure has the magic number 7 for the outer-inner tube relation.

The stability of the 7-1HMS structure was shown theoretically[18]. Further theoretical studies are expected to understand the gold HMS nanowires, of their chirality (pitch of the nanowire), and conductance.

The conductance of the HMS nanowires are intersting because of their chirality. Following to the elaborated experimental and theoretical works on carbon nanotubes[19], conductance is sensitive to the chirality. Recent theory[20] has predicted reduction of conductance by 2 ($2e^2/h$) , although the theory is meso-scopic.

5. A single strand of gold atoms

The finest nanowire should be a linear chain of atoms. A strand of gold atoms[3] was made by thinnig using electron beam technique as shown in Fig.3. The suspended gold strand has four atoms whose chain-end atoms bound to the bulk crystal. The chain-center atoms (two dark dots) bonded to these chain-end atoms have quite a large bond length reaching to 0.4 nm before rupture[14,21].

Several questions are given for such a long bond length; for the nature of bond, stability, and conductance. A theoretical result[22] suggest that a gold chain may change to an insulator for bond lengths longer than 0.3nm. Further experiments and theory have to be done systematically for complete understanding of bond nature, conductance[23]. Particularly interesting is to demonstrate spin conductance for strands of magnetic elements[24,25].

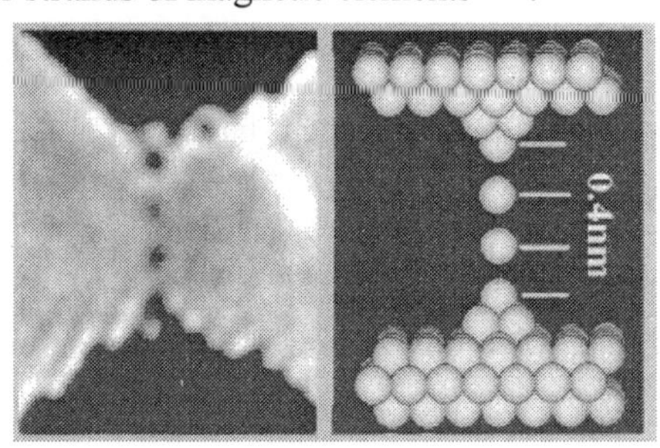

Fig.3 Transmission electron microscope image of a gold single strand(left panel) and its model(right panel). The strand has atomic bond distance of 0.4nm before its rupture. The strand bonds with gold crystal (upper and lower dark region of the panel) which has the[001] orientation.

6. Summary

Structure and conductance of metal QPC were studied by UHV electron microscopy with scanning tunneling microscopy. The efforts have confirmed that conductance quantization occurs when the QPC has a regular nanowire, and gold single strand has the conductance of ($2e^2/h$). At the same time , the efforts have revealed gold HMS nanowires having magic number 7, and a gold single strand having bond lengths of the order of 0.4nm.

To reveal nano-structures and their physical and chemical behavior, we need further experimental and theoretical efforts. Particularly important is UHV electron microscopy for developing nanoscience and technology, and without UHV condition we never reaches to right answer.

The work was done at "surface particle project" of ERATO, Japan Science Technology, and by the support of Grant-in-Aid for Specially Promoted Research from the Ministry of Education, Curture, Sports, Science and Technoology (MEXT).

References

1. J.I.Pascual, J.Mendez, J.Gomez-Herrero, A.M. Baro, N.Garcia and V.T.Binh, Phys. Rev. Lett. 71, 1852 (1993).
2. C.J.Muller, J.M. van Ruittenbeek and L.J.de Longh, Phys. Rev. Lett. 69, 140 (1992).
3. H.Yasuda and A.Sakai, Phys.Rev. B56, 1069 (1997).
4. J.L.Kosta-Kraemer, N.Garcia and H.Olin, Phys. Rev. B55, 12910 (1997).
5. B.Ludoph, M.H.Devoret, D.Esteve, C.Urbina and J.M.van Ruittenbeek, Phys. Rev. Lett. 82, 1530 (1999).
6. K.Hansen, E.Laesgaard, I.Stensgaard and F.Besenbaecher, Phys. Rev. B56, 2208 (1997).
7. J.L.Kosta-Kraemer, Phys. Rev. B55, R4875 (1997).
8. J.M.Krans, J.M.van Ruitenbeek, V.V.Fisun, I.K.Yanson and L.J.de Longh, Nature(London) 375, 767 (1995).
9. E.Scheer, N.Agrait, J.C.Cuevas, A.L.Yeyati, B.Ludoph, A.Martin-Rodero, G.R.Bollinger, J.M.van Ruittenbeek and C.Urbina, Nature(London) 394, 154 (1998).
10. Z.Gai, X.Li, B.Gao, R.G.Zhao, W.S.Yang and J.W.M.Frenken, Phys. Rev. B58, 2185 (1998).
11. J.L.Costa-Kramer, N.Garcia and H.Olin, Phys. Rev. Lett. 78, 4990 (1997).
12. J.L.Kosta^Craemer, N.Garcia, P.Garcia-
13. Mochales and P. A.Serena, M.I.Marques and A.Correia, Phys. Rev. B55, 5416 (1997).
14. M.R.Sorensen, M.Bandbyge and K.W.Jacobsen, Phys. Rev. B57, 3283 (1998).

15. H.Ohnishi, Y.Kondo and K.Takayanagi, Nature 395, 780 (1998).
16. R.Landauer, IBM J.Res.Dev. 1, 223 (1957): M.Buttiker, Y.Imry, R.Landauer, S.Pinhas, Phys.Rev.B31, 6207 (1985).
17. Y.Kondo and K.Takayanagi, Phys. Rev. Lett. 79, 3455 (1997).
18. Y.Kondo and K.Takayanagi, Science 289, 606 (2000).
19. E.Tosatti, S.Prestipino, S.Kostlmeier, A.Dal Corso, F.D.Di Tolla, Science 291, 288 (2001).
20. S.Iijima, Nature 354, 56 (1991).
21. M.Okamoto, T.Uda, K.takayanagi, Phys. Rev. B64, 033303 (2001).
22. H.Koizumi, Y.Ohshima, Y.Kondo, K. Takayanagi, Ultramicroscopy, 88 (2001) 17.
23. M.Okamoto and K.Takayanagi, Phys. Rev. B60, 7808 (1999).
24. J.A.torres, A.D.Corso, F.Ercolessi, J.J.Kohanoff, F.D.D. Tolla and J.M.Soler, Surface Sci. 426, L441 (1999).
25. F.Komori and N,Nakatsuji, J.Phys. Soc.Jpn. 68, 3786 (1999).
26. H.Imamura, N.Kobayashi, S.Takahashi and S.Maekawa, Phys. Rev. Lett. 84, 1003 (2000).

ATOMIC-SCALE STRUCTURES FABRICATED ON A HYDROGEN-TERMINATED SI SURFACE

T. HASHIZUME, S. HEIKE, M. FUJIMORI, M. KATO AND M. ISHIBASHI

Advanced Research Laboratory, Hitachi, Ltd., Hatoyama, Saitama, 350-0395, Japan

Atomic-scale structures on a hydrogen-terminated Si(100)-2x1-H surface are investigated by scanning tunneling microscopy/spectroscopy (STM/STS). Relaxation of atomic-scale dangling-bond (DB) structures can be well described by a Jahn-Teller distortion. When Ga atoms were deposited on the surface at 100 K, we observed a characteristic one-dimensional structure (Ga-bar structure). The Ga-bar structure is an STM image of a rapidly-migrating Ga atom confined in a linear potential well.

1 Introduction

Invention of scanning tunneling microscopy (STM)[1] and its use for atom manipulation[2] have made it possible to realize the idea of switching devices on the atomic and molecular scale as were proposed by Feynman.[3] The hydrogen-terminated Si(100)-2x1-H surface[4] shows semi-conductive band gap and is one of the promising substrates for atomic-scale device fabrication. Hydrogen atoms on the surface can be desorbed using the tunneling current of the STM and atomic-scale dangling-bond (DB) patterning has been demonstrated.[5,6]

We report on fabrication and characterization of atomic-scale structures on a hydrogen-terminated Si(100)-2x1-H surface. A row of fabricated DBs forms a one-dimensional atomic structure. Although this DB structure is chemically too reactive to be used in practical atomic-scale devices, we expect it to be an instructive example of an atomic-scale structure on an insulating or a semiconducting surface. We analyze our spectroscopy (STS) results based on the recent first-principles calculations.[7]

Thermally deposited Ga atoms on the surface are confined in a one-dimensional potential well between two adjacent dimer rows at a narrow range of temperatures near 100K, resulting in a continuous linear protrusion (Ga-bar structure). We point out that the height of the Ga-bar structure maps out the local variation of potential energy at individual adsorption sites.

2 Experimental

A Si(100) sample (P-doped, n-type, 7 to 18 mΩ·cm) was cut from a wafer and used as a substrate. An electrochemically etched ⟨111⟩ oriented single-crystal tungsten (W) wire was used as the STM tip. Details of surface and tip preparation were described previously. [6] Hydrogen atoms were extracted by moving the tip at a speed of 5 nm/s along a Si dimer row, typically at a sample bias voltage (V_s) of +2.8 V and a tunneling current of 0.9 nA. By heating a W filament with Ga metal to 600-650 °C, Ga atoms were evaporated *in situ*. The base pressure of the STM chamber was lower than 7×10^{-9} Pa during the STM observation. Experiments were carried out at various temperatures ranging from 30 to 150 K with deposited Ga of approximately 0.002 ML (1 ML is defined as the number of Si atoms on the bulk-terminated ideal Si(001) surface: 6.78×10^{14} atoms/cm^2).

3 Results and Discussion

We fabricated various-length DB structures composed of unpaired DBs, which are imaged off-center of the dimer row. In the cases of short DB structures, we observed interesting length-dependent charge redistribution. Figures 1(a) through (c) show gray-scale filled-

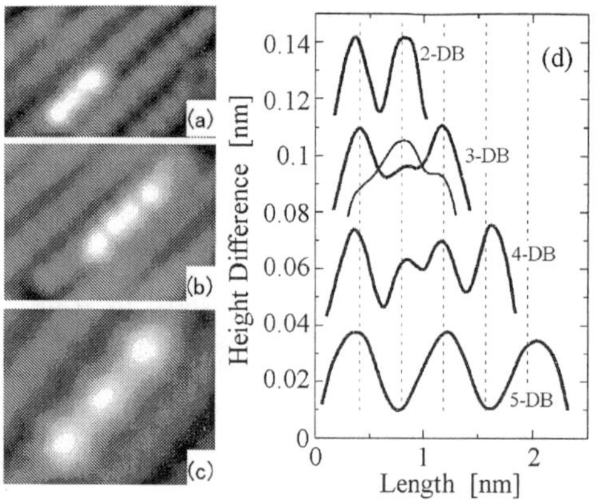

Figure 1. Filled-state STM images of (a) three-, (b) four-, and (c) five-DB structures. Curves in (d) show cross-sectional views obtained from the filled-state images and the thin line shows that from the empty-state image of the three-DB structure.

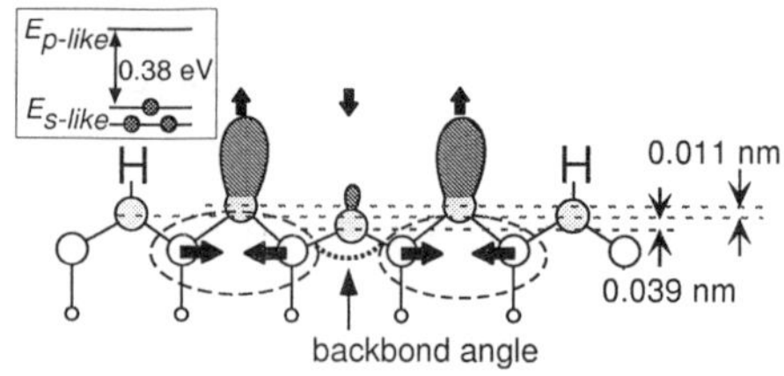

Figure 2. A ball-stick model of the three-DB structure showing displacement patterns of the first- and second-layer Si atoms indicated by the arrows. The circles denote the first- and second-layer Si atoms. The amount of charge in the DBs are schematically expressed by the ovals. The dotted ovals show the paired second-layer Si atoms. Dashed horizontal lines show the center of up and down Si atoms with DBs relative to the unrelaxed Si atoms. The inset shows the energy level diagram at the Γ point in wave-vector space associated with the three-DB structure. Filled circles stand for electrons. The second-layer Si atoms are displaced 0.0016 nm laterally to form pairs.

state STM images. Cross-sectional views of filled-state STM images in Fig. 1(d) reflect the charge redistribution.

Distinct alternate peaks are observed in the filled-state images for the odd-numbered DB structures (Figs. 1(a) and (c)). In the empty-state STM image of the three-DB structure, the center DB appears higher than the edge DBs (thin line in Fig. 1(d)). Theoretical calculation of the three-DB structure showed a remarkable qualitative agreement with the STM images.[8] Almost one unit charge of the center DB is transferred to the edge DBs. In contrast, the center DB has higher empty states.

In the charge redistribution process, pairing of the second-layer Si atoms is playing a crucial role (Fig. 2). When the second-layer Si atoms pair up, the backbond angle becomes smaller, which lowers the DB energy. The numerical calculations suggest that the DB character of the center Si atom becomes more p-like because the backbonds of the Si atom are more sp^2-like. In contrast, the DBs of the edge Si atoms become more s-like because the backbond of the Si atoms are more p^3-like. Since the s-like orbital with the energy of E_{s-like} is energetically more favorable than the p-like orbital (E_{p-like}), the center DB ends up with nearly empty charge. The electronic relaxation in the DB structure associated with the lattice distortion may be called as the Jahn-Teller distortion.[8]

In the filled-state image of the two-DB structure, the height difference between the DBs was small (Fig. 1(d)). However, the calculated result suggests a buckled atomic configuration, resulting in a non-equivalent charge distribution. The calculated flip-flop barrier is 48 meV, and a frequent flip motion, induced even at 100 K, results in the equivalent height of DBs observed by STM. The flip-flop can be explained by the exchange of a lone second-layer Si atom that is not forming a pair. The recombination of a pair in the second-layer Si atom effectively shifts the position of the lone second-layer Si atom. The motion is analogous to a soliton in polyacetylene[9] or phason on Si(100) surface.[10]

Figure 3(a) shows a typical gray-scale filled-state STM image of the Si(100)-2x1-H surface after the Ga deposition at 100 K. We observe a characteristic one dimensional (1D) structure, which we call a Ga-bar structure. The structure is formed parallel to the Si dimer rows and is located in the trough between two neighboring Si dimer rows.

By the first-principles calculations, we find that the adsorption sites of a Ga atom with the lowest potential energy are located in the trough between two adjacent dimer rows. We also find a metastable adsorption site with a potential energy higher by 221 meV at the pedestal site between two adjacent dimers. We evaluate the hopping frequency using a simple Arrhenius' equation. As a result of the difference in barrier height, the migration becomes anisotropic. By applying a voltage pulse with a large tunneling current on the Ga-bar structure (Vs = -3.0 V and tunneling current It = 0.7 nA for 200 ms, for example), the Ga atom moved out from the trough and the hydrogen-terminated surface was observed.

Figure 3(b) shows an STM image after removing the Ga-bar structure by a voltage pulse, revealing that the Ga-bar structure is terminated by the dihydride Si dimers (arrows) at both ends. Two dihydrides and a trough form a linear potential well for Ga-atom migration. We conclude that the Ga-bar structure is an image of a Ga atom, which is confined and migrating back and forth in a linear well of potential energy surface (PES).[11] We point out that the height of the Ga-bar structure maps out the local variation in potential energy at individual adsorption sites and 5 pm difference in height corresponds to approximately 1 meV in potential energy.[11]

Another example of the cross-section of the Ga-bar structure showed a local maximum around the middle of the Ga-bar structure, which is approximately 30 pm and corresponds to the adsorption energy difference of 6 meV. We found that the position of the local maximum coincides with the position of a missing dimer defect on the substrate 1.2 nm away from the Ga-bar structure (to the direction perpendicular to the dimer-row

Figure 3. (a) An STM image of a Ga-bar structure (7 nm x 3 nm, Vs = -2.0 V). (b) Same as (a) but after removing the Ga-bar structure by a voltage pulse (local dihydride species are shown by arrows).

direction). We believe that there is a local PES distortion in that region because of the defect. In this way, we can map out local adsorption energy variation by using the height variation of the Ga-bar structure. Detailed examination of this distortion will be discussed elsewhere.

A key step for measuring properties of the atomic structures is how we connect the atomic structures to the bulk electrodes.[12] We have been developing an SPM (scanning probe microscopy) based nanofabrication system that uses electron irradiation.[13,14] In this system, the exposure dose is controlled by varying the bias voltage between the tip and the substrate to keep a constant dose on the resist film. The SPM nanofabrication system has been applied to fabricate 10-nm-level uniform line-and-space resist patterns with high reproducibility.

We have also developed a resist-patterning method using an exposure control method of hybrid current-voltage control, which is a combination of the conventional constant-voltage and constant-current control methods. The method is superior to the conventional methods for drawing dot patterns. We have used the drawing method to fabricate complex resist patterns with resolution of 60 nm with a 50-nm thick resist film. We are currently developing a method for preparing fine electrodes connecting bulk electrodes and atomic structures. The results will be reported elsewhere.

4 Conclusions

In conclusion, we have reported atomic-scale DB and Ga structures on a hydrogen-terminated Si(100)-2x1-H surfaces. Dangling bonds (DBs) and DB structures are fabricated on the surface by STM to form DB patterns. Short DB structures made of unpaired DBs show the Jahn-Teller distortion. Migration of a Ga atom at 100K on the hydrogen-terminated Si(100)-2x1-H surface is imaged by STM as a continuous linear protrusion (Ga-bar structure). The height of the Ga-bar structure maps out the local variation of potential energy at individual adsorption sites.

5 Acknowledgement

This work was performed in collaboration with Y. Suwa, M. Ichimura, T. Onogi, Advanced Research Laboratory, Hitachi, Ltd., T. Hitosugi, S. Matsuura, S. Watanabe, The University of Tokyo, and Z.-Q. Li, K. Ohno, Y. Kawazoe, Institute for Materials Research, Tohoku University.

This study was partially sponsored by Special Coordination Funds of the Ministry of Education, Culture, Sports, Science and Technology of the Japanese Government.

References

1. G. Binnig, H. Rohrer, Ch. Gerber and E. Weibel, *Phys. Rev. Lett.* **49**, 57 (1982).
2. D. M. Eigler and E. K. Schweizer, *Nature* **344**, 524 (1990).
3. R. P. Feynman, *J. Microelectromechanical Systems* **1**, 60 (1992).
4. J. J. Boland, *Adv. Phys.* **42**, 129 (1993).
5. J. W. Lyding, T. C. Shen, J. S. Hubacek, J. R. Tucker and G. C. Abeln, *Appl. Phys. Lett.* **64**, 2010 (1994).
6. T. Hashizume, S. Heike, M. I. Lutwyche, S. Watanabe, K. Nakajima, T. Nishi and Y. Wada, *Jpn. J. Appl. Phys.* **35**, L1085 (1996).
7. S. Watanabe, Y. A. Ono, T. Hashizume, Y. Wada, J. Yamauchi and M. Tsukada, *Phys. Rev.* **52**, 10768 (1995); S. Watanabe, Y. A. Ono, T. Hashizume and Y. Wada,*Phys. Rev.* **54**, R17308 (1996).
8. T. Hitosugi, S. Heike, T. Onogi, T. Hashizume, S. Watanabe, Z. -Q. Li, K. Ohno, Y. Kawazoe, T. Hasegawa, and K. Kitazawa, *Phys. Rev. Lett.* **82**, 4034 (1999).
9. W. P. Su, J. R. Shrieffer and A. J. Heeger, *Phys. Rev. Lett.* **42**, 1698 (1979).
10. H. Shigekawa, K. Miyake, M. Ishida, K. Hata, H. Oigawa, Y. Nannichi, R. Yoshizaki, A. Kawazu, T. Abe, T. Ozawa and T. Nagamura, *Jpn. J. Appl. Phys.* **35**, L1081 (1996).
11. T. Hitosugi, Y. Suwa, S. Matsuura, S. Heike, T. Onogi, S. Watanabe, T. Hasegawa, K. Kitazawa and T. Hashizume, *Phys. Rev. Lett.* **83**, 4116 (1999).
12. M. Kato, M. Ishibashi, S. Heike and T. Hashizume, *Jpn. J. Appl. Phys.* **40**, 4317 (2001).
13. M. Ishibashi, S. Heike, H. Kajiyama, Y. Wada and T. Hashizume, *Appl. Phys. Lett.* **72**, 1581 (1998).
14. M. Ishibashi, S. Heike, M. Kato and T. Hashizume, *in Quantum Coherence and Decoherence - ISQM - Tokyo '01*, Y. A. Ono and K. Fujikawa, Eds. (World Scientific, 2002)(this issue) pp. .

NANOMETER-SCALE FABRICATION USING SCANNING PROBE LITHOGRAPHY

M. ISHIBASHI, S. HEIKE, M. KATO AND T. HASHIZUME

Advanced Research Laboratory, Hitachi, Ltd., Hatoyama, Saitama, 350-0395, Japan

We report a new form of exposure control method of scanning-probe nanolithography. It is a hybrid current-voltage control, which is a combination of the conventional constant-voltage and constant-current control methods. We have used the method to fabricate complex resist patterns with a dot-resolution of 60 nm using a 50-nm thick resist film. We found that the new control method is particularly suitable for drawing dot patterns.

1 Introduction

New devices that use functions of a molecule or an atomic wire have recently been investigated since a proposal by Feynman in 1959.[1] Such devices require a fabrication method that can fabricate electrodes and device structures of nanometer- to micrometer-level sizes. Scanning tunneling microscopy (STM)[2] and other scanning probe microscopy (SPM) are useful tools for atomic-scale and nanoscale fabrication as well as for surface characterization.

In particular, SPM nanofabrication using an organic resist film, scanning-probe (SP) lithography, is a useful technique for making sub 100-nm-level structures, because various methods to transfer patterns to the substrate, such as etching, lift-off, and plating, can be used. Since SP lithography using Langmuir-Blodgett film as a resist was first demonstrated,[3] SP lithography techniques, based on STM,[3,4,5] near-field scanning optical microscopy (NSOM),[6] and atomic force microscopy (AFM),[7-11] have been reported.

In the SP lithography, we use electron irradiation from the conductive AFM tip for patterning. A resist pattern fabricated by electron irradiation can have a steeper and higher-aspect-ratio cross-sectional shape, and can be fabricated faster than a pattern fabricated by the mechanical method.[12]

We have been developing an SP lithography system, in which the exposure dose is controlled by varying the bias voltage between the tip and the substrate to keep a constant current flowing through the resist film.[10,11] The SP lithography system has been applied to fabricate 10-nm-level uniform line-and-space resist patterns.

In this paper, we report a new form of SP lithography technique that can be used to fabricate various kinds of complex patterns with under-100-nm resolution. We use a raster scan writing method, and the resist patterns are fabricated by dot-array patterns using a hybrid constant-current and constant-voltage control method.

2 System configuration

A block diagram of the SP lithography system is schematically shown in Fig. 1. In the system, the tip-sample distance was controlled by an optical-beam reflection system used in a conventional contact-mode AFM. A current flow between the biased tip and the substrate through the resist film was used for exposure. A sample was mounted on a tube scanner, which controls x and y positions of the sample. The feedbacked z position was controlled by means of personal computer (PC). The exposure dose was also controlled through the PC.

We built a current/voltage (IV) converter using an OPA111 operational amplifier to measure the exposure current. As an SPM controller we used AD converters, DA

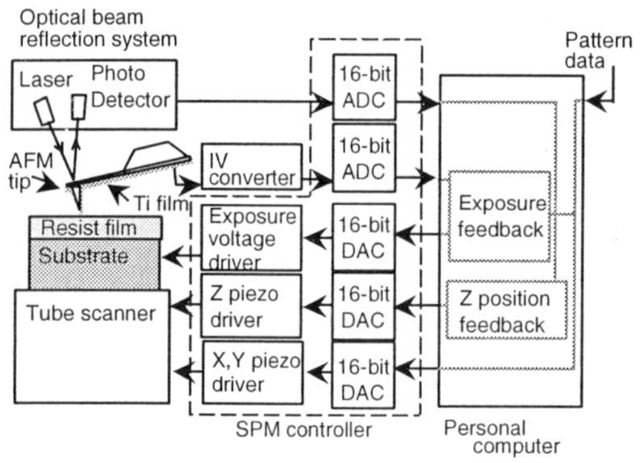

Figure 1. Block diagram of the SP lithography system used in the present study.

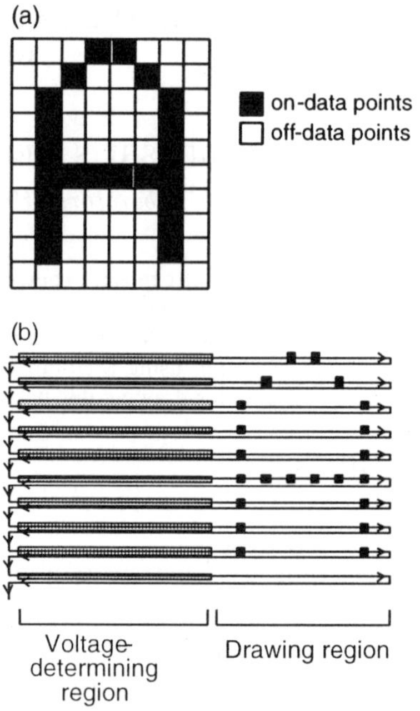

Figure 2. Raster scan writing using a hybrid current-voltage control: (a) an example of pattern data, and (b) tip trajectory (arrows) and exposure points (dark squares) of raster scan writing using the data of (a).

converters, an exposure voltage driver, and x, y, and z piezo drivers. A contact-mode AFM cantilever, coated with 20-nm-thick titanium by DC sputtering, was used as the tip for SP lithography.

In the constant-voltage method, a specified constant bias voltage is applied between the tip and the substrate at the on-exposure points. We used a 40-V exposure bias voltage for on-exposure and 20-V for off-exposure. In the constant-current method, the exposure bias voltage is varied to keep the tip-substrate current constant. We used a 30-pA exposure current for on-exposure and 3-pA exposure current for off-exposure.

In the hybrid current-voltage method, raster scan writing illustrated in Fig. 2 was used. The solid and open dots in Fig. 2(a) show an example of input data-point patterns. The arrows in Fig 2(b) show the tip trajectory and the solid dots show the exposure points of raster scan writing using the data of Fig. 2(a).

In this method, a bias voltage required for a specified current needed to make a latent image in the resist film is determined by applying the constant-current method for each raster-scanning line in the voltage-determining region (gray rectangles in the left half of Fig 2(b)) and the value is stored by the PC. Also the tip is raster scanned with a specified duration time at each data point.

When the tip is positioned over on-data points, the predetermined bias voltage is applied between the tip and substrate; when the tip is positioned over off-data points, less than the critical bias voltage is applied in the drawing region. For example, the on-exposure current was set to 30 pA, and the off-exposure voltage was set to 20 V lower than the on-exposure voltage, which is typically around 40 V for the 50-nm-thick resist film. The duration time for the tip to remain at each data point was varied from 1 to 10 ms.

3 Results and Discussion

When we used the constant-current method for rectangular solid-open array patterns, we found that the beginning of the each rectangular solid pattern did not have enough dose for making latent image in the resist film. The result indicates that the detected exposure current used for constant-current exposure feedback contained not only the current flowing in the resist film but also a charging current to the stray capacitance existing between cantilever and substrate.

If a pattern, such as a line pattern, does not require frequent exposure switching, the charging current is negligible because voltage variation in time is small. However, if a pattern, such as a dot pattern, requires frequent exposure switching, the charging current is not negligible. Moreover, constant-current control is unsuitable for quick exposure switching because it relies on closed-loop feedback. It is, therefore, difficult to fabricate dot array patterns when using constant-current control.

When we used the constant-voltage method for dot-array patterns, the dot size was not uniform during the raster scan. This suggests that the current flowing in the resist was not kept constant by the constant-voltage method because the tip shape changed during the raster scan. Thus, uniform dot pattern fabrication with high reproducibility using constant-voltage method may be difficult to achieve.

Figure 3(a) shows the tip trajectory and exposure points of a resist pattern formed with the hybrid current-voltage method. The tip duration time at each data point was set to 3 ms. The distance between exposure points are set to 200 nm to both the x and y directions. The SEM micrographs (Fig. 3(b) and (c)) show that uniform dots with diameter of 70 nm were fabricated over a wide area when the hybrid current-voltage method was used. This indicates that the current flowing through the resist was kept constant for the applied constant voltage because there was little change in the tip shape during each horizontal scan.

Rectangular resist patterns are also observed in the voltage-determining region.

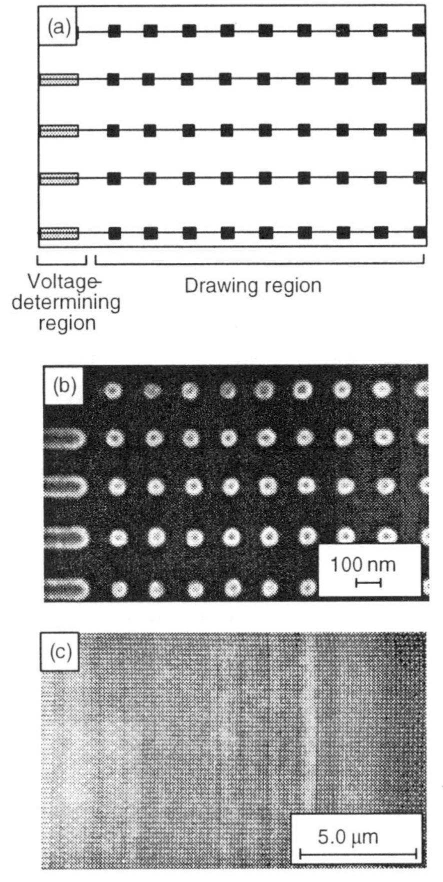

Figure 3. Dot patterning using hybrid current-voltage control: (a) tip trajectory and exposure points, and (b) and (c) SEM micrographs of formed resist pattern.

These patterns were formed because a current over a critical value flowed in the resist film when the voltage was being determined in this region. The hybrid current-voltage method also allows rapid switching of the exposure because it uses open-loop exposure control during pattern drawing. The hybrid current-voltage method is, therefore, suitable especially for making dot patterns.

We measured dot size dependence on the duration time of the tip with the hybrid current-voltage method. Four tips and a 50-nm-thick resist film were used in the experiment. The dot size became smaller as the duration time was shortened, and the mini-

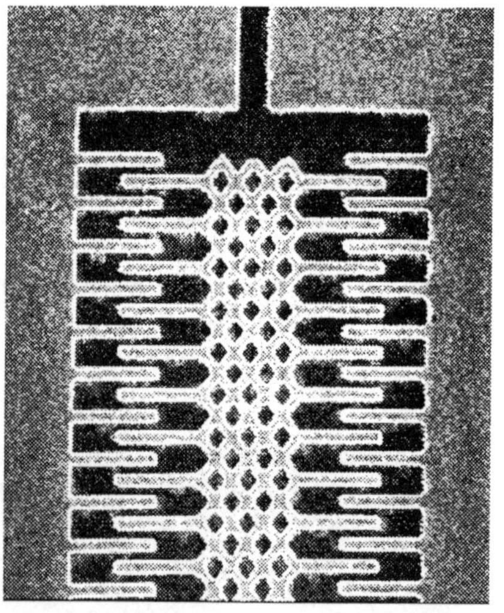

Figure 4. An SEM micrograph of a resist pattern formed using hybrid current-voltage control.

mum dot size of 60 nm was obtained at the duration time of 2 ms.

We performed raster-scan writing for a relatively complicated pattern data using the hybrid current-voltage method. The tip duration time at each data point was set to 3 ms, and the dot size formed was 70 nm. When the spacing between dots was reduced, neighboring dots overlapped and plane or line patterns with various width and shape were formed (Fig. 4).

4 Conclusions

A hybrid current-voltage method combining a constant-current and constant-voltage methods is developed for fabricating dot-array patterns using SP lithography. We used this method to successfully fabricate 50-nm-thick resist patterns with a resolution of 60 nm.

References

1. R. P. Feynman, *J. Microelectromechanical Systems* **1**, 60 (1992).
2. G. Binnig, H. Rohrer, Ch. Gerber and E. Weibel, *Phys. Rev. Lett.* **49**, 57 (1982).
3. M. A. McCord and R. F. W. Pease, *J. Vac. Sci. Technol.* **B4**, 86 (1986).
4. E. A. Dobisz and C. R. Marrian, *Appl. Phys. Lett.* **58**, 2526 (1991).
5. J. Chen, M. A. Reed, C. L. Asplund, A. M. Cassell, M. L. Myrick, A. M. Rawlett, J. M. Tour and P. G. Van Patten, *Appl. Phys. Lett.* **75**, 624 (1999) and references therein.
6. I. I. Smolyaninov, D. L. Mazzoni and C. C. Davis, *Appl. Phys. Lett.* **67**, 3859 (1995).
7. A. Majumdar, P. I. Oden, J. P. Carrejo, L. A. Nagahara, J. J. Graham and J. Alexander, *Appl. Phys. Lett.* **61**, 2293 (1992).
8. S. W. Park, H. T. Soh, C. F. Quate and S.-I. Park, *Appl. Phys. Lett.* **67**, 2415 (1995).
9. K. Wilder, H. T. Soh, A. Atalar and C. F. Quate, *J. Vac. Sci. Technol.* **B15**, 1811 (1997).
10. M. Ishibashi, S. Heike, H. Kajiyama, Y. Wada and T. Hashizume, *Appl. Phys. Lett.* **72**, 1581 (1998).
11. M. Ishibashi, N. Sugita, S. Heike, H. Kajiyama and T. Hashizume, *Jpn. J. Appl. Phys.* **38**, 2445 (1999) and references therein.
12. M. Wendel, S. Kühn, H. Lorenz, J. P. Kotthaus and M. Holland, *Appl. Phys. Lett.* **65**, 1775 (1994).

SPECTRAL STRUCTURE OF QUANTUM LINE WITH A DEFECT

TAKSU CHEON[1] TAMÁS FÜLÖP[2] IZUMI TSUTSUI[3]

[1] *Kochi University of Technology, Tosa Yamada, Kochi 782-8502, Japan*
E-mail: taksu.cheon@kochi-tech.ac.jp

[2] *Roland Eötvös University, Pázmány P. sétány 1/A, H-1117 Budapest, Hungary*
E-mail: fulopt@poe.elte.hu

[3] *High Energy Accelerator Research Organization (KEK), Tsukuba 305-0801, Japan*
E-mail: izumi.tsutsui@kek.jp

We study the spectral properties of one-dimensional quantum wire with a single defect. We reveal the existence of the non-trivial topological structures in the spectral space of the system, which are behind the exotic quantum phenomena that have lately been found in the system.

With the progress of the nanotechnology, it has become possible to manufacture quantum systems with desired specification.[1] The theoretical study of simple quantum system with nontrivial properties is now a legitimate and relevant subject in wider context outside of mathematical physics. It has been lately pointed out that one of such simple model systems of the idealized quantum wire with a single defect [2,3] possesses the properties such as strong vs weak coupling duality and spiral spectral anholonomy,[4,5] the features usually associated with the non-Abelian gauge field theories. Despite its simplicity, the model is a very generic one in the sense that it represents the long wave-length limit of arbitrary one-dimensional potential with finite spatial support. As such, probing those phenomena in its precise working is worthwhile, if only for its mathematical feasibility. That is exactly what we attempt in this paper.

We consider a quantum particle in one-dimensional line with a single defect placed at $x = 0$. In formal language, the system is described by the Hamiltonian

$$H = -\frac{\hbar^2}{2m}\frac{d^2}{dx^2}, \tag{1}$$

defined on proper domains in the Hilbert space $\mathcal{H} = L^2(\mathbf{R} \setminus \{0\})$. We ask what the most general condition at $x = 0$ is. We define the two-component vectors, [6]

$$\Phi = \begin{pmatrix} \varphi(0_+) \\ \varphi(0_-) \end{pmatrix}, \ \Phi' = \begin{pmatrix} \varphi'(0_+) \\ -\varphi'(0_-) \end{pmatrix}, \tag{2}$$

from the values and derivatives of a wave function $\varphi(x)$ at the left $x = 0_-$ and the right $x = 0_+$ of the missing point. The requirement of self-adjointness of the Hamiltonian operator (1) is satisfied if probability current $j(x) = -i\hbar((\varphi^*)'\varphi - \varphi^*\varphi')/(2m)$ is continuous at $x = 0$. In terms of Φ and Φ', this requirement is expressed as

$$\Phi'^{\dagger}\Phi - \Phi^{\dagger}\Phi' = 0, \tag{3}$$

which is equivalent to $|\Phi - iL_0\Phi'| = |\Phi + iL_0\Phi'|$ with L_0 being an arbitrary constant in the unit of length. This means that, with a two-by-two unitary matrix $U \in U(2)$, we have the relation,

$$(U - I)\Phi + iL_0(U + I)\Phi' = 0 . \tag{4}$$

This shows that the entire family Ω of contact interactions admitted in quantum mechanics is given by the group $U(2)$. In mathematical term, the domain in which the Hamiltonian H becomes self-adjoint is parametrized by $U(2)$ — there is a one-to-one correspondence between a physically distinct contact interaction and a self-adjoint Hamiltonian. We use the notation H_U for the Hamiltonian with the contact interaction specified by $U \in \Omega \simeq U(2)$.

We now consider following *generalized parity* transformations:

$$\mathcal{P}_1: \varphi(x) \to (\mathcal{P}_1\varphi)(x) := \varphi(-x), \tag{5}$$

$$\mathcal{P}_2: \varphi(x) \to (\mathcal{P}_2\varphi)(x) := i[\Theta(-x) - \Theta(x)]\varphi(-x)\ . \tag{6}$$

$$\mathcal{P}_3: \varphi(x) \to (\mathcal{P}_3\varphi)(x) := [\Theta(x) - \Theta(-x)]\varphi(x)\ . \tag{7}$$

These transformations satisfy the anti-commutation relation

$$\mathcal{P}_i\mathcal{P}_j = \delta_{ij} + i\epsilon_{ijk}\mathcal{P}_k. \tag{8}$$

Since the effect of $\mathcal{P}_i$ on the boundary vectors Φ and Φ' are given by $\Phi \xrightarrow{\mathcal{P}_i} \sigma_i\Phi$, $\Phi' \xrightarrow{\mathcal{P}_i} \sigma_i\Phi'$, where $\{\sigma_i\}$ are the Pauli matrices, the transformation $\mathcal{P}_i$ on an element $H_U \in \Omega$ induces the unitary transformation

$$U \xrightarrow{\mathcal{P}_i} \sigma_i U \sigma_i \tag{9}$$

on an element $U \in U(2)$. The crucial fact is that the transformation $\mathcal{P}_i$ turns one system belonging to Ω into another one with *same spectrum.* In fact, with any $\mathcal{P}$ defined by

$$\mathcal{P} := \sum_{j=1}^{3} c_j\,\mathcal{P}_j \tag{10}$$

with real c_j with constraint $\sum_{j=1}^{3} c_j^2 = 1$, one has a transformation

$$\mathcal{P}H_U\mathcal{P} = H_{U_\mathcal{P}} \tag{11}$$

where $U_\mathcal{P}$ is given by

$$U_\mathcal{P} := \sigma U \sigma \tag{12}$$

with

$$\sigma := \sum_{j=1}^{3} c_j\,\sigma_j. \tag{13}$$

One sees, from (11), that the system described by the Hamiltonians H_U has a family of systems $H_{U_\mathcal{P}}$ which share the same spectrum with H_U.

Let us suppose that the matrix U is diagonalized with appropriate $V \in SU(2)$ as

$$U = V^{-1}DV. \tag{14}$$

With the explicit representations

$$D = e^{i\xi}e^{i\rho\sigma_3} = \begin{pmatrix} e^{i\theta_+} & 0 \\ 0 & e^{i\theta_-} \end{pmatrix}, \tag{15}$$
$$\theta_\pm := \xi \pm \rho,$$

and

$$V = e^{i\frac{\mu}{2}\sigma_2}e^{i\frac{\nu}{2}\sigma_3}, \tag{16}$$

one can show easily that with $\sigma_V := e^{-i\frac{\nu}{2}\sigma_3}\, e^{-i\frac{\mu}{2}\sigma_2}\, e^{i\frac{\nu}{2}\sigma_3}\sigma_3 = \sigma_V^{-1}$, one has

$$U = \sigma_V D \sigma_V \tag{17}$$

which is of the type (12). One can therefore conclude that [A] the spectrum of the system described by H_U is uniquely determined by the *eigenvalue* of U, and [B] a point interaction characterized by U possesses the isospectral subfamily

$$\Omega_{iso} := \{H_{V^{-1}DV} | V \in SU(2)\}\,, \tag{18}$$

which is homeomorphic to the 2-sphere specified by the polar angles (μ, ν).

$$\Omega_{iso} = \{(\mu,\nu) | \mu \in [0,\pi], \nu \in [0, 2\pi)\} \simeq S^2. \tag{19}$$

There is of course an obvious exception to this for the case of $D \propto I$, in which case, Ω_{iso} consists only of D itself.

To see the structure of the spectral space, *i.e.* the part of parameter space $U(2)$ that determines the distinct spectrum of the system, it is convenient to make the spectrum of the system discrete. Here, for simplicity, we consider the line $x \in [-l, l]$ with Dirichlet boundary, $\varphi(-l) = \varphi(l) = 0$. One then has

$$V\begin{pmatrix} \varphi(0_+) \\ \varphi(0_-) \end{pmatrix} = \sin kl\Phi_0, \tag{20}$$
$$V\begin{pmatrix} \varphi'(0_+) \\ -\varphi'(0_-) \end{pmatrix} = k\cos kl\Phi_0,$$

with some common constant vector Φ_0. From (4), we obtain

$$1 + kL_0 \cot kl \cot\frac{\theta_+}{2} = 0, \tag{21}$$
$$1 + kL_0 \cot kl \cot\frac{\theta_-}{2} = 0.$$

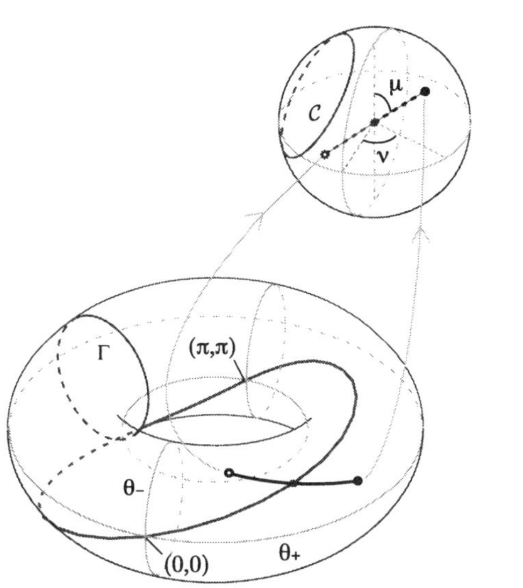

Figure 1. The parameter space $\{(\theta_+, \theta_-, \mu, \nu)\}$ is a product of the spectral torus T^2 specified by the angles (θ_+, θ_-) and the isospectral sphere S^2 specified by the angles (μ, ν).

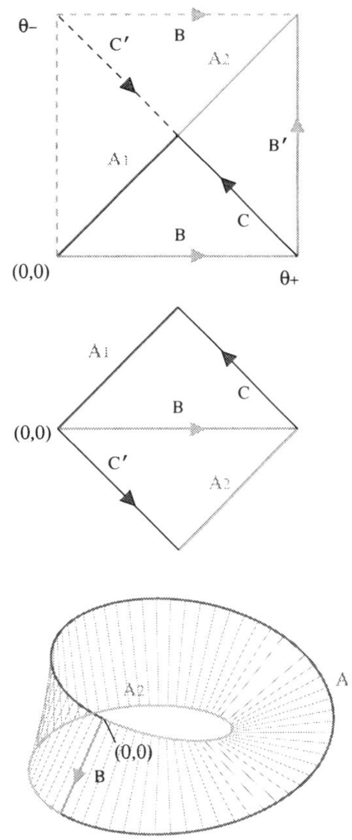

Figure 2. In the top figure, the distict spectral space Σ is the triangle surrounded by edges $A_1 + A_2$, B and B'. Since a subtriangle is spectrally identical to its isospectral image B–C'–A_2, Σ can be represented by the square A_1–C'–A_2–C in the middle figure. When the two spectrally identical edges C and C' are stitched together with the right orientation, we obtain the Möbius strip with boundary A_1–A_2 (the bottom figure).

This means that the spectrum of the system is effectively split into that of two separate systems of same structure, each characterized by the parameters θ_+ and θ_-. So the spectra of the system is uniquely determined by two angular parameters $\{\theta_+, \theta_-\}$. The entire parameter space $\Omega = \{\theta_+, \theta_-, \mu, \nu\}$ is a product of spectral space 2-torus

$$\begin{aligned} \Omega_{sp} &= \{(\theta_+, \theta_-) | \theta_+, \theta_- \in [0, 2\pi]\} \\ &\simeq T^2 = S^1 \times S^1, \end{aligned} \tag{22}$$

and the isospectral space $\Omega_{iso} = \{\mu, \nu\} \simeq S^2$ (See Fig. 1). Note, however, that this parameter space provides a double covering for the family of point inteactions $\Omega \simeq U(2)$ due to the arbitrariness in the interchange $\theta_+ \leftrightarrow \theta_-$. Accordingly, two systems with interchanged values for θ_+ and θ_- are isospectral. So the space of distinct spectra Σ is the torus $T^2 = \{(\theta_+, \theta-) | \theta_\pm \in [0, 2\pi)\}$ subject to the identification $(\theta_+, \theta_-) \equiv (\theta_-, \theta_+)$. Thus we have

$$\Sigma := \{Spec(H_U) | U \in \Omega\} = T^2/\mathbf{Z}_2, \tag{23}$$

which is homeomorphic to a Möbius strip with boundary (Fig. 2).

To relate the non-trivial topological structure found here and the exotic quantum phenomena we have alluded to in the introduction, the readers are referred to other publications.[7,8,9] Here we simply observe that the homotopy $\pi_1(T^2) = \mathbf{Z} \times \mathbf{Z}$ is behind the double spiral anholonomy,[4] and the isospectral family S^2 is the generalization of the duality[5] found earlier. Also, the existence of Dirac monopole in the isospectral parameter space, suggested by homotopy$\pi_2(S^2) = \mathbf{Z}$, has been found numerically[10] and experimentally[11].

This work has been supported in part by the Monbu-Kagakusho Grant-in-Aid for Scientific Research (Nos. (C)11640301 and (C)13640413).

References

1. *See, for example*, J.H. Thywissen *et. al.*, Euro. Phys. J. **D7**, 361 (1999).
2. P. Šeba, Czech. J. Phys. **B36**, 667 (1986).
3. S. Albeverio, F. Gesztesy, R. Høegh-Krohn and H. Holden, *Solvable Models in Quantum Mechanics* (Springer, Heidelberg, 1988).
4. T. Cheon, Phys. Lett. **A248**, 285 (1998).
5. T. Cheon and T. Shigehara, Phys. Rev. Lett. **82**, 2539 (1999).
6. T. Fülöp and I. Tsutsui Phys. Lett. **A264**, 366 (2000).
7. I. Tsutsui, T. Fülöp and T. Cheon, J. Phys. Soc. Jpn, **69**, 3473 (2000).
8. T. Cheon, T. Fülöp and I. Tsutsui, *to be published in* Ann. of Phys. (NY) (2001). LANL preprint quant-ph/0008123.
9. I. Tsutsui, T. Fülöp and T. Cheon, *to be published in* Journ. Math. Phys. (2001). LANL preprint quant-ph/0105066.
10. T. Cheon and T. Shigehara, Phys. Rev. Lett. **76**, 1770 (1996).
11. M. Brazovskaia and P. Pieranski, Phys. Rev. **E58**, 4076 (1998).

QUANTUM TRANSPORT IN TWO-DIMENSIONAL ELECTRON GAS IN ULTRA-SHORT PERIOD LATERAL SUPERLATTICES

YASUHIRO IYE, AKIRA ENDO, SHINGO KATSUMOTO

Insitute for Solid State Physics, University of Tokyo, Kashiwanoha, Kashiwa, Chiba 277-8581 Japan,

and

CREST Project, Japan Science and Technology Corporation, Mejiro, Toshima-ku, Tokyo 171-0031 Japan

E-mail: iye@issp.u-tokyo.ac.jp

YASUHIDE OHNO, SATOSHI SHIMOMURA AND SATOSHI HIYAMIZU

Graduate School of Engineering Science, Osaka University, Machikaneyama-cho, Toyonaka, Osaka 560-8531 Japan

Magnetotransport phenomena in unidirectional lateral superlattice (LSL) systems of short and ultra-short periods have been studied. Commensurability oscillation of magnetoresitance has been observed in the composite fermion (CF) regime near the Landau level filling $\nu = 3/2$, which is found to be consistent with the one expected for fully spin-polarized CFs subjected to spatial modulation of effective magnetic field. Quenching of spin gap at odd integer fillings has been observed in narrow quantum well samples with ultrashort period LSL potential. This is attributed to combined effect of suppression of exchange term due to the modulation potential and the quantum-well-thickness dependence of the g-factor. Reproducible quasipeirodic fluctuation of magnettoresistance has been observed in one of these samples. The phenomenon resembles the so-called universal conductance fluctuation in mesoscopic systems, although the present sample is of macroscopic size.

1 Introduction

Two-dimensional electron system (2DES) formed at semiconductor heterointerface, GaAs/AlGaAs in particular, offers an ideal experimental stage for investigation of various aspects of electronic transport – most notably integer and fractional quantum Hall effects (QHEs). Lateral superlattice, namely periodic structure built upon 2DES, adds a whole new class of physics associated with imposed periodicity, for example geometrical resonance effects, miniband formation and sophisticated Hofstadter butterfly spectra.

Mesoscopic periodic structures are usually created by use of microfabrication techniques such as electron beam lithography, wet or dry etching, and metal evaporation/liftoff process. The length scale definable by such techniques is limited by lithographic resolution and is typically $\sim$ 100nm. Efforts toward fabrication of still finer structures often involve sophisticated utilization of crystal growth characteristics. Examples are found in regrowth on cleaved side face of superlattice, step-flow growth on vicinal surfaces, and formation of self-organized quantum dot structures.

In this work, we investigate magnetotransport in unidirectional LSL systems based on GaAs/AlGaAs 2DES, with short and ultrashort periodicity. The samples of the former category have been fabricated by a standard lithographical method and those of the latter category by a special crystal growth technique. We show, in particular, that magnetotranport study of these LSL systems can shed light on the spin state of 2DES in quantizing magnteic fields

This paper is organized as follows. In the next section, the method of sample preparation and other experimental techniques are described. In section 3, the experimental results for the short period (a = 92nm) LSL are presented and the geometrical resonance effect in the vicinity of $\nu = 3/2$ is discussed. The composite fermion (CF) picture has proved extremely successful in interpreting behavior of 2DES in the vicinity of half integer fillings[3]. Although the CF picture at $\nu = 1/2$ has been extensively tested by numerous experiments, the corresponding problem at $\nu = 3/2$ is less explored. In particular, there seem to exist conflicting experimental results concerning the spin state of $\nu = 3/2$ CFs[4,5]. We address this issue from the geometrical resonance effect.

In section 4, the magnetotransport effects in ultrashort period (a = 12nm) LSL are discussed. Here, we focus on the effect of LSL potential modulation on the spin state in the quantum Hall regime. Since the electron g-factor in GaAs is small, $g = -0.44$, the bare Zeeman energy alone does not lead to spin-split Landau levels even in the typical magnetic field range where the QHE is studied. The occurrence of spin-split Landau levels, and hence the observability of odd integer quantum Hall states, is attributed to the exchange term in the total energy of 2DES[6]. The presence of potential modulation is expected to drastically change the situation. We present a systematic study of the spin gap at odd integer fillings and its evolution with quantum well thickness, which controls the strength of the potential modulation.

2 Experimental Method

Two types of LSL samples were fabricated and studied in the present work. One type was made by writing a line-and-space pattern of a period on resist coating on the surface of a GaAs/AlGaAs 2DES wafer by electron beam lithography as shown in Fig. 1(a)[1]. The carrier density and mobility were $n_e = 2.1 \times 10^{15}\text{m}^{-2}$ and $\mu = 76\text{m}^2/\text{Vs}$, respectively, and the 2DES plane resided 90nm below the top surface. A strain-induced weak potential modulation is generated by difference in thermal contraction between the resist material and the GaAs crystal. In order to detect the effect of modulation, a plane 2DES sample without any modulation structure was also fabricated on the same wafer.

The ultrashort period LSL quantum well wafers were grown at Osaka University utilizing step-bunching growth mode on the (775)B surface of GaAs[2]. By selecting suitable growth conditions, one can create well-ordered periodic faceting which results in a unidirectionally corrugated GaAs/AlAs interface with period a = 12nm, as depicted in Fig. 1(b). The strength of the potential modulation by this periodic modulation of the well thickness depends on the thickness L_w of the quantum well layer. Four wafers with different quantum well thicknesses ($L_w = 4, 5, 7$ and 10 nm) were grown. In order to measure the anisotropic transport, samples were prepared in an L-shaped Hall bar pattern. The carrier densities of the four samples were in the range of $n_e = 2.8 - 3.9 \times 10^{15}\text{m}^{-2}$. The mobilities in the direction of stripes (y-direction) fell in the range of $6.5 - 16\ \text{m}^2/\text{Vs}$. Those in the perpendicular direction (x-direction) are lower by one or two orders of magnitude, and are lower for smaller L_w.

Transport measurements were carried out by the standard low frequency ac technique with an excitation current low enough to ensure no heating effect at the lowest temperature (30 mK). Magnetotransport measurements were carried out in a 15 T superconducting solenoid and a top-loading dilution refrigerator equipped with a rotating sample stage.

3 Short Period ($a = 92$nm) LSL

3.1 *Low Field Magnetotransport*

Low field magnetoresistance of 2DES subjected to a periodic potential modulation

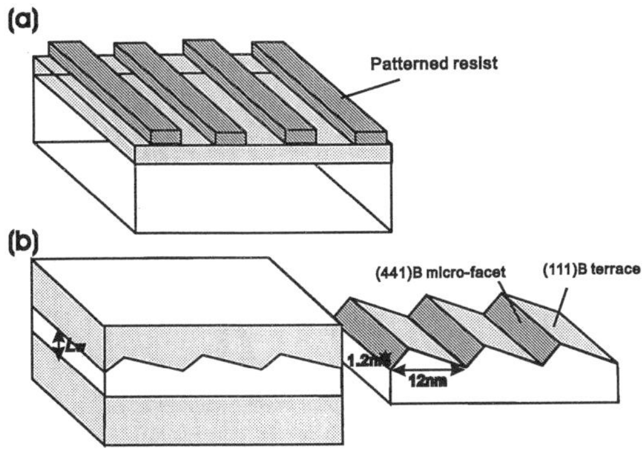

Figure 1. (a)Short period LSL fabricated by lithographic method. (b)Quantum well structure grown on (775)B surface exhibiting ultrashort period LSL structure.

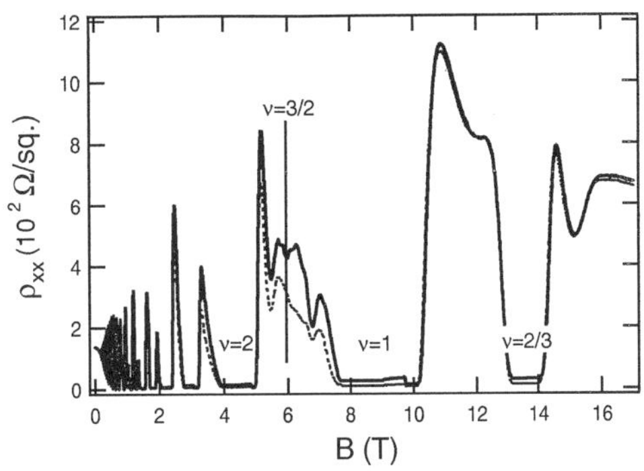

Figure 2. Magnetoresistance traces of a short period LSL sample in comparison with a plain 2DES sample.

exhibits oscillatory behavior. This phenomenon, known as commensurability oscillation (CO) or Weiss oscillation arises from geometrical resonance between the cyclotron motion of electrons and the modulation periodicity[7]. Minima in magnetoresistance occur periodically as a function of $1/B$, when the following commensurability condition is fulfilled.

$$\frac{2R_c}{a} = (n - \frac{1}{4}) \qquad n = 1, 2, 3, \cdots \qquad (1)$$

Here, $R_c = \hbar k_F/eB$ is the cyclotron radius of electrons at the Fermi energy. Analysis of the CO waveform furnishes estimate of the amplitude of the potential modulation[8]. For the present sample, the modulation amplitude was found to be $V_0 = 0.015$mV, which is about 0.2 % of the Fermi energy.

3.2 High Field Magnetotranport

Figure 2 shows the magnetoresitance traces of the modulated and plain 2DES up to 17 T at 30 mK. As evident from the figure, the traces for the modulated and for the plain 2DES almost coincide with each other except around 6 T. In this field range, which corresponds to $\nu = 3/2$ filling, the modulated 2DES exhibits extra structures which are interpreted as the CO features in the *low* field magnetotransport of CFs.

For the interpretation of the CO effect of CFs, the following points should be kept in mind. In the CF picture, spatial variation of electron density is equivalent to variation of the effective magnetic field for CFs. Therefore, although the modulation imposed to 2DES is purely electrostatic, it is seen by CFs as magnetic field modulation. The CO for magnetic modulation differs from that for electrostatic modulation in the oscillation phase[9]. Resistance minima for magnetic modulation occur at

$$\frac{2R_c}{a} = (n + \frac{1}{4}) \qquad n = 1, 2, 3, \cdots. \qquad (2)$$

Second point to note is that the value of k_F for a given density n_e of electrons depends on whether they are spin degenerate or not. The ordinary CO effect occurs at low magnetic field, so that electrons are spin degenerate and $k_F = \sqrt{2\pi n_e}$. If CFs are spin polarized, one should use $k_F = \sqrt{4\pi n_{CF}}$, n_{CF} being the density of the CFs.

3.3 Composite Fermions at $\nu = 3/2$

There are further complications specific to the $\nu = 3/2$ CFs as opposed to the $\nu = 1/2$ counterparts. The $\nu = 1/2$ CFs are composite particles each consisting of an electron plus two fictitious flux quanta, so that their density

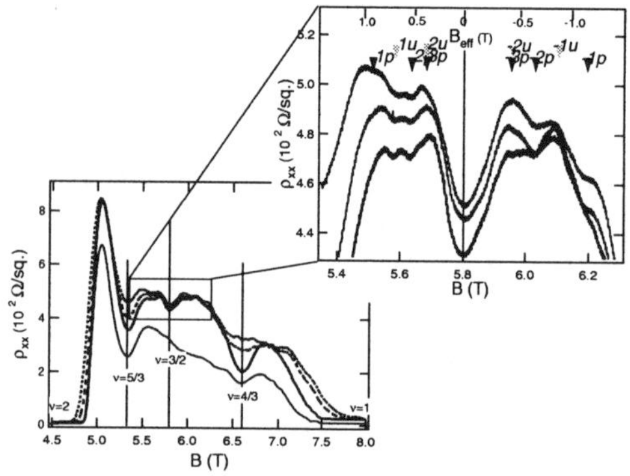

Figure 3. Magnetoresistance traces near $\nu = 3/2$ at three different temperatures. The figure on the right hand side is an expansion of the region in the vicinity of $\nu = 3/2$. The upper scale shows the effective magnetic field for the CFs. The arrows marked 1p, 2p, $\cdots$ and those marked 1u, 2u, $\cdots$ show the expected positions ofresistance minima of the commensurablitiy oscillation of CFs assuming fully spin-polarized and spin-unpolarized CFs, respectively.

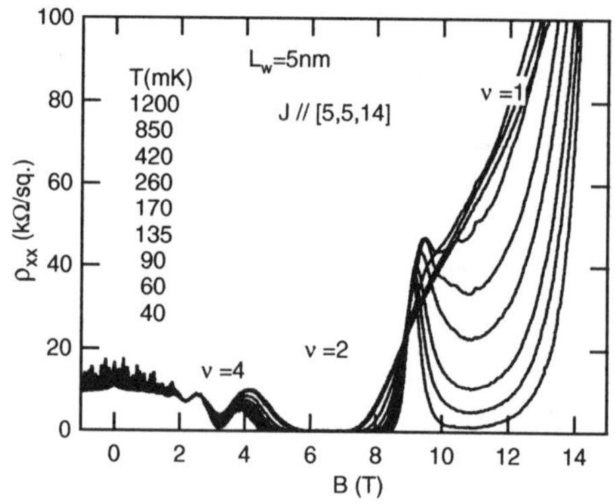

Figure 4. Magnetoresistance (ρ_{xx}) in a narrow quantum well sample (L_w = 5nm).

$n_{CF}^{(\nu=1/2)} is equal to the electron density n_e$, and the effective magnetic field seen by the CFs is $B_{eff}^{(\nu=1/2)} = B - B_{(\nu=1/2)}$. By contrast, the CF at $\nu = 3/2$ is formed by attaching two flux quanta to a *hole* from the $\nu = 2$ state. The density of the $\nu = 3/2$ CFs is given by

$$n_{\rm CF}^{(\nu=3/2)} = \frac{2eB}{h} - n_{\rm e}, \tag{3}$$

and the effective magnetic field reads

$$\begin{aligned} B_{\rm eff}^{(\nu=3/2)} &= B - 2n_{\rm CF}^{(\nu=3/2)}(\frac{h}{e}) \\ &= -3(B - B_{(\nu=3/2)}). \end{aligned} \tag{4}$$

Note here that the effective magnetic field for the n=3/2 CFs changes three times as fast as the external field.

Figure 3 shows the magnetoresistance traces in a narrow field range near $\nu = 3/2$ at three different temperatures. The black (gray) triangles indicate the expected positions of resistance minima of the magnetic Weiss oscillation assuming that the CFs are fully spin-polarized (spin-unpolarized). The observed resistance minima are in good agreement with the the solid triangle marks, suggesting that the $\nu = 3/2$ CFs are fully spin polarized.

4 Ultrashort Period (a = 12nm) LSL

4.1 QHE at even fillings

Figure 4 shows the traces of magnetoresistance in the sample (L_w = 5nm) with the current passed perpendicular to the stripes (ρ_{xx}). The magnetoresistance for the parallel current direction (ρ_{yy}) is similar in shape, although the absolute value differs by more than an order of magnitude.

The QHE features at odd integers are greatly suppressed in comparison with those at even integers. Those corresponding to $\nu = 3, 5, \cdots$ are not visible, and the activation energy Δ from the $\nu = 1$ state is as small as 0.2 K. The values of Δ for different filling factors are summarized in Fig. 5. It is evident that the values of Δ at odd fillings (spin gap) are one or two orders of magnitude smaller than those for even fillings (cyclotron gap) at comparable magnetic field.

$$E_g^{\rm even} = \frac{1}{2}\hbar\omega_c - \epsilon - \Gamma. \tag{5}$$

The slope of the $\Delta(B)$ at the even fillings is close to $\hbar\omega_c/2$, *i.e.* half the single particle cyclotron gap. The negative intercept at $B = 0$ becomes large in absolute value with decreasing L_w, that is, increasing modulation amplitude. It gives a measure of (half) the subband width associated with the LSL potential, with possible contribution from random potentials.

4.2 *QHE at odd fillings*

The results shown above indicate that the cyclotron gap at even fillings can be basically understood within the single particle picture. By contrast, many-body effect plays an essential role for the spin gap at odd fillings, as stated in the Introduction. The spin gap at an odd integer filling may be expressed as

$$E_g^{\text{odd}} = g'\mu_B B + \Delta E_{ex} - \epsilon - \Gamma. \quad (6)$$

Here, ΔE_{ex} represents the contribution from Coulomb exchange interaction which, in a simple picture, scales as $e^2/\ell \propto \sqrt{B}$ ($\ell = \sqrt{\hbar/eB}$ is the magnetic length.) The dependence of E_g^{odd} on L_w involves all these terms. The change in ϵ and Γ with L_w can be estimated from the activation energies at even fillings. The changes in the bare g-factor g' with L_w are caused by the finite penetration of electron wavefunction into the AlAs barrier layer. For narrow quantum wells, this is a substantial effect. The g-factor changes from the bulk GaAs value, $g = -0.44$, to a positive value as the quantum well width is reduced[10]. The many-body physics is contained in the exchange contribution ΔE_{ex}. How this term is suppressed by the modulation potential is what one hopes to calculate quantitatively[11].

In Fig. 5(b), the $\nu = 3$ QHE of the same sample is brought to different values of total magnetic field either by changing the electron density through persistent photocarriers or by tilting the sample. The fact that the B-dependence of E_g^{odd} is weaker than the bare Zeeman term (shown by the dotted line) with $|g| = 0.44$ suggests that the bare g-factor is close to zero in these narrow quantum wells.

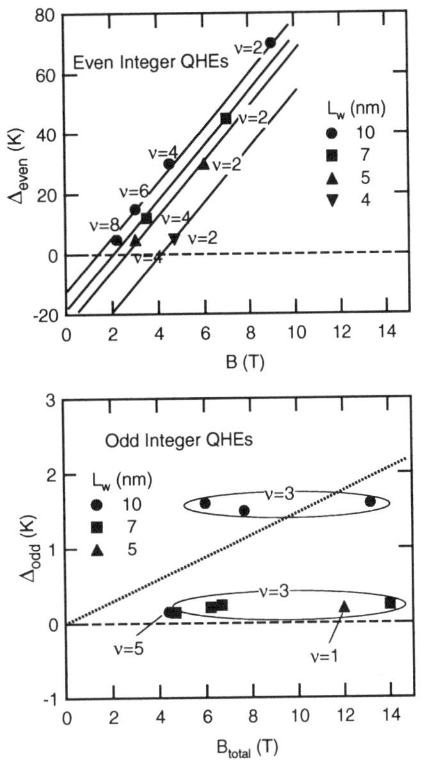

Figure 5. Measured values of activation energy for (a)even fillings and (b)odd fillings. The straight lines in (a) are drawn with the same slope as $\hbar\omega_c/2$. In (b), the data points for $\nu = 3$ (encircled) were taken on the same sample. The dotted line in (b) represents half the bare Zeeman splitting, $g\mu_B B/2$ with $|g| = 0.44$.

4.3 *Magnetoresistance Fluctuation*

Figure 6 shows the low field part of Fig. 4. The magnetoresistance traces exhibit quasiperiodic fluctuation at low temperatures, which is reproducible as long as the sample is kept cold. The fluctuation pattern changes after the thermal cycling to the room temperature. The characteristic (quasi-)period of the fluctuation (correlation field) ΔB_c is about 0.1 T if only major peaks are counted, and is about 0.05 T if smaller peaks

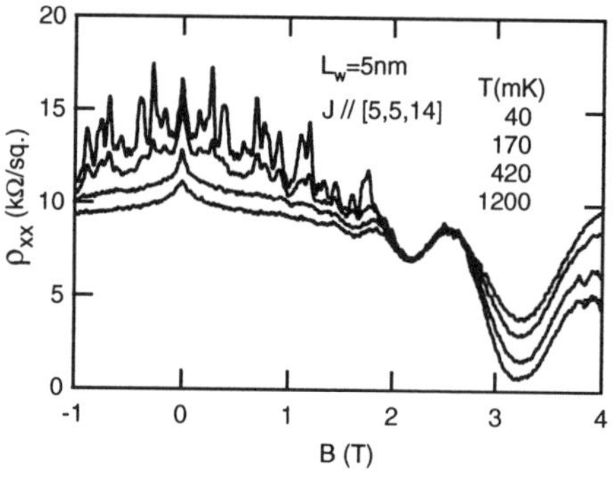

Figure 6. Reproducible magnetoresistance fluctuation observed in a Lw=5nm sample at low temperatures.

are included. Reproducible fluctuation was also observed in the resistivity parallel to the stripes, although the amplitude is smaller and the characteristic value of the correlation field is smaller. The amplitude of the conductance fluctuation at the lowest temperature is $0.1 - 0.5\ e^2/h$. Tilted field experiments have verified that the fluctuation pattern is governed by the field component perpendicular to the 2DES plane. The magnetoresistance fluctuation vanishes at higher temperatures. So far, the fluctuation phenomenon has been observed in one particular sample with Lw=5 nm but not in others.

The observed phenomenon has many features in common with the so-called universal conductance fluctuation (UCF) phenomena ubiquitously observed in mesoscopic systems[12]. However, the argument for the mesoscopic systems does not seem naively applicable to the present case which occurs in a macroscopic Hall bar sample. Other possible mechanisms for magnetoresistance variation, such as commensurability oscillation, Aharonov-Bohm type oscillation, or chaotic transport, do not seem to apply directly to the present case, either. The physical origin of the fluctuation in the present system is not identified at the moment.

Acknowledgments

This work was supported in part by a Grant-in-Aid for COE Research (No.12CE2004 "Control of Electronics by Quantum Dot Structures and Its Application to Advanced Electronics") and by a Grant-in-Aid for Scientific Research (No.13304025) from the Ministry of Education, Culture, Sports, Science and Technology (MEXT) Japan.

References

1. A. Endo, M. Kawamura, S. Katsumoto and Y. Iye, *Phys. Rev.* **B63** 113310-1 (2001).
2. M. Higashiwaki, M. Yamamoto, T. Higuchi, S. Shimomura, A. Adachi, Y. Okamoto, N. Sano and S. Hiyamizu, *Jpn. J. Appl. Phys.* **35** L606 (1996).
3. See, for example, S. Das Sarma, A. Pinzuk (eds.), *Perspective in Quantum Hall Effects* (Wiley, New York, 1997).
4. R.L. Willet and L.N. Pfeiffer, *Semicond. Sci. Technol.* **11** 1473 (1996)
5. R.R. Du, A.S. Yeh, H.L. Stormer, D.C. Tsui, L.N. Pfeiffer and K.W. West, *Phys. Rev. Lett.* **75** 3926 (1995).
6. T. Ando and Y. Uemura, *J. Phys. Soc. Jpn.* **37** 1044 (1974).
7. D. Weiss, K. von Klitzing, K. Ploog and G. Weimann, *Europhys. Lett.* **8** 179 (1989).
8. A. Endo, S. Katsumoto and Y. Iye, *Phys. Rev.* **B62** 16761 (2000).
9. S.Izawa, S. Katsumoto, A. Endo and Y. Iye, *J. Phys. Soc. Jpn.* **64** 706 (1995).
10. R.M. Hannak, M. Oestreich, A.P. Heberle, W.W. Ruehle and K. Koehler, *Solid State Commun.* **93** 313 (1995).
11. A.Manolescu and R.R.Gerhardts, *Phys. Rev.* **B51** 1703 (1995).
12. P.A. Lee and A.D. Stone, *Phys. Rev. Lett.* **55** 1622 (1985).

TRANSPORT PROPERTIES THROUGH ATOMIC-SIZE POINT CONTACTS OF MULTI-BAND METALS

K. UENO, M. ETO AND K. KAWAMURA
Faculty of Science and Technology, Keio University
3-14-1 Hiyoshi, Kohoku-ku, Yokohama 223-8522, Japan
E-mail: kueno@rk.phys.keio.ac.jp

We theoretically investigate the effect of the interband scattering on the transport through atomic-size point contacts of multi-band metals. The conductance is evaluated, taking account of the atomic structures by tight-binding models. In the case of single s band, the conductance is almost quantized in units of $2e^2/h$. In the multi-band model with broad and narrow s bands, the conductance quantization is also observed. With broad s and narrow p bands, on the other hand, the conductance is strongly suppressed with an increase in the interband scattering. This is because the hybridization between s and p orbitals is significantly different in the leads and in the contact region, which forbids the smooth transport of electrons between them.

1 Introduction

The conductance quantization through metallic point contacts has attracted a lot of interest. The quantization of conductance G in units of $2e^2/h$ has been observed experimentally for Au,[1,2,3,4] Ag,[2] Cu,[3,5,6] etc. The point contacts of magnetic metals have also been studied.[2,3,6] For Ni, the conductance quantization in units of e^2/h has been reported by T. Ono *et al.*,[6] when the magnetization is ferromagnetically saturated under a magnetic field.

Since the Fermi wavelength λ_F is of atomic size in metals, the atomic structures around the contacts should be taken into account to understand the quantization of G. For spin-polarized states, continuum models have been proposed,[7,8] which yield the quantized conductance in units of e^2/h. Although the calculated results are in accordance with the experimental results with Ni,[6] the applicability of the continuum models to metallic point contacts is not evident for the above-mentioned reason.

In this paper, we examine G through metallic point contacts theoretically, taking account of the atomic structures by tight-biding models. Other groups have investigated realistic and specific situations using tight-binding models [4,9,10,11] or by LDA calculations,[12,13] mainly for simple metals. In magnetic metals, broad s and narrow d bands coexist at the Fermi level, the effect of which has not been elucidated. We focus on the interband scattering effect on the transport, taking simple models with broad and narrow s bands, or broad s and narrow p bands. We surprisingly find that the interband scattering can strongly suppress the conductance in the latter case. Based on the calculated results, we discuss a possible mechanism for the quantization unit of e^2/h for Ni point contacts.[6]

2 Models and Calculation Method

We adopt tight-binding models of three-dimensional simple cubic lattice. We consider two geometries for the point contact: (I) the contact region consists of one atom, or (II) of 19 atoms with 3 atoms in the widest row (Fig. 1). The leads are semi-infinitely long in z direction and infinitely wide in x and y direction. The electron transfer is allowed only between neighboring atoms (solid lines in Fig. 1). The transfer integrals are assumed to be constant.[14] We neglect the electron-electron interaction. The spin degeneracy is always assumed and spin-flip processes are ignored.

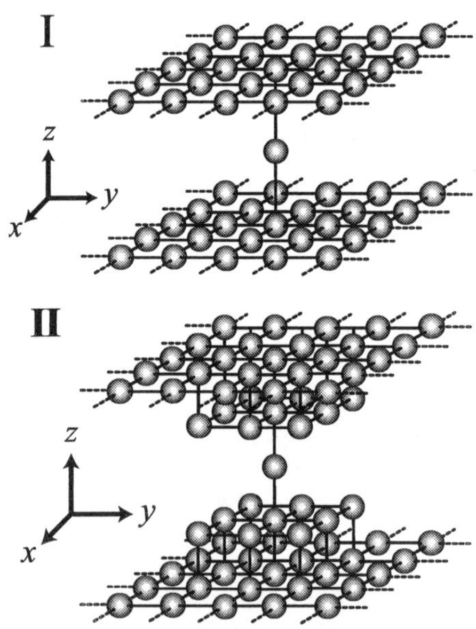

Figure 1. The geometries of the tight-binding models used for our calculations. The models consist of two leads and contact region which includes one atom (I), or 19 atoms (II). The leads are semi-infinitely long in z direction and infinitely wide in x and y direction. Spheres represent an atom which has single s orbital (model A), two s orbitals (model B), and s and p orbitals (model C).

We study three models with (A) single s band, (B) broad and narrow s bands, and (C) broad s and narrow p bands. In model A, an atom has a single s orbital. The transfer integral between the neighboring atoms is denoted by t_s. In model B, atoms have two s-like orbitals (s, s'). The transfer integrals are t_s ($t_{s'}$) for s (s') orbitals and $t_{s,s'}$ from s to s' orbital. In model C, s, p_x, p_y and p_z orbitals exist at each atom. The transfer integrals are t_s (t_p) for s (p) orbitals[15] and $t_{s,p}$ from s to p orbital.[16] The on-site energies are fixed at zero for all the orbitals. The conductance is calculated at zero temperature, using the Green function method.[17]

It should be noted that, in the experimental situation of point contacts, the conductance is measured while the size of the narrowest part at the contact changes. Instead, we calculate the conductance through a fixed geometry of the contact, as a function of the Fermi energy E_F.

3 Calculated Results

We begin with the transport properties in model A of single s band. In Fig. 2, G is shown as a function of E_F through the contact I (curve a) and II (curve b). The maximum of G is almost quantized to be $2e^2/h$. The quantization plateau of $G \approx 2e^2/h$ is more evident in the contact II than in contact I: As the atomic structure changes more gradually from the leads to the point contact, the electrons can be transported smoothly through the contact in the larger range of the energy. We conclude that the conductance is clearly quantized for single-band metals when the atomic structure is considered at the point contact in geometry II.

Now we discuss the effect of the interband scattering on the transport properties in the multi-band models. We present the calculated results only with geometry II. Figure 3 shows G as a function of E_F for model B. We change the transfer integral between s and s' orbitals; $t_{s,s'}/t_s = 0.0$ (curve a), 0.2 (curve b) and 0.4 (curve c). Despite the interband scattering by $t_{s,s'}$, G is still quantized in units of $2e^2/h$: $G \approx 4e^2/h$ in the energy range of $|E_F/t_s| < 0.1$, and $2e^2/h$ in the energy range of $0.2 < |E_F/t_s| < 1$. Both s and s' bands contribute to the conductance (two channels) in the former, whereas only the s band makes the main contribution (one channel) in the latter. (The energies in s and s' bands are $|E| \leq 6t_s$ and $|E| \leq 0.6t_s$, respectively.) The interband scattering does not seem to affect the conductance quantization in the case of two s-type bands.

For model C, the behavior of G is quite different from that in models A and B. In Fig. 4, the conductance G is shown as a function of E_F, changing the transfer integrals between s and p orbitals; $t_{s,p}/t_s = 0.001$ (curve a), 0.1 (curve b) and 0.5 (curve c). The p-p

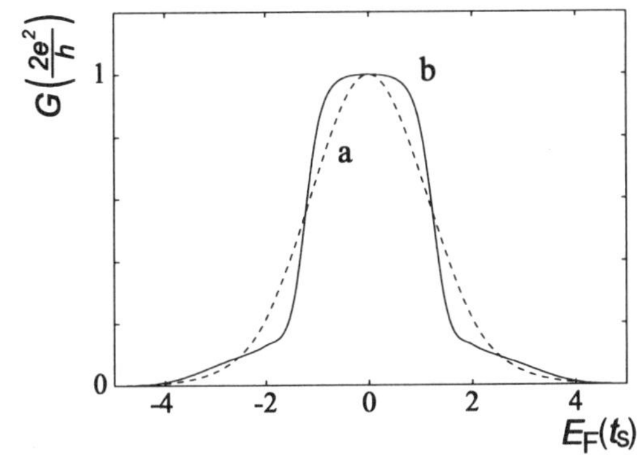

Figure 2. The conductance G as a function of the Fermi energy E_F of conduction electrons in model A. The geometry of atomic structure is (a) I and (b) II in Fig. 1.

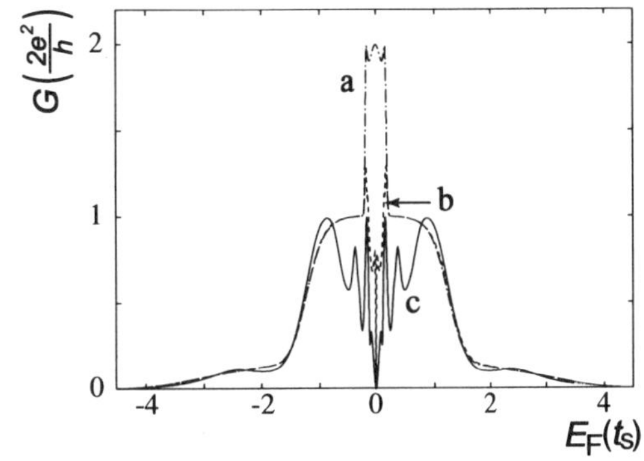

Figure 4. The conductance as a function of E_F in model C, with geometry II. The transfer integral between s and p orbitals is **a** $t_{s,p}/t_s = 0.0$, **b** 0.1 and **c** 0.5, whereas that for p orbitals is fixed at $t_p/t_s = 0.1$.

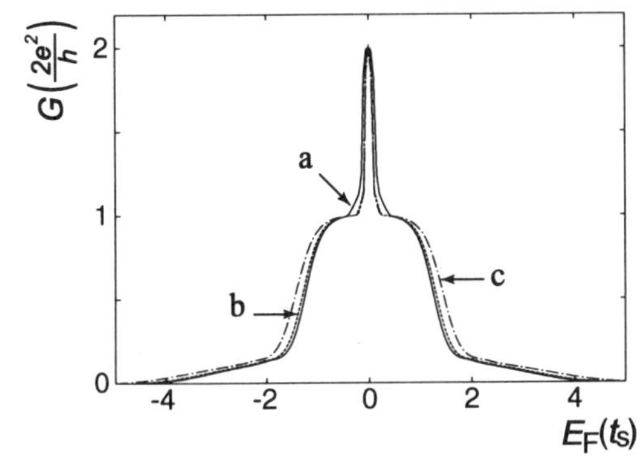

Figure 3. The conductance as a function of E_F in model B, with geometry II. The transfer integral between s and s' orbitals is **a** $t_{s,s'}/t_s = 0.0$, **b** 0.2 and **c** 0.4, whereas that for s' orbitals is fixed at $t_{s'}/t_s = 0.1$.

transfer integral is fixed at $t_p/t_s = 0.1$. When there is almost no interband scattering (curve a), G is quantized in units of $2e^2/h$: $G \approx 4e^2/h$ in the energy range of $|E_F/t_s| < 0.2$ (two channels of s and p_z bands) and $2e^2/h$ in the energy range of $0.2 < |E_F/t_s| < 1$ (one channel of s band: The p bands exist between $-0.2t_s$ and $0.2t_s$). With increasing the interband scattering (curves b and c), we find that (i) G is strongly suppressed in the energy range of p band, and (ii) a lot of sharp peaks of G appear in the same energy range. The reason for them is as follows. The transfer integral $t_{s,p}$ hybridizes the s orbitals with p orbitals. The hybridization can be quite different between in the Bloch states at the leads and in the electronic states around the contact region. This disturbs the smooth connection between them and, as a result, suppresses the conductance significantly. This also generates localized states in the contact region. The sharp peaks of G are attributed to the resonant tunneling through the localized states.

4 Conclusions and Discussion

We have theoretically studied the conductance through point contact of multi-band metals, taking into consideration the atomic structures by tight-binding models. For the single s band (model A), G is almost quantized in units of $2e^2/h$, when the connection of the contact region to the leads is sufficiently smooth. In the presence of broad s and narrow s' bands (model B), the conductance quantization is still seen. In the case of s and p bands (model C), however, the interband scattering crucially influences the transport properties: The conductance is strongly suppressed when s and p bands coexist at the Fermi level. This is because the hybridization between s and p orbitals is considerably different in the bulk Bloch states and in the electronic states around the contact. This forbids the smooth connection between them. The comparison of model B and C indicates that the anisotropy of the orbitals is important for the suppression of G by the interband scattering.

We find a lot of sharp peaks of G in model C, which are ascribed to the resonant tunneling through localized states around the point contact. However, they should be difficult to observe in usual experimental situations.[18] (i) At room temperatures, the electron energies are thermally smeared. When the resonant width is smaller than the thermal energy, the resonant tunneling cannot take place. (ii) In the experiment of point contact, the conductance is averaged over many samples. Since the localized levels are different from sample to sample, the sharp peaks of G should be averaged out.

Finally we discuss a possible mechanism of the the conductance quantization in units of e^2/h with Ni point contacts.[6] In Ni metal, $3d$ bands are fully occupied and $4s$ band is partly filled for electrons with majority spin, whereas both $3d$ and $4s$ bands appear at the Fermi level for electrons with minority spin. If spin-flip processes are negligible, the conductance of electrons with majority and minority spins can be evaluated separately. Our model A with single s band represents the electrons with majority spin. They yield the conductance quantization in units of e^2/h (half of the value in the previous section where the spin degeneracy is assumed). Our model C with broad s and narrow p bands might be applied to the electrons with minority spin. The interband scattering between s and d bands could suppress G for minority spin strongly, as in the case of s and p bands. In consequence the whole conductance is quantized in units of e^2/h. For the experimental situations, of course, we have to examine realistic structures of atoms with s and d bands. We have also to consider the electron-electron interaction particularly in narrow d bands, in further studies.

References

1. J. L. Costa-Krömer *et al.*, *Phys. Rev.* B **55**, 12910 (1997).
2. J. L. Costa-Krömer, *Phys. Rev.* B **55**, R4875 (1997).
3. K. Hansen *et al.*, *Phys. Rev.* B **56**, 2208 (1997).
4. E. Scheer *et al.*, *Nature* **394**, 154 (1998).
5. H. Ohnishi *et al.*, *Nature* **395**, 780 (1998).
6. T. Ono *et al.*, *Appl. Phys. Lett.* **75**, 1622 (1999).
7. H. Imamura *et al.*, *Phys. Rev. Lett.* **84**, 1003 (2000).
8. K. Nakanishi *et al.*, *Phys. Rev.* B **61**, 11278 (2000).
9. A. Levy Yeyati *et al.*, *Phys. Rev.* B **56**, 10369 (1997).
10. A. M. Bratkovsky *et al.*, *Phys. Rev.* B **52**, 5036 (1995).
11. J. C. Cuevas *et al.*, *Phys. Rev. Lett.* **80**, 1066 (1998).
12. N. Kobayashi *et al.*, *Jpn. J. Appl. Phys.* **38**, 336 (1999).
13. N. Kobayashi *et al.*, *Surf. Sci.* **433-435**, 854 (1999).
14. This is because the character of the atomic bonds should not be different significantly between in the bulk and around the contact.
15. We consider the transfer integral t_p between p_i orbitals in i direction only ($i = x, y, z$) and neglect in the other direction.
16. Although we consider s and p orbitals in model C, the situation is very different from that of Al.[4,11] We set the transfer integral for p orbitals to be one tenth of that for s orbitals, which are rather closer to the values for $3d$ and $4s$ orbitals in transition metals.
17. T. N. Todorov *et al.*, *J. Phys. Condens. Matter* **5**, 2389 (1993).
18. K. Ueno *et al.*, in preparation.

ENHANCED TUNNEL MAGNETORESISTANCE IN FERROMAGNETIC SINGLE ELECTRON TRANSISTOR

R. MATSUDA*, A. KANDA, Y. OOTUKA

Institute of Physics, University of Tsukuba, Tsukuba, 305-8571, Japan
CREST, JST, 4-1-8, Honcho, Kawaguchi, Saitama, 332-0012, Japan
* *School of Science, University of Tokyo, Hongo, Tokyo, 113-0033 Japan*

The tunnel magnetoresistance (TMR) of ferromagnetic single-electron transistors made of Ni and Co has been measured. We found that the TMR is enhanced by a factor about 10 between 4.2 K and 25 mK in almost all devices including those whose junction resistance is much higher than the quantum resistance. This indicates that the theories based on the higher-order tunneling are not sufficient to explain the enhancement.

1 Introduction

Since Jullier[1] found the so-called tunnel magnetoresistance effect (TMR) in a tunnel junction made of Fe and Co, electron transport in ferromagnetic tunnel junctions has attracted lots of interest from the aspects of both the basic research and the application. The TMR is caused because the symmetry on the spin direction is broken in ferromagnetic metals, and the resistance of the ferromagnetic tunnel junction changes depending on the relative orientation of magnetizations in both electrodes. While both magnetizations align to the direction of a magnetic field when the field is strong enough, they generally align differently at low fields. Moreover, the alignment should vary depending on the history of the field change. Thus, it gives rise to the hysteretic magnetoresistance. The TMR is characterized by the TMR ratio defined as

$$\gamma \equiv (R_{\mathrm{AP}} - R_{\mathrm{P}})/R_{\mathrm{P}}. \qquad (1)$$

Here, R_{P} (R_{AP}) denotes the resistance when magnetizations in the electrodes are parallel (antiparallel).

When one reduces the size of tunnel junctions, the tunneling of electron is known to be suppressed at low temperatures by the Coulomb blockade. This is because the single electron charging energy (E_{C}) that is necessary for an electron to tunnel into the other electrode is not fed from the reservoir when $k_{\mathrm{B}}T << E_{\mathrm{C}}$. The single-electron transistor (SET), a double junction system with a gate electrode, is a device which utilizes this phenomenon. By applying the gate voltage, one can tune the electrostatic potential of the island electrode and control the current through the device.

In 1997, Ono, Shimada and Ootuka[2] reported experiments on SET made of Ni and Co. They found two novel phenomena: one is the magneto Coulomb oscillation and the other is the enhancement of TMR. The former, that is, the periodic change of resistance as a function of external magnetic field is explained in the framework of the orthodox theory of single electron tunneling, when we takes the Zeeman energy of electron into account.[3] The latter is that the low-field magnetoresistance convincingly due to the TMR effect is largely enhanced when it is cooled to the Coulomb blockade regime. In their experiment, the TMR ratio that was about 4 % at 4.2 K grew up to 40 % at 20 mK, which is even larger than the maximum TMR ratio expected from spin polarization tunneling experiments.[4] Similar enhancement of TMR was found in two-dimensional array of ferromagnetic tunnel junction.[5] Later, there also appeared reports on the similar enhancement in granular films made of ferromagnetic metal and insulating material, where the electron transport is supposed to be due to tunneling between metallic small grains.[6,7]

According to the orthodox theory[8] of single electron tunneling, the resistance of SET at low temperature is essentially expressed as $R_0 \exp(E_C/k_BT)$, where R_0 being the sum of the tunnel resistance of two tunnel junctions, $E_C = e^2/2C$, and C the island capacitance. Although R_T changes to some extent by the TMR effect, we do expect the magnetic field does not affect the geometrical capacitance C. Thus, the simple argument based on the orthodox theory concludes that the TMR ratio should remain constant, and we need a theory beyond the orthodox theory to explain the enhancement.

There are several theoretical investigations on the mechanism of the TMR enhancement.[2,9,10,11,12] All of them are based on the contribution of higher-order tunneling. Although they predict the enhancement of TMR at low temperatures, the size of the enhancement does not seem large enough to explain the experiment. If the enhancement relates to the higher-order tunneling, the enhancement should vary depending on the tunnel resistance. Namely, the enhancement should be evident only for conductive junctions. In this paper, we report our recent experimental results on TMR in ferromagnetic SET having various tunnel resistance.

2 Experiment

We fabricated Ni/Co/Ni-SET by following procedure. To begin with, gold contact pads 100nm thick were made on a Si substrate adopting the photo-lithography and vacuum deposition. Next we spin-coated two layers of resist between which evaporated thin germanium film is interposed. After small SET patterns were drawn and developed in an upper resist layer through the standard electron-beam lithography process, we transferred the patterns to the Ge film by CF_4 reactive ion etching. Then, the lower resist around the patterns was removed by O_2 plasma etching to form suspended Ge mask. Ni and Co films 30-45 nm thick were deposited by double angle evaporation in a vacuum chamber at 10^{-8}torr. We adopted two sorts of tunnel barriers, Al_2O_3 and NiO. In the former case, thin Al_2O_3 film was vacuum-deposited in low-pressure O_2 atmosphere (10^{-6} torr). In the latter case, NiO barrier was formed by plasma oxidation of Ni electrodes. Finally the films deposited onto the resist were lifted-off in organic solvent. Figure 1 shows a SEM image of a device. The size of tunnel junction is about 0.1 μm $\times$ 0.1 μm.

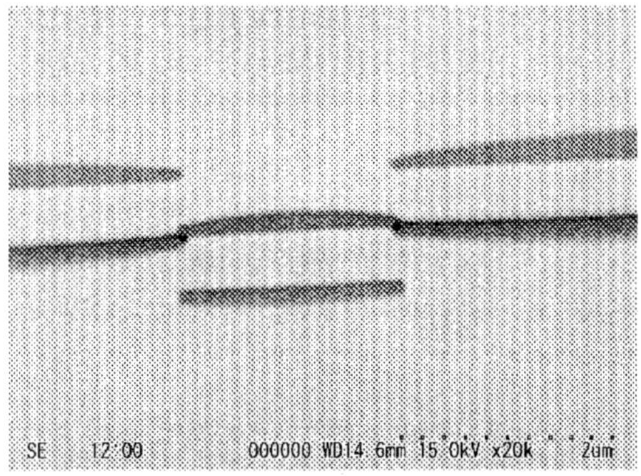

Figure 1. SEM image of a Ni/Co/Ni-SET

Measurement was done at temperatures down to 20 mK using a dilution refrigerator. A magnetic field up to 2 T was applied to the direction parallel to the long axis of the electrodes. The conductance was measured in the dc- or ac- constant voltage mode. The current through the device was amplified by a current amplifier and detected by a digital voltmeter or a lock-in amplifier. The tunnel resistance of a junction (R_T) was determined from the resistance at 4.2 K assuming that the two junctions had the same tunnel resistance. The single-electron charging energy E_C was obtained from the offset-voltage in the I-V characteristics at the lowest temperature. During the TMR measurement, the bias voltage was kept smaller than $2E_C/e$ to assure the zero-bias condition.

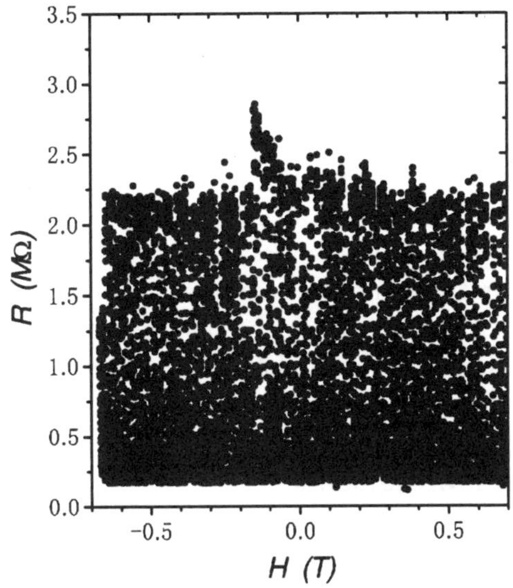

Figure 2. Magnetoresistance of a Ni/Co/Ni-SET at $T = 26$ mK
The magnetic field was swept from 0.7 T to −0.7 T.

3 Results and Discussions

We measured eight devices, tunnel resistances of which ranged from 27 kΩ to 3.1 MΩ. The charging energies, E_C, were in the range from 20 to 80 μeV. The TMR ratios measured at 4.2 K ranged from 2.2 % to 7.5 %. We found a tendency that the SET with Al_2O_3 barriers had larger TMR ratio than that with oxidized nickel barriers. As bulk NiO is antiferromagnetic, there can be some magnetic interaction between tunneling electron and barrier, which could cause spin-flip during the tunneling and degrade the TMR ratio.

Fig.2 shows a typical result of TMR measurement at low temperatures. The device has tunnel resistance $R_T = 39$ kΩ, and the TMR ratio of 2.2 % at 4.2 K. We measured the resistance while sweeping the magnetic field H from 0.7 T to −0.7 T and scanning the gate voltage V_G rather quickly. In the figure, we plot all the data in R-H plane. The structure seen in $0 > H > -0.2$ T is the TMR. The TMR ratio is about 22 % at 26 mK, which is evidently much larger than the value at 4.2 K. Thus the TMR of this device is enhanced by a factor 10.

There are several theoretical investigations on the TMR enhancement. In the orthodox theory, only the sequential tunneling is taken into account. At low temperature, however, the sequential tunneling is strongly suppressed and the contribution of the higher-order tunneling can be important. Because current carried by n-th order tunneling process is proportional to ${R_T}^{-n}$, the change in R_T gives rise to a much larger change in the current, resulting in the large TMR ratio.[2,9] This can be argued from another aspect. We discussed the renormalization of the charging energy E_C due to the quantum fluctuation of charge.[10] The E_C that characterizes the temperature dependence of resistance really changed in the magnetic field. Both theories predict the enhancement of the TMR ratio at low temperature to some amount. But quantitatively, the prediction of the enhancement does not seem large enough to explain the experimental results: The second order tunneling[9], cotunneling, is apparently not sufficient to explain the enhancement of factor 10 observed in previous[2] and present experiments. The theory of the quantum fluctuation of charge predicts the renormalization of the charging energy much smaller than the experiment.[10]

If the enhancement relates to the higher-order tunneling, the enhancement should be large for conductive junctions. Figure 3 shows how the size of the TMR ratio varies as a function of temperature. In order to compensate the variation of the TMR ratio and of the charging energy among devices, we plot the TMR ratio normalized to that at 4.2 K against the temperature normalized to E_C. The temperature shown in the figure is the electron temperature that is determined from the temperature dependence of resistance. In Fig.3, we depict results for three devices the tunnel resistances of which are about 39 kΩ,

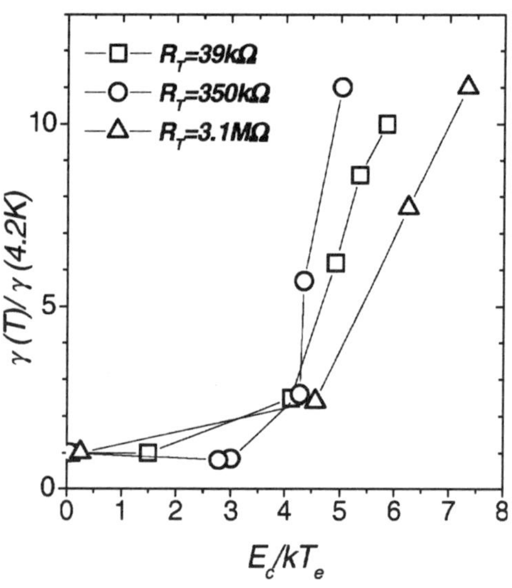

Figure 3. Temperature dependence of TMR ratio The TMR ratio γ is normalized to that at 4.2 K, and the electron temperature T_e is normalized to the single electron charging energy E_C.

350 kΩ, and 3.1 MΩ. As the quantum resistance $R_Q = h/2e^2 = 12.9$ kΩ, they correspond to $R_T/R_Q \approx 3$, 27, and 240. We see that similar degree of enhancement is seen in all devices including those with $R_T >> R_Q$. Thus, the prediction based on the higher-order tunneling fails to explain the experiment in the R_T-dependence, and we need another explanation for the enhancement.

Wang and Brataas[12] calculated the magnetoresistance of a ferromagnetic SET in the strong tunneling case where the junction resistance is smaller than the quantum resistance, and showed that there is a large enhancement of the TMR in the strong tunneling regime at a very low temperature compared with the weak coupling case. The most conductive junction we have measured had R_T = 27 kΩ, which is a little larger than R_Q. The TMR enhancement of this device was similar to those shown in Fig.3. The problem how the enhancement behaves in the strong tunneling regime is an interesting subject that should be clarified experimentally.

In conclusion, we have investigated the enhancement of TMR of Ni/Co/Ni-SETs whose tunnel resistances range from 27 kΩ to 3.1 MΩ. We found that the TMR is enhanced by a factor of about 10 between 4.2 K and 25 mK, and the saturation of the enhancement in SET with $R_T >> R_Q$, which is expected by the theories based on the higher-order tunneling, has not been observed. It means an unidentified mechanism other than the higher-order tunneling is responsible for the TMR enhancement.

Acknowledgments

We thank K.Ono and H.Shimada for helpful advice.

References

1. M. Jullier, Phys. Lett. **54A**, 225 (1975).
2. K. Ono, H. Shimada and Y. Ootuka, J. Phys. Soc. Jpn. **66** 1261 (1997).
3. H. Shimada, K.Ono and Y.Ootuka, J. Phys. Soc. Jpn. **67** 1359 (1998).
4. R. Meservey, and P. M. Tedrow, Phys. Rep. **238**, 173 (1994).
5. K. Ono, H. Shimada and Y. Ootuka, J. Phys. Soc. Jpn. **65** 3449 (1996).
6. S. Mitani, S. Takahashi, K. Takanashi *et al.*, Phys. Rev. Lett. **81** 2799 (1998).
7. K. Yakushiji, S. Mitani, K. Takahashi *et al*, Appl. Phys. Lett. **78**, 515 (2001).
8. C. W. Beenakker, Phys. Rev. B **44**, 1646 (1991).
9. S. Takahashi, S. Maekawa, Phys. Rev. Lett. **80** 1758 (1998).
10. K. Ono, H. Shimada and Y. Ootuka, J. Phys. Soc. Jpn. **67** 2852 (1998), Y. Ootuka, R. Matsuda, K. Ono, and H.S himada, Physica **B 280** 394 (2000).
11. S. Iwabuchi, T. Tanamoto, R. Kitakawa, Physica B. **249-251** 276 (1998).
12. X. H. Wang and A.Bratters, Phys. Rev. Lett. **83**, 5138 (1999).

OBSERVATION-DEPENDENT DECOHERENCE IN AN AHARONOV-BOHM RING

K. KOBAYASHI*, H. AIKAWA*, S. KATSUMOTO*†, AND Y. IYE*†

*Institute for Solid State Physics, University of Tokyo, Chiba 277-8581, Japan

†CREST, Japan Science and Technology Corporation, Mejiro, Tokyo 171-0031, Japan

E-mail: knsk@issp.u-tokyo.ac.jp

We found that the probe-arrangement greatly affects quantum coherence in mesoscopic samples. The present result poses important questions on pictures on quantum decoherence and intrinsic decoherence at absolute zero.

1 Introduction

Quantum decoherence in solids, which is the biggest obstacle for the realization of quantum computers, is now one of the most important issues in fundamental physics. Mohanty *et al.*[1] measured temperature dependence of phase-coherence time (τ_ϕ) of various systems and alleged that there is intrinsic decoherence even at absolute zero originated from zero-point fluctuation (ZPF) of the electromagnetic environment. This allegation opened debate on the origin of decoherence at the lowest temperatures[2].

In this work, we report an experiment, which suggests that the quantum coherence in mesoscopic samples strongly depends on the measurement configuration. Coherent transport in mesoscopic conductors imposed us to reconsider the meaning of "probes" for the measurements, i.e., in the mesoscopic regime, we cannot define distinct boundary between sample and probes. Hence the Landauer-Büttiker (LB) formalism[3] for such conductors, in which the probes are equally treated, has achieved great success in analyzing experiments[4]. When analyzing this class of mesoscopic transport phenomena, one usually assumes τ_ϕ to be independent of measurement configuration. However, we feel that this assumption is not immune to scrutiny. For example, Buks *et al.*[5] proved that a current that flows close to a quantum dot causes decoherence. Similarly if one exchanges current probes in a mesoscopic sample, that would result in some variations of potential fluctuation or shot-noise fluctuation and cause some difference in decoherence.

2 Experiment

We prepared an AB ring shown in Fig.1 (a) by wet-etching a two-dimensional electron gas system at an AlGaAs/GaAs heterostructure with mobility 9×10^5 cm^2/Vs and sheet carrier density 3.8×10^{11} cm^{-2}. The electron mean free path (l_e) $\sim$ 8 μm is larger enough than the length of one arm of the ring $L \sim 2$ μm, ensuring that the motion of electrons in the ring is quasi-ballistic. Two samples (#1 and #2) with almost same geometry were measured, yielding almost the same results. The probe configuration of the present devices allowed us to perform two distinct types of measurement. One is a conventional four-terminal measurement: electric current I_{14} is applied between the terminal 1 and 4, and then the voltage V_{23} between 2 and 3 is measured as shown in Fig. 1 (b), giving the resistance $R_{14,23}$. The other setup shown in Fig. 1 (c) is referred as a non-local setup henceforth, yielding the signal $R_{12,43}$ (current: 1 $\rightarrow$ 2, voltage: 4 $\rightarrow$ 3). To minimize the difference due to extrinsic effects and instrumental noise, the same instrumental setup was used for both measurements.

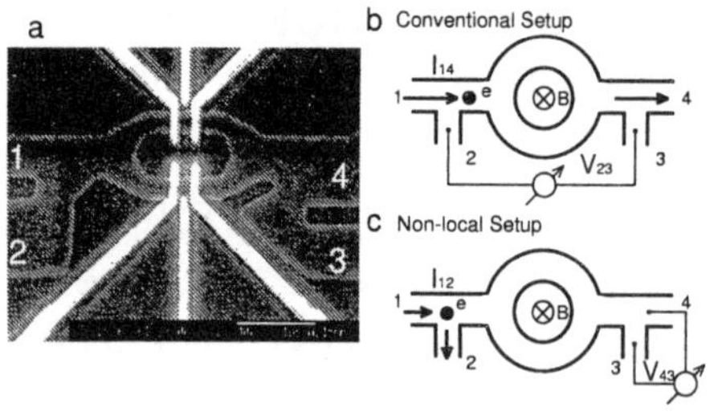

Figure 1. (a) Scanning electron micrograph of the AB ring. The Au/Ti gates on top of the arms of the AB ring were kept open for the experiments discussed in this paper. (b) The conventional four-terminal setup (current: $1 \to 4$, voltage: $2 \to 3$). (c) The non-local four-terminal setup (current: $1 \to 2$, voltage: $4 \to 3$). All the measurements were performed between 30 mK and 4.2 K using a dilution refrigerator. The standard lock-in techniques were used with typically $I_{14} = 3$ nA and $I_{12} = 15$ nA for the conventional and non-local setup, respectively, which we have carefully checked to be sufficiently small to be free from any heating effect.

3 Results and Discussions

The upper part of Fig. 2 (a) shows the conventional resistance $R_{14,23}$ obtained at 30 mK. The AB oscillation is superposed on the magnetoresistance coming from the conductance channels which do not encircle the ring. The oscillation period is $\Delta B_{AB} = 3.1 \pm 0.5$ mT, which agrees with the one expected from the ring size. The lower part of Fig. 2 (a) shows the extracted AB component (ΔR_{AB}), whose ratio is about two percent of the total resistance of the ring ~ 1 kΩ. Fig. 2 (b) represents the signal $R_{12,43}$ obtained for the non-local setup and its AB component ΔR_{AB}. The AB signal relative to the total resistance is greatly enhanced and it averages up to 20 % of the total signal with its maximum ~ 75 % at $B \sim -0.017$ T. Such an AB amplitude is the largest among all the AB experiments ever reported.

This unique property of the non-local measurement can be explained by the LB formula. According to the formula[3], $R_{14,23} = (h/2e^2)(T_{21}T_{34} - T_{24}T_{31})/D$ and $R_{12,43} = (h/2e^2)(T_{41}T_{32} - T_{42}T_{31})/D$, where T_{ij} (≥ 0) is a transmission coefficient from terminal i to j ($i, j = 1 \ldots 4$, $i \neq j$) summed over the relevant conductive channels and the denominator D is a function including all the T_{ij}'s. To include the AB effect, T_{ij} in a single conductive channel case can be expressed as $T_{ij} = \alpha_{ij} + \beta_{ij}\cos(2\pi\Phi/\Phi_0 + \Delta)$, where α_{ij} and β_{ij} are functions of B, temperature T, and electron energy E. The bias voltage is so small in our case that E can be replaced by the Fermi energy E_F. Φ_0 is a flux quantum and Φ is magnetic flux surrounded by the electron path. Δ represents a geometrical phase difference that an electron wave acquires along its path around the ring. When there exist multiple conductive channels, T_{ij} should be the sum over the incoming and outgoing channels. Combined with this expression and the above LB formula for $R_{14,23}$ and $R_{12,43}$, it can be shown that $R_{12,43}$ shows very large AB amplitude. The essence is that the term $T_{21}T_{34}$ in the numerator of $R_{14,23}$ is expected to be large in magnitude but contains little AB component, because both T_{21} and T_{34} are dominated by channels which barely enter the ring. By contrast, the two terms in the numerator of $R_{12,43}$ pick up the AB component more effectively. Thus, $R_{12,43}$ shows enhanced AB effect even when $|\beta_{ij}/\alpha_{ij}| \ll 1$ for all the T_{ij}'s, while the ratio of the AB component relative to the total signal in $R_{14,23}$ remains of the order of this small value. More quantitative simulations with realistic T_{ij}'s can be performed to see such enhanced AB effect.

We have seen that the non-local measurement yields quite different result from the conventional one in terms of the relative AB component. Still, as shown above, the difference is interpreted within the LB framework. However, there is another aspect which seems to suggest a fundamental difference between the two configurations. Figures 3 (a) and (b) show the resistance obtained for the two setups, respectively, at different temperatures. In Fig. 3 (c), we plot the temperature depen-

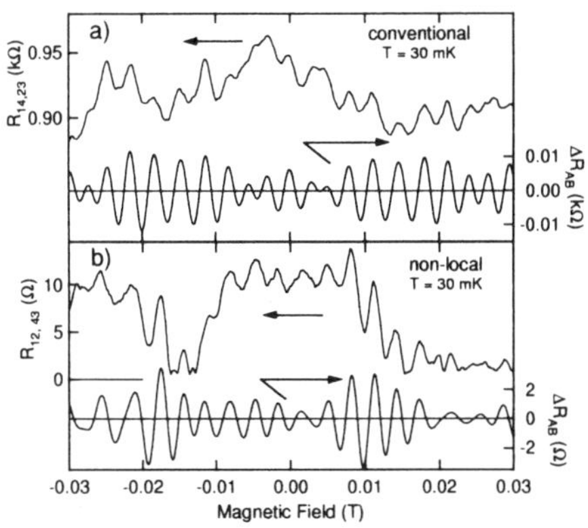

Figure 2. (a) Typical AB oscillation obtained in the conventional setup. (Bottom) The AB component ΔR_{AB} digitally extracted from $R_{14,23}$ by means of the Fourier analysis. (b) The counterpart of (a) obtained in the non-local setup.

dence of the FFT intensity of the AB component derived from $R_{14,23}$ and $R_{12,43}$. Each of them is normalized to the value at 30 mK, which is justified as the total resistances is only slightly temperature dependent. Remarkably, the AB oscillation in the non-local setup survives at higher temperatures than that conventionally obtained. The data can be fitted to $\exp(-aT)$ with a as a fitting parameter. We obtained $a = 0.72$ and 1.0 K^{-1} for the two samples in the non-local configuration, while $a = 2.3$ and 2.5 K^{-1} for the conventional case.

There are two main factors that diminish the AB amplitude. One is the thermal broadening of the electron wave packets. The other is the decoherence which contributes as $\beta_{ij} \propto \exp(-\tau_L/\tau_\phi(T))$ where $\tau_L = L/v_f$ (v_f: Fermi velocity). Since these effects are presumed to be included commonly in all the T_{ij}'s, the temperature dependence of the AB amplitude should be the same between both measurements as long as the same T_{ij}'s in the LB formula are adopted in both measurements. Because the thermal averaging in the ballistic regime is expected to occur in the time scale $\tau_{th} \sim \hbar/k_BT$, τ_ϕ^{-1} is replaced by $\tau_\phi^{-1} + \tau_{th}^{-1}$ and this τ_{th} would contribute to the AB degradation as $\exp(-bT)$ where $b \sim 1$ K^{-1}. Our observation, therefore, indicates that $\tau_\phi \propto T^{-1}$ with very different coefficients between the two setups. Judging from this temperature dependence and the fact that the AB effect is observable even at 4.2 K in the non-local setup, the coherence is supposed to survive at higher temperature in the non-local setup than in the conventional setup.

Recently, several studies on the decoherence in the ballistic AB ring were reported. Cassé *et al.*[6] explained the temperature dependence of the AB amplitude as due to the thermal averaging. Hansen *et al.*[7] showed that the thermal averaging alone cannot explain the AB degradation and $\tau_\phi \propto T^{-1}$, being consistent with our observation. Theoretically, Seelig and Büttiker[8] proposed that $\tau_\phi \propto T^{-1}$ due to the charge fluctuations caused by a nearby capacitor. Here, we propose that, while decoherence occurs mainly through the electron-electron interaction in both setups, the observed configuration-dependent decoherence is due to the difference of the current path in the conventional and non-local setups. In the former setup, the momentum of the electron is almost conserved in traversing the ring from the terminal 1 to 4 since the motion of an electron is quasi-ballistic. In the latter setup, however, the momenta of the electron incoming from 1 and that outgoing from 2 are opposite in direction. Possibly, it gives rise to the suppression of charge fluctuation in the ring, and hence the coherence of electron might be preserved in the non-local setup.

Here two remarks are in order. Our observation of $\tau_\phi \propto T^{-1}$, which holds for two different setups with different coefficients, is not compatible with the "intrinsic decoherence" due to ZPF[1]. Second, at this moment, it is not clear how much current flows around the ring in the non-local setup. Although no net current flows across the ring, current can flow along the conductive channel mu-

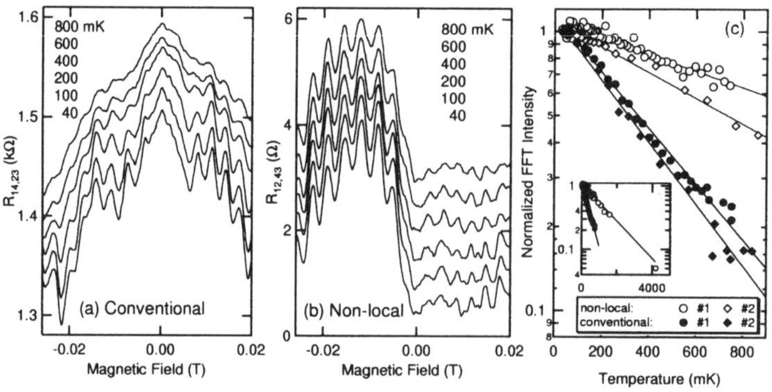

Figure 3. (a) The conventional AB oscillation of the sample #1 taken at different temperatures. The data at $T \geq 100$ mK are incrementally shifted upward by 0.02 kΩ. (b) The counterpart of (a) for the non-local measurement. The data at $T \geq 100$ mK are incrementally shifted upward by 0.5 Ω. (c) Temperature dependence of the FFT intensity of the AB amplitude for the two types of measurement. The results for the two samples (sample #1 and #2) are shown. The solid lines are fitted to the exponential decay function of temperature. The inset shows the result up to 4.2 K for the sample #1. Consistency between two different samples strongly indicates that this observation is not sample-specific but due to the measurement configuration.

tually between terminals as the LB formula presumes. This problem looks as delicate as the one about the current in the edge channel in the quantum Hall effect.

4 Conclusion

Applying the conventional and non-local measurements to the AB ring, we have observed that the decoherence is dependent on the probe configurations. While τ_ϕ behaves as $\propto T^{-1}$ in both setups, coherence survives at higher temperature in the non-local setup. We propose that the finite current across the ring may cause significant decoherence in the conventional setup. Our observation also indicates that the existence of intrinsic decoherence still disputable. The above result is not at all inconsistent with the LB formula, which has been successfully applied widely, but suggests that coherence is dependent on how we measure, the point that has been overlooked thus far.

Acknowledgments

This work is supported by Grant-in-Aid for Scientific Research and by Grant-in-Aid for COE Research ("Quantum Dot and Its Application") from the Ministry of Education, Science, Sports, and Culture.

References

1. P. Mohanty *et al.*, Phys. Rev. Lett. **78,** 3366 (1997).
2. D. S. Golubev and A. D. Zaikin, Phys. Rev. Lett. **81,** 1074 (1998); *ibid.* **82,** 3191 (1999); I. L. Aleiner, B. L. Altshuler, and M. E. Gershenson, *ibid.* **82,** 3190 (1999).
3. M. Büttiker, Phys. Rev. Lett. **57,** 1761 (1986).
4. For example, M. L. Roukes *et al.*, Phys. Rev. Lett. **59,** 3011 (1987).
5. E. Buks *et al.*, Nature **391,** 871 (1998).
6. M. Cassé *et al.*, Phys. Rev. B **62**, 2624 (2000).
7. A. E. Hansen *et al.*, Phys. Rev. B **64** 45327 (2001).
8. G. Seelig and M. Büttiker, cond-mat/0106100.

OSCILLATION PHENOMENA IN HIGH ENERGY PHYSICS: CP VIOLATION IN B-MESON DECAYS AND LONG BASELINE NEUTRINO OSCILLATION

KENZO NAKAMURA

High Energy Accelerator Research Organization (KEK)
Oho, Tsukuba, Ibaraki 305-0801, Japan
E-mail: kenzo.nakamura@kek.jp

B-factory experiments in search for CP violation in neutral B-meson decays and long baseline neutrino oscillation experiments are among the most important particle physics experiments either ongoing or in preparation. Quantum-mechanical oscillation phenomena play the central role in these experiments. Very recently, large CP violation has been discovered in neutral B-meson decays by the Belle experiment at KEK and the BaBar experiment at SLAC. The first long baseline neutrino oscillation experiment, K2K, is being conducted between KEK and Super-Kamiokande at Kamioka over a distance of 250 km. Preliminary results are in favor of muon-neutrino oscillation. This article focuses on the experiments in Japan, Belle and K2K.

1 Introduction

The most popular and extensively studied quantum-mechanical oscillation phenomenon in particle physics has been K^0-$\bar{K}^0$ oscillation. It should also be noted that CP violation was first discovered in decays of long-lived neutral K mesons, and until very recently it has been observed only in these decays. Recently, however, high-luminosity e^+e^- colliders called B factories enabled particle physicists to explore B^0-$\bar{B}^0$ oscillation in detail and to search for CP violation in neutral B-meson decays.

Within the framework of the standard model, CP violation in the quark sector is explained by the three-generation quark mixing represented by the CKM (Cabibbo-Kobayashi-Maskawa) matrix. The CKM matrix is a 3×3 unitary matrix parametrized by the three mixing angles and a CP-violating phase. The current knowledge of the CKM matrix predicts large CP-violation effects in B-meson decays. The primary purpose of the B-factory experiments is to explore them precisely and to investigate if the standard model reasonably accounts for the results, or physics beyond the standard model should be invoked.

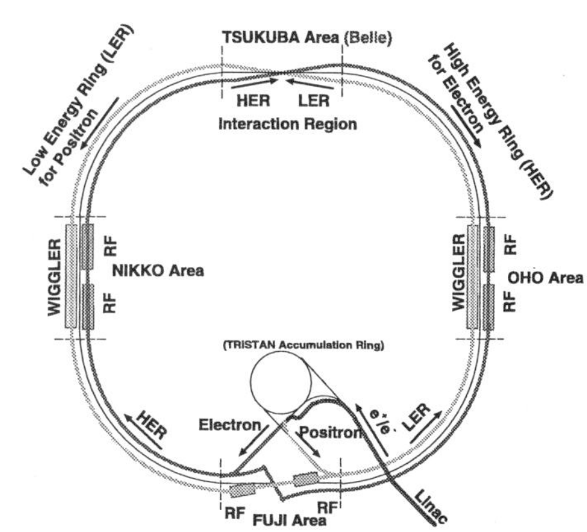

Figure 1. Schematic of the KEK-B asymmetric e^+e^- collider. Electrons and positrons collide only at the Tsukuba area where the Belle detector is located.

Recently, the Belle experiment[1] at KEK and the BaBar experiment[2] at SLAC have reported the discovery of large CP violation in neutral B-meson decays. In this paper, the Belle experiment and its CP results are presented in section 2.

In the lepton sector, compelling evidences for neutrino oscillation, and, therefore, finite neutrino mass have been obtained recently in atmospheric-neutrino and solar-neutrino observations. In general, the three-generation neutrino mixing is represented by the MNS (Maki-Nakagawa-Sakata) ma-

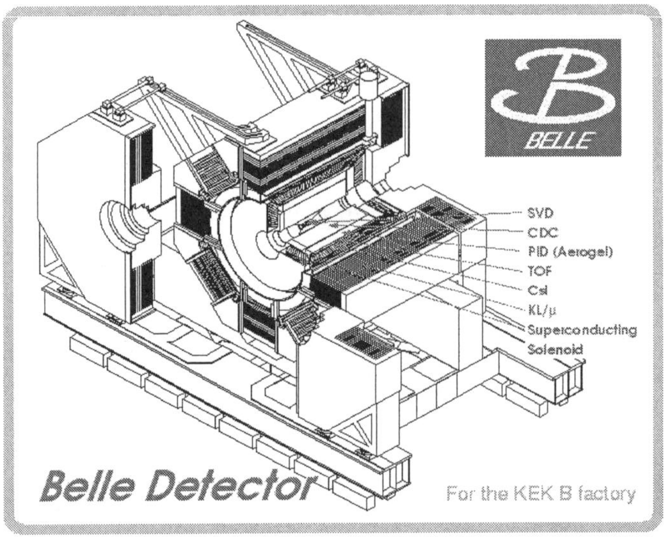

Figure 2. Cutaway view of the Belle detector.

trix which has the same parametrization as the CKM matrix. However, as Δm^2 relevant to the solar-neutrino oscillation ($< 2 \times 10^{-4}$ eV^2) is much smaller than Δm^2 relevant to the atmospheric-neutrino oscillation ($\sim 3 \times 10^{-3}$ eV^2), these oscillations are decoupled and experimental results can be analyzed in terms of two-neutrino oscillations. (Note, however, that CP violation is a genuine three-generation phenomenon: there is no CP violation for two-neutrino oscillations.)

The evidence of atmospheric-neutrino oscillation obtained by the Super-Kamiokande experiment[3] indicates $\nu_\mu \to \nu_\tau$ oscillation. This oscillation mode can be investigated with better precision using accelerator-produced muon-neutrino beam with energy E_ν of order GeV and a baseline length of $L \sim$ several hundred km, since the best sensitivity to Δm^2 is given by Δm^2 (eV^2) $\sim E$ (GeV)/L (km).

The K2K (KEK-to-Kamioka) experiment[4] is the first accelerator long baseline neutrino-oscillation experiment. Its primary purpose is to confirm the muon-neutrino oscillation found by Super-Kamiokande[3] and to determine the oscillation parameters with better accuracy. In section 3, the K2K experiment and its preliminary results are described. Other long baseline neutrino-oscillation experiments, MINOS and OPERA, are under preparation in United States (Fermilab to Soudan mine) and in Europe (CERN to Gran Sasso in Italy), respectively, but they are expected to turn on in 2005.

2 CP Violation in B-Meson Decays

The best strategy for the B-factory experiments to discover CP violation in the B^0 decays has been considered to utilize the decay into $J/\psi K_S^0$, $J/\psi K_L^0$, and similar states, which are CP eigenstates. If an initially pure B^0 is produced, it mixes with a $\bar{B}^0$ through the second-order weak interactions, causing B^0-$\bar{B}^0$ oscillation. CP violation arises due to the quantum-mechanical interference of the two amplitudes corresponding to the $B^0 \to f$ and $\bar{B}^0 \to f$ decays to the same CP eigenstate f. Time-dependent CP-violating asymmetry $A(t) = \frac{\Gamma(\bar{B}^0 \to f) - \Gamma(B^0 \to f)}{\Gamma(\bar{B}^0 \to f) + \Gamma(B^0 \to f)} = -\xi_f \sin 2\phi_1 \sin \Delta \mathrm{mt}$ is the relevant quantity to be measured, since the time-integrated rates are the same for $B^0 \to f$ and $\bar{B}^0 \to f$ and do not show CP violation. Here, Γ is a decay rate at a proper time t after production, ξ_f is the CP eigenvalue of f, Δm is the mass difference between the two B^0 mass eigenstates. $A(t)$ is proportional to $\sin 2\phi_1$, where the angle ϕ_1 is a measure of the magnitude of CP violation. The problem is how to prepare the initially pure B^0 and $\bar{B}^0$.

In the Belle Experiment, B^0-$\bar{B}^0$ pairs are obtained as the decay products of Υ(4S)'s. Since the pair is born with a constraint of the Υ(4S) quantum numbers, $J^{PC} = 1^{--}$, the B^0-$\bar{B}^0$ state evolves coherently until one of the mesons decays. If the flavor of the meson that decayed at time t_1 can be tagged, the other meson's flavor can be known at t_1. In this way, a pure B^0 or $\bar{B}^0$ can be prepared at time t_1. Suppose that the prepared meson of known flavor decayed at time t_2. Now, the time-dependent CP-violating asymmetry $A(t_2 - t_1)$ should be measured. This method, therefore, requires measurement of $\Delta t = t_2 - t_1$.

If Υ(4S)'s are produced at a symmetric e^+e^- collider, they are produced at rest. In this case the B mesons from Υ(4S) decay can travel only 30 μm on average. This is too short a distance to measure with electronic detectors. The solution is an asymmetric collider of 3.5 GeV e^+ and 8.0 GeV e^- beams. This collider is called KEK-B (Fig. 1). Υ(4S)'s produced at KEK-B is moving in the direction of the e^- beam. The decay B mesons are boosted in the same direction, and travel distances measurable with a silicon vertex detector. The B^0 and $\bar{B}^0$ mesons are nearly at rest in the Υ(4S) center-of-mass system. Therefore, Δt can be determined from the spatial separation of the two decay vertices. Figure 2 schematically shows the Belle detector. It has a silicon vertex detector (SVD) as the innermost component.

Recently, the Belle Collaboration reported[1] the result with an integrated

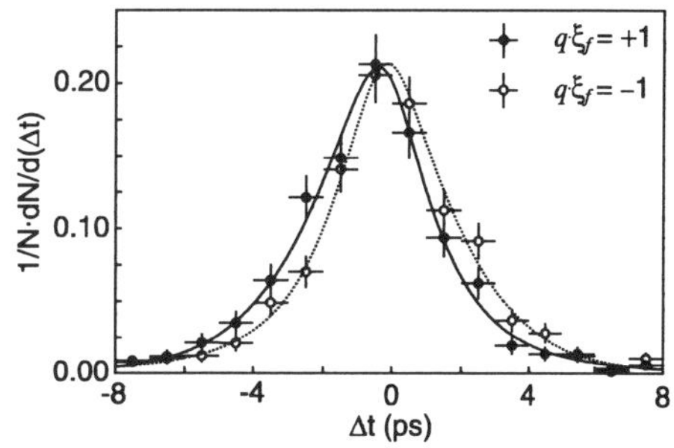

Figure 3. Δt distributions for the events with $q\xi_f = +1$ (filled points) and $q\xi_f = -1$ (open points). The results of the global fits with $\sin 2\phi_1 = 0.99$ are shown as solid and dashed curves, respectively.

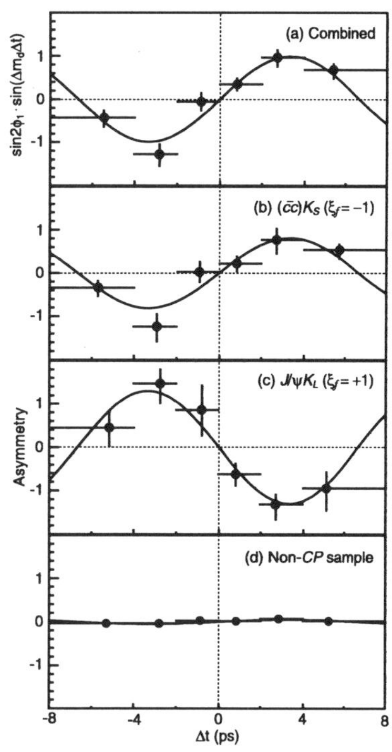

Figure 4. (a) The asymmetry obtained from separate fits to each Δt bin for the full data sample. The curve is the result of the global fit. The corresponding plots for the (b) $(c\bar{c}K_S$ ($\xi_f = -1$), (c) $J/\psi K_L$ ($\xi_f = -1$), and (d) B^0 control samples are also shown. The curves are the results of the fits applied separately to the individual data samples.

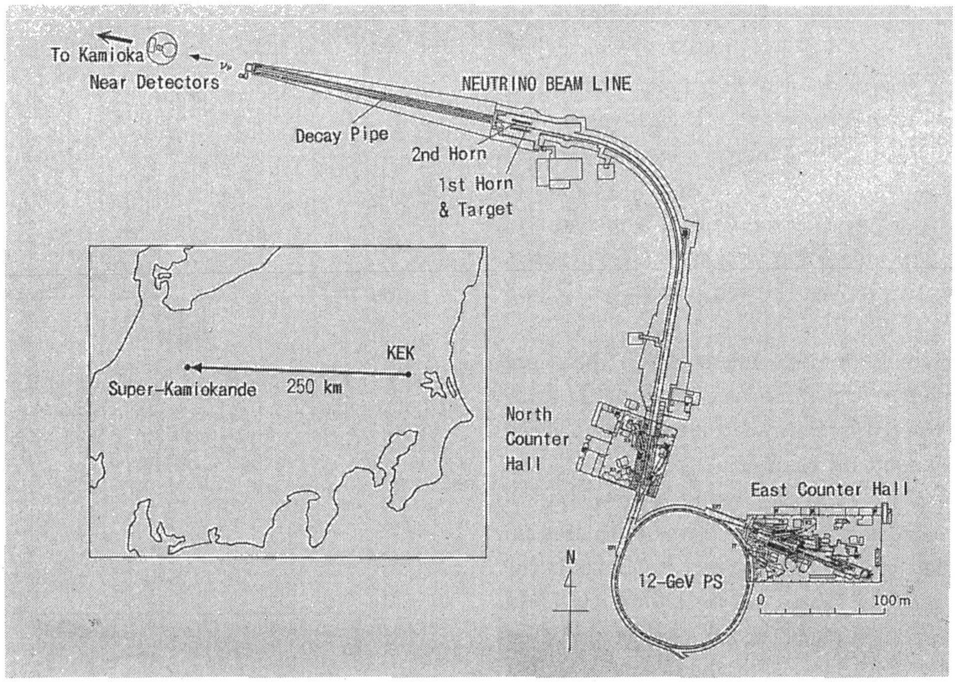

Figure 5. Layout of the KEK neutrino beam line. An insert shows the locations of KEK at Tsukuba and Super-Kamiokande at Kamioka.

liminosity of 29.1 fb^{-1}. We do not discuss details of the event reconstruction and event selection here. The interested readers are refered to Ref.[1] Figure 3 shows the Δt distributions for the events with $q\xi_f = +1$ and $q\xi_f = -1$. Here, $q = +1$ (-1) is assigned for the events with the tag side meson being B^0 ($\bar{B}^0$). Without CP violation these two distributions should be the same. These results clearly show CP violation. Figure 4 shows the asymmetry A(t) for (a) the combined data sample, (b) ($c\bar{c}K_S$ ($\xi_f = -1$), (c) $J/\psi K_L$ ($\xi_f = -1$), and (d) non-CP eigenstate control samples. From these data the Belle Collaboration obtained sin $2\phi_1 = 0.99 \pm 0.14$(stat) ± 0.06(sys), concluding that there is a large CP violation in the neutral B-meson system. A competing experiment, BaBar, at SLAC also reported[2] sin $2\phi_1 = 0.59 \pm 0.14$(stat) ± 0.05(sys).

The Belle[1] and BaBar[2] results are consistent to within the experimental uncertainties and established large CP violation in neutral B-meson decays.

3 Long Baseline Neutrino Oscillation

K2K, the first long baseline neutrino-oscillation experiment[4] with a baseline distance of hundreds of kilometers aims at confirming neutrino oscillation $\nu_\mu \to \nu_x$, in the Δm^2 range of $10^{-2} \sim 10^{-3}$ eV^2, suggested by Super-Kamiokande in the atmospheric neutrino observations.[3] ν_x may be ν_τ or ν_s. (ν_s is a sterile neutrino. The Super-Kamiokande's atmospheric-neutrino observation, however, disfavors[5] $\nu_\mu \to \nu_s$.)

Figure 5 shows a layout of the KEK neutrino beam line and the locations of KEK and Super-Kamiokande. A muon-neutrino beam with an average energy of 1.3 GeV is produced by the KEK 12-GeV proton synchrotron and shot to Super-Kamiokande

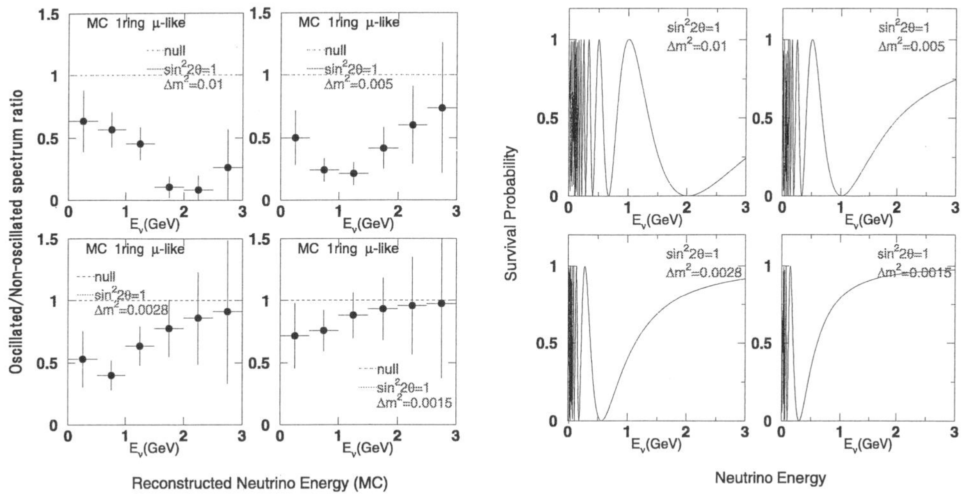

Figure 6. The four panels on the left show the "oscillated/non-oscillated spectrum ratio" expected to be observed in Super-Kamiokande for the oscillation parameters indicated in the individual panels. These are the results of the simulation study assuming 10^{20} 12-GeV protons impinging on the pion production target. Single-ring μ-like events are selected and the neutrino energy is reconstructed by assuming quasi-elastic (QE) scattering. Because of the non-QE background, the ratio does not reach 0 where the oscillation causes the maximum effect. Nevertheless, for $\Delta m^2 \geq 0.0028$ eV^2, a dip is clearly seen at the expected neutrino energy where the survival probability becomes 0. Each of the four panels on the right shows the survival probability which is calculated by assuming the same neutrino-oscillation parameters as used for the corresponding panel on the left.

which is located $\sim$ 1000 m underground in the Kamioka mine in Gifu Prefecture, about 250 km west to KEK. In the K2K Experiment, $\nu_\mu \to \nu_\tau$ is studied only in the disappearance mode because of the τ production threshold ($\sim$ 4 GeV).

The expected signal of neutrino oscillation is the reduction of the muon-neutrino flux, namely, $N_{\text{obs}}/N_{\text{exp}} < 1$ where N_{obs} is the number of muon-neutrino events observed in Super-Kamiokande, while N_{exp} is that expected for no neutrino oscillation. N_{exp} should be determined from the neutrino flux measured by the near detector located at about 300 m from the production target. Another method is a comparison of the neutrino energy spectrum measured by the near detector and that measured by the far detector. The latter method has a better sensitivity to the neutrino oscillation.

The goal of the experiment is to observe $\sim$ 200 charged-current events (in the case of no neutrino oscillation) in the 22000-ton fiducial volume of Super-Kamiokande with 10^{20} protons on the production target. Figure 6 shows some results of simulation studies of what is expected with this statistics.

The experiment started in June 1999. Up to April, 2001, 44 charged-current events, time-correlated with the KEK neutrino beam, were detected. This is compared to $63.9^{+6.1}_{-6.6}$ events expected for no neutrino oscillation. The probability of observing 44 events or less as a result of the statistical fluctuation of the expected $63.9^{+6.1}_{-6.6}$ events is less than 3%. These 44 events are further classified into 1-ring μ- and e-like events and multiring events. The results are listed in Table 1 and compared with the expected number of events with hypotheses of no neutrino oscillation and neutrino oscillation with $\Delta m^2 = 3$, 5, and 7 $\times 10^{-3}$ eV^2 and $\sin^2 2\theta = 1$. As can be seen from this table, the observed results are consistent with $\Delta m^2 = 3 \times 10^{-3}$ eV^2.

Table 1. K2K results up to the end of April, 2001. The observed number of events fully contained in the Super-Kamiokande's fiducial volume are compared with the expected number of events with hypotheses of no neutrino oscillation and neutrino oscillation with $\Delta m^2 = 3$, 5, and 7 $\times 10^{-3}$ eV^2 and $\sin^2 2\theta = 1$. The fully contained events are classified into the 1-ring events (third row) and the multiring events (sixth row). The 1-ring events are further classified into the μ-like events (fourth row) and the e-like events (fifth row). The uncertainties associated with the expected number of events are given only for the case of no oscillation.

Event Type	Observed number of events	Expected number of events			
		No oscillation	Δm^2 (eV^2)		
			3×10^{-3}	5×10^{-3}	7×10^{-3}
FC (22.5 kt)	44	$63.9^{+6.1}_{-6.6}$	41.5	27.4	23.1
1-ring	26	38.4 ± 5.5	22.3	14.1	13.1
μ-like	24	34.9 ± 5.5	19.3	11.6	10.7
e-like	2	3.5 ± 1.4	2.9	2.5	2.4
multiring	18	25.5 ± 4.3	19.3	13.3	10.0

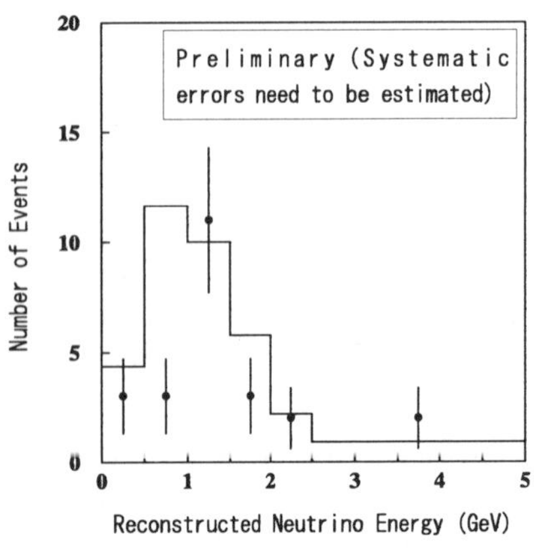

Figure 7. The reconstructed energy distribution obtained from the 24 fully-contained 1-ring μ-like events which were observed in the Super-Kamiokande's 22.5 kton fiducial volume in time correlation with the KEK neutrino beam. The bin size is 0.5 GeV except for the highest energy bin where the $2.5 \leq E_\nu \leq 5$ GeV interval is shown as one bin. The histogram shows the expected ν_μ energy spectrum for the case of no neutrino oscillation, normalized to the expected number of events (34.9). The systematic errors associated with this spectrum have yet to be estimated.

Figure 7 shows a preliminary result of the spectrum comparison. Though the systematic errors associated with the spectrum measured by the near detector should be estimated yet, the observed number of events in the 0.5 - 1 GeV bin is much less than the expected number for no neutrino oscillation. For $\Delta m^2 = 3 \times 10^{-3}$ eV^2, the oscillation minimum is expected at $\sim$ 0.6 GeV. Therefore, the preliminary spectrum data are also consistent with $\Delta m^2 = 3 \times 10^{-3}$ eV^2.

To conclude this section, the preliminary K2K results are in favor of muon-neutrino oscillation, and consistent with the Super-Kamiokande's atmospheric neutrino observation.

References

1. K. Abe *et al.*, *Phys. Rev. Lett.* **87**, 091802 (2001).
2. B. Aubert *et al.*, *Phys. Rev. Lett.* **87**, 091801 (2001).
3. Y. Fukuda *et al.*, *Phys. Rev. Lett.* **81**, 1562 (1998).
4. S.H. Ahn *et al.*, *Phys. Lett.* B **511**, 178 (2001).
5. S. Fukuda *et al.*, *Phys. Rev. Lett.* **85**, 3999 (2000).

DYNAMIC OBSERVATION OF VORTICES IN HIGH-T_c SUPERCONDUCTORS

Akira TONOMURA

*Advanced Research Laboratory, Hitachi, Ltd., Hatoyama, Saitama*350-0395, Japan.
and
CREST, Japan Science and Technology Corporation (JST), Kawaguchi, Saitama 332-0012, Japan.
E-mail: tonomura@harl.hitachi.co.jp

Quantized magnetic vortices inside high-T_c superconductors were observed using the phase shift of electron waves passing through the vortex magnetic fields with our recently developed 1-MV field-emission electron microscope, which has the brightest electron beam yet attained. We were able to investigate the microscopic pinning mechanism of columnar defects. which are produced by irradiation of high-energy [heavy ions] and are regarded as one of the most effective pinning centers in high-T_c superconnductors. In particular, we determined under which conditions individual vortex lines are trapped along tilted columnar defects in Bi-2212 thin films, which results in a strong pinning effect.

1. Introduction

When a magnetic field is applied to a type-II superconductor, magnetic flux penetrates the superconductor in the form of thin filaments. These filaments are called "vortices", because the magnetic flux is formed by a vortex supercurrent. These vortices hold the key to the practical application of superconductors as electrical conductors. This is because they begin to move when a current is applied due to the Lorentz force exerted on them by the current, eventually breaking down the superconductivity. Therefore, a dissipation-free current cannot be obtained unless the vortices are somehow pinned.

Although extensive efforts have been made, mainly by trial and error, to develop practical superconducting materials with large critical currents, the microscopic mechanism of vortex pinning has not yet been fully explained, as vortices and pinning centers are both too tiny to be observed directly. Techniques using the Bitter method[1)], scanning tunneling microscopy (STM)[2)], tiny Hall devices[3)], and tiny SQUIDs[4)], and other methods have been developed to observe individual vortices, but these techniques detect only vortices near the surface of a superconductor, not those inside the superconductors.

We have long attempted to directly and dynamically observe both individual vortices and pinning centers inside superconductors by using electron microscopes. In 1989, the magnetic lines of force of vortices leaking from the surface of a superconducting lead film were observed[5)] in the form of holographic interference electron micrographs[6)]. This method allows us to observe the magnetic lines of force but not the material defects.

With a new transmission method using Lorentz microscopy, individual vortices and also defects in niobium thin films were observed in an out-of-focus image[7)]. With this method, the microscopic vortex pinning phenomena, which determine the critical current of superconductors, were directly observed through a microscope. The high penetration power and the high brightness of the electron beam of our recently developed 1-MV field-emission electron microscope[8)] have enabled us to observe vortices even inside high-T_c superconductors[9)].

2. Experimental Apparatus

Electric or magnetic fields localized in a microscopic region present phase objects to an illuminating electron beam. These phase objects cannot be observed as in-focus electron micrographs but can be observed as holographic interference micrographs or Lorentz micrographs. However, since weak phase objects such as superconducting vortices can be observed only by using a highly collimated electron beam, we have continued to develop brighter electron beams which enable a collimated illumination. In fact, every time we obtained a brighter beam, new possibilities opened up. With our 200-kV microscope we can measure an electron phase shift as small as 1% of the electron wavelength,[10)] enabling us to carry out experiments on quantum-mechanical phenomena such as the single-electron build-up of an interference pattern[11] and the Aharonov-Bohm (AB) effect[12, 13, 14]. With our 300-kV microscope, we can directly observe the dynamics of vortices in superconductors.[7]

Our next challenge is to clarify the unusual behaviors of vortices which reflect the effect of the layered structure in high-T_c superconductors. Although the practical use of high-T_c superconductors is expected to change our world drastically, progress, though steady, has been slow, because the vortices are very easily moved. For these vortices to be directly observable with our new transmission method, the film has to be ten times thicker because the magnetic radius of vortices, or the penetration depth, in high-T_c superconductors is ten times larger than in Nb film. Since we are unable to observe such a thick film with our 300-kV microscope, we developed a 1-MV field-emission microscope[8].

The microscope, schematically illustrated in Fig. 1, is divided into three parts in order to minimize the vibration of the microscope column as well as to greatly stabilize the high voltage. The 1-MV high voltage is produced by a Cockcroft-Walton generator using an upright tank. This high voltage is transmitted through a cable to a middle tank, where electrical systems are installed to control the field-emission gun. The effects of the AC ripples of these power supplies are completely confined within their own tanks, and the ripple values are greatly reduced during transmission through the cables.

This three-tank system is indispensable from the standpoint of a mechanical-vibration-free system in which the tiny electron source (50 Å in diameter) does not move relative to the microscope column even by only a fraction of its diameter The developed microscope has the highest beam brightness, $2\times10^{10}A$ / cm^2, and the shortest lattice resolution, below 0.5 Å, yet obtained.

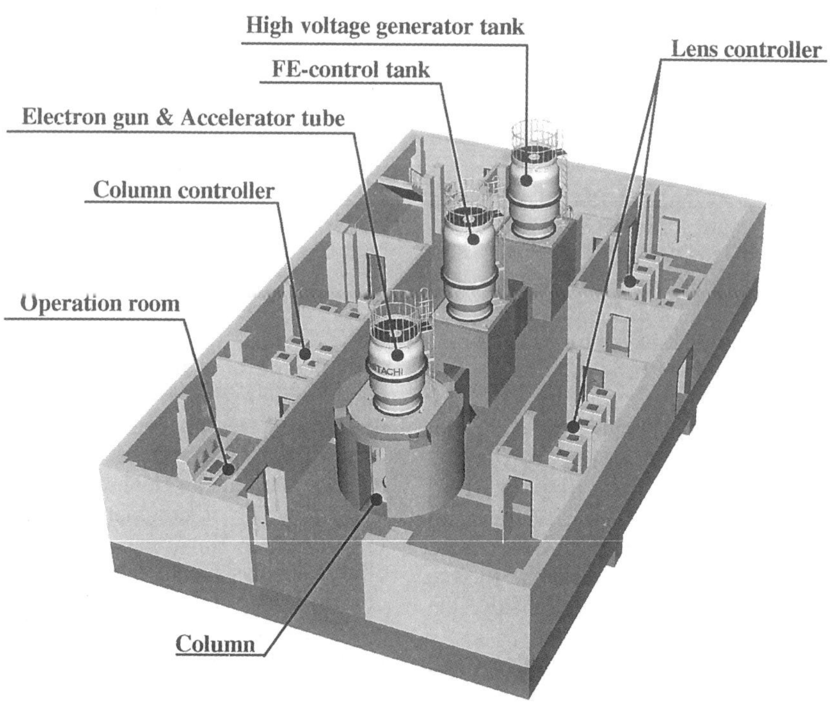

Figure 1. 1-MV holography electron microscope.

3. Experimental Method

3.1 Lorentz microscopy

A schematic of the experimental arrangement for the Lorentz microscopy is shown in Fig. 2. A film sample, 4000 Å thick, was prepared by cleaving a single crystal of Bi-2212 (transition temperature T_c = 85 K). A magnetic field of up to 100 Oe was applied to the film, and a collimated electron beam was applied incident to the film, which was tilted at 30°. We observed the vortices as defocused Lorentz micrographs. The electrons transmitted through the film were phase-shifted due to the AB effect of the vortex magnetic field, so they contained information about the spatial distribution of the magnetic flux, mainly inside the film. This phase distribution was transformed into an intensity distribution by image defocusing, so the magnetic flux distribution, *i.e.*, the vortex-line arrangement, inside the film could be determined from Lorentz micrographs taken at appropriate defocusing distances.

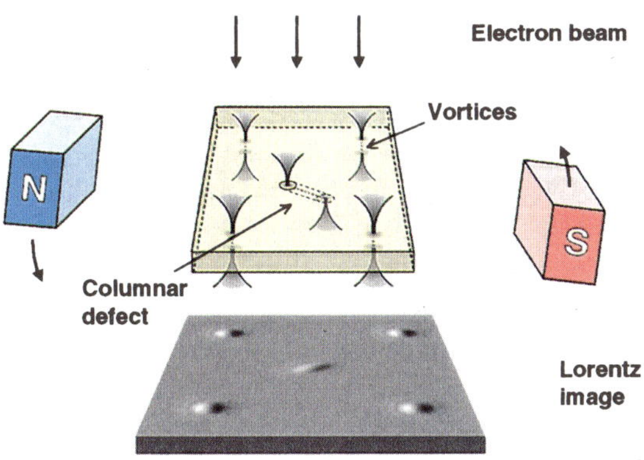

Figure 2. Experimental set-up for Lorentz microscopy of vortices in superconducting thin film with tilted columnar defects. Some vortex lines are trapped at columns, while others penetrate the film perpendicularly. A collimated electron beam was applied incident to the film, which was tilted at 30°. The phase distribution of the transmitted beam reflected the vortex magnetic field inside the film, so the Lorentz image of a vortex, *i.e.*, the coherently defocused image, differed depending on whether it was perpendicular or tilted.

3.2 Columnar defects

We irradiated the sample with parallel 240-MeV Au^{15+} ions to produce tilted columnar defects at 70°. Such columnar defects are one of the most effective pinning centers in high-T_c layered superconductors. A vortex line in such superconductors tends to split into vortex "pancakes" in each layer, and these pancakes move around, especially at high temperatures and in high magnetic fields. Columnar defect can trap even such vortex lines.

However, the trapping of a vortex line at a columnar defect had never been directly observed because previous method detects only vortices near the superconductor surfaces. We therefore attempted to distinguish between vortex lines trapped at columnar defects and those untrapped, since different Lorentz images should be formed if the distributions of vortex magnetic flux differ.

4. Experimental Results

4.1 Trapped vortices

Fig. 3 shows a Lorentz micrograph of vortices in a Bi-2212 thin film when a magnetic field was applied in the direction of tilted columnar defects. The tiny spots, each consisting of bright and dark contrasting features, are images of the vortices. Careful examination shows that some are circular images and the others, indicated by arrows, are elongated images with weaker contrast.

By using image simulation investigating the correspondence of the vortex images to the column images, we found that the circular images indicate vortex lines penetrating the film perpendicularly to the film plane and that the elongated images indicate vortex lines trapped along the tilted columnar defects. The origin of these images is explained by comparing an electron micrograph and Lorentz micrograph in Fig. 4. The tiny thin lines, 100 Å thick and 1 μm long in the electron micrograph (Fig. 4(a)) indicate projected images of tilted columnar defects. When these images were defocused, the column images became blurred and, eventually, they disappeared completely by spreading out. However, when they were defocused even further, new images appeared, as shown in the Lorentz micrograph (Fig. 4(b)). They are the vortex images. The reason the vortex images appeared when the images were defocused is that the images were produced by the phase contrast. The elongated images indicated by the arrows correspond to vortex lines trapped at tilted columnar defects, since they appear exactly at the columnar defects. The circular images appeared in regions without columnar defects.

We then investigated whether vortex lines remained trapped along columnar defects, even when the magnetic field direction was very different from that of the columns. When the magnetic-field directions were $\theta = 0°$, and ±70° (columnar defect direction: $\theta_\phi = 70°$), all the vortex lines were found to remain trapped along columns at $T > 19$ K.

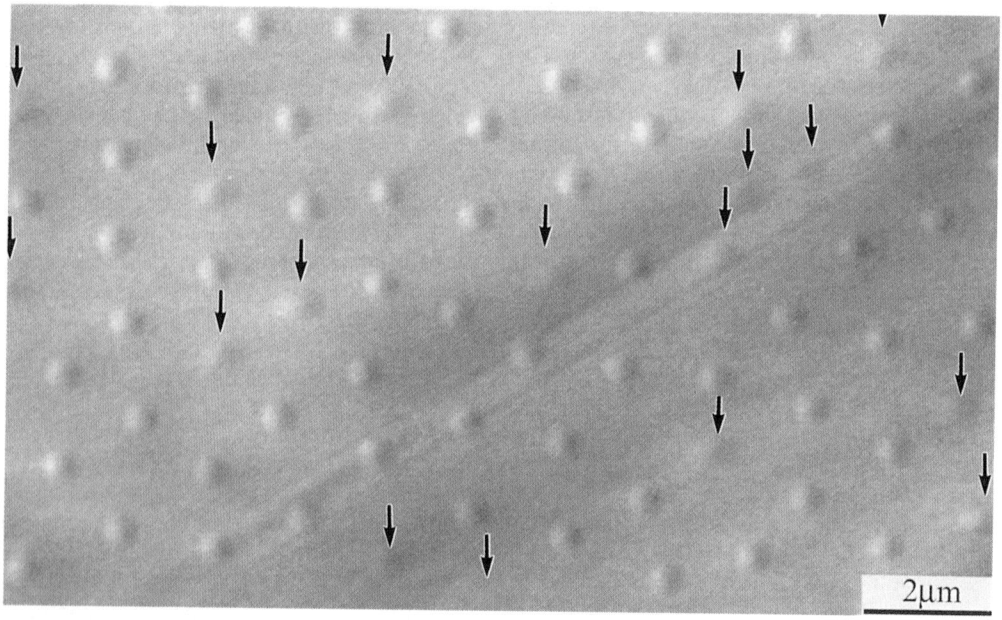

Figure 3. Lorentz micrograph of vortices in Bi-2212 thin film with columnar defects tilted at 70°. Tiny spots are images of vortices. Careful examination shows that some are circular images and the others are elongated images with weaker contrast.

Figure 4. Comparison of vortices in Bi-2212 thin film, and an in-focus image of columnar defects:
(a) electron micrograph, (b) Lorentz micrograph. Due to the low density of columnar defects, some vortices were trapped at columnar defects. The images of untrapped vortex lines perpendicular to the film plane are circular spots having bright and dark regions. Vortex images located at columnar defects are elongated spots (indicated by arrows) with lower contrast, since these vortex lines were trapped at columnar defects tilted by 70°, as illustrated in Fig. 2.

4.2 *Temperature dependence of vortex-line arrangement*

When *T* was increased above 19K, field-cooled vortex lines were found to be trapped at tilted columnar defects, regardless of the direction of the applied magnetic field. When *T* was decreased to 12-14 K, we found to our surprise that the vortices trapped at the columnar defects began to stand up perpendicularly to the film plane, one after another. We also found that this change was not reversible; instead, hysteresis was observed when increasing and decreasing the temperature: when *T* was increased, the perpendicular vortices at 7 K did not begin to tilt at 12-14 K, but at 15-19 K. These transition temperatures were rather scattered for different trapped vortex lines.

To investigate the reason why vortex lines stand up perpendicularly to the film plane, we observed the vortex movements at various temperatures. We found that the movements varied with the sample temperature. At $T = 7$ K, the vortices were strongly pinned by the background pinning. When a strong driving force was applied to them by changing the value of *H*, the vortices migrated very slowly. All the vortices moved slowly and uniformly as if there were no columnar defects.

The migration speed increased with *T*. Above 16 K, some vortices began to be trapped at defects, while others migrated by passing around the trapped vortices. The migrating vortices sometimes stopped at defects. Then after a while they began to migrate again. When *T* was increased above 25 K, the vortices began to be trapped at specific points. By comparing the vortex images and in-focus images, we found that these points were located at columnar defects. When a stronger driving force was applied, these trapped vortices were depinned from the columns and began hopping.

5. Discussion

The observed static and dynamic characteristics of vortices in Bi-2212 thin films with columnar defects can be consistently interpreted as follows. Above 25 K, columnar-defect pinning dominates, so even when a magnetic field is applied from any direction, the vortex lines are firmly trapped, even along greatly tilted columns. When a driving force is applied to such vortices, they hop from one defect to another. When *T* is decreased below 25 K, the hopping movement gradually changes to migration movement, since the background collective-pinning due to more abundant atomic-size defects of other types (for example, oxygen defects) increases relative to the columnar-defect pinning. Here, the vortices, when driven, intermittently stop and migrate. Since the migration speed increases with the rise in *T*, the migration is thought to be caused by thermally activated depinning of a vortex line from many atomic-size defects, one by one. The trapped vortex lines are tilted along columns. When *T* is decreased below 12 K, the trapped vortex lines begin to stand up perpendicularly to the film plane, just like untrapped vortices. At around 7 K, all the vortices migrate or drift slowly, as if they were moving in a viscous medium containing no pinning centers, because the pinning of columnar defects is almost completely hidden by the background pinning, though there are still some exceptional vortices strongly trapped at multiple defects.

The behavior of vortices in Bi-2212 thin films was thus microscopically clarified by direct observation.

References

1) U. Essman, H. Träuble, Phys. Lett. A 24 (1967) 526-527.
2) H. F. Hess, R. B. Robinson, J. V. Waszczak, Phys. Rev. Lett. 64 (1990) 2711-2714.
3) A. M. Chang, H. D. Hallen, L. Harriott, H. F. Hess, H. L. Kao, J. Kwo, R. E. Miller, R. Wolfe, J. van der Ziel, T. Y. Chang, Appl. Phys. Lett. 61 (1992) 1974-1976.
4) L. N. Vu, M. S. Wistrom, D. J. Van Harlingen, Appl. Phys. Lett. 63 (1993) 1693-1695.
5) T. Matsuda, S. Hasegawa, M. Igarashi, T. Kobayashi, M. Naito, H. Kajiyama, J. Endo, N. Osakabe, A. Tonomura, R. Aoki, Phys. Rev. Lett. 62 (1989) 2519-2522.
6) A. Tonomura, "Electron Holography"- 2nd Edition, Springer, Heidelberg, 1999.
7) K. Harada, T. Matsuda, J. Bonevich, M. Igarashi, S. Kondo, G. Pozzi, U. Kawabe, A. Tonomura, Nature 360 (5 November 1992) 51-53.
8) T. Kawasaki, T. Yoshida, T. Matsuda, N. Osakabe, A. Tonomura, I. Matsui, K. Kitazawa, Appl. Phys. Lett. 76 (2000)

1342-1344
9) A. Tonomura, H. Kasai, O. Kamimura, T. Matsuda, K. Harada, Y. Nakayama, J. Shimoyama, K. Kishio, T. Hanaguri, K. Kitazawa, M. Sasase, S. Okayasu, Nature 412 (2001) 620-622.
10) A. Tonomura, T Matsuda, T. Kawasaki, J. Endo, N. Osakabe, Phys. Rev. Lett. 54 (1985) 60-62.
11) A. Tonomura, J. Endo, T. Matsuda, T. Kawasaki, H. Ezawa, Amer. J. Phys. 57 (1989) 117-120.
12) A. Tonomura, T. Matsuda, R. Suzuki, A. Fukuhara, N. Osakabe, H. Umezaki, J. Endo, K. Shinagawa, Y. Sugita, H. Fujiwara, Phys. Rev. Lett. 48 (1982) 1443-1446.
13) A. Tonomura, H. Umezaki, T. Matsuda, N. Osakabe, J. Endo, Y. Sugita, Phys. Rev. Lett. 51 (1983) 331-334.
14) A. Tonomura, N. Osakabe, T. Matsuda, T. Kawasaki, J. Endo, S. Yano, H. Yamada, Phys. Rev. Lett. 56 (1986) 792-795.
15) L. Civale, A. D. Marwick, T. K. Worthington, M. A. Kirk, J. R. Thompson, L. Krusin-Elbaum, Y. Sun, J. R. Clem, F. Holtzberg, Phys. Rev. Lett. 67, (1991) 648-651.

PRECISION OPTICAL FREQUENCY METROLOGY USING PULSED LASERS

TH. UDEM, R. HOLZWARTH, M. ZIMMERMANN, AND T. W. HÄNSCH
Max-Planck Institut für Quantenoptik, 85748 Garching, Germany
E-mail: Thomas.Udem@mpq.mpg.de

Femtosecond laser frequency comb techniques are vastly simplifying the art of measuring the frequency of light. A single mode-locked femtosecond laser is now sufficient to synthesize hundreds of thousands of evenly spaced spectral lines, spanning much of the visible and near infrared region. The mode frequencies are absolutely known in terms of the pulse repetition rate and the carrier-envelope phase slippage rate, which are both accessible to radio frequency counters. Such a universal optical frequency comb synthesizer can serve as a clockwork in atomic clocks, based on atoms, ions or molecules oscillating at optical frequencies.

1 Introduction

For more than a century, precise optical spectroscopy of atoms and molecules has played a central role in the discovery of the laws of quantum physics, in the determination of fundamental constants, and in the realization of standards for time, frequency, and length. The advent of highly monochromatic tunable lasers and techniques for non-linear Doppler-free spectroscopy in the early seventies had a dramatic impact on the field of precision spectroscopy[1,2]. Today, we are able to observe extremely narrow optical resonances in cold atoms or single trapped ions, with resolutions ranging from 10^{-13} to 10^{-15}, so that it might ultimately become possible to measure the line center of such a resonance to a few parts in 10^{18}. Laboratory experiments searching for slow changes of fundamental constants would then reach unprecedented sensitivity. A laser locked to a narrow optical resonance can serve as a highly stable oscillator for an all-optical atomic clock[3] that can satisfy the growing demands of optical frequency metrology, fiber optical telecommunication, or navigation. However, until recently there was no reliable optical "clockwork" available that could count optical frequencies of hundreds of THz. Most spectroscopic experiments still rely on a measurement of optical wavelengths rather than frequencies. Unavoidable geometric wavefront distortions have so far made it impossible to exceed an accuracy of a few parts in 10^{10} with a laboratory-sized wavelength interferometer. To measure optical frequencies, only a few harmonic laser frequency chains have been built during the past 30 years which start with a cesium atomic clock and generate higher and higher harmonics in non-linear diode mixers, crystals, and other non-linear devices[5,6,7,8]. Phase-locked transfer oscillators are needed after each step, so that such a chain traversing a vast region of the electromagnetic spectrum becomes highly complex, large, and delicate, and requires substantial resources and heroic efforts to build and operate. Most harmonic laser frequency chains are designed to measure just one single optical frequency.

In 1998, our laboratory has introduced a revolutionary new approach that vastly simplifies optical frequency measurements. We could demonstrate that the broad comb of modes of a mode-locked femtosecond laser can be used as a precise ruler in frequency space[9,10]. This work has now culminated in a compact and reliable all-solid-state frequency "chain" which is actually not really a chain any more but requires just a single mode-locked laser[11,12,13,14,3]. As a universal optical frequency comb synthesizer it provides the long missing simple link between optical and microwave frequencies. For the first time, small scale spectroscopy laboratories have now access to the ability to measure or synthesize any optical frequency with extreme precision. Femtosecond frequency comb techniques have since begun to rapidly gain widespread use, with precision measurements in Cs[9], Ca[15,16,4], CH_4[4], H[17], Hg^+[15,4], I_2[12,18,4,19], Yb^+[4,20], Sr^+[4] and In^+[21,4].

The same femtosecond frequency comb techniques are also opening new frontiers in ultrafast physics. Control of the phase evolution of few cycle light pulses, as recently demonstrated[13,22], provides a powerful new tool for the study of highly non-linear phenomena that should depend on the phase of the carrier wave relative to the pulse envelope, such as above threshold ionization, strong field photoemission, or the generation of soft x-ray attosecond pulses by high harmonic generation.

In the first experiment of its kind, we have applied the frequency comb of a mode-locked femtosecond laser to measure the frequency of the cesium D_1 line[9] that provides an important link for a new determination of the fine structure constant[23] α. More recently, we have measured the absolute frequency of the hydrogen 1S-2S two-photon resonance in a direct comparison with a cesium atomic fountain clock to within 1.9 parts in 10^{14}, thus realizing one of the most accurate measurement of an optical frequency to date[15,4]. During the past few years, precision spectroscopy of hydrogen has yielded a value for the Rydberg constant that is now one of the most accurately known fundamental constant[24].

2 Optical Frequency Differences

While it has been extremely difficult in the past to measure an absolute optical frequency, a small frequency difference or gap between two laser frequencies can be measured rather simply by superimposing the two laser beams on a photodetector and monitoring a beat signal. The first experiments of this kind date back to the advent of cw He-Ne-lasers in the early sixties[25]. Modern commercial fast photodiodes and microwave frequency counters make it possible to directly count frequency differences up to the order of 100 GHz. Since the gap between the high frequency endpoint of a traditional harmonic laser frequency chain and an unknown optical frequency can easily amount to tens or hundreds of THz, there has long been a strong interest in methods for measuring much larger optical frequency differences.

Motivated by such problems in precision spectroscopy of atomic hydrogen, we have previously introduced a general, although perhaps not very elegant solution for the measurement of large optical frequency gaps with the invention of the optical frequency interval divider (OFID) which can divide an arbitrarily large frequency difference by a factor of precisely two[26,27]. An OFID receives two input laser frequencies f_1 and f_2. The sum frequency $f_1 + f_2$ and the second harmonic of a third laser $2f_3$ are created in non-linear crystals. The radio frequency beat signal between them at $2f_3 - (f_1 + f_2)$ is used to phase-lock the third laser at the exact midpoint. With a divider chain of n cascaded OFIDs, the original frequency gap can be divided by a factor of 2^n. To measure an absolute optical frequency rather than a frequency gap the determination of the interval between the laser frequency f and its own second harmonic $2f$ was suggested[26]: $f = 2f - f$.

Frequency intervals up to several THz can also be measured with passive optical frequency comb generators[28,29]. These devices where then proposed to significantly shorten an OFID chain of the $2f - f$ type[30].

3 Femtosecond Light Pulses

The periodic pulse train of a mode-locked laser can be described in the frequency do-

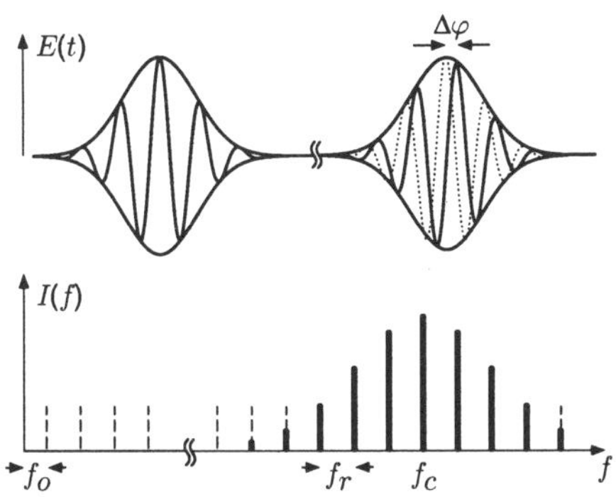

Figure 1. Two consecutive pulses of the pulse train emitted by a mode locked laser and the corresponding spectrum (right). The pulse to pulse phase shift $\Delta\varphi$ results in a frequency offset because the carrier wave at f_c moves with the phase velocity v_p while the envelope moves with the group velocity v_g.

main as a comb of equidistant modes, so that such a laser can serve as an active OFCG. More than twenty years ago, the frequency comb of a mode-locked picosecond dye laser has first been used as an optical ruler to measure the sodium 4d fine structure splitting[31]. This route was further pursued in the seventies and eighties[32,33], but the attainable bandwidths were never sufficiently large to make it a widespread technique for optical frequency metrology. Broadband femtosecond Ti:sapphire lasers have existed since the beginning of the nineties. Our experiments at Garching[10,14] as well as recent experiments at NIST[35] have shown conclusively, that such lasers can play a crucial role in this field.

To understand the mode structure of a fs frequency comb and the techniques applied for its stabilization one can look at the idealized case of a pulse circulating in a laser cavity with length L as a carrier wave at f_c and an envelope function $A(t)$. This function defines the pulse repetition time T, and the pulse repetition frequency $f_r = T^{-1}$ by demanding $A(t-T) = A(t)$ where $T = 2L/v_g$ with cavity mean group velocity v_g (see Fig. 1). Assuming the periodicity of the envelope function the electric field at a given place (e.g. at the output coupler) can be written as

$$\begin{aligned} E(t) &= Re\left(A(t)e^{-2\pi f_c}\right) \\ &= Re\left(\sum_q A_q e^{-2\pi(f_c+qf_r)}\right) \end{aligned} \tag{1}$$

where A_q are Fourier components of $A(t)$. This equation shows that the resulting spectrum consists of a comb of laser modes that are separated by the pulse repetition frequency. Since f_c is not necessarily an integer multiple of f_r the modes are shifted from being exact harmonics of the pulse repetition frequency by an offset that is chosen to be smaller than f_r:

$$f_n = nf_r + f_o \qquad n = \text{a large integer} \tag{2}$$

This equation maps two radio frequencies f_r and f_o onto the optical frequencies f_n. While f_r is readily measurable, it is not easy to access f_o unless the frequency comb contains more than an optical octave[34]. In the time domain the frequency offset is obvious because the group velocity differs from the phase velocity inside the cavity and therefore the carrier wave does not repeat itself after one round trip but appears phase shifted by $\Delta\varphi$ as shown in Fig. 1. The offset frequency is then calculated from $f_o = \Delta\varphi/2\pi T$[36,34]. Note that such a fs frequency comb has two free parameters which are the repetition frequency f_r and the offset frequency $f_o < f_r$. Depending on the application one or both of them may be phase locked.

4 Femtosecond Combs as Frequency Rulers

At the high peak intensities of femtosecond laser pulses non-linear effects due to the $\chi^{(3)}$ non-linear susceptibility are considerable even in standard silica fibers. The output spectrum of a femtosecond laser can be broadened significantly via self phase modulation in an optical fiber. Provided the fiber action is the same for every pulse, we expect the above argument that led to the mode structure to remain valid for the broadened frequency comb. To test this property we have used used an OFID and locked the two input lasers, separated by as much as 44 THz, to modes that where created by the fiber. We then measured a beat note between the output of the OFID at the average of the input frequencies and a nearby mode of the comb and found[37] that the modes are equally spaced at the level of a few parts in 10^{18}. Note that the coherence between the pulses is obviously preserved.

For most applications it is desirable to phase-lock both of the comb's degrees of freedom simultaneously. A piezo driven folding mirror changes the cavity length and leaves $\Delta\varphi$ approximately constant as the additional path in air has a negligible dispersion. A mode-locked laser that uses two intracavity prisms to produce the negative group velocity dispersion ($\partial^2\omega/\partial k^2$) necessary for Kerr-lens mode-locking provides us with a means for independently controlling the pulse repetition rate. We use a second piezo-transducer to slightly tilt the mirror[34] at the dispersive end of the cavity about a vertical pivot that corresponds to the mode f_m. We thus introduce an additional phase shift $\Delta\Phi$ proportional to the frequency distance from f_m, which displaces the pulse in time and thus changes the round trip group delay. In the frequency domain one could argue that the length of the cavity stays constant for the mode f_m while higher (lower) frequency modes experience a longer (shorter) cavity (or vice versa, depending on the sign of $\Delta\Phi$). In the case where only dispersion compensation mirrors are used to produce the negative group velocity dispersion one can modulate the pump power or manipulate the Kerr lens by slightly tilting the pump beam[14]. Although the two controls (i.e. cavity length and pump power or tilt) are not independent they affect the round trip group delay T and the round trip phase delay differently and which allows us to control both, f_o and f_r.

5 Absolute Optical Frequencies

For the absolute measurement of optical frequencies one has to determine frequencies of several 100 THz in terms of the definition of the SI second represented by the cesium ground state hyperfine splitting of 9.2 GHz. Extending our principle of determining large frequency differences to the intervals between harmonics or subharmonics of an optical frequency leads naturally to the absolute measurement of optical frequencies. In the most simple case this is the interval between an optical frequency f and its second harmonic[26] $2f$. But of course other intervals can be used as well.

6 Frequency Combs Spanning more than an Octave

The first absolute measurement of an optical frequency with a fs frequency comb[11] has inspired further rapid advances in the art of frequency metrology. In collaboration with P. St. Russell, J. Knight and W. Wadsworth from the University of Bath (UK) we have used novel microstructured photonic crystal fibers[38] to achieve further spectral broadening of femtosecond frequency combs. The remarkable dispersion characteristics attainable with these fibers including zero group velocity dispersion well below 800 nm and the high peak intensities associated with the short pulses and the small core size, enables one to observe a range of unusual non-linear optical effects[39], including very effective spectral broadening to more than an optical octave even with the moderate output power from the laser oscillator. Similar experiments have been reported by S. Cundiff, J. Hall and coworkers in Boulder using a fiber fabricated at Lucent Technologies[12,13].

With an octave wide spectrum we can directly access the interval between an optical frequency f and its second harmonic $2f$ as shown in Fig. 2. This allows the direct com-

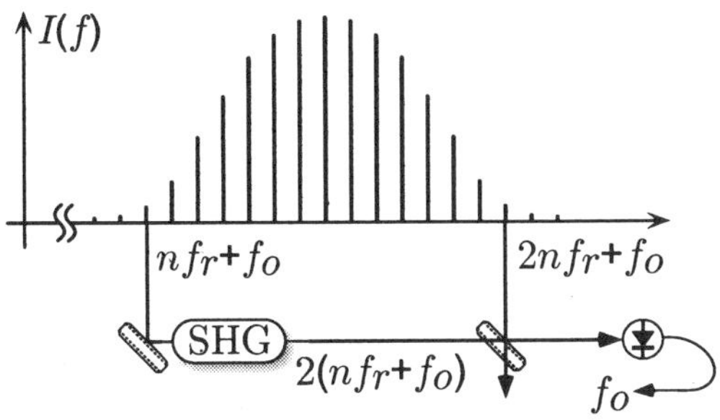

Figure 2. To obtain the frequency offset f_o the "red" portion of an octave spanning comb is frequency doubled (SHG) and a beat note with the "blue" wing is observed.

parison of radio and optical frequencies without the need of any optical frequency interval dividers or further non-linear steps[11,12,13,14]. Such an optical frequency synthesizer directly relates all optical frequencies f_n contained within the comb to the radio frequencies f_r and f_o which may be locked to an atomic clock or a GPS receiver. Our $2f - f$ optical frequency synthesizer is based on a 25 fs Ti:sapphire high repetition rate ring laser (GigaOptics, model GigaJet). While the ring design makes it almost immune to feedback from the fiber, the high repetition rate increases the available power per mode. The set-up requires only 1 square meter on our optical table with the potential for further miniaturization. At the same time it supplies us with a reference frequency grid across much of the visible and infrared spectrum with comb lines that may be distinguished with a wavemeter. This makes it an ideal laboratory tool for precision spectroscopy and a compact solid state system for all optical clocks[3].

To check the integrity of the broad frequency comb and evaluated the overall performance of the $2f - f$ optical synthesizer we compared it with an earlier set-up[11] that was based on bridging the frequency gap between $3.5f$ and $4f$. After averaging all data we found[14] an agreement within 5.1×10^{-16}. A similar testing was recently performed by the NIST group[35]. No systematic effects were visible in these experiments.

7 Outlook

Other important applications of this frequency domain technique in the time domain where the carrier offset slippage frequency is an important parameter and needs to be controlled for the next generation of ultrafast experiments[40]. In collaboration with F. Krausz (Vienna technical university) we have applied fs comb techniques to control the phase evolution of ultra-short pulses lasting for only a few optical cycles[22]. Future applications of precise optical frequency measurements also include the search for variations in the fundamental constants and the test of CPT invariance with anti-hydrogen now underway at CERN. We believe that the development of accurate optical frequency synthesis marks only the beginning of an exciting new period of ultra-precise physics. Finally we would like to thank our collaborators, without their help the work presented here would not have been possible.

References

1. *Non-linear Spectroscopy, Proceedings of the International School of Physics "Enrico Fermi"*, ed. N Bloembergen (North Holland Publ. Co., 1977).
2. *Frontiers in Laser Spectroscopy, Proceedings of the International School of Physics "Enrico Fermi"*, ed. T W Hänsch and M Inguscio (North Holland Publ. Co., 1995).
3. S A Diddams *et al.*, Science **293**, 825 2001.
4. See contributions of S A Diddams *et al.*, J C Bergquist *et al.*, H R Telle *et al.*, E E Curtis *et al.*, F Riehle *et al.*, P V Pokasov *et al.*, A Yu Nevsky *et al.* K Sugiyama *et al.*, S N Lea *et al.*, Chr Tamm *et al.*, P Dubé *et al.* and Th Becker *et al.*,

in the Proceedings of the *6th Symposium on Frequency Standards and Metrology*, St.Andrews/Scottland, September 9-14, 2001.
5. Th Udem *et al.*, Phys. Rev. Lett. **79**, 2646 1997.
6. H Schnatz *et al.*, Phys. Rev. Lett. **76**, 18 1996.
7. C Schwob *et al.*, Phys. Rev. Lett. **82**, 4960 1999.
8. J E Bernard *et al.*, Phys. Rev. Lett. **82**, 3238 1999.
9. Th Udem, J Reichert, R Holzwarth, and T W Hänsch, Phys. Rev. Lett. **82**, 3568 1999.
10. Th Udem, J Reichert, R Holzwarth, and T W Hänsch, Opt. Lett. **24**, 881 1999.
11. J Reichert, M Niering, R Holzwarth, M Weitz, Th Udem, and T W Hänsch, Phys. Rev. Lett. **84**, 3232 2000.
12. S A Diddams *et al.*, Phys. Rev. Lett. **84**, 5102 2000.
13. D Jones *et al.*, Science **288**, 635 2000.
14. R Holzwarth, Th Udem, T W Hänsch, J C Knight, W J Wadsworth, and P St J Russell, Phys. Rev. Lett. **85**, 2264 2000.
15. Th Udem *et al.*, Phys. Rev. Lett. **86**, 4996 2001.
16. J Stenger *et al.*, Phys. Rev. A **53**, 021802(R) 2001.
17. M Niering *et al.*, Phys. Rev. A **84**, 5456 2000.
18. A Yu Nevsky *et al.*, Opt. Commun. **192**, 263 2001.
19. R Holzwarth *et al.*, Appl. Phys. B **73**, 269 2001.
20. J Stenger *et al.*, Opt. Lett. **26**, 1589 (2001).
21. J von Zanthier *et al.*, Opt. Lett. **25**, 1729 2000.
22. A Apolonski *et al.*, Phys. Rev. Lett. **85**, 740 2000.
23. J M Hensley, A Wicht, B C Young, and S Chu, proceedings of th 15th International Conference on laser spectroscopy ICOLS XV (2001).
24. P J Mohr, and B N Taylor, Rev. Mod. Phys. **72**, 351 2000.
25. A Javan *Fundamental and Applied Laser Physics, Proceedings of the Esfahan Symposium*, ed. M. S. Feld, A. Javan, and N. A. Kurnit (John Wiley & Sons, New York 1973).
26. T W Hänsch *Proceedings of The Hydrogen Atom*, ed. T W Hänsch (Springer, Berlin 1989).
27. H R Telle, D Meschede, and T W Hänsch, Opt. Lett. **15**, 532 1990.
28. M Kourogi, B Widiyatomoko, Y Takeuchi, and M Ohtsu, J. Qaunt. Electr. **31**, 2120 1995.
29. L R Brothers, D Lee, and N C Wong, Opt. Lett. **19**, 2451994.
30. N Nakagawa, M Kourogi and M Ohtsu, Appl. Phys. B **57**, 425 1993.
31. J N Eckstein, A I Ferguson, and T W Hänsch, Phys. Rev. Lett. **40**, 847 1978.
32. V P Chebotayev, and V A Ulybin Appl. Phys. **50**, 1 1990.
33. D M Kane, S R Bramwell, and A I Ferguson, Appl. Phys. **39**, 171 1986.
34. J Reichert, R Holzwarth, Th Udem, and T W Hänsch, Opt. Commun. **172**, 59 1999.
35. S A Diddams, L Hollberg, L S Ma, and L Robertson Opt. Lett. in press.
36. D J Wineland *et al. Proceedings of The Hydrogen Atom*, ed. T W Hänsch (Springer, Berlin 1989).
37. R Holzwarth PhD thesis, Ludwig-Maximilians-Universität, Munich, Germany (2000).
38. J C Knight, T A Birks, P St J Russell, and D M Atkin, Opt. Lett. **21**, 1547 1996.
39. W J Wadsworth *et al.*, Electron. Lett. **36**, 53 2000.
40. G G Paulus *et al.*, Nature **414**, 182 2001.

INTERFEROMETRIC GRAVITATIONAL WAVE DETECTOR IN JAPAN

NOIKATSU MIO* AND THE TAMA COLLABORATION

*Department of Advanced Materials Science, University of Tokyo
7-3-1 Hongo, Bunkyo-ku, Tokyo 113-0033, Japan
E-mail: norikatsu@k.u-tokyo.ac.jp

A gravitational wave (GW) is a ripple of the space-time propagating with the speed of light; its existence was predicted by Einstein in his theory of general relativity. It is now very promising that huge laser interferometers which are being constructed all over the world can really detect GWs coming from catastrophic astrophysical events such as coalescence of binary neutron stars and supernova; we can expect the detection within several years. In Japan, there is a project, called TAMA, in which a 300-m interferometer has been constructed and is being improved. The current status of the TAMA project as well as other projects for GW detection is reported.

1 Introduction

A gravitational wave (GW) is a ripple of the space-time propagating with the speed of light; its existence was predicted by Einstein in his theory of general relativity.[1] Since the gravitational interaction is extremely weak, the direct detection of the GW by a detector on the earth has not been realized so far even though the 40-year continuous effort has been devoted. However, it is now very promising that huge laser interferometers which are being constructed all over the world can really detect GWs coming from catastrophic astrophysical events such as coalescence of binary neutron stars and supernova. These interferometers have the sensitivity to detect GWs of the amplitude (h) less than 10^{-21}; the amplitude of the GW is equivalent to the space strain or relative displacement. If we consider the distance between the sun and the earth (1.5×10^{11} m), the induced displacement by the GW of $h = 10^{-21}$ is on the same order of the Bohr radius; this shows the difficulty of the detection of the GW. Since GWs are generated by the coherent motion of a significant amount of mass, the information carried by the GW is quite different from that by electromagnetic waves. Thus, the detection of GWs offers the new information of the Universe; this means the birth of the GW astronomy.

2 Interferometric GW detector

A schematic of an interferometric GW detector is shown in Fig. 1.[2] All of optical elements such as mirrors and a beam splitter are suspended as a pendulum. These objects behave as a free mass for an incident GW of which the frequency is much higher than that of the pendulum resonance. The equivalent displacement induced by the GW (amplitude h with an appropriate polarization) is approximately given by

$$\delta L \sim \frac{1}{2} hL, \tag{1}$$

for each arm (L is the length of the arm). Since the displacements in x and y arms are opposite in sign, owing to the nature of the GW, the signal appearing in the change of the fringe is doubly enhanced. The optimum path length is determined as the turn-around time of the light in the arm is the same as the half period of the GW. For instance, if one wish to detect a GW of 1 kHz, the arm length should be chosen as 75 km. Since it is impossible to find a place for such an interferometer, the light should be instead bounced back and forth, many times between two mirrors set in each arm; this can be achieved by using a Fabry-Perot (FP) cavity. When the cavity is used, the equivalent optical path length is given by $2L\mathcal{F}/\pi$ where L and $\mathcal{F}$ are the real length and the finesse of the cav-

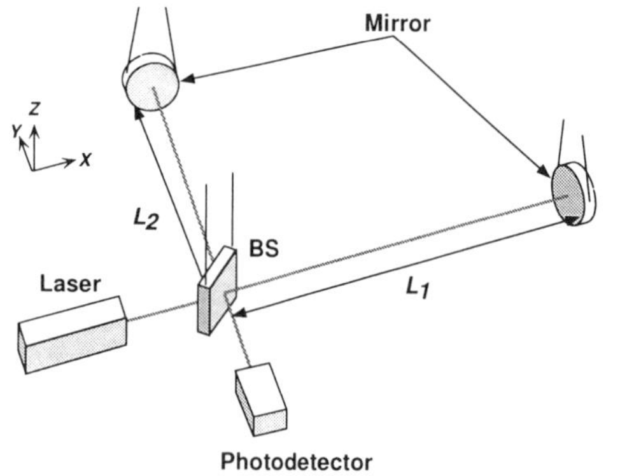

Figure 1. Schematic view of an interferometric gravitational wave detector; this is a Michelson interferometer, where all of the optical elements such as mirrors and a beam splitter are suspended as a pendulum. Both arms have nearly same length ($L_1 \sim L_2 \sim L$).

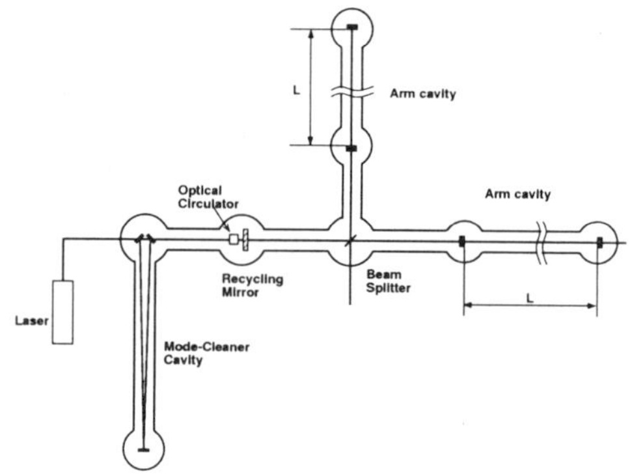

Figure 2. Schematic diagram of the detector; this is a Michelson interferometer where Fabry-Perot cavities are installed in both arms. A stable laser with high output power is used as a light source. A mode cleaner cavity is set between the laser and the interferometer for reducing the distortion of the laser beam.

ity. With high quality mirrors which are now available, it is easy to obtain sufficient path length without loosing the light power.

Figure 2 is a schematic diagram of the detector. A stable laser with high output power is used as a light source. Currently, a laser-diode (LD) pumped Nd:YAG laser is used because of its high efficiency and stability. The power of the laser must be high to reduce the shot noise. For the initial phase of the detector projects, lasers of about 10-W output power have been developed. To obtain the single frequency laser oscillation, injection-locking[3,4] is usually used. Since the fluctuations of the intensity and frequency must be quite small, the elaborated stabilization systems are being developed.[5]

The laser light is introduced to an optical cavity, called mode cleaner (MC), which is used for clearing the spatial mode of the laser light and reducing the beam pointing and geometry fluctuations.[6] The ring-type cavity is used in order to remove the backward light reflected from the MC.

After the MC, the light is led to a main interferometer; this is a huge Michelson interferometer where Fabry-Perot cavities are installed in both arms. The real length of the cavity is 3-4 km in order to realize the sensitivity sufficient for the detection of GWs; the longer cavity requires smaller bounce number thus is less sensitive to the mirror vibration induced by external disturbances or thermal noise force. The interference fringe between the reflected light beams is observed in order to obtain the information on the GW.

There is a mirror, called recycling mirror, between the MC and the main interferometer. This mirror reflects the light coming from the main interferometer; the phase of the reflected light is tuned to coincide with that of the incident light from the laser. By using this mirror, the light power inside the interferometer can be enhanced; this results in the reduction of the shot noise. This scheme is called as power recycling. The enhancement factor in the light power is recycling gain; the recycling gain of 10-100 is expected. If the gain of 100 is achieved, the equivalent power of 1 kW can be obtained with a 10-W laser; this is enough for detecting GW of $h \sim 10^{-21}$.[7,8,9]

3 Noise source of the interferometer

The fundamental noise sources of the detector are as follows: seismic noise ($\sim$ 10Hz), thermal noise (10Hz$\sim$ 100Hz) and shot noise (100Hz$\sim$).

The typical spectrum of seismic noise at the frequency f is considered as[10]

$$\delta x_{\rm seismic} = \frac{10^{-7}}{(f/1\,{\rm Hz})^2}{\rm m}/\sqrt{\rm Hz}. \qquad (2)$$

In order to reduce the noise, there are so many researches on the vibration isolation system.[5]

Thermal noise originates from two parts: pendulum motion and mirror elastic vibration.[11] The noise spectrum can be estimated by means of the fluctuation dissipation theorem with the information on the mechanical loss; the loss can be characterized by the Q-value of the resonance. The thermal noise spectrum at the temperature T is expressed as

$$\delta x_{\rm thermal} \propto \sqrt{\frac{T}{Q}}. \qquad (3)$$

Thus, high-Q material search has been performed so far; silica and sapphire show high Q, larger than 10^7.[12,13,14] The most direct way to reduce the thermal noise is to cool the system down to cryogenic temperature. The test system has shown that the mirror can be cooled to 30 K by heat conduction of sapphire wires.[15]

The shot noise is due to the quantum nature of the light; the number of the incident photons randomly fluctuates with Poisson statistics for the coherent state light. This noise appears at the readout stage of the interferometer. The minimum detectable phase change with the light source of wavelength λ and power P is given by

$$\delta\phi_{\rm shot} = \sqrt{\frac{4\pi\hbar c}{\lambda\eta P}\Delta f}, \qquad (4)$$

where η and Δf are the quantum efficiency at the photon-electron conversion and the bandwidth of the measurement, respectively. Here, c is the speed of light.

4 Big projects

In the world, several projects are going on to construct large GW detectors: LIGO, VIRGO, GEO and TAMA. The LIGO project is to construct two detectors in the US; the sites for the detectors are Hanford, Washington and Livingston, Louisiana. It has been approved in 1991 and started in 1992.[16] These detectors have 4-km long vacuum pipes which have been completed; several interferometers will be installed in these vacuum systems. At the LIGO Hanford Observatory which has a 2-km arm interferometer and 4-km one, the shorter interferometer has been completed and now is being finely tuned to obtain the best performance. LIGO will start the observation at the beginning of 2003. After 4-year observation, the next phase, called LIGO II will be started. For this advanced detector, LIGO scientific collaboration (LSC) has been organized; many research institutes and universities join this forum.

The VIRGO project is the collaborative project between France and Italy to construct an interferometer whose arm length is 3km; the site is in Pisa, Italy.[17] The outstanding feature of this detector is the design for detecting a GW of very low frequencies around 10Hz. In order to enhance the sensitivity at such a frequency, an elaborated vibration isolation system is under development in Pisa. The French group is responsible for the optical system design, as well as for the development of low-loss optics and a stabilized laser of high power. The construction of the vacuum system will be finished at the middle of 2002 and the start of the operation is scheduled in 2003.

GEO is the German and UK collaboration which is constructing a 600-m long interferometer at Hannover in Germany.[18] The project has adopted a particular optical design, called dual recycling[19] and introduced advanced technologies for achieving the similar sensitivity of larger ones. The detector will be completed in 2001 and will perform a test run; this is planned as a coincidence run with LIGO.

TAMA is the Japanese project and described in detail in the next section.[a]

5 TAMA project

TAMA is the project to construct a 300-m FP-type detector (named TAMA300) at the Mitaka campus of national astronomical observatory (NAO); this project started in 1995 and will continue to the end of March, 2002. The purpose of this project is to show the feasibility for the full scale interferometer. Since the size of the detector is small, the attainable sensitivity is worse than those of other detectors. However, we decided that the required specifications should be same as those for full-scale interferometers. The continuous operation is also planned to show the fact that the interferometer can really work for a long time as an observatory.

The design of TAMA300 is based on that described in the previous section. As a light source, we used an LD-pumped Nd:YAG laser with injection locking; its output power was 10W, obtained by an end-pumping scheme.[4] The mirrors were made of monolithic synthetic silica; the surface for the reflection was super-polished and coated by ion-beam sputtering.[20,21] These technologies are indispensable for the interferometric gravitational wave detectors. The vacuum system consisted of chambers, ducts and pumping systems. The inner surface of the chambers and the ducts were treated by electro-chemical buffing; the pressure could be kept on the order of 10^{-6} Pa without baking.[22]

The mirrors and other critical optical components were suspended as a double pendulum with an eddy-current damping system.[23] They were set on vibration isolation stacks; the active vibration isolation system which supports the stack has been introduced to obtain further vibration isolation.

There were many control systems for the length and alignment of the cavities, the fringe point of the Michelson interferometer, the frequency and intensity of the laser and so on as shown in Fig 3. The most significant point is to reduce the fluctuation without introducing any extra noises. Now, a computer-controlled locking system is installed for long term operations.

TAMA300 is now the only one interferometer which can be used for observations. Since the completion of the Phase I system (without recycling) in 1999, the various improvements have been done. Figure 4 shows the latest sensitivity with the history of the improvement. The highest one is $5 \times 10^{-21} /\sqrt{\text{Hz}}$ at 700 Hz; this is the best sensitivity in the world.[24] The operation is now quite stable; the interferometer can be operated for one day without loosing the locked state. We have performed the data taking run (DTR) several times.[24,25] The data-recorded time was longer than 300 hours from the first DTR to the fifth DTR. The sixth DTR is being performed from 1st August 2001 to 20th September 2001; data for 1000 hours will be acquired during this DTR.

6 Future plan in Japan

Since the scale of TAMA is small, we cannot expect the detection of extra-Galactic GWs; for the GW astronomy in Japan, we need to construct the full-scale interferometer. Thus, we are now proposing the project, called LCGT;[26] in this project we plan to construct a 3-km interferometer in Kamioka mine where the seismic noise is quite small

[a] TAMA Web page is http://tamago.mtk.nao.ac.jp.

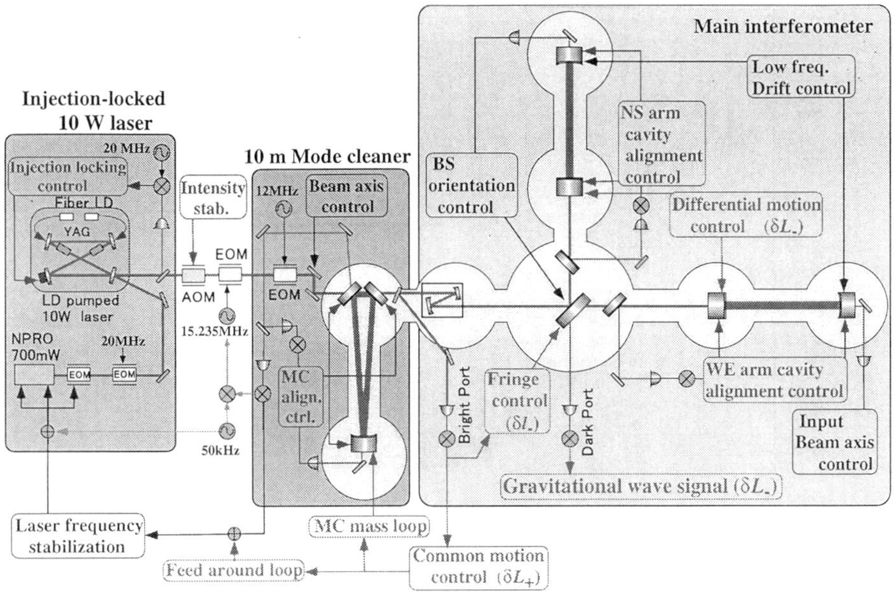

Figure 3. Schematic diagram of the control system of TAMA300. There are many control systems for the length and alignment of the cavities, the fringe point of the Michelson interferometer, the frequency and intensity of the laser.

because of the underground site. We are also willing to introduce a cryogenic technique in order to reduce the effect due to the thermal noises. Although the budget for the project has not been approved, the R&D projects are progressed. A test facility has been completed and several experiments are going on at the campus of Institute of Cosmic Ray Research (ICRR), University of Tokyo. Also, the 20-m prototype interferometer which was built at NAO[9,21] has been moved to Kamioka mine and is being operated with improved sensitivity.

7 Summary

Within a few years, several interferometers will be operated for detecting GWs. We are strongly expecting that the gravitational wave extends our understanding of the Universe to its horizon. The detection of the GW is no longer a dream. This is just about to appear in front of us.

Acknowledgments

The authors would appreciate Koji Arai and Masaki Ando for offering the figures in this article.

References

1. Kip Thorne in *300 years of Gravitation*, eds. S. W. Hawking and W. Israel (Cambridge Univ. Press, Cambridge, 1987).
2. A. Giazotto, *Phys. Rep.* **182**, 365 (1989).
3. R. Barillet et al., *Meas. Sci. Technol.* **7**, 162 (1996).
4. S. T. Yang et al., *Opt. Lett.* **21** 1676 (1996).
5. see articles in *Gravitational Wave Detection II* eds. S. Kawamura and N. Mio (Universal Academy Press, Tokyo, 2000)
6. Rüdiger et al., *Opt. Acta*, **28**, 641 (1981).
7. D. Schnier et al., *Phys. Lett.* A **225**, 210

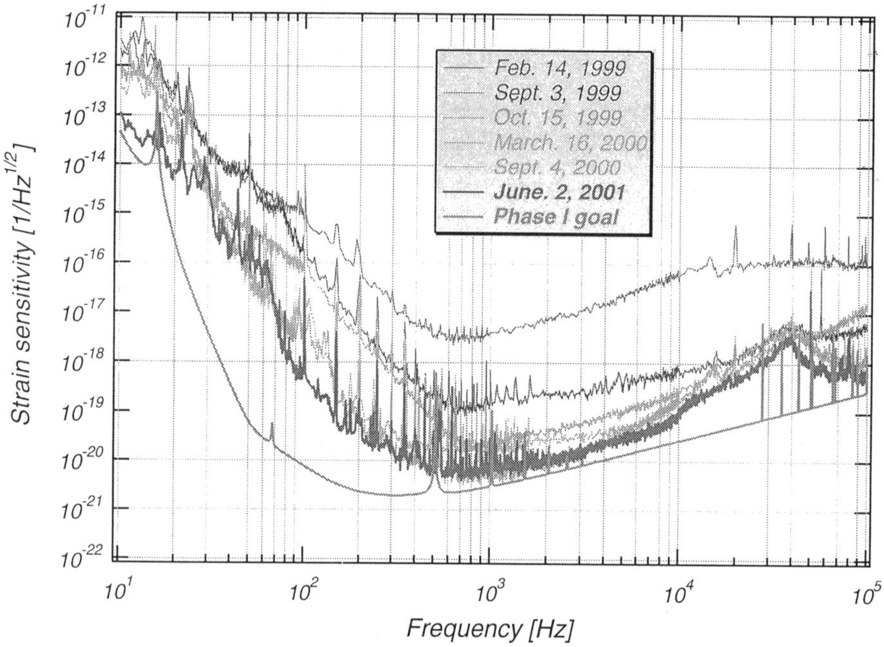

Figure 4. Latest sensitivity of TAMA300. The noise spectrum measured by TAMA300 is expressed as the sensitivity for the amplitude of the gravitational wave. The history of the improvement is also shown. Within two years, the noise level has been reduced by two orders of magnitude.

(1997).

8. P. Fritschel et al., *Phys. Rev. Lett.* **80**, 3181 (1998).
9. S. Sato et al., *Appl. Opt.-OT* **38**, 4616 (2000).
10. D. Shoemaker et al., *Phys. Rev.* D **38**, 423 (1988).
11. P. R. Saulson, *Phys. Rev.* D **42**, 2437 (1990).
12. T. Uchiyama et al., *Phys. Lett.* A **261**, 5 (1999).
13. K. Numata et al., *Phys. Lett.* A **276**, 37 (2000).
14. K. Numata et al., *Phys. Lett.* A **284**, 162 (2001).
15. T. Uchiyama et al., *Phys. Lett.* A **242**, 211 (1998).
16. A. Abramovici et al, *Science* **256**, 325 (1992); http://www.ligo.caltech.edu.
17. E. Majonara, in *Gravitational Wave Detection II* eds. S. Kawamura and N. Mio (Universal Academy Press, Tokyo, 2000) p.15; http://www.virgo.infn.it.
18. B. Willke et al., in *Gravitational Wave Detection II* eds. S. Kawamura and N. Mio (Universal Academy Press, Tokyo, 2000) p.25; http://www.geo600.uni-hannover.de.
19. G. Heinzel et al., *Phys. Rev. Lett* **81**, 5493 (1998).
20. A. Ueda at al., *Opt. Rev.* **3**, 369 (1996).
21. S. Sato et al., *Appl. Opt.-OT* **38**, 2880 (1999).
22. Y. Saito et al., *Vacuum* **53**, 353 (1999).
23. K. Tsubono et al., *Rev. Sci. Instrum.* **64**, 2237 (1993).
24. M. Ando et al., *Phys. Rev. Lett.* **86**, 3950 (2001).
25. H. Tagoshi et al., Phys. Rev. D **63**, 062001 (2001).
26. K. Kuroda et al., *Int. J. Mod. Phys.* D **8**, 557 (1999) .

OBSERVATION OF HANBURY BROWN-TWISS CORRELATIONS FOR FREE ELECTRONS

HARALD KIESEL AND FRANZ HASSELBACH

Institut für Angewandte Physik der Universität Tübingen
Auf der Morgenstelle 10, D-72076 Tübingen, Germany
E-mail: franz.hasselbach@uni-tuebingen.de

The astronomers Hanbury Brown and Twiss (HBT) were the first to observe in 1956 that the fluctuations in the counting rate of photons originating from uncorrelated point sources become, within the coherently illuminated area, slightly enhanced compared to a random sequence of classical particles[1,2,3]. This at a first glance mysterious formation of correlations in the process of propagation turns out to be a consequence of quantum interference between two indistinguishable photons and Bose-Einstein statistics[4]. The latter requires that the composite wave function is a symmetrized superposition of the two possible paths. For fermions, by virtue of the Pauli principle, no two particles are allowed to be in the same state. The corresponding antisymmetrized two particle wave function excludes overlapping wave trains, i.e., simultaneous arrivals of two fermions at contiguous, coherently illuminated detectors are forbidden. These anticorrelations have been observed for the first time for a beam of *free* electrons in spite of the low mean number (degeneracy) of $\sim 10^{-4}$ electrons per cell in phase space. With respect to the low degeneracy, our experiment is the fermionic twin of HBT's experiment with photons.

1 Introduction

Correlations between successive detections of bosons resp. fermions are observed even if the emission of the two particles cannot physically influence one another because, e.g., the emission sites have a space-like separation. In such a radiation field correlation builds up in the process of propagation. At points sufficienly far from the sources the field becomes highly correlated. This obervation of HBT in 1956 seemed so strange to a considerable portion of the physics community that they declared it as physically absurd[a], although the van Cittert Zernike theorem of partial coherence which expresses the field correlations (coherence) at two points in an optical field emerging from incoherent planar sources was formulated nearly two decades before.

In essence, HBT have shown that measuring field correlations (2nd order coherence) provides another — apart from observing interference fringes — sometimes more handy and/or powerful method for measuring the absolute value of the degree of coherence. E.g., with their stellar *intensity* interferometer HBT were able to measure the diameter of Sirius, undisturbed by fluctuations of the refractive index of the atmosphere[5]. The reason why (anti)correlations between *free* photons (electrons) in coherent beams are so difficult to observe and why the fundamentally different behaviour of free bosons and fermions does not at all become obvious in light and electron optics — in spite of the extreme difference between Bose-Einstein and Fermi-Dirac statistics[b] — is due to the very low occupation numbers (degeneracy) in phase space. Only recently, with the advent of high brightness field electron emitters with degeneracies reaching 10^{-4} an experiment for free fermions seemed to become feasible[6,7]. The first fermionic HBT experiments in semiconductor devices where *highly degenerate* beams of electrons are conveniently available[8,9] have been successfully performed in 1999.

[a] A lively description of the situation at that time is given in R. Hanbury Brown, 'The Intensity Interferometer', Taylor and Francis, New York 1974, p. 7.

[b] While any number of photons is allowed in one cell, for electrons by Pauli's exclusion principle the occupation of a cell is one resp. two for antiparallel spins.

2 Experimental procedure

The coincidence method chosen in the present 'antibunching' experiment is an alternative to HBT's correlation procedure for gathering information about correlations in a stream of particles: Two detectors are coherently illuminated by an electron field emitter. According to the Pauli principle no two electrons are allowed to be in the same quantum state, i.e., to arrive at both detectors simultaneously[c]. In other words, if our detectors had a time resolution corresponding to the coherence time T_c which is on the order of 10^{-14} s, no coincidences would be observed. The time resolution of our fast coincidence counter of T_r=26 ps was about three orders of magnitude less. For incoherent illumination of the detectors the electrons behave like classical particles. Due to the insufficient time resolution a certain random coincidence rate is observed. When we change the illumination from incoherent to coherent we expect a reduction of the random coincidences (by a factor T_c/T_r of about 10^{-3} in the present experiment) due to the fact that within the first 10^{-14} s after arrival of an electron no second one is allowed to arrive. This reduction is a signature of antibunching.

3 Experimental set-up

Our experimental set-up corresponds to HBT's stellar interferometer: The tiny effective virtual source of an electron field emitter illuminates via magnifying quadrupoles two small collectors (Fig. 1a, right hand side, rhs). With increasing magnification the effective lateral distance of the collectors decreases and their illumination changes from incoherent to totally coherent. In turn a continuous increase of anticorrelations of arrival times of the electrons is expected.

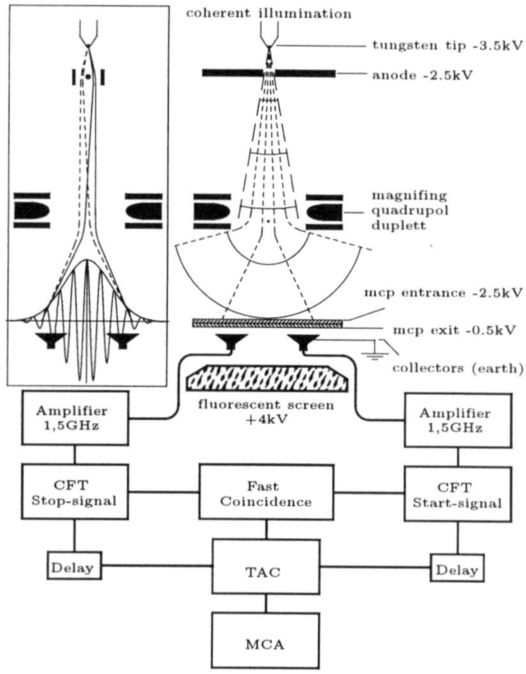

Figure 1. Electron optical set-up (top) and fast coincidence electronics (bottom) to measure electron anticorrelations. The spherical segments emerging from the cathode represent single coherence volumina. Between electron source and quadrupole a biprism (inset on the left) is inserted temporarily to check the coherence of illumination of the collectors.

The quadrupoles produce an elliptically shaped beam of coherent electrons. For geometrical reasons less coherent electrons are missing the collectors compared to stigmatic magnification. This reduces the measuring time T_M largely. The collectors, 4 mm in diameter (impedance 50 Ohms) inserted between the exit of the two cascaded channel plates (mcp) and the fluorescent screen, collect the electron avalanches initiated by single electrons. Into the actual electron optical set-up[d] an electron biprism (inset on the left hand side, lhs, of Fig. 1) is integrated. It allows to examine the state of coherence of illumination of the collectors[e] by observing the overlap of the fringes with the shadows of

[c]The fluctuations in orthogonal polarizations resp. spin directions are independent. Therefore, when the experiment is performed with unpolarized light/electron beams, enhancement resp. suppression of the extra fluctuations is reduced by a factor 1/2 compared to polarized beams.

[d]Its construction principles may be found, e.g., in [10].
[e]At least in the direction perpendicular to the fringes.

the collectors on the fluorescent screen. By this means the magnification factors which are necessary for coherent, partially coherent and incoherent illumination of the collectors are determined. The antibunching experiments are performed without biprism in the beam path at these predetermined magnifications.

The very short electron avalanches leaving the channel plates (rise time and width ~.5 ns) are transferred coaxially from the collectors via microwave amplifiers (bandwidth 1.5 GHz) to modified Constant Fraction Trigger (CFT) modules which extract timing signals with low time-jitter and -walk. A first coincidence circuit preselects events within a time window of ±3 ns and opens the gate of a Time to Amplitude Converter (TAC). The time spectra are accumulated by a Multi Channel Analyzer (MCA). By an additional delay of 3 ns in the stop channel of the time to amplitude converter, arrival time zero shows up in the center of the time spectra. Stop signals arriving prior to the start signal are displayed on the negative time axis.

The emission of electrons is a Poissonian process, therefore — for incoherent illumination — the probability of arrival of an electron after an elapsed time $\tau = t_0 - t_1$ is given by $P(\tau) = e^{-\overline{n}\tau}$ where $\overline{n}$ is the average counting rate. I.e., given a start signal at time t_0, the probability for the next stop signal at t_1 depends exponentially on the counting rate $\overline{n}_{\text{stop}}$ in the stop channel.

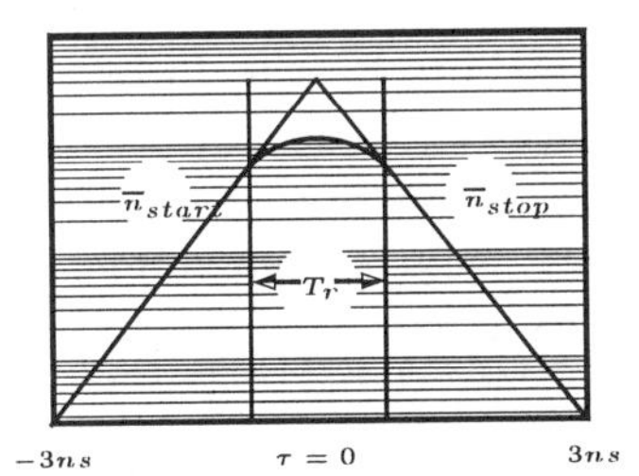

Figure 2. Time spectra in semi-logarithmic representation expected for Poissonian processes, see text.

In semi-logarithmic representation this probability is represented by a straight line with a slope given by $-\overline{n}_{\text{stop}}$ (Fig. 2, rhs). Correspondingly, stop signals arriving prior to the start signal are displayed on the negative time axis with a slope of $\overline{n}_{\text{start}}$ (Fig. 2, lhs). The peak of the real time spectrum is rounded within the resolution time window T_r of the fast coincidence.

Features of our cold <100> oriented tungsten field emitter are: Extraction voltage 900V, total current 1.5μA, energy width ΔE_{FWHM} of .3eV which corresponds to a standard deviation ΔE of .13eV (calculated under the assumption of a Gaussian energy distribution), virtual source diameter ~36nm, brightness $4.4 \cdot 10^7 \frac{\text{A}}{\text{cm}^2 \cdot \text{sr}}$, coherence time T_c=3.25·10^{-14}s, coherent particle current 4.7·$10^9$1/s, degeneracy 1.6·10^{-4}.
In spite of a vacuum of 10^{-10}mbar and a constant emission current during the data accumulation times T_{M} (see Table 1), the counting rates of the detectors sometimes increased by more than a factor of 2 or fell below .5 of the desired value. In these cases a reduction resp. an increase in coherence of illumination of the collectors took place. Therefore, time spectra with counting rates outside these limits were discarded. $T_{\text{M}_{\text{eff}}}$ is the data accumulation time resulting after subtraction of times of unstable emission.

4 Results

In order to prove antibunching a total of four spectra were accumulated for about 30 hrs each. The first for incoherent illumination of the detectors, the following for partially, totally and the last again for partially coherent illumination. The evaluation procedure of the experimental time spectra consisted of smoothing and normalizing the spectra followed by visually superimposing the incoherent with the coherent resp. partially coherent spectra. The results are summarized in Fig. 3 and Table 1. The antibunching signal S, i.e., the missing coincidences ΔN in the coherent and partially coherent versus the incoherent spectrum becomes visible as a flattening of the peak in Fig. 3 within the time resolution window of ± 13 ps. The measured relative reduction in coincidences amounts to $S_{rel} = 1.26 \cdot 10^{-3}$ with a signal to noise ratio S/N of 3. As expected, the reduction in coincidence rate and the signal to noise ratio are smaller for partially coherent illumination.

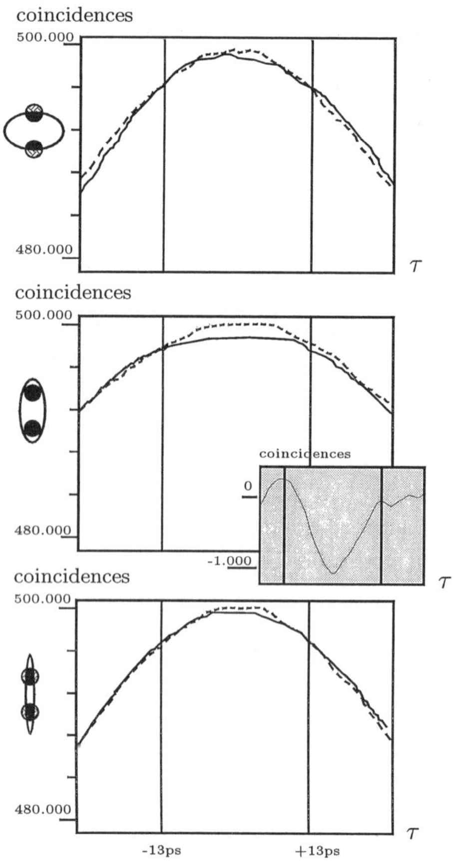

Figure 3. Antibunching as a function of coherence of illumination of the collectors. The coincidence rate for incoherent illumination (dashed lines) is compared to that for partially coherent on top, coherent in the middle and again partially coherent illumination on the bottom (full lines). In the inset $N_{coh.} - N_{incoh.}$ is given. In the pictograms the ellipses represent the coherently illuminated areas, the circles the collector areas. The coherently illuminated parts of the collectors are marked in black.

5 Conclusion

Antibunching or — in the parlance of interferometry — interference between a system

Table 1. Summary of results. S, S_{rel} are the absolute, resp. relative coincidence reduction, S/N the signal to noise ratio, T_M the total measuring time and $T_{M_{eff}}$ the effective measuring time after subtraction of times of instable emission of the field emitter.

	illumination		
	part. coh.	coh.	part. coh.
$S = \Delta N$	4.567	16.942	8.292
$S_{rel}/10^{-3}$	0.61	1.26	0.34
S/N	2.2	3.0	1.4
T_M/min	1402	1690	4531
$T_{M_{eff}}$/min	1402	1642	4450

consisting of two particles resp. second order coherence has been observed for *massive free fermions* for the first time. The experimental technique opens a gateway to new fundamental tests of quantum mechanics and statistics, e.g., observation of quantum statistics on interference phenomena and experimental tests of interaction of fields and potentials with charged two-fermion systems[6].

Acknowledgement

We thank the Deutsche Forschungsgemeinschaft for financial support.

References

1. R. Hanbury Brown, R.G. Twiss, Phil. Mag. **45** (1954), 663.
2. R. Hanbury Brown, R.Q. Twiss, Nature **177** (1956), 27.
3. R.Q. Twiss, A.G. Little, R. H. Brown, Nature **178** (1957), 1447.
4. E.M. Purcell, Nature **178** (1956), 1449.
5. R. Hanbury Brown, Twiss R.Q., Nature **178** (1958), 1046.
6. M.P. Silverman, Phys. Lett. **A120** (1987), 442-446 and references therein.
7. T. Kodma et al., Phys. Rev. **A57**(1998), 2781.
8. M. Henny et al., Science **284** (1999), 296.
9. W.D. Oliver, J. Kim, C. Liu, Y. Yamamoto, Science **284**(1999), 299.
10. F. Hasselbach, Z. Phys. B – Condensed Matter – **71** (1988), 152.

JAMIN-TYPE INTERFEROMETER FOR COLD NEUTRONS USING ETALON PLATES

M. KITAGUCHI*, H. FUNAHASHI,

Department of Physics, Kyoto Univ., Kyoto 606-8502, Japan

**The Institute for Physical and Chemical Research(RIKEN), 2-1 Hirosawa, Wako, Saitama 351-0198, Japan*

E-mail: kitaguch@nh.scphys.kyoto-u.ac.jp, hal@nh.scphys.kyoto-u.ac.jp

AND M. HINO

Research Reactor Institute, Kyoto Univ., Kumatori, Osaka 590-0494,Japan

E-mail: hino@rri.kyoto-u.ac.jp

The authors have proposed and tested a new type of multilayer cold-neutron interferometer based on a pair of etalons. The range of experimental application of conventional multilayer cold-neutron interferometer was limited due to the small spatial separation between the two coherent beams. Using etalons with an air gap of 20μm in spacing we have observed interference fringes with the contrast of $(3.5 \pm 0.7)\%$. The present results have demonstrated the feasibility of developing a cold neutron interferometer with a large path separation to carry out high precision measurements and new types of experiment.

1 Introduction

Neutron interferometry is a powerful technique for studying fundamental physics. Generally, an interferometer consists of a splitter, a phase shifter, and an analyzer (Figure 1). A splitter divides the incident wave into two coherent components, a phase shifter produces the relative phase between the two components, and an analyzer recombines them. Interference fringes are given as a function of the relative phase. The relative phase is written as

$$\Delta\phi = 2\pi\frac{m\lambda L}{h^2}\Delta E, \tag{1}$$

where m is neutron mass, λ is neutron wavelength, L is interaction path length, and ΔE is energy difference between the two beams. A large dimensional interferometer for long wavelength neutrons has the advantage to increase the sensitivity to small interactions ΔE. Though such a kind of interferometer was realized by using multilayer mirrors[1] , the beam separation was extremely small. The aim of our development is to increase the spatial beam separation of a multilayer interferometer in order to broaden the applicability of neutron interferometry.

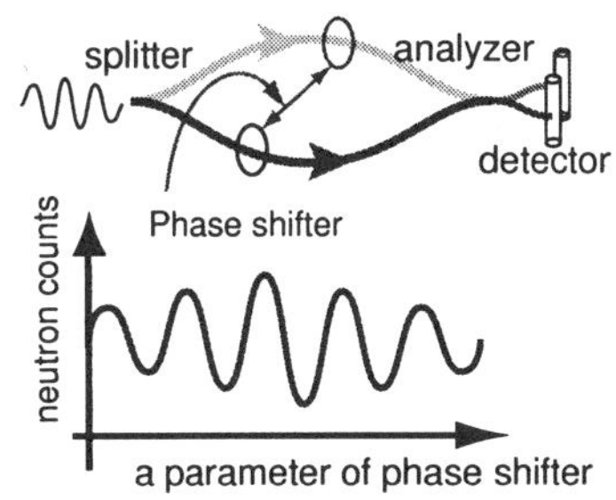

Figure 1. The concept of an interferometer. Neutron counts oscillate with a parameter of the phase shifter.

2 Cold neutron interferometer using etalon plates

2.1 Neutron spin interferometer

We have an improve *neutron spin interferometer* which is based on the cold neutron interferometer using multilayer mirrors [1] and neutron spin echo method [2]. The spin interferometer enables us to carry out high precision experiments due to its high contrast [3]. It contains a pair of *spin splitters*. The spin splitter is a multilayer which consists of a magnetic mirror on top, a gap layer, and a non-magnetic mirror. The spin splitter separates spatially up and down spin components into two parallel waves. The waves superpose each other spatially on the second spin splitter. Magnetic field provided by phase-shifter coil gives the relative phase between the two waves. There are some remarkable experiments using the neutron spin interferometer, for example, double Stern-Gerlach experiments [4] and delayed choice experiments [5]. Because the gap layer of a conventional spin splitter is fabricated with vacuum evaporation method, the gap is not thick enough to separate the two beams spatially. The range of application of the cold-neutron spin interferometer is limited due to the small beam separation. A cold neutron interferometer with large spatial separation enables us to carry out high precision measurements and new types of experiment. For example, we can insert some devices into the gap between paths of the interferometer. We can also measure more precisely the interaction through which the relative phase $\Delta\phi$ in eq.(1) depends on the area enclosed by beam path.

2.2 Spin splitters based on etalons

We used *etalons* in order to enlarge the spatial separation between two parallel waves in neutron spin interferometer. An etalon consists of two parallel planes which are very smooth. We can purchase special etalons with planes smooth enough to be used as substrates of neutron mirrors (RMS roughness is less than 3Å). By depositing a magnetic mirror and a non-magnetic mirror on the parallel planes of an etalon we can produce a spin splitter which provide a large separation of the two beams as shown in Fig. 2. Figure 3 shows photograph of the present etalons. The etalon spacing is 20μm . The mono-layer of Permalloy45 ($Fe_{55}Ni_{45}$) with 800Å thickness and the mono-layer of nickel with 800Å thickness are made as a

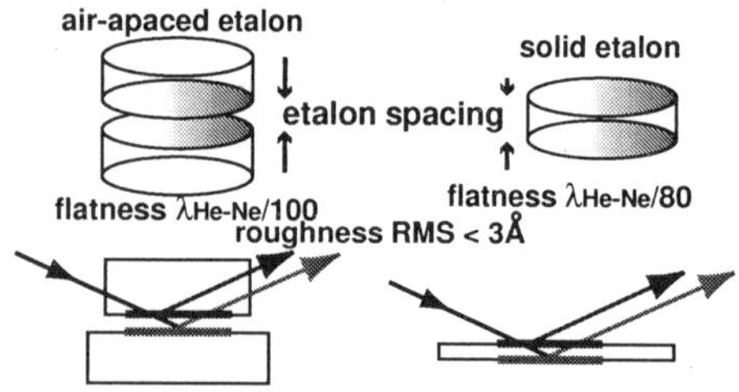

Figure 2. Etalon. Etalons with neutron mirrors can be used as spin splitters.

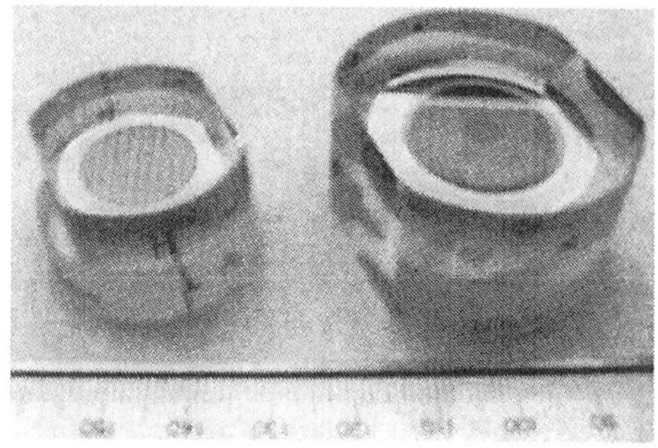

Figure 3. Etalons. The first etalon(left) has a diameter of 30mm. Clear aperture is 20mm in diameter. The second etalon(right) has a diameter of 42mm. Clear aperture is 30mm in diameter. These etalons have air gap spacing of 20μm.

magnetic mirror and non-magnetic mirror respectively. Figure 4(a) shows the reflectivity of these mirrors by numerical simulation according to the optical potential model [6]. In this simulation the nuclear and magnetic potentials for Permalloy45 are 220neV and 96.5neV and the nuclear potentials for nickel is 243.4neV. At an appropriate incident angle, the magnetic mirror reflects only the spin up component and the non-magnetic mirror reflects the spin down component which is transmitted through the magnetic mirror. Figure 4(b) shows the measurement of reflectivity for up and down spin neutron from the etalons. Figure 4 shows that 0.9 degree is a suitable incident angle.

2.3 Experimental setup

Figure 5 illustrates the experimental setup. The experiment was performed using the cold neutron beam line 'MINE' at the JRR-3M reactor in JAERI. The beam had a wavelength

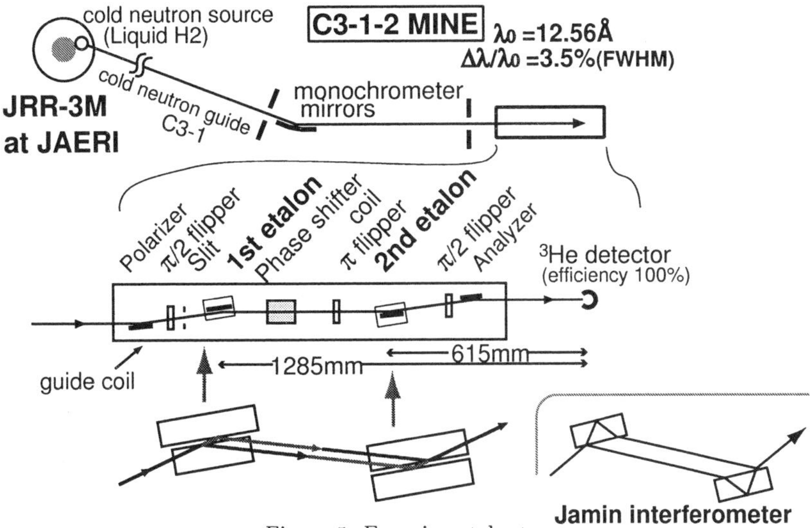

Figure 5. Experimental setup.

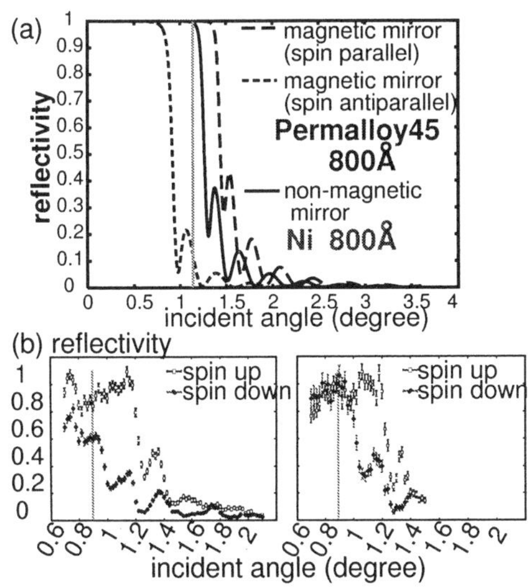

Figure 4. (a) The reflectivity of mirrors on etalons by numerical simulation. At an appropriate incident angle indicated by gray line, the layer of Permalloy45 reflects only spin up component and the layer of nickel reflects transmitted spin down component. (b) The measurement of reflectivity for up and down spin neutron from the etalons.

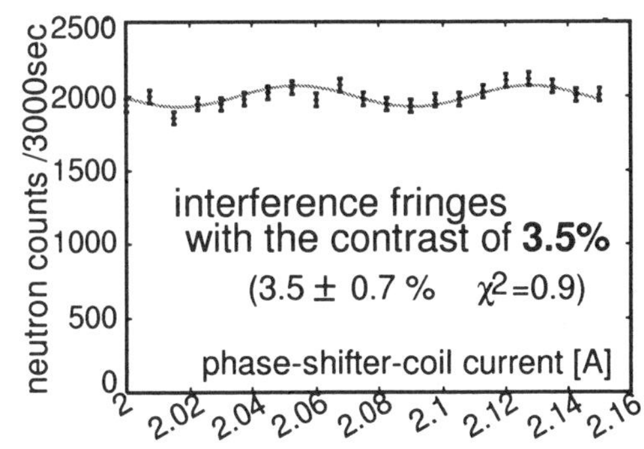

Figure 6. Experimental result. We have observed the interference fringes with the contrast of $(3.5 \pm 0.7)\%$.

of 12.6Å and a bandwidth of 8.5% in FWHM. In the viewpoint of geometrical optics the pair of etalons is equivalent to a *Jamin interferometer* , which is the oldest type of interferometer of visible light.

2.4 Results and discussion

We have observed interference fringes with the contrast of $(3.5 \pm 0.7)\%$ (Figure 6). The reduced χ^2 value of least-square fit was 0.9. The results of five experimental runs were consistent with each other. The interference fringes shown in Fig.6 is a sum of all five runs.

There are three main reasons for the loss of contrast. The first is finite beam polarization, which limits a maximum contrast. It measured 90% without the pair of etalons. The second is insufficient reflectivity of the non-magnetic mirror of the first etalon. The low reflectivity makes the unbalance of the intensity between the two beams, which decreases the contrast of fringes. The decreased contrast from the maximum contrast of 90% was estimated at 87% . The third is misalignment concerning the tilting angle of the second etalon from the first etalon. For the dispersion of wavenumber of beam σ_k and the spatial dislocation L between the superposed two optical paths, the contrast of interference fringes is given by [7]

$$\exp(-\frac{1}{2}(\sigma_k L)^2). \tag{2}$$

The increase of the spatial dislocation decreases the contrast. The tilting angle of the second etalon causes the transverse dislocation. The beam divergence of 0.23 degree composed by the experimental collimation determined the transverse-wavenumber dispersion σ_k of 0.002Å^{-1} and gave the decrease of the contrast by using eq.(2). We did not have a tilting-angle control technique in this experiment. The the tilting angle of 0.17 degree and the unbalanced reflectivity can be given to explain the decrease of the contrast to 3.5% from the maximum contrast of 90% . Recently, we have established a new alignment technique using a survey instrument based on a laser. Using this technique the tilting angle is adjusted within 0.025 degree, which will enable us to observe contrast of 15% even with the etalon spacing of 100μm.

3 Summary

The present results have demonstrated the feasibility of development of a large dimensional interferometer with a large spatial path separation for long-wavelength neutrons using etalons. We are continuing on the test experiments using a new alignment technique and new air-spaced etalons with etalon spacing of 20μm and 100μm thickness. We plan to increase the etalon air gap spacing up to 1mm.

The enlargement of path separation enables us to carry out high precision measurements and new types of experiment which have never been accomplished by other interferometers developed so far. We also plan some applications of the present type of interferometer, for example, high precision measurement of the topological Aharonov-Casher effect [8] and a gravitationally-induced quantum interference experiment [9].

Acknowledgments

The authors gratefully acknowledge fruitful discussions with Prof. T. Ebisawa, Prof. S. Tasaki, and Prof. H. M. Shimizu. This work was supported by the inter-university program for common use JAERI and KUR, and financially by a Grant-in-Aid for Scientific Research (No.12740149). One of the authors (M.K.) was supported by the Junior Research Associates Program of RIKEN.

References

1. H.Funahashi et al., *Phys. Rev.* A **54**, 649 (1996).
2. F. Mezei, *Z. Phys.* **255**, 146 (1972).
3. T. Ebisawa et al., *Phys. Rev.* A **57**, 4720 (1998).
4. Y. Otake et al., *Proc. Int. Symp. Foundations of Quantum Mechanics* (ISQM-Tokyo '98) pp.323-326 (1999).
5. T. Kawai et al., *Nucl. Inst. and Meth.* A **410**, 259 (1998)
6. S. Yamada et al., *Annu. Rep. Res. Reactor Inst. Kyoto Univ.* 11, 8 (1978)
7. Y. Morikawa et al., *Il Nuovo Cimento* vol. 105 B, N. 5, 507 (1990).
8. A. Cimmino et al., *Nucl. Inst. and Meth.* A **440**, 579 (2000).
9. S. A. Werner et al., *Physica* B **151**, 22 (1988); K. C. Littrell et al., *Phys. Rev.* A **56**, 1767 (1997).

PROPOSITION OF A NANOSCALE COLLIDER EXPERIMENT TO EXAMINE THE LOSS OF VISIBILITY DUE TO ENTANGLEMENT

KOUICHI TOYOSHIMA*, TAKASI ENDO AND YUTAKA HIRAYOSHI

Faculty of Science and Engineering, University of Saga, Saga 840-8502, Japan
E-mail: toyo@cc.saga-u.ac.jp

Loss of interference takes place due to entanglement and resembles the collapse of the wave packet. We propose a simple collision experiment to examine this effect with apparatuses already available.

1 Loss of visibility due to entanglement

Entanglement is one of the key ideas of the modern quantum mechanics, such as quantum teleportation. We will remind you that it may reveal another interesting phenomena, lack of interference of particles from a coherent source, and we claim that it can be experimentally verified with existing techniques.

It is believed that the coherence of the matter wave arises if (1) the source size is small and (2) variation of the energy is small. Coherence brings about high fringe visibility when interfered in some way. However, the particle beam does not exhibit interference at all if they undergo an interaction with another particle before, making an entangled pair with it [1]. It doesn't matter however small the size of this interaction region and the energy spread of the emerging particles are. Here we obtain an exotic particle beam which shows no apparent wavelike behavior.

Suppose the particles 1, 2 in states a, $\bar{\mathrm{a}}$ and b, $\bar{\mathrm{b}}$ are coupled as

$$|\Psi(\boldsymbol{r}_1,\boldsymbol{r}_2)\rangle = |\mathrm{a}\rangle_1|\bar{\mathrm{a}}\rangle_2 + |\mathrm{b}\rangle_1|\bar{\mathrm{b}}\rangle_2 \,.$$

Then, if we observe only the particle 1, we must calculate $\langle\Psi|\Psi\rangle$ with integrating over the position of the particle 2.

$$\begin{aligned}&\langle\Psi(\boldsymbol{r}_1)|\Psi(\boldsymbol{r}_1)\rangle\\ &= |\phi_{\mathrm{a}}(\boldsymbol{r}_1)|^2\,\langle\bar{\mathrm{a}}\,|\,\bar{\mathrm{a}}\rangle + |\phi_{\mathrm{b}}(\boldsymbol{r}_1)|^2\,\langle\bar{\mathrm{b}}\,|\,\bar{\mathrm{b}}\rangle\\ &\quad + 2\mathrm{Re}\,[\,\phi_{\mathrm{a}}^*(\boldsymbol{r}_1)\phi_{\mathrm{b}}(\boldsymbol{r}_1)\langle\bar{\mathrm{a}}\,|\,\bar{\mathrm{b}}\rangle\,] \qquad (1)\end{aligned}$$

where ϕ_{a} or ϕ_{b} is a single particle wavefunction for particle 1. We assume the orthonormality of the wavefunction of the particle 2,

$$\langle\bar{\mathrm{a}}\,|\,\bar{\mathrm{a}}\rangle = \langle\bar{\mathrm{b}}\,|\,\bar{\mathrm{b}}\rangle = 1 \text{ and } \langle\bar{\mathrm{a}}\,|\,\bar{\mathrm{b}}\rangle = 0\,.$$

So that, $\langle\Psi(\boldsymbol{r}_1)|\Psi(\boldsymbol{r}_1)\rangle = |\phi_{\mathrm{a}}(\boldsymbol{r}_1)|^2+|\phi_{\mathrm{b}}(\boldsymbol{r}_1)|^2$ with no interference term.

Such a pair can be prepared as a momentum-entangled pair of electrons simply by a collision, as shown in Fig. 1. If we are on the center-of-mass system and the collision region can be limited to a very small volume, then we will have a very strange particle source which should have the temporal and spatial coherence but shows no interference effect at all.

2 Experimental feasibility to obtain non-interfering coherent beam

The above condition can be realized by colliding the thin electron beams of same energy at a nearly head-on geometry. If we take 180 degree crossing angle we have no energy spreading over the scattering angle, but at this geometry the collision volume cannot be defined, since we assume DC beams. So we take several degrees less than 2π for the crossing angle. Main cause of the energy spreading of the scattered particles is the spread of angles of the incoming beams. To minimize this effect, the beams from the cathodes are truncated within small solid angles, and the observation angle is taken as close as possible to the both beam axes. Mölenstedt biprism

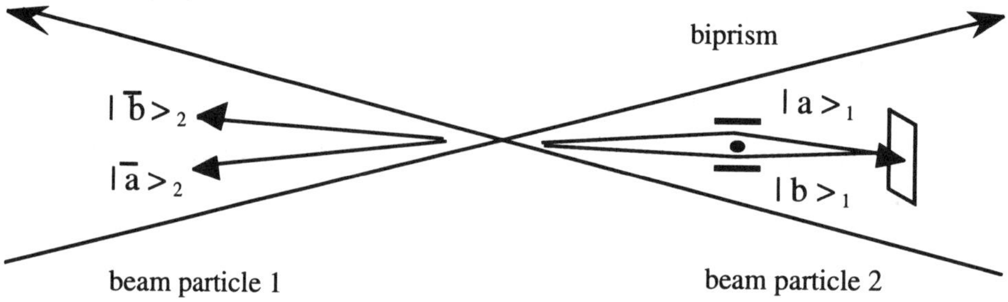

Figure 1. Proposed experimental arrangement.

is used as a sensitive interferometer.

We denote path lengths in Fig.2 as follows:

$$l_1 = \overline{A_1P_1Q}, \quad l_2 = \overline{A_1P_2Q}$$

$$l_1' = \overline{A_2P_1Q}, \quad l_2' = \overline{A_2P_2Q}$$

$$\Delta l_1 = l_1 - l_1', \quad \Delta l_2 = l_2 - l_2'$$

To ensure that the loss of visibility or interference is not due to the spreads of energy or of source volume, $\Delta l = |l_1 - l_2|$ and $\Delta l' = |l_1' - l_2'|$ should be smaller than the longitudinal coherence length, and $\Delta' l = \Delta l_1 - \Delta l_2$ should be small so that the particles originated from different points of the source, e.g. A_1 and A_2 in Fig. 2, should undergo nearly the same phase differences between the two paths around the biprism. These requirements are expressed as follows

$$\Delta k\, \Delta l \ \text{ or } \ \Delta k\, \Delta l' < 1 \tag{2}$$

$$k\, \Delta' l < 1 \tag{3}$$

where k and Δk are wave number and its spread of the scattered electrons.

Denoting the size of the biprism ($\overline{P_1P_2}$) by d and the deviation angle from the axis to the source by α , Δl and $\Delta l'$ is approximated by $d \sin\alpha$. If we use $s_{\max}$ for the size of the source viewed from the biprism and L for the distance from the source to the biprism, $\alpha = s_{\max}/L$.

Small crossing angle of two beams (denoted by 2β) is chosen and biprism is set inside of this angle, as shown in Fig.1. Then $s_{\max} \cong W$ (W is the width of the beam) and therefore

$$\Delta l, \ \Delta l' \cong Wd/L\,. \tag{4}$$

Next let us evaluate $\Delta' l$. Using the parameters defined in Fig.3,

$$\Delta l_1 = s\cos\theta_1, \quad \Delta l_2 = s\cos\theta_2\,,$$

$$\Delta' l = |s(\cos\theta_1 - \cos\theta_2)| \cong s\sin\theta_2 \Delta\theta\,.$$

Considering the smallness of β and that the collision volume is elongated along the biprism axis, $\Delta' l$ is evaluated to be $W\Delta\theta$.

Now we are prepared for the numerical evaluation. We assume the energy and size of the focus to be 5 keV ($k = 3.62 \times 10^{11}$ [m^{-1}]) and 10 nm respectively. The angular spread is limited to 0.02 rad (1.15 degree) by strongly collimating the beam. For β we take 3.5 degree. For the biprism, let us assume L = 10 mm and d = 1 mm. With this geometry spreads of energy due to the spread in angle of the colliding beams are 18.2eV and 3.7eV for 0.02 rad of $\Delta\beta_1$ and $\Delta\beta_2$ respectively.

With these parameters the quantities of (2) and (3) are estimated as follows:

$$\Delta k\, \Delta l = 0.66, \quad k\, \Delta' l = 3.62 \times 10^{-3}\,.$$

Condition for the energy (wave number) spread is rather marginal but we should say that it is within our technical effort.

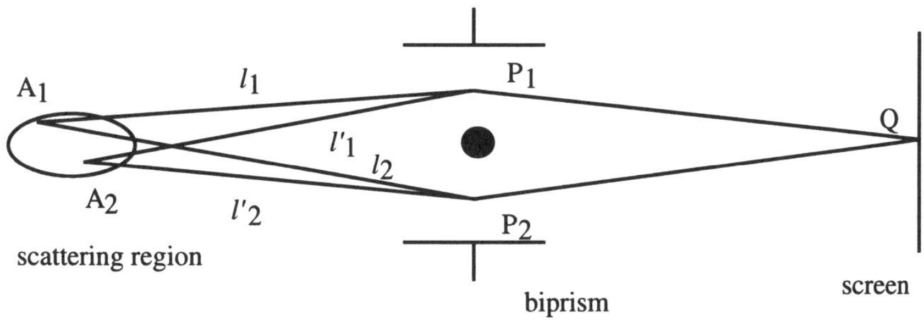

Figure 2. Definition of parameters of the interferometer.

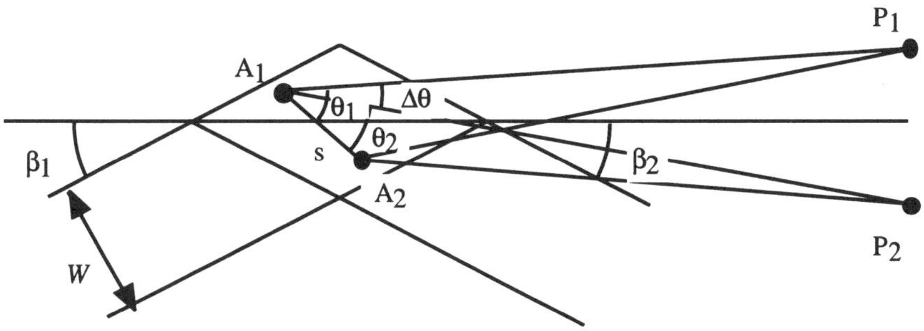

Figure 3. Definition of parameters for the collision region.

3 Yield and "calibration"

The luminosity of the collision region is roughly estimated to be

$$L = \frac{I^2}{e^2 W v \sin 2\beta}$$

where v is the velocity of the electron and I is the beam intensity in ampere.

Assuming a moderate intensity of 10μA and an attenuation factor of 1.5×10^{-3} due to collimation, we obtain the luminosity of 1.97×10^{23} [$\mathrm{m}^{-2}\mathrm{s}^{-1}$]. Using the Coulomb scattering cross section and assuming the solid angle into the biprism to be 10^{-2}sr, then we obtain the rate of a few counts per second.

To exclude the possibility that the lack of fringe is not due to the source size or the energy spread, we must check the apparatus is capable to observe fringes if it would appear. This may be possible by taking coincidences with partner particles. If we observe simultaneously the particle 2 at a certain position after recombining the two states ($|\bar{\mathrm{a}}\rangle_2$ and $|\bar{\mathrm{b}}\rangle_2$ in Fig.1), we will get a non-zero value in place of $\langle \bar{\mathrm{a}} | \bar{\mathrm{b}} \rangle$ in eq. 1 .

References

1. Anton Zeilinger, Review of Modern Physics, **71** S288 (1999) .

QUANTUM INFORMATION ASPECT OF BLACK HOLE

A. HOSOYA

Department of Physics
Tokyo Institute of Technology
Oh-Okayama, Meguro-ku, Tokyo 152
Japan

It is pointed out that quantum information theory has a nice application to black hole physics, because the quantum state is an entangled state of the particles inside and outside of the black hole just like the Einstein-Podolsky-Rosen pair. We show in particular that the increase of the generalized entropy by quantum process outside the horizon of a black hole is more than the Holevo bound of mutual information between a message prepared by an agent located outside the horizon and that received by an observer at infinity.

1 Introduction

Recently there has been substantial development in the field of quantum computation,teleportation and cryptography, which are main topics of the symposium. However, the concept of quantum information is so fundamental that it should not be restricted to applications to various quantum technology but should be incorporated in basic science. Here I will give an application of quantum information theory to black hole physics. Probably before going into details it would be appropriate to give a very brief introduction of black hole thermodynamics and explain why quantum information is relevant for a black hole.

A heavy star of mass more than several times the solar mass gravitationally collapses to a black hole. Near the black hole the gravitational force is so strong that even light cannot escape from a black hole. The marginal region is called the event horizon of the black hole. For a spherical black hole of mass M, the marginal radius is $2MG/c^2$ with G and c being the Newton constant and the light velocity.(Hereafter we take the unit $G = c = 1$) It is around 3 km for an object of solar mass.

Classically nothing can go out of a black hole. However, as Hawking [1] showed in 1974, there is a quantum effect that particles are emitted from the black hole of mass M in a black body spectrum with the temperature $T_{BH} = \frac{\hbar}{8\pi M}$ in the above units, numerically around $10^{-7}K$ for a solar mass black hole. Roughly speaking the mechanism of the Hawking radiation is the following. The strong gravitation force near the horizon creates a pair of particles, one of which travels to infinity while the other falls into the black hole. The pair is quantum mechanically correlated a lá Einstein-Podlsky-Rosen (EPR) pair, which plays a key role in quantum information theory.[2]

More explicitly,the quantum state of the matter in the black hole spacetime is the Hartle-Hawking state,

$$|\psi>_{HH}=\sum_n \sqrt{c_n}|n>_B |n>_A, \qquad (1)$$

with $c_n = e^{-\frac{\omega n}{T_{BH}}}/Z$ being the Boltzmann factor. $Z = \sum_n e^{-\frac{\omega n}{T_{BH}}}$ and T_{BH} is the Hawking temperature defined above.This state is an entangled state of the n-particles inside $|n>_B$ and outside $|n>_A$ of the black hole. We trace over the inaccessible B-state to obtain a mixed state for the observer outside, $\rho_A = tr_B(|\psi>_{HH}<\psi|) = \sum_n c_n|n>_A<n|$, the canonical density operator.[3]

2 The generalized second law in black hole thermodynamics

There is an apparent paradox called "Wheeler's demon" that we can evade the second law of thermodynamics by disposing of a garbage entropy into a black hole. According to the generalized second law, however, we have to pay the price for it. The black hole becomes bigger when the entropy is thrown in. More precisely, it says that the sum of the black hole entropy $S_{BH} = \frac{1}{4\hbar}A$ and the ordinary matter entropy S_{matter}; $S_{tot} = S_{BH} + S_{matter}$, does not decrease, where $A = 16\pi M^2$ is the area of the horizon of the black hole of mass M. There are plenty of evidence to support it. Example are a gedanken experiment and a general argument based on the EPR like entanglement of the particle states of inside and outside the event horizon.[4]

It is historically interesting to point out that Bekenstein[5] proposed the generalized second law in the black hole spacetime prior to the discovery of the Hawking radiation [1] on the basis of information theoretical argument in a gedanken experiment and opened up the black hole thermodynamics, one of the most fascinating fields of theoretical physics. [3]It has been shown that there is an almost complete parallelism between black hole physics and thermodynamics from the zero-th to the third laws.

3 The Holevo bound

Information can be processed(sent, received,disposed etc.) only by physically performing measurements. We say, for example, that the information prepared by an agent outside a black hole is retrieved by another agent at infinity if the two experimental outcomes are correlated. A quantification of the correlation is the mutual information defined below. In what follows we will show that the increase of the generalized entropy is bounded below by the Holevo bound[6,7], which in turn is the upper bound of the classical mutual information.

Imagine a quantum experiment outside of the black hole. The state will change in general as $\rho \to \rho' = \sum_\alpha A_\alpha \rho A_\alpha^\dagger = \sum_\alpha p_\alpha \rho'_\alpha$, with $\sum_\alpha A_\alpha^\dagger A_\alpha = 1$. The transition is a trace preserving POVM (positive operator valued measure), where $p_\alpha = tr(A_\alpha \rho A_\alpha^\dagger)$ is the probability to get the result $\alpha.\rho'_\alpha$is the normalized density operator. We suppose that the experiment is local and an isothermal process due to the Unruh effect of the accelerated system with the temperature $\tilde{T}(r) = \frac{T_{BH}}{\chi(r)}$, the blue shifted temperature from the Hawking temperature T_{BH} of the cavity surrounding the black hole at infinity. The first law of black hole physics is

$$\Delta S_{BH} = \frac{\Delta W}{T_{BH}}, \tag{2}$$

where ΔW is the work to be done by an agent at infinity needed for the quantum experiment. The ordinary thermodynamics tells us that the work ΔW needed in the isothermal process is more than or equal to the difference of the free energy:

$$\begin{aligned} &\Delta W \geq \Delta F \\ \Delta F &= \textstyle\sum_\alpha p_\alpha[E_\alpha - \tilde{T}(S'_\alpha - logp_\alpha)]\chi - (E_0 - \tilde{T}S_0)\chi \\ &= [S_0 - (\textstyle\sum_\alpha p_\alpha S'_\alpha - \sum_\alpha p_\alpha logp_\alpha)]T_{BH}. \end{aligned}$$

The last equality holds because the internal energy does not change ;$E_0 = \sum_\alpha p_\alpha E_\alpha$ in an isothermal process. S_0,S' and S'_α are defined by $S_0 = S(\rho_0)$,$S' = S(\rho')$ and $S'_\alpha = S(\rho'_\alpha)$,where $S(\rho) = -tr(\rho log\rho)$ is the expression for the von Neumann entropy for a general state ρ. The quantity$-\sum_\alpha p_\alpha logp_\alpha$ represents the entropy of the detector.

Combining the first law of black hole physics and the second law of thermodynamics above, we obtain $\Delta S_{BH} = S'_{BH} - S_{BH} \geq S_0 - \sum_\alpha p_\alpha(S'_\alpha - logp_\alpha)$ We write it in a more illuminating way as

$$S'_{tot} - S_{tot} \geq S' - \sum_\alpha p_\alpha S'_\alpha, \tag{3}$$

where S' is the matter entropy after the measurement and $S'_{tot} = S'_{BH} + S' - \sum_\alpha p_\alpha log p_\alpha$ is the generalized entropy including the detector entropy. The right hand side is the famous Holevo bound :

$$S(\sum_\alpha p_\alpha \rho'_\alpha) - \sum_\alpha p_\alpha S(\rho'_\alpha) \geq I', \quad (4)$$

where I' is the mutual information which represents the correlation between the two measurements. Precisely, with $\{E_j\}$ being the orthogonal projection summing to unity which corresponds to the observation by the second agent at infinity and should be distinguished from the previous POVM's,we have

$$I'(E) = -\sum_{j,\alpha} p_\alpha p(j|\alpha) log \frac{p(j)}{p(j|\alpha)}, \quad (5)$$

where $p(j|\alpha) = tr(E_j \rho'_\alpha)$ is the conditional probability to obtain the outcome j when the state ρ'_α is prepared and $p(j) = \sum_\alpha p(\alpha)p(j|\alpha)$ is the average probability to obtain j. The above expression can be interpreted as the uncertainty of the first measurement minus its uncertainty after the second measurement,i.e., the knowledge about the prepared sate,$\sum_\alpha p_\alpha \rho'_\alpha$obtained by the measurement $\{E_j\}$ at infinity. The equality can be achieved for some projection $\{E_j\}$ if and only if the components of ρ'_α's are mutually commutable. In this case ρ'_α's can be simultaneously diagonalized so that we can choose, for example, that $A^\dagger_\alpha A_\alpha = E_j$ as the best that the second agent can do. We obtain in this optimal case $I'(E) = -\sum_\alpha p_\alpha log p_\alpha$. This is nothing but the Shannon information entropy stored by the first measurement.

4 Summary and Discussion

We have shown that the increase of the generalized entropy by quantum process outside the horizon of a black hole is more than the Holevo bound of mutual information which would be obtained by quantum observation;

$$\Delta S_{tot} \geq I'. \quad (6)$$

What we have used as physics are the energy conservation in the black hole physics and the second law of ordinary thermodynamics. Our result may be interpreted that the harder experimentalist works to understand the universe, the more the black region of ignorance or garbage grows!

However,there remains a long standing problem: loss of information of the initial state by evaporation of a black hole.[8] From our view point , it is crucial to clarify the meaning of "information" to resolve this paradox.[9]

References

1. S.W. Hawking,*Commun.Math.Phys.* **43** 199 (1975).
2. For a compact review,V.Vedral,*Role of Relative Entropy in Quantum Information Theory*, quant-phys/0102094(2001).
3. R.M.Wald,*Quantum Field Theory in Curved Spacetime and Black Hole Thermodynamics*, Chicago University Press(1994).
4. J.D.Bekenstein,*Phys.Rev.***D23** 287 (1981),W.G.Unruh and R.M.Wald,*Phys.Rev.* **D25** 942 (1982), **D27** 2271 (1983), V.P.Frolov and D.Page, *Phys.Rev.Lett.* **71**(24) 3902 (1993), A.Hosoya, A.Carlini and T.Shimomura*Phys.Rev.* **D63**104008 (2001).
5. J.D.Bekenstein,*Phys.Rev.***D7** 2333 (1973),**D9** 3292 (1974),**D12** 3077(1975).
6. A.S.Holevo, *Probl. Peredachi Inform.* **9**, 177 (1973).
7. A. Peres, "Quantum Theory:Concepts and Methods", (1995) Kluwer Academic Publishers.
8. D.N.Page, *Proc.5th Canadian conference on general relativity* ed. R.Mann amd R. McLenagham(Singapore:World Scientific) (1994).
9. A. Hosoya and A. Carlini, in preparation (2001).

DECOHERENCE OF ANOMALOUSLY-FLUCTUATING STATES OF FINITE MACROSCOPIC SYSTEMS

AKIRA SHIMIZU,* TAKAYUKI MIYADERA,** AND AKIHISA UKENA
Department of Basic Science, University of Tokyo, 3-8-1 Komaba, Tokyo 153-8902, Japan

In quantum systems of a macroscopic size V, such as interacting many particles and quantum computers with many qubits, there exist pure states such that fluctuations of some intensive operator $\hat{A}$ is anomalously large, $\langle \delta \hat{A}^2 \rangle = \mathcal{O}(V^0)$, which is much larger than that assumed in thermodynamics, $\langle \delta A^2 \rangle = \mathcal{O}(1/V)$. By making full use of the locality, we show, starting from Hamiltonians of macroscopic degrees of freedom, that such states decohere at anomalously fast rates when they are weakly perturbed from environments.

1. INTRODUCTION

We consider a quantum system, which extends spatially over a *macroscopic but finite* volume V. Such a system includes, for example, many particles confined in a box of volume V, and a system composed of N two-level systems (qubits), for which $V \propto N \gg 1$. Such a system, in general, has pure states such that fluctuations of some intensive operator,

$$\hat{A} = \frac{1}{V} \sum_{x \in V} \hat{a}(x), \tag{1}$$

where $\hat{a}(x)$ is an operator at point x, is anomalously large;

$$\langle \delta \hat{A}^2 \rangle = \mathcal{O}(V^0). \tag{2}$$

We call such a pure state an 'anomalously fluctuating state' (AFS), because $\langle \delta \hat{A}^2 \rangle$ is much larger than that assumed in thermodynamics, $\langle \delta A^2 \rangle = \mathcal{O}(1/V)$.

In *closed* quantum systems, AFSs appear naturally, as explained in section 2. Experimentally, however, it is rare to encounter AFSs. This apparent contradiction may be explained by the fact that real physical systems are not completely closed: Interactions with environments would destroy AFSs very quickly. Effects of environments have been discussed intensively in studies of 'macroscopic quantum coherence' [1], and of quantum measurement [2]. These previous studies assumed that the principal systems of interest were describable by a *small number* of collective coordinates, which interact *non-locally* with some environment(s). However, justification of these assumptions is not clear. Although such models might be applicable to systems which have a non-negligible energy gap to excite 'internal coordinates' in the collective coordinates, there are many systems which do not have such an energy gap. Moreover, the results depended strongly on the choices of the coordinates and the form of the nonlocal interactions, so that general conclusions were hard to draw. In this work, we study the decoherence rates of AFSs and normally-fluctuating states (NFSs), starting from microscopic Hamiltonians of macroscopic degrees of freedom, by making full use of the locality: interactions must be local (Eq. (3)), and macroscopic variables must be averages over a macroscopic region (Eq. (1)). To express the locality manifestly, we use a local field theory throughout this work.

2. ANOMALOUSLY-FLUCTUATING STATES

AFSs generally appear in, e.g., (i) finite systems which will exhibit symmetry breaking if V goes to infinity, and (ii) quantum computers with many qubits.

In case (i), we can find states (of finite systems) which approach a symmetry-breaking vacuum as $V \to \infty$. We call such a state a pure-phase vacuum (PPV). It has a finite expectation value $\langle \hat{M} \rangle = \mathcal{O}(V^0)$ of an order parameter $\hat{M}$, and has negligible fluctuations $\langle \delta \hat{A}^2 \rangle = \mathcal{O}(1/V)$ for *any* intensive operator $\hat{A}$ (including $\hat{M}$) [3–8].

Hence, PPVs are NFSs. In a mean-field approximation, PPVs have the lowest energy. However, it is known that the exact lowest-energy state of a finite system (without a symmetry-breaking field) is generally the *symmetric* ground state (SGS), for which $\langle \hat{M} \rangle = 0$ [3–8]. PPVs have a higher (or, in some special cases, equal) energy than the SGS. The SGS is composed primarily of a superposition of PPVs with different values of $\langle \hat{M} \rangle$ [3–8]. As a result, it has an anomalously large fluctuation of $\hat{M}$; $\langle \delta \hat{M}^2 \rangle = \mathcal{O}(V^0)$. Therefore, if one obtains the exact lowest-energy state (e.g., by numerical diagonalization) in case (i), it is generally an AFS.

In case (ii), various states appear in the course of a quantum computation. Some state may be an NFS, for which $\langle \delta A^2 \rangle = \mathcal{O}(1/V)$ for *any* intensive operator A. This means that correlations between distant qubits are weak [8,10]. Properties of such states may be possible to emulate by a classical system with local interactions, because entanglement is weak. We therefore conjecture that other states – AFSs – should appear in some stage of the computation for a quantum computer to be much faster than classical computers. In fact, we confirmed this conjecture in Shor's algorithm for factoring [10].

We stress that AFSs are peculiar to *quantum* systems of macroscopic but *finite* sizes, and thus are very interesting. In (local) classical theories, a state such that $\langle \delta A^2 \rangle = \mathcal{O}(V^0)$ is possible only as a mixed state. In (local) quantum theory of infinite systems, any pure states (including excited states) are NFSs [3,4], because all AFSs for a finite V become mixed states in the limit of $V \to \infty$ [3,4,7].

3. INTERACTING MANY BOSONS

We first consider interacting many bosons confined in a uniform box of volume V with the periodic boundary conditions [6,7,11]. Since the Hamiltonian $\hat{H}$ commutes with the number of bosons $\hat{N}$, there exist simultaneous eigenstates of $\hat{N}$ and $\hat{H}$. We denote the lowest-energy state for a given value of N as $|N, \mathrm{G}\rangle$. This is the SGS, for which $\langle \hat{\Psi} \rangle = 0$, where $\hat{\Psi}$ denotes the intensive order parameter $\hat{\Psi} \equiv (1/V) \sum_{x \in V} \hat{\psi}(x)$, and $\hat{\psi}(x)$ is the boson operator at point x in the box. On the other hand, $|N, \mathrm{G}\rangle$ has the long-range order, $\langle \hat{\psi}^\dagger(x) \hat{\psi}(x') \rangle = \mathcal{O}(V^0)$ for $|x - x'| \sim V^{1/3}$ [7,11]. We can easily show that it has an anomalously-large fluctuation of $\hat{\Psi}$; $\langle \delta \hat{\Psi}^\dagger \delta \hat{\Psi} \rangle = \mathcal{O}(V^0)$, which shows that $|N, \mathrm{G}\rangle$ is an AFS. On the other hand, by superposing $|N, \mathrm{G}\rangle$ of various values of N [12], we can construct a state $|\alpha, \mathrm{G}\rangle$, for which $\langle \hat{\Psi} \rangle = \mathcal{O}(V^0)$ and $\langle \delta A^2 \rangle = \mathcal{O}(1/V)$ for *any* intensive operator A [7]. Namely, $|\alpha, \mathrm{G}\rangle$ is a PPV, hence is an NFS. Although it is not an eigenstate of $\hat{H}$, $|\alpha, \mathrm{G}\rangle$ does not collapse for a macroscopic time [6].

Since the energy of $|N, \mathrm{G}\rangle$ is lower (by $\mathcal{O}(V^0)$) than that of $|\alpha, \mathrm{G}\rangle$ for the same value of $\langle \hat{N} \rangle$ [6], there seems to be nothing against the realization of $|N, \mathrm{G}\rangle$ if the system is closed. However, most real systems are not completely closed, and weak perturbations from environments can alter the situation dramatically. In fact, effects of interactions with an environment, which was assumed to be a huge room that has initially no bosons, on these states was studied in Ref. [11], where it was shown that $|N, \mathrm{G}\rangle$ decoheres much faster than $|\alpha, \mathrm{G}\rangle$, as bosons escape from the box into the environment. Namely, the SGS (which is an AFS) is much more fragile than the PPV (an NFS), for interacting many bosons in a leaky box. This may be the first example in which the SGS and PPVs are identified *and* the fragility of the SGS as well as the robustness of PPVs are shown, for a non-trivial interacting many-particle system.

4. GENERAL SYSTEMS

We next consider a *general* finite system of a *large* V, interacting with a *general* environment E, with a *local* interaction,

$$\hat{H}_{\mathrm{int}} = \lambda \sum_{x \in V_{\mathrm{C}}} \hat{a}(x) \otimes \hat{b}(x), \qquad (3)$$

where $\hat{a}(x)$ and $\hat{b}(x)$ are local operators (*any* functions of the fields and their conjugate momenta at point x) of the principal system and E, respectively, *at the same point* $x \in V_C$. Here, V_C ($\subseteq V$) is a 'contact region' between the principal system and E. Since we are interested in the case of weak perturbations from E, we assume that the coupling constant λ (≥ 0) is small.

The initial ($t = 0$) state, $\rho(0) = |\phi\rangle\langle\phi|$, of the principal system is either an AFS or an NFS, and the initial state of the total system is assumed to be the uncorrelated product, $\rho_{tot}(0) = \rho(0) \otimes \rho_E$, where ρ_E is a time-invariant state of E. We are interested in the time evolution of the reduced density operator, $\rho(t) \equiv \mathrm{Tr}_E[\rho_{tot}(t)]$, which generally evolves from the pure state $|\phi\rangle\langle\phi|$ into a mixed state. As a measure of the purity, we evaluate the 'linear entropy' defined by $S_{lin}(t) \equiv 1 - \mathrm{Tr}[\rho(t)^2]$. If $|\phi\rangle$ is translationally invariant [13], both spatially and temporally, we find to $\mathcal{O}(\lambda^2)$ that [8]

$$S_{lin}(\phi, t) \geq \frac{\lambda^2}{\hbar^2} g_{00} \langle\phi|\delta\hat{A}^\dagger \delta\hat{A}|\phi\rangle t, \qquad (4)$$

where $\hat{A} \equiv (1/V)\sum_{x\in V} \hat{a}(x)$, and g is a positive matrix defined by the time correlation of $\hat{b}$ of E;

$$g_{k_1 k_2} \equiv \frac{1}{2}\int_{-\infty}^{\infty} ds \langle \hat{b}^\dagger_{k_1} \hat{b}_{k_2}(s)\rangle. \qquad (5)$$

Here, $b_k \equiv \sum_{x\in V_C} b(x) \mathrm{e}^{-ikx}$, where the sum is not over the entire region of E, but over V_C. Since the rhs of Eq. (4) is proportional to t, we can interpret it divided by t as a *lower bound of the decoherence rate*, which we denote γ. It is proportional to the fluctuation of the intensive operator $\hat{A}$ composed of $\hat{a}(x)$ which constitutes H_{int}. As discussed in the next section, in real physical systems many terms would exist in $\hat{H}_{\mathrm{int}}$; $\hat{H}_{\mathrm{int}} = \hat{H}^{[1]}_{\mathrm{int}} + \hat{H}^{[2]}_{\mathrm{int}} + \cdots$, where $\hat{H}^{[\ell]}_{\mathrm{int}} = \lambda^{[\ell]} \sum_{x\in V_C^{[\ell]}} \hat{a}^{[\ell]}(x) \otimes \hat{b}^{[\ell]}(x)$. It may be possible to construct a state which is *exactly robust* (i.e., does not decohere at all) against *one* of $\hat{H}^{[\ell]}_{\mathrm{int}}$'s. Such a state, however, would be fragile against another $\hat{H}^{[\ell]}_{\mathrm{int}}$. It is therefore important to judge the fragility and robustness against *all* local interactions.

To γ of an AFS, some of $\hat{H}^{[\ell]}_{\mathrm{int}}$'s can give an anomalously large contribution, because, by definition, $\langle\phi|\delta\hat{A}^{[\ell]\dagger}\delta\hat{A}^{[\ell]}|\phi\rangle = \mathcal{O}(V^0)$ for some $\hat{A}^{[\ell]}$, hence

$$\gamma^{[\ell]} = (\lambda^{[\ell]2}/\hbar^2) g^{[\ell]}_{00} \times \mathcal{O}(V^0), \qquad (6)$$

which can be anomalously large. To see this, we estimate the prefactor g_{00}. Let V_E^{corr} be the size of the region in which $\int_{-\infty}^{\infty} dt \langle b^*(x) b(0,t)\rangle$ is correlated in E. We can roughly estimate that $g_{00} \propto V_C^2$ when $V_E^{\mathrm{corr}} > V_C$, whereas $g_{00} \propto V_C V_E^{\mathrm{corr}}$ when $V_E^{\mathrm{corr}} < V_C$ [8]. In either case, $\gamma^{[\ell]}$ becomes anomalously large, in the sense that $\gamma^{[\ell]}$ is proportional to the *square* (or product) of a macroscopic volume(s), if the contact region V_C and the correlation region V_E^{corr} are macroscopically large. Hence, AFSs are fragile in some environment [9].

To γ of an NFS, on the other hand, *none* of $\hat{H}^{[\ell]}_{\mathrm{int}}$'s can give such an anomalously large contribution, because $\langle\phi|\delta\hat{A}^{[\ell]\dagger}\delta\hat{A}^{[\ell]}|\phi\rangle = \mathcal{O}(1/V)$ for any $\hat{A}^{[\ell]}$. Hence, NFSs are less fragile than AFSs, except when $V_E^{\mathrm{corr}[\ell]}$ is microscopically small for all ℓ. Here, we have used a moderate word "less fragile" instead of a stronger word "robust" because, for NFSs of general systems, $\gamma^{[\ell]}$ can be proportional *linearly* to a macroscopic volume [9], unlike the case of section 3.

5. EFFECTIVE THEORIES

Usually, we are only interested in phenomena in some energy range ΔE. Hence, it is customary to analyze a physical system by an effective theory which correctly describes the system only in ΔE. In some cases, the degrees of freedom N of the effective theory can become small even for a macroscopic system when, e.g., a non-negligible energy gap exists in ΔE because then the number of quantum states in ΔE can be small. Some SQUID systems are such examples. We here exclude such systems, and concentrate on systems whose N is a macroscopic number, because otherwise the difference between $\mathcal{O}(1/N)$ and $\mathcal{O}(N^0)$ would be irrelevant. The effective theory can be constructed from an elementary dynamics by an appropriate renormalization process. In this process, in general, many interaction terms would be generated in the effective interaction $H_{\rm int}$ [14]. Hence, it seems rare that $H_{\rm int}$ does not have any term which takes the form of Eq. (3) such that $\langle \delta A^\dagger \delta A \rangle = \mathcal{O}(N^0)$ for the AFS under consideration.

If N were small, one could drop terms with small $\lambda^{[\ell]}$. Since N is large, however, such an approximation can be wrong, because, as discussed in the previous section, its contribution may be enhanced by an anomalously large factor and become relevant to AFSs, however small $\lambda^{[\ell]}$ is.

6. IMPLICATIONS

We finally discuss implications of the above results, for (i) and (ii) of section 2.

In case (i), our results suggest a new origin of symmetry breaking in finite systems [11]. Although symmetry breaking is usually described as a property of infinite systems, it is observed in finite systems as well. Our results suggest that although a PPV (which is an NFS) has a *higher* energy than the SGS (an AFS), the former is realized because the latter is extremely fragile in environments. This scenario, may be called 'environment-induced symmetry breaking' after Zurek's 'environment-induced superselection rule' [2], may be the physical origin of symmetry breaking in finite systems.

In case (ii), our results show that the decoherence rate can be estimated by fluctuations of intensive operators, which depend strongly on the number of qubits N and the natures of the states of the qubits [15]. This may become a key to the fight against decoherence in quantum computation.

In both cases, we stress that the *approximate* robustness against *all* local interactions (between the principal system and environments) would be more important than the *exact* robustness against a *particular* interaction, because, as discussed in section 4, many types of interactions would coexist in real physical systems, and the exact robustness against one of them could imply fragility to another.

---

* E-mail: shmz@ASone.c.u-tokyo.ac.jp

** Present address: Department of Information Sciences, Science University of Tokyo.

[1] See, e.g., proceedings of the previous ISQM conferences.

[2] W. H. Zurek, Phys. Rev. **D24**, 1516 (1981).

[3] D. Ruelle, *Statistical Mechanics: Rigorous Results* (Benjamin, Reading, 1969).

[4] R. Haag, *Local Quantum Physics* (Springer, Berlin, 1992).

[5] P. Horsh and W. von der Linden, Z. Phys. **B72**, 181 (1988).

[6] A. Shimizu and T. Miyadera, Phys. Rev. **E64**, 056121 (2001).

[7] A. Shimizu and T. Miyadera, to appear in J. Phys. Soc. Jpn. **76-1** (2002).

[8] T. Miyadera and A. Shimizu, quant-ph/0109065.

[9] More general and detailed analyses will be described in A. Shimizu, in preparation.

[10] A. Ukena and A. Shimizu, unpublished.

[11] A. Shimizu and T. Miyadera, Phys. Rev. Lett. **85**, 688 (2000)

[12] Conversely, $|N, \mathrm{G}\rangle$ is a superposition of $|\alpha, \mathrm{G}\rangle$ over all values of the angle of α.

[13] This assumption is not satisfied by general AFSs and NFSs. We here apply Eq. (4) to some classes of them for which the assumption is approximately satisfied.

[14] We expect that by an appropriate renormalization process $H_{\rm tot}$ can be made local in the relevant space-time scale.

[15] This fact was first pointed out by G. M. Palma, K.-A. Suominen and A. K. Ekert, Proc. Roy. Soc. Lond. A (1996) **452**, 567. In contrast to the present theory, they argued the dependence of the decoherence rate on the *Hamming distance*, assuming *non-interacting* qubits and a *non-local* system-environment interaction.

A COARSE-GRAINED POSITION OPERATOR AND DECOHERENCE IN MACROSCOPIC SYSTEMS

A. NAGASATO, T. AKAMINE, T. YONEDA AND A. MOTOYOSHI

Department of Physics, Kumamoto University, Kumamoto 860-8555, Japan

E-mail: tyoneda@aps.cms.kumamoto-u.ac.jp, motoyosi@aster.sci.kumamoto-u.ac.jp

The behavior of macroscopic system is investigated with a coarse-grained position operator based on the SO(3,1), whose macroscopic nature induces decoherence leading to classical behavior.

1 Introduction

It is an important problem how the transition from quantum to classical system occurs. This problem is deeply connected with the problems of measurement. In the processes of measurement, decoherence plays a crucial role. Therefore, more detailed analyses of the origin and the process of decoherence are necessary for further establishment of the foundation of quantum mechanics. They are simultaneously interesting for the light they may shed on the problem of recent new applications of quantum mechanics for quantum information processing, since decoherence is an obstacle for them.

Machida and Namiki[1] showed that macroscopic system as an open system with the uncertainty of the number of atoms induces the reduction of wave packet. One of the present authors (A.M.)[2] have discussed another description of this macroscopic nature and proposed a coarse-grained position operator based on the O(4).

In this paper, making use of a more reasonable coarse-graind position operator based on the SO(3,1), the origin of decoherence and the transition from quantum to classical system are investigated. In comparison with other theories of open system, our operator has wide applicability for any macroscopic system if it is described by Hamiltonian. In fact, we can discuss generalized master equation[3] and make a modification of Green model[4] against Furry's criticism with this operator.

2 Coarse-grained Position Operator

We can consider that a macroscopic size L has finite random fluctuation ΔL.[1] An example of coarse-grained position operator having this macroscopic nature is[5]

$$\hat{x}_j = i\hbar\{\frac{\partial}{\partial p_j} - \frac{\langle(\Delta L)^2\rangle}{\hbar^2} p_j \sum p_k \frac{\partial}{\partial p_k}\}, \quad (1)$$

where $\langle(\Delta L)^2\rangle$ denotes the averaged value of $(\Delta L)^2$. This operator can be constructed in parallel with Ref. 2. Making use of the projective co-ordinates, we may define that

$$p_i \equiv \alpha_L^{-1/2} \cdot \xi_i/\xi_5, \quad (2)$$

where $[p_i, p_j] = 0$, $\alpha_L \equiv \langle(\Delta L)^2\rangle/\hbar^2$ and $\alpha_L p^2 \equiv \Gamma$ is a dimensionless constant. We may consider a restricted p-space $p^2 = p_1^2 + p_2^2 + p_3^2 \leq \alpha_L^{-1}$. This restricted p-space is embedded in a fictitious (3+1) flat ξ-space by Eq.(2), and constructs a 3-dimensional world with constant curvature. Making use of the infinitesimal generators which leaves $\xi_\alpha\xi^\alpha$ invariant

$$G^{\alpha\beta} = i(\xi^\alpha \frac{\partial}{\partial\xi^\gamma} g^{\gamma\beta} - \xi^\beta \frac{\partial}{\partial\xi^\delta} g^{\delta\alpha}),$$

$$\alpha, \beta, \gamma, \delta : 5, 1, 2, 3, \; g_{55} = -g_{ii} = 1, \quad (3)$$

we can define a quantum mechanical operator $\hat{x}_i$ as a coarse-grained macroscopic position operator $\hat{x}_i \equiv -\hbar\alpha_L^{1/2} G_{5i}$. The commutation relations for the generators $G_{\alpha\beta}$ are given by

$[G_{\alpha\beta}, G_{\gamma\delta}] = -i(g_{\alpha\gamma}G_{\beta\delta}+g_{\beta\delta}G_{\alpha\gamma}-g_{\alpha\delta}G_{\beta\gamma}-g_{\beta\gamma}G_{\alpha\delta})$, from which we have $[\hat{x}_i, \hat{x}_j] = \hbar^2\alpha_L\cdot[G_{5i}, G_{5j}] = -i\hbar^2\alpha_L \cdot G_{ij}$. Making use of the variables ξ and p, the commutation relation for $[\hat{x}_i, p_j]$ is given by

$$i\hbar(\delta_{ij} - \frac{\xi_i\xi_j}{\xi_5^2}) = i\hbar(\delta_{ij} - \alpha_L p_i p_j), \qquad (4)$$

and commutation relation for $[\hat{x}_i, \hat{x}_j]$ is given by

$$\begin{aligned} &- \hbar^2\alpha_L(\xi_i\frac{\partial}{\partial\xi_j} - \xi_j\frac{\partial}{\partial\xi_i}) \\ =& - \hbar^2\alpha_L(p_i\frac{\partial}{\partial p_j} - p_j\frac{\partial}{\partial p_i}), \qquad (5)\end{aligned}$$

respectively. Finally, we obtain

$$\begin{aligned}\hat{x}_j &= i\hbar\alpha_L^{1/2}(\xi_5\frac{\partial}{\partial\xi_j} + \xi_j\frac{\partial}{\partial\xi_5}) \\ &= i\hbar\{\frac{\partial}{\partial p_j} - \alpha_L p_j\sum p_k\frac{\partial}{\partial p_k}\}, \qquad (6)\end{aligned}$$

as the solution of the above commutation relations. This operator furnishes Eq.(1). Microscopic limit as a geometrical point is simply given by $\Delta L \to 0$. In this limit, usual position operator and commutation relations can be recovered. On the other hand, we may consider that macroscopic distance between two points smaller than $\langle(\Delta L)^2\rangle^{1/2}$ cannot be discriminated and macroscopic limit is given by $\langle(\Delta L)^2\rangle/L^2 = \epsilon_L^2 \to 0$. This means that the fluctuation ΔL can be ignored in this limit in comparison with its macroscopic size L. This limit shows that $\alpha_L p^2 = \Gamma = \langle(\Delta L)^2\rangle \cdot p^2/\hbar^2 \to 1$ in the present representation of the macroscopic nature, because of that this limit can be rewritten into $\langle(\Delta L)^2\rangle = \epsilon_L^2 L^2 \Rightarrow \Gamma = \epsilon_L^2 L^2 \cdot p^2/\hbar^2 \to 1$ and $\epsilon_L^2 L^2 \to \hbar^2/p^2 \Rightarrow L^2 \gg \hbar^2/p^2$ for $\epsilon_L \to 0$. The commutation relations $[\hat{x}_i, p_i]$ tends to zero in this limit. Therefore, the operator $\hat{x}_i$ becomes classical physical quantity x_i. Consequently the commutation relation $[\hat{x}_i, \hat{x}_j] \Rightarrow [x_i, x_j]$ also becomes zero. It should be noted that these relations are obtained without to take the limit $\hbar \to 0$. When the dimensionless parameter $\Gamma = \alpha_L p^2$ tends to 1 and the macroscopic size L is sufficiently larger than de Broglie wave length $\lambda = h/p$ accompaning with the motion of the center of mass, deviation of the fluctuation $\langle(\Delta L)^2\rangle$ of the macroscopic size L becomes negligible. Therefore, we can discriminate different two macroscopic points separated over several de Broglie wave length. Hence, we can define the distance of two separated macroscopic points.

3 The Transition from Quantum to Classical

We may consider crystal lattice consisting of sufficiently large number N atoms as an example of the macroscopic system. The Hamiltanian of this system can be written by

$$H = \sum_{i=1}^{3N}\frac{p_i^2}{2m} + \sum_{ij}U_{ij}x_i x_j, \qquad (7)$$

in the harmonic approximation. This Hamiltonian becomes into the form

$$H = \sum_{i=1}^{N}\left(\frac{p^2}{2m} + \frac{1}{2}Kx^2\right)_i, \qquad (8)$$

by the choice of normal co-ordinates. In this system, the macroscopic nature given in the preceding section can be reduced to a fluctuation $\langle(\Delta\ell)^2\rangle$ of the position of atoms. In virtue of the central limit theorem, we have $\langle(\Delta L)^2\rangle = N\cdot\langle(\Delta\ell)^2\rangle$, and obtain

$$\begin{aligned}\tilde{x}_j &= i\hbar\{\frac{\partial}{\partial p_j} - \frac{\langle(\Delta\ell)^2\rangle}{\hbar^2}p_j\sum p_k\frac{\partial}{\partial p_k}\}, \\ \alpha_\ell &\equiv \frac{\langle(\Delta\ell)^2\rangle}{\hbar^2}, \quad \alpha_\ell p^2 \equiv \gamma \ll 1, \qquad (9)\end{aligned}$$

for x_{op}. In order to make up this system with Hamiltonian (8) to have the macroscopic nature, we need to replace $H(p,x)$ by $\tilde{H}(p,\tilde{x})$ in p-representation. Hence, to investigate the behavior of this system is essentially equivalent to solve following eigenvalue equation

$$(p^2/2m + K\tilde{x}^2/2)\Psi = E\Psi,$$

$$\Psi = \Phi(p)\cdot Y_{lm}(\theta,\phi). \tag{10}$$

We can introduce a dimensionless independent variable $z = 1/\alpha_\ell p^2$, then Eq.(10) becomes[5]

$$4\alpha_\ell(z-1)^2 z\frac{d^2\Phi}{dz^2} + 2\alpha_\ell(z-1)^2\frac{d\Phi}{dz} + \{\frac{2E}{K\hbar^2} - \frac{1}{mK\hbar^2\alpha_\ell z} - \alpha_\ell l(l+1)z\}\Phi = 0. \tag{11}$$

This differential equation becomes solvable by the transformation $\Phi = (1-z^{-1})^{-2A}\Phi_1$ with $A = 1/8 \pm 1/4\sqrt{1/4 + 1/mK\hbar^2\alpha_\ell^2}$. We now find a solution for Φ_1 in the form $\Phi_1 = \sum a_n(z-1)^{s+n}$. The boundary condition that $z^{2A}(z-1)^{s-2A}$ does not diverge in $\Phi = z^{2A}(z-1)^{s-2A}\sum a_n(z-1)^n$, exclusivelly allows only upper sign in the index A. Denoting $K\langle(\Delta\ell)^2\rangle/2 \equiv V_\ell$ and $(\hbar\omega/V_\ell)^2 \equiv \lambda$, from the boundary condition that Φ_1 must be polynomial for positive integers n, l and $n = l = 0$, we obtain

$$|E_{n,l}|/V_\ell = (2n + l + 3/2)\sqrt{1+\lambda} + 2\{(2n+1)\cdot(n+l+1) - 1/4\}, \tag{12}$$

for the eigenvalues of E in Eq.(10).[5] Equation (12) becomes $|E| \simeq (2n+l+3/2)\hbar\omega$ for $V_\ell \ll \hbar\omega$, which furnishes quantum mechanical energy levels of a three-dimensional isotropic harmonic oscillator in spherical polar co-ordinates. Similarly, Eq.(12) simply becomes $|E| \simeq [4n(n+l+2) + 3(l+1)]V_\ell$ for $V_\ell \gg \hbar\omega$. For $n = l = 0$, Eq.(12) gives zero point energy $|E| = 3\hbar\omega/2$ for $V_\ell \ll \hbar\omega$, and zero point fluctuation of energy $|E| \simeq 3K\langle(\Delta\ell)^2\rangle/2$ for $V_\ell \gg \hbar\omega$.

The wave functions of this system also can be obtained from Eq.(11). The function Φ_1 is given by the following hypergeometric function

$$\Phi_1 = a_0 F(a,b,c;1-z),\ a_0 = \text{const}, \tag{13}$$

where $a = -(n+l+1/2)$, $b = -n$ and $c = a + b - \sqrt{1+\lambda}/2$. Finally, we have

$$\Phi_{n,l}(p) = a_0 z^{2A}(z-1)^{s-2A}F(a,b,c;1-z), \tag{14}$$

for $\Phi(p)$ in Eq.(10).

On the other hand, in exactly quantum mechanical case ($\alpha_\ell = 0$), it is known that the wave function of this system is given by

$$\Phi_{q;n,l}(p) = \left\{\frac{2\beta^3\Gamma(n+l+3/2)}{n!\{\Gamma(l+3/2)\}^2}\right\}^{1/2} \cdot\rho^l e^{-\rho^2/2}F(-n, l+3/2;\rho^2), \tag{15}$$

where $\rho = \beta p$ and $\beta^2 = 1/\sqrt{mK\hbar^2}$.

For example, the first three functions $\Phi_{n,l}(p)$ are

$$\Phi_{0,0}(\ p\ ) = a_0(1-\alpha_\ell p^2)^{-(1+\sqrt{1+\lambda})/4},$$

$$\Phi_{0,1}(\ p\ ) = a_0(1-\alpha_\ell p^2)^{-(1+\sqrt{1+\lambda})/4} \cdot\left(\frac{\alpha_\ell p^2}{1-\alpha_\ell p^2}\right)^{1/2},$$

$$\Phi_{1,0}(\ p\ ) = a_0(1-\alpha_\ell p^2)^{-(1+\sqrt{1+\lambda})/4} \cdot\left(\frac{\alpha_\ell p^2}{1-\alpha_\ell p^2} + \frac{3}{5+\sqrt{1+\lambda}}\right), \tag{16}$$

for $2n + l = 0$ (n=0, l=0), $2n + l = 1$ (n=0, l=1) and $2n + l = 2$ (n=1, l=0). Corresponding quantum mechanical wave functions $\Phi_{q;n,l}(p)$ are

$$\Phi_{q;0,0}(p) = \left(\frac{2\beta^3}{\pi^{1/2}}\right)^{1/2}\cdot e^{-\beta^2p^2/2},$$

$$\Phi_{q;0,1}(p) = \left(\frac{8\beta^3}{3\pi^{1/2}}\right)^{1/2}\cdot e^{-\beta^2p^2/2}\cdot\beta p,$$

$$\Phi_{q;1,0}(p) = \left(\frac{2\beta^3}{3\pi^{1/2}}\right)^{1/2}\cdot e^{-\beta^2p^2/2} \cdot(1-\frac{2}{3}\beta^2p^2). \tag{17}$$

The lack of exponential factor in the function $\Phi_{n,l}(p)$ in comparison with the wave function $\Phi_{q;n,l}(p)$ is due to the broadness of potential energy originated from the fluctuation. Three energy eigenvalues $E_{n,l}$ for states in Eq.(16) are

$$|E_{0,0}| = \frac{3}{2}(1+\sqrt{1+\lambda})V_l,$$
$$|E_{0,1}| = \frac{3}{2}(1+\sqrt{1+\lambda})V_l + 3V_l,$$
$$|E_{1,0}| = \frac{7}{2}(1+\sqrt{1+\lambda})V_l + 8V_l, \quad (18)$$

in comparison with quantum mechanical energy levels $E_{q;n,l}$, which are given by $E_{q;0,0} = 3\hbar\omega/2$, $E_{q;0,1} = 5\hbar\omega/2$ and $E_{q;1,0} = 7\hbar\omega/2$. If the fluctuation of energy were larger than twice of zero point energy ($V_l \geq 3\hbar\omega$), common factor $1+\sqrt{1+\lambda}$ in Eqs. (16), (18) reads

$$1+\sqrt{1+\lambda} \simeq 2. \quad (19)$$

Therefore, this common factor in Eq.(16) can be approximated by

$$(1-\alpha_\ell p^2)^{-(1+\sqrt{1+\lambda})/4} \simeq 1 + \frac{1}{2}\alpha_\ell p^2, \quad (20)$$

because of that $\alpha_\ell p^2 \ll 1$. This approximation offers

$$\Phi_{0,0}(p) \simeq a_0(1+\frac{1}{2}\alpha_\ell p^2),$$
$$\Phi_{0,1}(p) \simeq a_0(1+\frac{1}{2}\alpha_\ell p^2)\sqrt{\alpha_\ell}p,$$
$$\Phi_{1,0}(p) \simeq a_0(1+\frac{1}{2}\alpha_\ell p^2)(\frac{1}{2}+\alpha_\ell p^2), \quad (21)$$

for Eq.(16), which are unnormalizable. Simultaneously, even if $V_l \sim 3\hbar\omega$, wide energy levels ($9\hbar\omega$ for $2n+l=0$) are covered by the fluctuation of energy. Therefore, considerable quantum mechanical energy levels become indistinguishable, which offers classical property.

The function $\Phi_{n,l}(p)$ approaches to the following form with increasing $(2n+l)$

$$\Phi_{n,0}(p) \simeq a_0(1+\frac{1}{2}\alpha_\ell p^2)\cdot e^{-Bn}\cdot e^{\frac{Bn}{\alpha_\ell p^2}}, \quad (22)$$

which gives the position probability density for the classical harmonic oscillator and B is a positive number of order unity $B = 1/2 \sim 4$.

4 Conclusion and Discussions

The transition from quantum to classical system is investigated with a coarse-grained position operator. Essential difference between $\alpha_\ell = 0$ and $\alpha_\ell \neq 0$ is important, the former gives quantum mechanical case, the latter gives classical property. In fact, we cannot obtain Eq.(16) from Eq.(17) through any limitting procedure. In addition, naturally, it does not permitted to take the limit $\alpha_\ell \to 0$ in Eq.(11). Only the energy of the system is continuously connected through Eq.(12). Introducing an extra-dimension, we have get a unified description of quantum and classical mechanics. This result is obtained without to take the limit $\hbar \to 0$.

References

1. S. Machida and M. Namiki, *Prog.Theor.Phys.* **63**,1457,1833 (1980); M. Namiki and S. Pascazio, *Phys.Rep.* **232**, 301 (1993).
2. A. Motoyoshi, *Nuovo Cim.* B **108**, 373, 385(1993); *Prog.Theor.Phys.* **91**, 625(1994).
3. T. Yoneda, A. Nagasato, T. Akamine, A. Motoyoshi, *Phys.Lett.* A **280**, 271 (2001).
4. T. Yoneda, A. Nagasato, T. Akamine, A. Motoyoshi, *Nuovo Cim.* B, **116**, 73 (2001).
5. T. Yoneda, A. Nagasato, T. Akamine, A. Motoyoshi, *Nuovo Cim.* B, in print (2001).

ENTANGLEMENT OF AN ASSEMBLY OF N IDENTICAL TWO LEVEL ATOMS

D. ULAM-ORGIKH AND M. KITAGAWA

Graduate School of Engineering Science, Osaka University
1-3 Machikaneyama-cho, Toyonaka, Osaka 560-8531 Japan;
CREST, Japan Science and Technology Corporation
E-mail: uka@laser.ee.es.osaka-u.ac.jp kit@qc.ee.es.osaka-u.ac.jp

We consider an assembly of N indistinguishable two-level atoms, for which a concept of spin-squeezing and entanglement measures are well defined. We explicitly show that a spin-squeezed state has pairwise entanglement and a squeezing parameter serves as a measure of entanglement in that case.

It is well known that both entangled states and squeezed states are valuable resources of quantum correlations. Entanglement being an essential ingredient in quantum information science is originated from the Hilbert space structure and the superposition principle, while squeezing is based on the uncertainty principle — another vital principle of quantum mechanics. Thus, it is of interest to study their relationship.

We deal in this Contribution with an assembly of N indistinguishable two-level atoms for which both entanglement measure and spin-squeezing are well defined. The time evolution of the assembly is governed by a Hamiltonian which is left unchanged under any permutation of the atomic labels. Thus, we introduce the following notation for permutation symmetric state: $|\Phi_s(N,k)\rangle = \sum_{\text{perm}} |1^{\otimes k}0^{\otimes(N-k)}\rangle$, where $\sum_{\text{perm}}$ describes a summation over all distinct permutations of $N-k$ zero's and k one's. A state we are dealing with may be written as

$$|\Psi\rangle = \sum_{k=0}^{N} c_k |\Phi_s(N,k)\rangle, \tag{1}$$

since the symmetric states are orthogonal $\langle\Phi_s(N,k)|\Phi_s(N,k')\rangle = \binom{2s}{k}\delta_{k',k}$. Moreover, the atomic assembly can also be considered as N identical spin-1/2 or qubit system, and one may define collective angular momentum operators as $J_k = \frac{1}{2}\sum_{i=1}^{N}\sigma_{k,i}$ $(k = x,y,z)$, where $\sigma_{k,i}$ is a Pauli operator for ith spin-1/2.

The eigenstates $|J,m\rangle$ of the total angular momentum operator $\hat{J}^2$ and J_z are equivalent to the normalized permutation symmetric states: $|J,m\rangle = \binom{2s}{J-m}^{-1/2}|\Phi_s(2J,J-m)\rangle$ $(m = -J,-J+1,\ldots,J)$. In other words the state Eq. (1) is restricted to the constant $J = N/2$ angular momentum subspace of maximum multiplicity $2J+1$.

In this subspace, a Clebsch-Gordan composition of spin-J into the spin-J_1 and spin-$(J-J_1)$ can be written in the following simple form:

$$|\Phi_s(N,k)\rangle = \sum_{l=0}^{\min\{2J_1,k\}} |\Phi_s(2J_1,l)\rangle \otimes |\Phi_s(N-2J_1,k-l)\rangle, \tag{2}$$

and turns out to be a useful tool.

For instance, with $J_1 = 1$ we can easily write the reduced density matrix $\rho_{AB} = Tr_{CD\ldots}|\Psi\rangle\langle\Psi|$ as :

$$\rho_{AB} = \begin{pmatrix} \langle\varphi_0|\varphi_0\rangle & \langle\varphi_0|\varphi_1\rangle & \langle\varphi_0|\varphi_1\rangle & \langle\varphi_0|\varphi_2\rangle \\ \langle\varphi_1|\varphi_0\rangle & \langle\varphi_1|\varphi_1\rangle & \langle\varphi_1|\varphi_1\rangle & \langle\varphi_1|\varphi_2\rangle \\ \langle\varphi_1|\varphi_0\rangle & \langle\varphi_1|\varphi_1\rangle & \langle\varphi_1|\varphi_1\rangle & \langle\varphi_1|\varphi_2\rangle \\ \langle\varphi_2|\varphi_0\rangle & \langle\varphi_2|\varphi_1\rangle & \langle\varphi_2|\varphi_1\rangle & \langle\varphi_2|\varphi_2\rangle \end{pmatrix}, \tag{3}$$

where $\langle\varphi_j|\varphi_i\rangle = \sum_{k=0}^{N-2} c_{k+i}c^*_{k+j}\binom{N-2}{k}$ $(i,j = 0,1,2)$, which has at most three eigenvalues due to its symmetry property. Using the following correlation sums
$\sum_{k=0}^{2J} k\binom{2J}{k}c_k c^*_{k-1} = \langle J_+\rangle$,
$\sum_{k=0}^{2J} k(k-1)\binom{2J}{k}c_k c^*_{k-2} = \langle J_+^2\rangle$,
$\sum_{k=0}^{2J} k\binom{2J}{k}|c_k|^2 = \langle(J - J_z\rangle$,

$\sum_{k=0}^{2J} k^2 \binom{2J}{k} |c_k|^2 = \langle (J - J_z)^2 \rangle,$
$\sum_{k=0}^{2J} k^2 \binom{2J}{k} c_k c_{k-1}^* = \langle J_+(J - J_z) \rangle$
$= \quad \frac{1}{2} \{ (N+1) \langle J_+ \rangle - \langle J_+ J_z + J_z J_+ \rangle \},$
one can express the matrix elements via mean values of the first and second order momenta:

$$\langle \varphi_0 | \varphi_0 \rangle = \frac{J(J-1) + \langle J_z^2 \rangle + (2J-1)\langle J_z \rangle}{N(N-1)},$$
$$\langle \varphi_1 | \varphi_1 \rangle = \frac{J^2 - \langle J_z^2 \rangle}{N(N-1)},$$
$$\langle \varphi_2 | \varphi_2 \rangle = \frac{J(J-1) + \langle J_z^2 \rangle - (2J-1)\langle J_z \rangle}{N(N-1)},$$
$$\langle \varphi_0 | \varphi_1 \rangle = \frac{(J - \frac{1}{2})\langle J_+ \rangle + \frac{1}{2}\langle J_z J_+ + J_+ J_z \rangle}{N(N-1)},$$
$$\langle \varphi_0 | \varphi_2 \rangle = \frac{\langle J_+^2 \rangle}{N(N-1)},$$
$$\langle \varphi_1 | \varphi_2 \rangle = \frac{(J - \frac{1}{2})\langle J_+ \rangle - \frac{1}{2}\langle J_z J_+ + J_+ J_z \rangle}{N(N-1)}.$$

It is obvious that extreme mean values of the momenta are the rotational invariants and entanglement measures should be defined by them. As for spin-squeezing in the sense of quantum correlations it also should be defined by these invariants.

In literature, many definitions of spin-squeezing are used. A definition based on uncertainty principle of momentum operators is not invariant under rotation and thus does not necessarily reflect quantum correlations in contrast to the boson squeezing. Thus we use the following definition[1]: a spin is regarded as squeezed only if the minimum variance $\langle \Delta J_\perp^2 \rangle$ of a spin component perpendicular to the direction $\hat{n}$ of the mean spin vector is smaller than the standard quantum limit (SQL) $\frac{J}{2}$ of the coherent spin state (CSS), i.e., the state is spin-squeezed iff $\xi_\perp = \langle \Delta J_\perp^2 \rangle / (J/2) < 1$. The idea is that quantum correlations are needed to overcome SQL and this definition is intuitively supposed to be used only for the maximum multiplicity subspace in which CSS belongs.

It is important that this definition also implies negative pair-wise correlations[2]. We show in this contribution that these correlations are indeed quantum by using some measure of entanglement – concurrence[3,4,5]. The concurrence is defined as $C_{\text{mix}} = \max\{0, \lambda_1 - \lambda_2 - \lambda_3 - \lambda_4\}$ for a mixed state of a pair of spins AB, where λ_1, λ_2, λ_3, and λ_4 are the square roots of the eigenvalues of the product $\rho_{AB}(\sigma_y \otimes \sigma_y)\rho_{AB}^*(\sigma_y \otimes \sigma_y)$ in decreasing order. Here the asterisk denotes complex conjugation. For a pure state, it reduces simply to $C_{\text{pure}} = 2\sqrt{\det \rho_A}$, where $\rho_A = Tr_B \rho_{AB}$.

Although there is no widely accepted definition of concurrence or entanglement for more qubits, the permutation symmetry of the state allows one to apply the above definition in the following sense[5]: (1) The entanglement between one of the spin with the rest $C_{\text{rest}} = C_{\text{pure}}$ with $\rho_A = Tr_{BCD\ldots}\rho$, i.e., by considering all other spins except spin A as a single object. (2) Pair-wise entanglement $C_{\text{pair}} = C_{\text{mix}}$ with $\rho_{AB} = Tr_{CDE\ldots}\rho$. We refer to Coffman *et. al.*[5] for details and just note that C_{rest} is valid only for a pure state while C_{pair} may also be used for mixed state. A residual entanglement or tangle was also defined as $\tau = C_{\text{rest}}^2 - (N-1)C_{\text{pair}}^2$.

Now it is easy to see that concurrence C_{rest} of permutation symmetric pure state is defined by the mean spin value as

$$C_{\text{rest}} = \sqrt{1 - (\langle J_{\hat{n}} \rangle / J)^2}, \tag{4}$$

by using

$$\rho_A = Tr_B \rho_{AB} = \frac{1}{J} \begin{pmatrix} J + \langle J_z \rangle & \langle J_- \rangle \\ \langle J_+ \rangle & J - \langle J_z \rangle \end{pmatrix}$$

and $\langle J_+ \rangle \langle J_- \rangle = \langle J_x \rangle^2 + \langle J_y \rangle^2$.

As for pair-wise entanglement, let us consider the case of $\langle J_z J_+ + J_+ J_z \rangle = 0$, which includes states symmetric under the exchange 0 and 1. Under the rotation of the frame of reference both entanglement and squeezing should not change. Therefore, by applying appropriate rotations one may choose as: $\langle J_x \rangle = \langle J_y \rangle = 0$, $\langle J_z \rangle = \langle J_{\hat{n}} \rangle$. Then by means of a virtual rotation around $\hat{n}$, the extrema of perpendicular variances are found in the form

$$\langle J_{\substack{\max \\ \min}}^2 \rangle = \tfrac{1}{2} \{ \langle J_x^2 + J_y^2 \rangle \pm \sqrt{[\langle J_x^2 - J_y^2 \rangle]^2 + \langle J_x J_y + J_y J_x \rangle^2} \},$$

or we have $\langle J_x^2 + J_y^2 \rangle = \langle J_{\max}^2 + J_{\min}^2 \rangle$ and $|\langle J_+^2 \rangle| = \langle J_{\max}^2 - J_{\min}^2 \rangle$.

By denoting $A = (J - 2\langle J^2_{\min}\rangle)/(J(2J-1))$ and $B = (J^2 - \langle J^2_{\hat{n}}\rangle - [\langle J^2_{\max} - J^2_{\min}\rangle^2 + 4(J^2 - \langle J^2_{\max}\rangle)(J^2 - \langle J^2_{\min}\rangle)]^{1/2})/(J(2J-1))$, the concurrency may be written as:

$$C_{\text{pair}} = \max\{0, |A-B| - (A+B)\} = \begin{cases} \max\{-2A, 0\} & \text{when } A < B \\ \max\{-2B, 0\} & \text{when } B < A \\ |A-B| & \text{when } A+B=0. \end{cases} \quad (5)$$

One may readily check that $A < 0$ iff $K^{i,j}_{\min,\min} < 0$, and $B < 0$ iff $K^{i,j}_{\hat{n},\hat{n}} < 0$ by using pair-spin correlation coefficient $K^{i,j}_{\mu,\mu} = \langle\sigma_{\mu,i}\sigma_{\mu,j}\rangle - \langle\sigma_{\mu,i}\rangle\langle\sigma_{\mu,j}\rangle = [2\langle J^2_\mu\rangle - J - (2J-1)\langle J_\mu\rangle^2/J]/[J(2J-1)]$. Thus, we see the equivalence between the negative pair-spin correlation ($K^{i,j}_{\mu,\mu} < 0$) and quantum pair-wise entanglement ($C_{\text{pair}} > 0$) in this case. For instance, the following simple relation holds

$$C_{\text{pair}} = -K^{i,j}_{\min,\min} = \frac{1-\xi_\perp}{N-1} \quad (6)$$

for spin-squeezed state $\xi_\perp < 1$. The state with $A+B=0$ is a new kinds of minimum uncertainty state in the sense that the equality $[J^2 - \langle J^2_{\min}\rangle][J^2 - \langle J^2_{\max}\rangle] = \frac{1}{4}(N-1)^2\langle J_{\hat{n}}\rangle^2$ holds and the concurrence becomes $C_{\text{pair}} = (|1-\xi_{\min}|)/(N-1)$.

We skip a discussion for more general case due to the limited space and apply the results for some interesting states.

Example. 1 (*Dicke state* $|J,m\rangle$). It is well known that an atomic Dicke state $|J,m\rangle$ displays strong atom-atom correlations[6] and spontaneous emission of the super-radiance state ($m \approx 0$) has quantum origin because the state has no quadruple momentum responsible for classical radiation. We see all Dicke states are indeed quantum correlated (Fig. 1) by finding the concurrencies:

$$C_{\text{rest}} = \sqrt{(J+m)(J-m)/J^2}$$

and

$$C_{\text{pair}} = \frac{\sqrt{(J+m)(J-m)}}{J(2J-1)}\left[\sqrt{(J+m)(J-m)} - \sqrt{(J+m-1)(J-m-1)}\right].$$

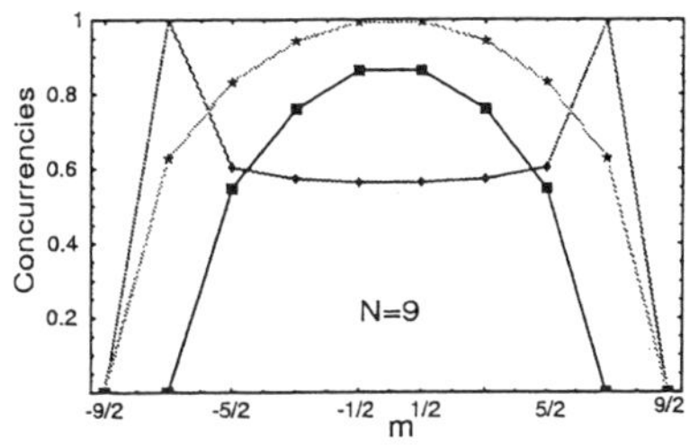

Figure 1. Entanglement measures for a Dicke state $|J,m\rangle$; ★, concurrence for entanglement between one of the spins with the rest C_{rest}; ♦, normalized pair-wise concurrence $(N/2)C_{\text{pair}}$; ■, residual entanglement τ. Super-radiant Dicke state ($m \approx 0$) has indeed maximum concurrence C_{rest} or entanglement.

The super-radiance state has the maximum entanglement C_{rest}, thus justifying the definition for C_{rest}. Dicke state possesses also pair-wise entanglement $C_{\text{pair}} > \frac{1}{N}$, but not spin-squeezed $\langle J^2_{\max}\rangle = \langle J^2_{\min}\rangle \geq J/2$.

Example. 2 (*A symmetric superposition of Dicke states* $|\varphi\rangle = \frac{1}{\sqrt{2}}[|J,m\rangle + |J,-m\rangle]$). These states are maximally entangled in the sense that $C_{\text{rest}} = 1$ and can be considered as a generalization of N-Cat like state (Fig. 2). It is surprising that only the states with $m \in [-\sqrt{N}/2, \sqrt{N}/2]$ have pair-wise entanglement, while the others close to the N-Cat state $\frac{1}{\sqrt{2}}(|J,J\rangle) + |J,-J\rangle)$ do not, as can be seen from $C_{\text{pair}} = \max\left\{0, \frac{J-2m^2}{J(2J-1)}\right\} \leq 1/(N-1)$. It can be easily checked that this state is also the minimum uncertainty one ($A+B=0$) and not spin-squeezed $\langle J^2_{\max}\rangle = \langle J^2_{\min}\rangle > J/2$.

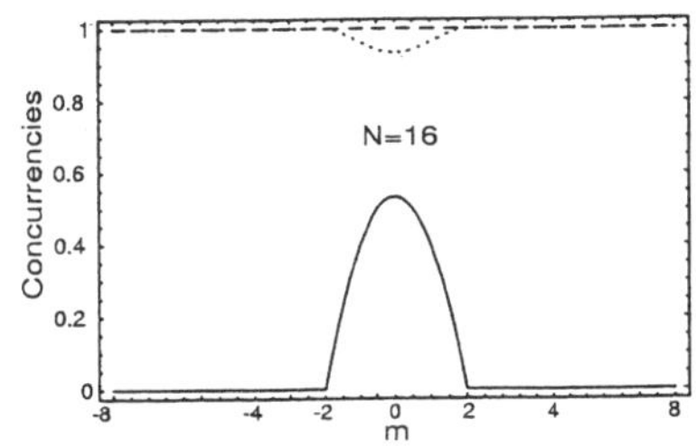

Figure 2. Concurrencies for a symmetric superposition state $|J,m\rangle + |J,-m\rangle$. Solid-line: C_{pair}, Dashed-line: C_{rest}, Dotted-line: τ. All states are maximally entangled in the sense $C_{\text{rest}} = 1$, while pair-wise entanglement appears only for the states with $m \in [\sqrt{N}/2, \sqrt{N}/2]$.

Example. 3 (*Spin-squeezing models*). We plot the evolution of entanglement of one-axis twisting $H_{\text{oa}} = \hbar\chi J_z^2$ and two-axis twisting $H_{\text{ta}} = \hbar\chi(J_+^2 - J_-^2)$ models of spin-squeezing with an initial CSS. Here $\nu = \chi t$ is a squeezing evolution parameter and t is an evolution time. It is important that spin-squeezing or negative pair-wise entanglement which is capable to overcome SQL in the presence of decoherence[2] occurs at the starting stage of the squeezing evolution in both models as can be seen from Figs. 3-4. For both models, we simply have one-to-one correspondence $\xi_\perp = 1 - (N-1)C_{\text{pair}}$, except the singular case of $N = 3$ for two-axis twisting ($C_{\text{pair}} = \frac{|1-\xi_\perp|}{N-1}$), which is a minimum uncertainty state and plotted in Fig. 5.

In summary, we have shown that the definition of spin squeezing[1] implies quantum negative pair-wise correlations for indistinguishable particles. We have found simple relations between concurrence and physical quantities, which hint some physical motivation for concurrency[4,5]. The further study using the extreme mean values for higher momenta to find possible measures of multipartite entanglement of pure and mixed states would be interesting.

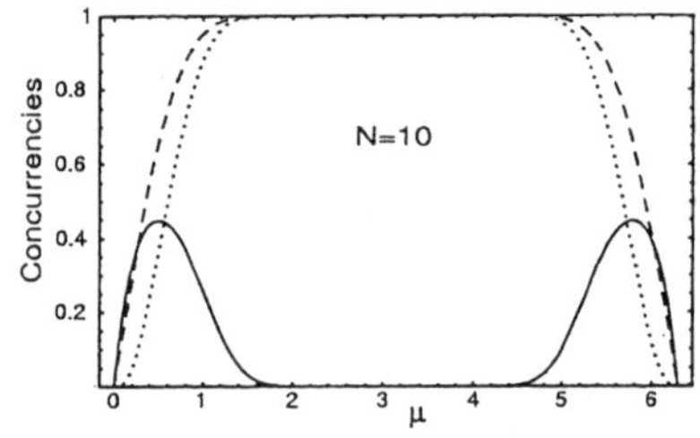

Figure 3. Entanglement of one-axis twisting vs squeezing evolution parameter μ. Solid-line: C_{pair}, Dashed-line: C_{rest}, Dotted-line: τ. A state is said to be spin-squeezed when $\xi_\perp = 1 - (N-1)C_{\text{pair}} < 1$.

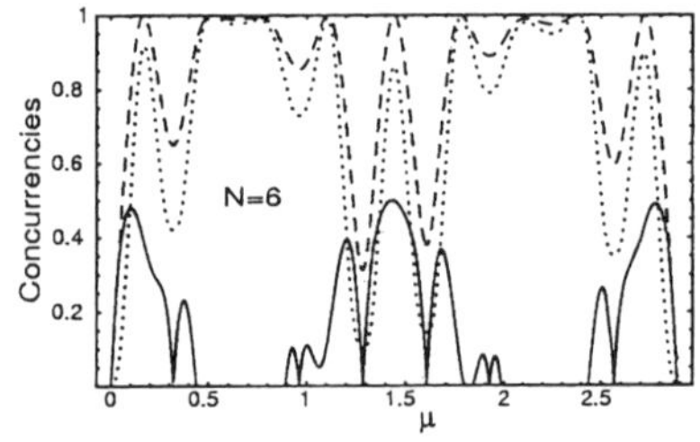

Figure 4. Evolution of entangled measures for two-axis twisting Hamiltonian. Solid-line: $C_{\text{pair}} = \frac{1-\xi_\perp}{N-1}$, Dashed-line: C_{rest}, Dotted-line: τ. A pair-wise negative correlation or spin-squeezing always appears at the beginning short stage of evolution, although the evolution picture becomes more chaotic for large N.

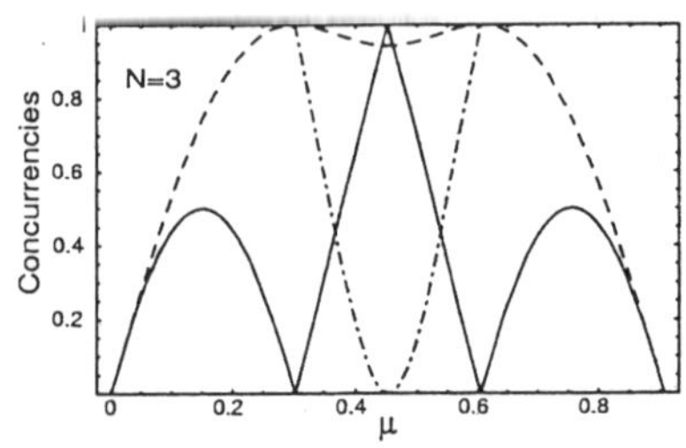

Figure 5. Entanglement of two-axis twisting vs μ. The non-symmetric singular case of $N = 3$. Solid-line: C_{pair}, Dashed-line: C_{rest}. It is not spin-squeezed $\langle J^2_{\text{min}}\rangle \geq J/2$ in the middle region of $\mu \in [\frac{\pi}{6\sqrt{3}}, \frac{\pi}{3\sqrt{3}}]$. We have also plotted a negative correlation coefficient (Dash-dotted line) $\langle \Delta J^2_{\vec{n}}\rangle(4/J^2)/(1 - \langle J_{\vec{n}}\rangle^2 J^{-2})$.

References

1. M. Kitagawa and M. Ueda, Phys. Rev. Lett. **67**, 1852 (1991); Phys. Rev. A **47**, 5138 (1993).
2. D. Ulam-Orgikh and M. Kitagawa, Spin squeezing and decoherence limit in Ramsey spectroscopy, Phys. Rev. A. , Accepted for publication (2001).
3. S. Hill and W.K. Wootters, Phys. Rev. Lett. **78**, 5022 (1997).
4. W.K. Wootters, Phys. Rev. Lett. **80**, 2245 (1998).
5. V. Coffman, J. Kundu, and W.K. Wootters, Phys. Rev. A. **61**, 052306 (2000).
6. R.H. Dicke, Phys. Rev. **93**, 99 (1954).

NEUTRON SCATTERING BY ENTANGLED SPATIAL AND SPIN DEGREES OF FREEDOM

E. B. KARLSSON

Department of Physics, Uppsala University, P.O. Box 530, S-75121 Uppsala, Sweden
E-mail: erk@fysik.uu.se

S. W. LOVESEY

Rutherford Appleton Laboratory, Chilton, Didcot, Oxon OX11 0QX, UK
E-mail: s.w.lovesey@rl.ac.uk

A theoretical interpretation of a set of neutron scattering experiments on protons and deuterons is presented in terms of quantum entanglement of spatial and spin degrees of freedom. A soluble model for the scattering of pairs of particles is developed, and it is applied to hydrogen in different condensed matter environments. Decoherence limits the lifetime of quantum entanglement to times of the order of 10^{-15} s.

1 Proton entanglement in condensed matter

Quantum entanglement of atoms is known to exist for extended times (up to microseconds) in very well isolated systems, such as atomic traps or r.f. cavities, but it is lost very rapidly as a result of interaction with an environment. The influence of decoherence[1] is expected to depend on the square of the size of the quantum subsystem under consideration, as well as on the frequency and momentum transfer in system-environment interaction events. For typical condensed-matter environments it can be estimated, even for systems as small as simple molecules, that decoherence is so fast that entanglement is virtually unobservable by standard observation techniques.

In recent experiments, published by Chatzidimitriou-Dreismann *et al.*[2,3] and by Karlsson *et al.*,[4,5] the effective duration of an observation τ_{sc} for protons in water, polymers, and metallic hydrides, has been reduced down to the sub-femtosecond range by suitably choosing the experimental conditions in neutron scattering. All these experiments show anomalously low proton cross-sections for $\tau_{sc} < 10^{-15}$ s. It will be shown here that this is to be expected if spatial and spin degrees of freedom of the particles are entangled and the duration of a neutron scattering event is shorter than the decoherence time, τ_{coh}, for the entanglement.

It is known that thermal neutron scattering experiments (with $\tau_{sc} \approx 10^{-13}$ s) show cross-sections of molecular hydrogen, H_2, which are different from the sum of the cross-section of two individual protons. This is due to correlations in the two-particle wavefunction caused by the indistinguishability of protons. This is observed, not only for molecular hydrogen in gaseous form (where environmental interactions are relatively small) but also in liquefied H_2 at low temperatures. For molecules, like CH_4, small effects of proton-proton correlations still remain[6] when studied in thermal neutron scattering, but more complicated molecules show cross-sections that are well explained in terms of distinguishable nuclei.

The new experiments use neutrons of higher energy (20–100 eV) in Compton scattering. At these energies the struck proton is highly excited, which makes it necessary to reformulate the standard calculation of the neutron cross-section.[7,8]

2 A soluble model for neutron Compton scattering by indistinguishable particles

As a starting point we have obtained without approximation the cross-section for Compton scattering by two identical particles, α and β referred to sites 1 and 2, whose spatial and spin degrees of freedom are correlated because of their indistinguishability. The initial state is ($\zeta = (-1)^J$)

$$\frac{1}{\sqrt{2}}\{\phi_1(\mathbf{R}_\alpha)\phi_2(\mathbf{R}_\beta) + \zeta\phi_1(\mathbf{R}_\beta)\phi_2(\mathbf{R}_\alpha)\}\,\chi_M^J(\alpha,\beta). \quad (1)$$

Here $\chi_M^J(\alpha,\beta)$ is the initial spin state with total angular momentum J and projection M. The neutron scattering operator has the form

$$V = b_\alpha \exp(i\mathbf{k}\cdot\mathbf{R}_\alpha) + b_\beta \exp(i\mathbf{k}\cdot\mathbf{R}_\beta), \quad (2)$$

where b is the scattering length operator. The specific features of Compton scattering are reflected in the following Ansatz for the state of the pair after the neutron encounter (having spin states $J'M'$ and $\zeta' = (-1)^{J'}$),

$$\frac{1}{\sqrt{2}}\{\exp(i\mathbf{p}'\cdot\mathbf{R}_\alpha)\psi(\mathbf{R}_\beta) + \zeta'\exp(i\mathbf{p}'\cdot\mathbf{R}_\beta)\psi(\mathbf{R}_\alpha)\}\,\chi_{M'}^{J'}(\alpha,\beta) \quad (3)$$

Here, one particle remains in a state of low energy, described by $\psi(\mathbf{R})$, whereas the other is in the form of a plane wave. At this stage their quantum correlation is not yet broken and the identity α or β of the struck particle will be settled only after the decoherence has occurred. It is assumed that the coherence length of the neutron wave-packet is not too short so as to exclude spatial coherence.

It was shown in Ref. 7 that the spatial integrals are reduced such that the matrix element $\langle f|V|i\rangle$ can be written

$$\begin{aligned}\langle f|V|i\rangle &= K(\mathbf{p})\frac{1}{2}\chi_{M'}^{J'}(\alpha,\beta)[b_\alpha + \zeta\zeta' b_\beta] \\ &\quad\cdot\chi_M^J(\alpha,\beta)[T_2 + \zeta\exp(-i\mathbf{p}\cdot\mathbf{d})T_1] \\ &= K(\mathbf{p})F(J'M',JM) \quad (4)\end{aligned}$$

where $K(\mathbf{p}) = \int d\mathbf{R}\exp(-i\mathbf{R}\cdot\mathbf{p})\phi_1(\mathbf{R})$ and $T_j = \int d\mathbf{R}\psi^*(\mathbf{R})\phi_j(\mathbf{R}), (j = 1,2)$. Here $\mathbf{p}$ and $\mathbf{p}' = \mathbf{p}+\mathbf{k}$ are the initial and final wave-vectors, respectively, of the struck particle, with $\mathbf{k}$ the momentum transfer in the process. The last equality in (4) defines $F(J'M',JM)$. The spatial phase factor $\exp(-i\mathbf{p}\cdot\mathbf{d})$ arises in the momentum wavefunction created from $\phi_2(\mathbf{R}) = \phi_1(\mathbf{R}-\mathbf{d})$ and the momentum distribution of the struck particles is contained in the expression $K(\mathbf{p})$. The Ansatz correctly reproduces the energy-dependence of scattering with intensity centred about the recoil energy. The scattering length operators in (2) have the form $b = A + B\mathbf{s}\cdot\mathbf{I}$, where $\mathbf{s}$ is the neutron spin operator and $\mathbf{I}$ the spin operator for the individual scattering particles, making up the correlated pair.

When initial and final states are orthogonal, only terms with $J' \neq J$ are contributing, and the scattered signal can be written

$$\begin{aligned}|F(J'M',JM)|^2 &= \langle J'M'|b_\alpha|JM\rangle^2 \\ &\quad\cdot|T_2 + \zeta\exp(-i\mathbf{p}\cdot\mathbf{d})T_1|^2. \quad (5)\end{aligned}$$

To obtain the actual intensity, $\langle J'M'|b|JM\rangle^2$ is averaged over the projections of the initial total spin and summed over the values of the projection of the final total spin.

The cross-section per particle is

$$\frac{1}{2}[I(I+1)/4]B^2\{|T_1|^2 + |T_2|^2\}. \quad (6)$$

Recognizing $\pi I(I+1)B^2/4$ as the incoherent cross-section σ_{inc} the cross-section per particle is reduced by a factor

$$f = \frac{1}{4}(\sigma_{\text{inc}}/\sigma)\{|T_1|^2 + |T_2|^2\} \quad (7)$$

when $J' \neq J$. Since $|T_1|^2$ and $|T_2|^2$ both can be shown to be $\leq 1/2$, the entanglement in scattering, by spatial and spin degrees of freedom, is seen to lead to a reduction for protons of more than 75% of the cross-section $\sigma = \sigma_{\text{coh}} + \sigma_{\text{inc}}$. It is notable that this is a purely quantum mechanical effect since the overlap of final and initial states is zero and cannot lead to additional correlations. If the superposition in (3) is reduced to a mixture, each particle acts as a local scatterer with $f = 1$.

3 Applications

Let us consider protons and set $I = 1/2$. Relaxing the condition $J' \neq J$ and allowing all (J, J') consistent with the angular momentum coupling gives,

$$f = \frac{1}{4}\left\{\frac{1}{2}\left|T_1 + \exp(i\mathbf{p}\cdot\mathbf{d})T_2\right|^2 + \frac{3}{2}\left|T_1 - \exp(i\mathbf{p}\cdot\mathbf{d})T_2\right|^2\right\}. \tag{8}$$

This is equal to $f = 1/4$ when the local vibrations $\mathbf{p}$ are perpendicular to the vector $\mathbf{d}$ connecting the two protons. Isotropic vibrations have 2/3 of the intensity $\perp$ $\mathbf{d}$, which results in $f = 1/3$.

For the metal hydrides,[4,5,9] the measured reduction factors for the proton cross-sections fall in the range $(f)_{\rm obs} \approx 0.6$–$0.8$ for the shortest observation times while the polymer polystyrene,[9] has $(f)_{\rm obs} \approx 0.8$. The fact that the observed reduction is less than calculated in the two-particle model can be explained by an entanglement that is not maximal since more than two protons can be involved, as well as by an insufficient neutron coherence length. The published data for partially deuterated water[2] can be fitted in absolute terms (Fig. 1) by our model if it is assumed that HD combinations are not quantum correlated and DD correlation effects are small ($f \approx 1$, see below). In water, the intramolecular (two-particle) H-correlations dominate, and furthermore, d is small (1.4 Å). The H-vibrations in the water molecule are such that the product $\mathbf{p}\cdot\mathbf{d}$ makes $f_{\rm HH} \approx 3/8$.

Corresponding data for deuterated hydrides are also presented in Refs. 4 and 9. In general, the anomalies are much smaller than for the protons, with less than 10% reduction compared to standard values. The much weaker degree of entanglement observed for deuterons is in keeping with the expected decrease of quantum overlap with increasing mass.

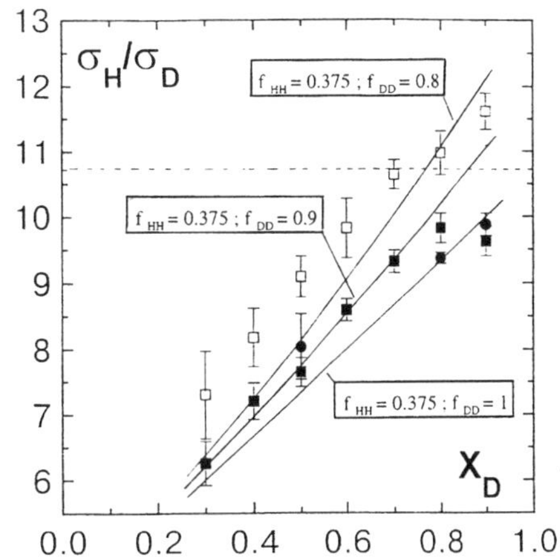

Figure 1. Ratios of H to D scattering intensities in partially deuterated water (X_D being the deuterium fraction) as measured by Chatzidimitriou-Dreismann *et al.*[2] using Au-foils (full circles) and U-foils (squares). The lines are predictions from the present theory, using $f_{\rm HH} = 3/8$ and $f_{\rm DD} = 1$, 0.9 and 0.8.

4 Decoherence

A second unique feature of the neutron Compton scattering technique is that the data can be analysed in a time-differential manner in the range below and around 10^{-15} s. This was first pointed out and implemented in Ref. 4 and is based on the relation between the duration $\tau_{\rm sc}$ and the transferred momentum $k(\theta)$ at different scattering angles θ,

$$\tau_{\rm sc}(\theta) \approx \frac{M}{k(\theta)\langle p^2\rangle^{1/2}}. \tag{9}$$

Figure 2 shows published data for protons in an Nb-hydride[4] (the same is valid for Pd-hydride[9]) with the cross-section reaching standard, individual proton values for $\tau_{\rm sc} > 10^{-15}$ s. The polymer and water experiments[2,3] cover a lower time range within which the cross-section reduction is

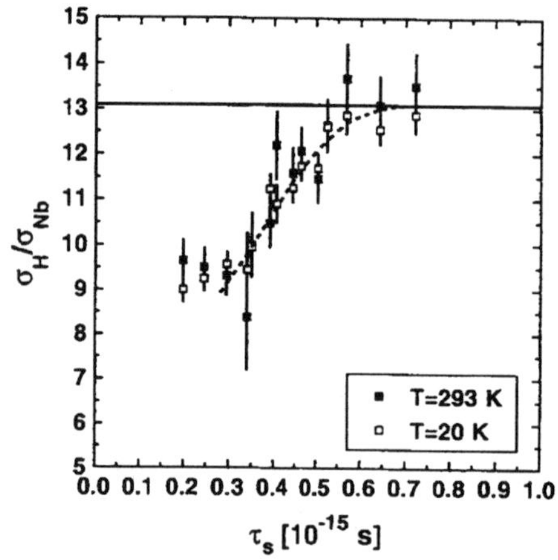

Figure 2. Ratio of H to Nb scattering intensities in Nb-hydride, as a function of the duration of a scattering event. The full line is the ratio 13.1 expected without proton entanglement in scattering.

fairly constant.

The Y-hydrides, reported in this Symposium, [5] show a time-dependence but do not reach full values in the time range observed. The return to normal values with increasing observation time is interpreted as an effect of decoherence where phase relations are lost and the wavefunction (3) is reduced to a mixture. With such a reduced final state the two-particle model predicts the classical result for individual particles. It is unlikely that the normal thermally excited processes in the environment are the main decoherence mechanism in this case, since the cross-section reductions are insensitive to temperature variations.[4] More probably, the short decoherence time in the metal hydrides is caused by the strong excitations when the struck particle starts it trajectory. If left in thermal equilibrium it is likely that forces that create and destroy entanglement would be in balance over times longer than 10^{-15} s.

5 Conclusions

The cited experiments, together with the theoretical analysis,[7,8] indicate that closely lying protons are entangled in scattering over times at least of the order of 10^{-15} s. The entanglement may, as in the theoretical model sketched above, be a result of the indistinguishability of the particles. Alternatively, entanglement might be due to the coupling of the protons to other common degrees of freedom. The existence of local quantum entanglement at this time scale may be of importance for the initial stages of chemical reactions involving hydrogen in condensed matter.

References

1. E. Joos and H. D. Zeh, Z. Phys. B **59**, 223 (1985).
2. C. A. Chatzidimitriou-Dreismann, T. Abdul-Redah, R. M. F. Streffer, and J. Mayers, Phys. Rev. Lett. **79**, 2839 (1997).
3. C. A. Chatzidimitriou-Dreismann, T. Abdul-Redah, and J. Sperling, J. Chem. Phys. **113**, 2784 (2000).
4. E. B. Karlsson, C. A. Chatzidimitriou-Dreismann, T. Abdul-Redah, R. M. F. Streffer, B. Hjörvarsson, J. Öhrmalm, and J. Mayers, Europhys. Lett. **46**, 617 (1999).
5. E. B. Karlsson, T. Abdul-Redah, T. J. Udovic, B. Hjörvarsson, and C. A. Chatzidimitriou-Dreismann, this Symposium.
6. A. C. Zemach and R. J. Glauber, Phys. Rev. **101**, 129 (1956).
7. E. B. Karlsson and S. W. Lovesey, Phys. Rev. A **61**, 062714 (2000).
8. E. B. Karlsson and S. W. Lovesey, Physica Scripta **65**, 112 (2002).
9. T. Abdul-Redah, R. M. F. Streffer, C. A. Chatzidimitriou-Dreismann, B. Hjörvarsson, E. B. Karlsson and J. Mayers, Physica B **276–278**, 824 (2000).

SHORT-LIVED PROTON ENTANGLEMENT IN YTTRIUM HYDRIDES

E. B. KARLSSON[1], T. ABDUL-REDAH[2], T. J. UDOVIC[3], B. HJÖRVARSSON[4] AND C. A. CHATZIDIMITRIOU-DREISMANN[2]

[1]*Department of Physics, Uppsala University, P.O. Box 530, SE-75121 Uppsala, Sweden*
[2]*Institute of Chemistry, Stranski Lab., Sekr. ER 1, TU Berlin, D-10623 Berlin, Germany*
[3]*Materials Science and Engineering Laboratory, NIST, Maryland 20899, USA*
[4]*Department of Materials Physics, Royal Institute of Technology, SE-10044 Stockholm, Sweden*

Previous experiments on $NbH_{0.8}$ and $PdH_{0.6}$ have shown large anomalies in the cross sections for protons, when studied by neutron Compton scattering. Here, these investigations are extended to the metallic hydrides YH_2, YH_3, YD_2, YD_3, and $Y(H_xD_{1-x})_3$. Considerably reduced cross sections for hydrogen are observed both in YH_2 and YH_3, but only minor ones for YD_2 and YD_3. The scattering time depends on the neutron scattering angle, which allows a time-differential analysis where the time window lies around one femtosecond. The anomalies persist longer in YH_2 and YH_3 than in $NbH_{0.8}$ and $PdH_{0.6}$. The reduced cross sections are interpreted as a result of quantum entanglement between protons, surviving for a few fs in the solids.

1 Introduction

Neutron Compton scattering (NCS) is a method where the interaction takes place within a very short time, of the order of 10^{-16} to 10^{-15}s. NCS measurements can also be analysed in a time-differential manner. They offer unique possibilities to study quantum correlations and entanglement in systems where decoherence is fast. There are now several experiments which have shown that protons exhibit anomalously low cross sections when observed in neutron Compton scattering.

The first one was made by Chatzidimitriou-Dreismann et al. on various H_2O-D_2O mixtures,[1] where, for the lower D-concentrations, the ratio of scattering cross-sections of H and D, σ_H/σ_D, was found to be about 30% lower than expected from the tabulated cross sections. Another experiment was performed by Karlsson et al.[2] on a set of Nb-hydrides, $Nb(H_{1-x}D_x)_{0.8}$, with x = 0, 0.28, 0.54, 0.81 and 0.96, where σ_H/σ_{Nb} and σ_D/σ_{Nb} were determined, again showing strongly reduced values for σ_H/σ_{Nb}. A measurement on $PdH_{0.6}$ by Abdul-Redah et al.[3] gave similar results as those obtained for $NbH_{0.8}$. More recently, NCS experiments have been performed by Chatzidimitriou-Dreismann et al. on solid hydrocarbon polymers[4], organic liquids (submitted) and amphiphilic molecules (submitted), showing a deficiency of about 20% in H-cross section, when compared to that of carbon analyzed simultaneously in the same sample.

Our theoretical interpretation is that these anomalous cross-sections are due to short-lived quantum entanglement (QE) of the protons. QE of protons exists in the H_2 molecules, which makes thermal neutron cross sections of both para- and ortho-hydrogen widely different from that of the sum of two individual protons. Thermal neutrons with their long interaction times (of the order of 10^{-13}s) do not show any anomalies when they scatter on protons in normal condensed matter environments, but Compton scattered neutrons may still do so if the decoherence times τ_{dec} of QE are longer than ca. 10^{-15}s.

QE effects are expected to disappear as soon as environment-induced decoherence sets in. In some of the NCS experiments[2,3] on hydrogen in solids, such a return to normal cross-sections has been observed when the duration of the scattering was increased above 10^{-15}s. In the present experiments these investigations are extended to metal hydrides of higher H- (and D-) concentrations.

2 Experiments

Yttrium hydrides were chosen as test materials since their structure and physical properties have been well studied in earlier experiments [5]. In YH_2, the H atoms occupy the tetrahedral sites of an fcc Y sublattice. The corresponding hydrogen atom density is ≈ 57 atoms/nm^3 with near-neighbor H-H distances of 2.6 Å. In YH_3, two-thirds of the H atoms are located in distorted tetrahedral sites while one-third are arranged in trigonal-type sites in the vicinity of the metal basal planes of an hcp Y sublattice. In this case, the corresponding hydrogen atom density is ≈ 78 atoms/nm^3 with smaller near-neighbor H-H distances in the range of 2.1 – 2.6 Å.

Yttrium hydrides with the following compositions: (*i*) $Y(H_{1-x}D_x)_2$ (x=0 and 1) and (*ii*) $Y(H_{1-x}D_x)_3$ (x=0, 0.2 and 1) were prepared at NIST, by quantitatively controlled reaction of high-purity Y (99.99 atomic %) with H_2 and D_2 gases. Neutron scattering measurements were made at the Electron-Volt Spectrometer (EVS) at ISIS, Rutherford-Appleton Laboratory, UK.

In a Compton experiment, neutrons of $E \approx 20 - 150$ eV energy are scattered by the nuclei in the sample. For scattering on hydrogen (H or D) momentum and energy transfers are large and the protons (deuterons) are recoiled out of their positions in the molecule or in the crystal where they are situated. In the EVS instrument outgoing neutrons are selected within a narrow energy interval, $E_1 = 4.91 \pm 0.14$ eV by the use of Au-197 resonance foils. Time-of-flight (TOF) neutron spectra are recorded, as described in Ref. 6. The inset of Fig. 1 shows examples of TOF spectra taken for YH_3 and YD_3 at one particular scattering angle, $\theta = 71°$. These spectra have been normalized according to a standard procedure based on the impulse approximation[6] and the peak areas A_H, A_D and A_Y should be proportional to $N_H\sigma_H$, $N_D\sigma_D$ and $N_Y\sigma_Y$.[1] However, the Y-signal can not be separated from that of the Al-container, since the Al-peak falls only slightly below the Y-peak in TOF spectra taken at forward scattering angles ($\theta < 90°$). Therefore, the angular range of the detectors was extended to $\theta = 140°$, where these peaks can be reasonably well resolved. The ratio A_Y/A_{Al} determined from the back scattering spectra (with $\theta > 90°$) was then used to extract the Y-fraction from the composite high mass peaks for $\theta < 90°$.

The forward scattering is detected by two detector banks, each composed of 8 Li-doped glass scintillators, which together cover an angular range from about 40 to 80 degrees. The σ_H/σ_Y (and σ_D/σ_Y) ratios were derived for each individual detector and compared with the tabulated cross section ratios.[7] The results reveal that the measured σ_H/σ_Y ratios are far below the standard value $81.7/7.70 = 10.6$, while the deviation from $7.63/7.70 = 0.99$ for σ_D/σ_Y is only marginal (data not shown). Both hydrides YH_3 and YH_2 exhibit the same anomalies. Furthermore, there is a tendency towards lower σ_H/σ_Y values at the higher scattering angles; cf. Fig. 1.

Strong dependences on scattering angle were noticed for the Nb-H and Pd-H systems.[2,3] In these works it was also shown that the dependence on scattering angle can be expressed as a dependence on the duration of the scattering process, i.e. the scattering time τ_{sc}, which for the neutron Compton scattering can be defined by[8]

$$\tau_{sc} = \frac{M}{\hbar q(\theta)\langle p_q^2\rangle^{1/2}} \tag{1}$$

Here $\langle p_q^2\rangle^{1/2}$ is the momentum spread (in Å^{-1}) in the local vibration of the scattering particle and $q(\theta)$ the angular dependent transferred momentum in Å^{-1}. Introducing

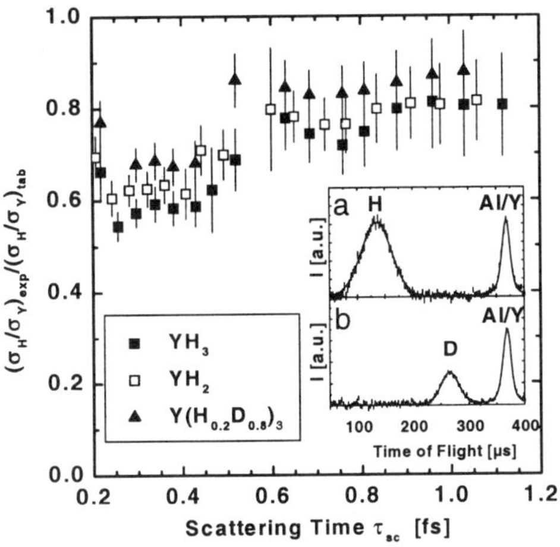

Figure 1. Experimentally determined ratios $(\sigma_H/\sigma_Y)_{exp}$ normalized with the tabulated one $(\sigma_H/\sigma_Y)_{tab}$ of YH_3, YH_2, and $Y(H_{0.2}D_{0.8})_3$ as a function of scattering time τ_{sc}. There is a strong deviation of the ratio σ_H/σ_Y from the tabulated value $(\sigma_H/\sigma_Y)_{tab.} = 10.6$. Examples of TOF spectra for a scattering angle of $\theta = 71$ degrees are shown in the inset: a) TOF spectrum of YH_3; b) TOF spectrum of YD_3 (accumulated over ca. 12 hrs). Intensities are given in arbitrary units.

the angular dependence explicitly, it is possible to write, for scattering on protons (with mass M),

$$\tau_{sc} = \frac{1}{tg(\theta)}\sqrt{\frac{M}{2E_1\langle p_q^2\rangle}} \tag{2}$$

With the actual values of the momentum spread, $\langle p_q^2\rangle^{1/2} \approx 4$ Å^{-1} (derived from the same EVS-spectra), and the range of angles used here, the data for H-scattering correspond to a time range $(0.2-1.2)\times 10^{-15}$ s.

Fig. 1 shows the ratios between observed (σ_H) and tabulated[7] $(\sigma_H)_{table}$ total H-cross sections, assuming that the "anomaly" is present only in σ_H and not in σ_Y. They are plotted as function of the scattering time τ_{sc} and can be compared with the result of the Nb-H and Pd-H data.[2,3] A slight τ_{sc}-dependence is visible. A general trend seems to be that the cross section reduction exhibits the same features for all yttrium hydride samples investigated, within present experimental error. Note also that the experiments in water[1] and polystyrene[4], which show no dependence on scattering angle, correspond to an essentially shorter time range owing to the larger values of $\langle p_q^2\rangle^{1/2}$ in these materials.

3 Interpretation and Discussion

For an explanation of the strong anomalies in NCS on hydrogen, it is essential to note that this technique offers a uniquely short experimental time window[1] as compared to most other neutron scattering arrangements (where usually holds $\tau_{sc} > 10^{-14}$s). It has been pointed out[9] that, in our case, it may still be possible to observe interference phenomena involving two or more protons since the coherence length $l_{coh} = \lambda^2/2\Delta\lambda$ is set by the energy sharpness of the resonance foil used to select the outgoing neutrons. The energy $E_1 = 4.91$ eV with its sharpness ± 0.14 eV corresponds to an estimated longitudinal coherence length of 2.5 Å,[9] which is of the same order of magnitude as the H-H distances in the metal hydrides (and considerably longer than the H-H distance in pure water). Because of this l_{coh} and the very short τ_{sc} realized at EVS, there exist therefore possibilities to observe effects of very short-lived quantum correlations between the scattering particles.[9]

The specific cross-sections for the H_2 molecule, mentioned in the Introduction, are due to the exchange correlations set up by the indistinguishability of the particles. The present experimental situation differs from thermal scattering on H_2 in that one of the two particles in the final state is, in the end, emitted in the form of a plane wave, but the symmetry rules are still expected to influence the interferences in the scattering from nearby H-atoms,[9] as long as their local quantum correlations remain within the (very short) scattering time in the NCS process.

A solvable two-particle model for particle entanglement effects in NCS has been put forward by Karlsson and Lovesey,[9,10] based on the rules for identical particles and the assumption that the local quantum correlations are not broken until after the scattering process has taken place. The recoiling particle or normal thermal processes have had no time to induce excitations in the solid or liquid in question. The strongly reduced cross section at the shortest observation times and the gradual transition towards normal cross sections (observed[2,3] for longer times in Nb-H and Pd-H) can be explained qualitatively by this model. Another theoretical model, in which, however, exchange correlations due to the possible indistinguishability of scattering particles play no dominant role, was proposed by Chatzidimitriou-Dreismann et al.[4,11] This is based on the basic van Hove formalism of neutron scattering, taking explicitly into account short-lived multi-particle entanglement. In was shown that entanglement and decoherence, which are caused by the ubiquitous many-body interactions in condensed matter, can explain the reduced cross sections density of H, if the decoherence time τ_{dec} is roughly of the order of magnitude of τ_{sc}.[4]

Fig. 1 shows that it is possible to follow the process of decoherence in the subfemtosecond range. For the yttrium hydrides studied here, the standard individual proton cross-sections are not reached at 1 fs as for the Nb-and Pd-hydrides mentioned,[2,3] but the latter had a considerably lower hydrogen content than the present ones and therefore less H-H overlap. The considered effect is set by the competition between the entanglement-creating forces and decoherence, caused by interactions with the environment.

Without going into the details of the mechanisms by which quantum correlations are lost during the scattering process, it seems possible to state that the experiments discussed in this report have shown the following: If hydrogenous systems are observed over sufficiently short times ($\leq 10^{-15}$ s), QE of nearby H-atoms is important, and it appears to survive decoherence.[4,9-11] This is a piece of information of interest for questions concerning the "system-environmental interactions" and the "quantum/classical boundary", as well as for the dynamics of various stages of chemical reactions involving hydrogen in condensed matter.

References

1. C. A. Chatzidimitriou-Dreismann, T. Abdul-Redah, R. M. F. Streffer and J. Mayers, *Phys. Rev. Lett.* **79,** 2839 (1997).
2. E. B. Karlsson, C. A. Chatzidimitriou-Dreismann, T. Abdul-Redah, R. M. F. Streffer, B. Hjörvarsson, J. Öhrmalm and J. Mayers, *Europhys. Lett.* **46,** 617 (1999).
3. T. Abdul-Redah, R. M. F. Streffer, C. A. Chatzidimitriou-Dreismann, B. Hjörvarsson, E. B. Karlsson and J. Mayers, *Physica* B **276-278,** 824 (2000).
4. C. A. Chatzidimitriou-Dreismann, T. Abdul-Redah and J. Sperling, *J. Chem. Phys.* **113,** 2784 (2000).
5. (a) T. J. Udovic et al., *Phys. Rev.* B **50,** 15739 (1994); (b) T. J. Udovic et al., *J. Phys. Chem. Solids* **57,** 423 (1996); and papers cited therein.
6. J. Mayers, T. M. Burke and R. J. Newport, *J. Phys. Condensed Matter* **6,** 641 (1994).
7. V. F. Sears, *Neutron News* **3,** 26 (1992).
8. G. I. Watson, *J. Phys. Condens. Matter* **8,** 5955 (1996)
9. E. B. Karlsson and S. W. Lovesey, Technical Report RAL-TR-2001-021.
10. E. B. Karlsson and S. W. Lovesey, *Phys. Rev.* A **61,** 062714 (2000).
11. C. A. Chatzidimitriou-Dreismann, T. Abdul-Redah and J. Sperling, *J. Alloys and Compounds*, in press.

SHORT-LIVED QUANTUM ENTANGLEMENT OF PROTONS AND DISSOCIATION OF C-H BONDS IN CONDENSED MATTER — A NEW EFFECT

C. A. CHATZIDIMITRIOU-DREISMANN AND T. ABDUL-REDAH

Institut für Chemie, Stranski-Laboratorium, Sekr. ER 1, Technische Universität Berlin, Str. d. 17. Juni 112, D-10623 Berlin, Germany

E-mail: dreismann@chem.tu-berlin.de, abdul-redah@chem.tu-berlin.de

In earlier neutron Compton scattering (NCS) experiments on H_2O/D_2O mixtures at $T \approx 298$ K we observed, for the first time, a striking "anomalous" decrease of the ratio σ_H/σ_D of the total scattering cross sections of H and D. This "anomaly" was found to depend strongly on the H/D composition of the liquid. Extending recent NCS results obtained from solid polystyrene, we present here new results concerning the quantum dynamics and dissociation of C-H bonds (at $T \approx 298$ K) in: (*a*) liquid benzene and C_6H_6/C_6D_6 mixtures; (*b*) fully protonated and partially deuterated polystyrene; and (*c*) liquid mixtures of H-acetone (CH_3COCH_3) and D-acetone (CD_3COCD_3). The considered NCS effect was given a theoretical explanation based on short-lived protonic quantum entanglement (QE) and decoherence. The variety of the new results suggests that, in the short-time scale of the NCS experiment, protonic quantum dynamics is strongly correlated with that of electronic degrees of freedom participating in the various chemical bonds.

1 Introduction

Quantum dynamics in condensed matter may lead to a novel kind of short-lived quantum entanglement (QE) between particles. This effect was also predicted to cause the appearance of "anomalously" reduced cross-section densities of protons in scattering experiments.[1] Motivated by these investigations, we have proposed and carried out (since 1995) various neutron Compton scattering[2] (NCS) experiments on liquid H_2O/D_2O mixtures,[3] D_2O-solutions of urea,[4] various metal-hydrogen systems[5] (in collaboration with E. B. Karlsson), polystyrene,[6] amphiphilic molecules,[7] and most recently in other molecular systems containing C and H; see below. These NCS-results provide further strong evidence for the existence of our predicted new QE effect[1,3] between protons at ambient conditions. In these experiments, the measured NCS cross-section density of H is reduced significantly, e.g. by ca. 20%. According to standard theory,[2] and due to the large energy and momentum transfers involved, the characteristic time scale of the NCS process is very short; see below and Refs. 3–7.

New experimental NCS results are presented below, which indicate a wide applicability of the effect in various physical, chemical and biological systems. Possible theoretical interpretations[6,8] are mentioned.

2 Experimental

Our NCS experiments were carried out with the eVS instrument of the ISIS neutron spallation source, Rutherford Appleton Laboratory, U.K. For experimental details, see e.g. Refs. 3-6. The eVS instrument might be viewed as an "atom mass spectrometer": each atom should give rise to one corresponding peak in the measured time-of-flight (TOF) spectrum, cf. Fig. 1. According to standard NCS theory,[2] the integral intensity A_X of each peak must be proportional to the number density N_X of the corresponding atom X in the sample and to the associated total scattering cross section σ_X. As a consequence, for H and any other atom X, the relation

$$\frac{A_H}{A_X} = \frac{N_H \sigma_H}{N_X \sigma_X} \tag{1}$$

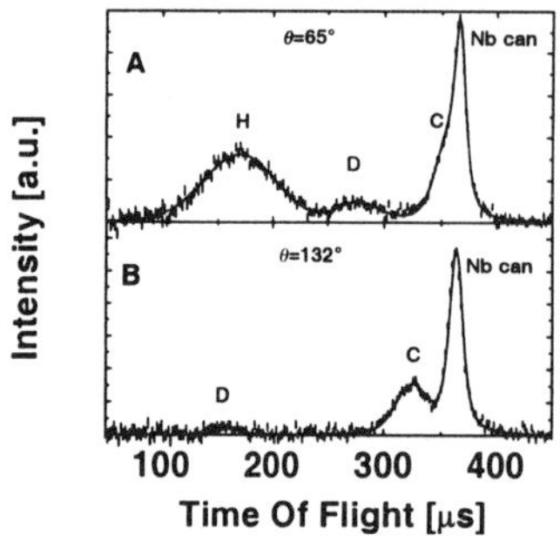

Figure 1. A TOF spectrum of a liquid C_6H_6/C_6D_6 mixture with H:D=1:1 in a Nb can, at $T \approx 298$ K. Full line: fitted TOF spectrum. For the scattering angle $\theta = 65°$, (**A**), the C and Nb peaks overlap, but for $\theta = 132°$, (**B**), they are well resolved.

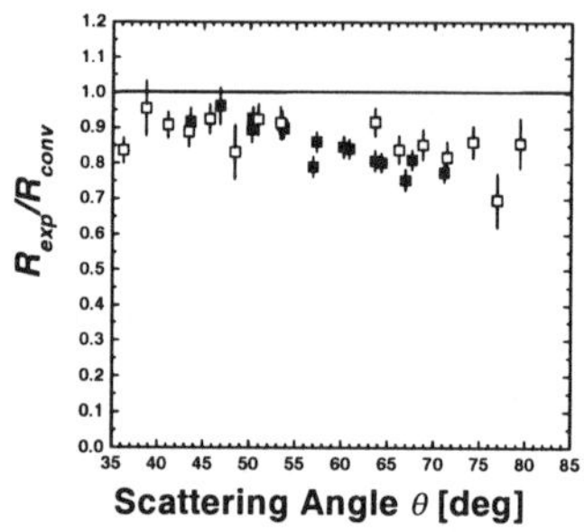

Figure 2. The ratios $R_{exp}(H,C)/R_{conv}(H,C)$ for a C_6H_6/C_6D_6 mixture (H:D=0.500:0.500) (full squares) and a partially deuterated polystyrene sample (H:D=0.544:0.456) (open squares), vs. scattering angle θ, corresponding to $0.05\,\text{fs} < \tau_{sc} < 0.6\,\text{fs}$.

holds strictly within standard NCS theory.[3] Its validity is immediately subject to experimental test since N_H/N_X is known by sample preparation and/or stoichiometry.

The energy transfer was in the range 3–150 eV, and the chemical C-H bonds are broken for scattering angles $\theta > 45°$.[6] According to general NCS theory,[2] the characteristic scattering time τ_{sc} of the neutron-proton collision in our experiments is about 0.05–0.6 femtoseconds.[5–7]

3 Results

Benzene. Our first NCS experiment concerns the detection of the considered effect in C_6H_6 and various C_6H_6/C_6D_6 mixtures. The liquids were put in a special annular metallic can made of Nb. The TOF spectra exhibit the following features. For small scattering angles θ the recoil peaks of C overlap with the peak of the Nb nuclei of the metallic can. To determine the correct intensity of the C recoil peak, 15 detectors were positioned in the "backward" regime ($\theta > 90°$), where, due to the larger momentum transfer, a well visible separation of the maxima of the C and Nb recoil peaks is achieved; see Fig. 1. This increases the reliability of the results considerably.

The anomalous QE effect is clearly visible, cf. Fig. 2 (full squares). From the measured peak areas A_H and A_C, the ratio $R_{exp}(H,C) = A_H/A_C$ has been determined and compared with $R_{conv}(H,C) = N_H\sigma_H/N_C\sigma_C$, i.e., the right-hand side of Eq. (1). The latter denotes the expected value of this ratio according to conventional NCS theory.[2] For all samples investigated, the values of R_{exp} seem to exhibit a weak dependence on the scattering angles and the associated[2] scattering times τ_{sc}. These results reveal an anomalous decrease of R_{exp} with respect to the conventional value of R_{conv}, which amounts to about 20%.

We emphasize that the data analysis procedure used already incorporates the well known transformation from the "free atom" to the "bound atom" cross section.[2–7]

Polystyrene. Our second NCS experiment is on the following solid samples of fully protonated and partially deuterated polystyrene:
(1) $[-CH_2CHC_6H_5-]_n$,
(2) $[-CD_2CDC_6H_{4.97}D_{0.03}-]_n$,
(3) $[-CD_2CDC_6H_{4.35}D_{0.65}-]_n$.
The isotopic compositions of (2) and (3) are H:D = 0.621:0.379 and H:D = 0.544:0.456, respectively. The scattering geometry was the same as in our previous works.[3,6] Since the polystyrene samples are solids (flat foils of

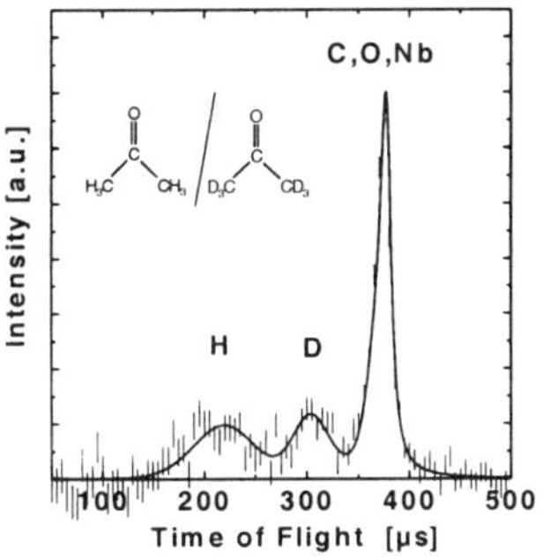

Figure 3. A TOF spectrum (accumulated over ca. 8 hrs) of a liquid H-acetone/D-acetone mixture with D mole fraction $X_D = [D]/([H]+[D]) = 0.8$ at $T \approx 298$ K as measured at scattering angle $\theta = 57°$.

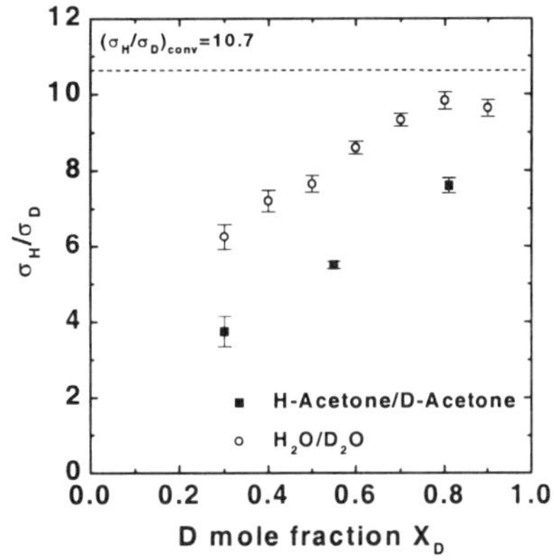

Figure 4. The experimentally determined ratios (σ_H/σ_D) (full squares) of the H-acetone/D-acetone mixtures at different fractions X_D=[D]/([H]+[D]). For comparison the (σ_H/σ_D) ratios from our first NCS experiment on liquid H_2O/D_2O mixtures (Ref. 3) are shown (open circles). $T \approx 298$ K.

ca. 0.2 mm thickness), no metallic container is necessary. This fact greatly facilitates the data analysis procedure, which yields the measured ratio $R_{exp}(H,C) = A_H/A_C$. For the fully protonated polystyrene, the atomic ratio H:C is 1:1. Using the conventional "bound atom" cross-sections,[9] i.e. $\sigma_H = 81.67$ barns and $\sigma_C = 5.564$ barns, one has $R_{conv}(H,C)$=14.7. The values of $R_{conv}(H,C)$ for the partially deuterated samples follow analogously.

The striking result of this experiment is that, in all samples investigated, R_{exp} is found to be anomalously decreased by ca. 15-20%, on the average. For illustration, see Fig. 2 (open squares), in which the results of the aforementioned sample (3) are shown. Furthermore, for all samples investigated, the ratios R_{exp}/R_{conv} seem to exhibit a weak dependence on the scattering angle θ, and thus also on the scattering time τ_{sc}.[6]

Summarizing the above experiments, we have found that the NCS results of polystyrene and benzene samples having similar H:D composition are essentially the same — within present experimental error. The weak θ- (and thus also τ_{sc}-) dependence of R_{exp} indicated in Fig. 2 is in clear contrast to the very strong τ_{sc}-dependence of σ_H/σ_{Metal} in some metallic hydrides.[5] The anomalously reduced $R_{exp}(H,C)$ exhibits a rather weak dependence on the H:D composition of the samples [to be published].

Acetone. Very recently, three H-acetone/D-acetone mixtures have been studied by NCS, using the annular Nb-can mentioned above. For an example of a TOF spectrum, see Fig. 3. Due to the current reconstruction of the eVS-Vesuvio instrument of ISIS, the observations were limited only to scattering angles $55° < \theta < 72°$. Therefore it was not possible to determine reliably the peak area of Nb and its relation to the peak areas of C and O; cf. Fig. 3 with Fig. 1A. For this reason, and since no significant θ-dependence of the "anomalous" effect was observed, the ratio $R_{exp} = A_H/A_D$ of the H and D peaks was averaged over the detectors and compared with conventional theory — as was also done in our first NCS experiment on H_2O/D_2O mixtures.[3]

Preliminary NCS results from acetone are shown in Fig. 4. They show the following unexpected features: The experimentally determined ratios σ_H/σ_D are very strongly reduced with respect to the conventional value[9] 81.7/7.6=10.7. This "anomaly" is much more pronounced (roughly by a factor of two!) than in the corresponding H_2O/D_2O mixtures.[3] The strong dependence of the considered "anomaly" on the D mole fraction X_D (see Fig. 4) leads to the conclusion that, in acetone, it is also of intermolecular origin.

4 Additional Remarks

It is important to note that the protons (deuterons) in C-H (C-D) bonds are tightly localized and do not exchange their positions. In contrast, a fast H/D exchange in H_2O/D_2O mixtures is well known, which also leads to the formation of HDO molecules. Thus, the positional exchange of protons is not the main reason for the observations.

Contradicting the basic Eq. (1), our NCS effect has no conventional interpretation.[3–8] To date there exist two proposed theoretical interpretations[6,8] of it, in both of which protonic QE plays a fundamental role: (*i*) In the theoretical model of Ref. 8, exchange correlations between pairs of identical particles are crucial. In contrast, (*ii*) in our theoretical model,[6] spatial QE and its decoherence — concerning protonic and electronic degrees of freedom — are considered to be crucial, and spin-QE plays herein a less significant role.

Let us give here a short theoretical outline. In the limits of incoherent and impulse approximations[2] characterizing NCS, the dynamic structure factor reads[2]

$$S(\mathbf{q},\omega) = \langle \delta(\omega - \omega_r - \mathbf{q}\cdot\hat{\mathbf{p}}/m)\rangle \qquad (2)$$

$\hbar\mathbf{q}$ and $\hbar\omega$ are the momentum and energy transfers from the neutron to the scattering nucleus, respectively. $\hbar\omega_r$ is the recoil energy of the struck nucleus with mass m and $\hat{\mathbf{p}}$ is its momentum operator. $\langle ...\rangle \equiv \sum W_n \langle n|...|n\rangle$ is the appropriate combined quantal and thermodynamic average (W_n: classical probabilities; $|n\rangle$: many-body quantum states) related with the condensed matter system. A typical matrix element of $S(\mathbf{q},\omega)$ is then

$$s_{nn} = Tr[\rho_1\, \delta(\omega - \omega_r - \mathbf{q}\cdot\hat{\mathbf{p}}/m)] \qquad (3)$$

where ρ_1 is the relevant one-body reduced density operator obtained from the exact N-body operator $\rho_n = |n\rangle\langle n|$. Due to the fundamental decoherence effect, ρ_1 in the position representation, $\{|x\rangle\}$, is known to exhibit an additional time dependence (where $\Lambda > 0$ is the so-called localization rate, cf. Ref. 10):

$$\langle x|\rho_1(t)|x'\rangle = \rho_1(x,x',0)\cdot \mathrm{e}^{-\Lambda|x-x'|^2 t} \qquad (4)$$

which contains a "reduction factor" given by the exponential. The experimentally relevant quantity is then given by the time average

$$\overline{s_{nn}} = \frac{1}{\tau_{sc}}\int_0^{\tau_{sc}} s_{nn}(t)\, dt \qquad (5)$$

which is "anomalously reduced" due to decoherence.[6,10] Thus, the same holds also for $S(\mathbf{q},\omega)$, and the associated total cross-section, given by ω-integration of $S(\mathbf{q},\omega)$.

References

1. C. A. Chatzidimitriou-Dreismann et al., *Phys. Rev. Lett.* **75,** 3008 (1995).
2. (a) V. F. Sears, *Phys. Rev.* B **30,** 44 (1984). (b) G. I. Watson, *J. Phys.: Condens. Matter* **8,** 5955 (1996).
3. C. A. Chatzidimitriou-Dreismann, T. Abdul-Redah, F. Streffer and J. Mayers, *Phys. Rev. Lett.* **79,** 2839 (1997).
4. C. A. Chatzidimitriou-Dreismann, T. Abdul-Redah and F. Streffer, *Ber. Bunsenges. Phys. Chem.* **102,** 544 (1998).
5. (a) E. B. Karlsson et al., *Europhys. Lett.* **46,** 617 (1999). (b) T. Abdul-Redah et al., *Physica* B **276-278,** 824 (2000).
6. C. A. Chatzidimitriou-Dreismann, T. Abdul-Redah and J. Sperling, *J. Chem. Phys.* **113,** 2784 (2000).
7. C. A. Chatzidimitriou-Dreismann, T. Abdul-Redah and B. Kolarić, *J. Am. Chem. Soc.* **123,** (Dec. 2001), in press.
8. E. B. Karlsson and S. W. Lovesey, *Phys. Rev.* A **61,** 062714 (2000).
9. V. F. Sears, *Neutron News* **3,** 26 (1992).
10. E. Joos and H. D. Zeh, *Z. Phys.* B **59,** 223 (1985).

IS IT POSSIBLE TO DETERMINE THE TIME WHEN THE AHARONOV-BOHM EFFECT OCCURS

T. KAUFHERR

School of Physics and Astronomy, Tel Aviv Univrsity, Tel Aviv 69978, Israel
E-mail: trka@post.tau.ac.il

Y. AHARONOV

School of Physics and Astronomy, Tel Aviv Univrsity, Tel Aviv 69978, Israel
and
Department of Physics, University of South Carolina, Columbia, S.C. 29208

We show that the Fourier transform of the velocity distribution of a charged particle moving in the field free zone surrounding a solenoid changes abruptly when the particle passes by the solenoid. This change can be attributed to the exchange of a conserved gauge invariant quantity between the particle and the solenoid, and is responsible for the Aharonov-Bohm effect.

1 Introduction

The question when the Aharonov-Bohm (A-B) effect [1] occurs has always been open. The effect is usually discussed in terms of the shift in the interference pattern of a charged particle as a result of the non vanishing line integral of the vector potential around a solenoid. Below, we show that the change in the probability distribution of the velocity, which is due to a non local interaction with the magnetic flux of the solenoid, occurs at a definite time. This non local interaction with the magnetic flux involves an exchange of a gauge invariant conserved quantity.

2 Modular Momentum

Quantum interference, unlike classical interference, is a manifestation of an underlying dynamics. To see this, consider a set up consisting of a charged particle moving in the vicinity of a capacitor, but prevented from entering the electric field. Consider first the position and momentum of the particle. The Heisenberg equations of motion are,

$$\dot{\vec{x}} = \frac{i}{\hbar}\,[H,\vec{x}] = \frac{\vec{p}}{m}\,,$$

$$\dot{\vec{p}} = \frac{i}{\hbar}\,[H,\vec{p}] = e\vec{E}\,. \tag{1}$$

Since the electric field vanishes in the region where the particle can be found, the average position and momentum of the particle are unaffected by the electric field. Also the averages of higher moments of position and momentum remain unaffected. So far the situation in quantum theory is similar to the classical one. But consider the displacement operator

$$f(p_x) = e^{\frac{i}{\hbar}p_x L}\,. \tag{2}$$

$$\frac{df}{dt} = \frac{i}{\hbar}\,[H,f] = \frac{ie}{\hbar}\Big(\int_x^{x+L} E\,dx'\,\Big)\,f\,, \tag{3}$$

where x denotes the position of the particle . So we have discovered a quantity that has changed. Compare this with the classical analog of $f(p_x)$. Writing

$$f(p_x) = e^{i2\pi\frac{p_x}{p_0}}\,, \tag{4}$$

where $p_0 = \frac{h}{L}$, it satisfies

$$\frac{df(p_x)}{dt} = -\{H, f(p_x)\} = eE\,\frac{\partial f}{\partial p_x} = i\frac{2\pi}{p_0}\,eEf, \tag{5}$$

where $\{\}$ denote Poisson brackets. It too does not change unless the particle actually comes in contact with the field. The displacement operator $f(p_x)$ is a function of modular momentum $p_x(mod\ p_0)$ [2] defined by

$$p_x(mod\ p_0) = p_x - Np_0 \, , \tag{6}$$

and so that its eigenvalues satisfy

$$0 \leq p_x(mod\ p_0) \leq p_0 \, . \tag{7}$$

N is an operator having integer eigenvalues. It is instructive to consider this question also in the Schrödinger representation. Consider a particle in the state

$$\Psi_\alpha = \Psi_1 e^{i\alpha} + \Psi_2 \, , \tag{8}$$

where Ψ_1 and Ψ_2 are non-overlapping wave packets, and $\Psi_2(x) = \Psi_1(x+L)$. α represents the change in the wave function of the charged particle due to its interaction with the capacitor. The average values of x and p_x do not depend on α nor do averages of any polynomial in x or p_x . Yet [a]

$$\left\langle e^{i2\pi \frac{p_x}{p_0}} \right\rangle = \frac{1}{2} e^{i\alpha} \, . \tag{9}$$

The implication of this is that the modular momentum has changed. We proceed to show that modular momentum has its own conservation law. Consider a collision between two systems 1 and 2 such as the charged particle and capacitor mentioned above, and let

$$\pi_1 = \cos(2\pi \frac{p_1}{p_0})$$
$$\pi_2 = \cos(2\pi \frac{p_2}{p_0}) \, , \tag{10}$$

where p_1, p_2 are the components of the momentum in any given direction. Conservation of momentum in that direction, i.e.,

$$p_1 + p_2 = p_1' + p_2' \tag{11}$$

then implies that also $C = \cos 2\pi \frac{p_1'+p_2'}{p_0}$ is conserved, which leads to the conservation law (see fig.1)

[a] Note that because Ψ_α is a non-analytic function of x, $\left\langle e^{i2\pi \frac{p_x}{p_0}} \right\rangle = \left\langle \sum_n \frac{(2\pi i)^n}{n!} \left(\frac{p_x}{p_0}\right)^n \right\rangle \neq \sum_n \frac{(2\pi i)^n}{n!} \left\langle \left(\frac{p_x}{p_0}\right)^n \right\rangle$.

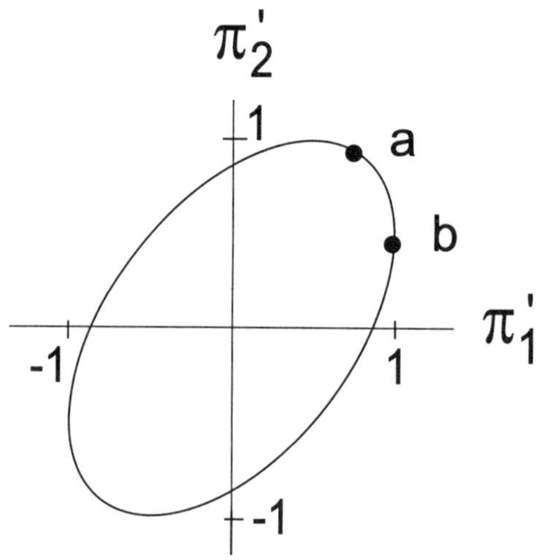

Figure 1. Conservation law for modular momentum. Due to a collision, the system moves from one point of the ellipse to another. For example, from point a to point b.

$$1 - C^2 = (\pi_1')^2 + (\pi_2')^2 - 2C\,\pi_1'\,\pi_2' \, , \tag{12}$$

which, in general, describes an ellipse. During a collision, modular momentum is exchanged under the constraint (12). Obviously, $\Delta\pi_2$ need not in general be equal to $-\Delta\pi_1$ (where $\Delta\pi_1 = \pi_1' - \pi_1, \Delta\pi_2 = \pi_2' - \pi_2$.) Instead, the systems translate from one point of the ellipse to another.

3 The effect of the vector potential on the velocity distribution

With the aid of the modular variables discussed above we can gain new insight into the nonlocal effects on a charged particle moving outside a solenoid. Consider a charged particle prepared in a superposition of two wave packets

$$\Psi = \Psi_1 + \Psi_2 \, , \tag{13}$$

where $\Psi_2(y) = \Psi_1(y+L)$. We also assume that Ψ_1, Ψ_2 occupy non-overlapping

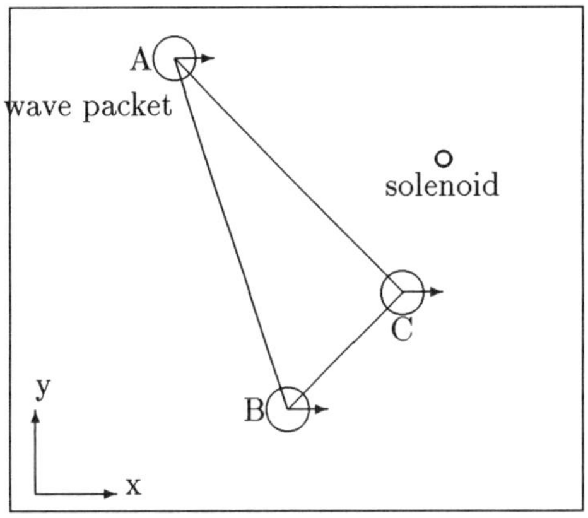

Figure 2. Particle in a superposition of three wave packets moving in the x-direction. Only the distribution of the velocities in the directions AB and AC will change when the respective lines cross the solenoid, while the distribution of the velocity in the direction BC does not change during the motion.

regions in space. The particle passes by a thin solenoid enclosing a flux Φ, but sufficiently far from it, so that its overlap with the solenoid is negligible. Thus it always moves in the field free region surrounding the solenoid. We want to show that the distribution of the velocity in the direction of the line connecting the two wave packets changes when this line crosses the solenoid in such a way that the solenoid lies in the interval between the packets (see fig. 2 and the explanation underneath). Without loss of generality we can assume that the line connecting the two packets is in the y direction. Then, let us choose the gauge

$$\begin{aligned} A_x &= \delta(x)\theta(y)\Phi \\ A_y &\equiv 0\,. \end{aligned} \tag{14}$$

where $\theta(y)$ is the step function. We choose this gauge so that before and after the particle passes by the solenoid $\vec{p} = m\vec{v}$, since the vector potential vanishes in that region. The Fourier transform of the velocity distribution is then (we take $\hbar = 1$)

$$\int P_r(mv_y)e^{imv_yL}dv_y = \int P_r(p_y)e^{ip_yL}dp_y = \;=<\Psi|e^{ip_yL}|\Psi>=\frac{1}{2} \tag{15}$$

since e^{ip_yL} is a displacement operator. When Ψ_1 has crossed the potential line, the particle's state is

$$\Psi_\alpha = \Psi_1 e^{i\alpha} + \Psi_2\,, \tag{16}$$

where $\alpha = \frac{e\Phi}{c\hbar}$ (the A-B effect). Consequently, the Fourier transform of the velocity distribution has changed by

$$\delta\left\langle e^{imv_yL}\right\rangle = \frac{1}{2}\left(e^{i\alpha} - 1\right)\,. \tag{17}$$

Since the velocity is gauge invariant, the geometry of where this change is occurring must also be gauge independent.

Still, let us consider the situation in the Coulomb gauge where

$$\vec{A} = -\frac{\Phi}{2\pi r}\hat{e}_\varphi\,. \tag{18}$$

We compare the value of the Fourier transform of the velocity distribution just before the line that connects the wave packets has crossed the solenoid, when $x = -\varepsilon$, with its value immediately afterwards, when $x = \varepsilon$. We find that it has changed by,

$$\delta\left\langle e^{im\vec{v}\cdot\vec{L}}\right\rangle = \frac{1}{2}\left(e^{-i\frac{e}{c}\oint\vec{A}\cdot d\vec{l}} - 1\right) = \frac{1}{2}\left(e^{i\frac{e\Phi}{c}} - 1\right), \tag{19}$$

in agreement with (17), and where we have used

$$e^{i(\vec{p}-\frac{e}{c}\vec{A})\cdot\vec{L}} = e^{-i\frac{e}{c}\int_{\vec{r}}^{\vec{r}+\vec{L}}\vec{A}\cdot d\vec{l}}\,e^{i\vec{p}\cdot\vec{L}}\,. \tag{20}$$

Thus, the change in the Fourier transform of the velocity distribution occurs when the line that connects the wave packets crosses the solenoid, even though in this gauge there is no sudden change in the relative phase between the packets.

4 The effect of the vector potential on the angular velocity distribution

The exchange of modular angular velocity occurs on a circle. To demonstrate this, we refer to the same set up as before, i.e., a charged

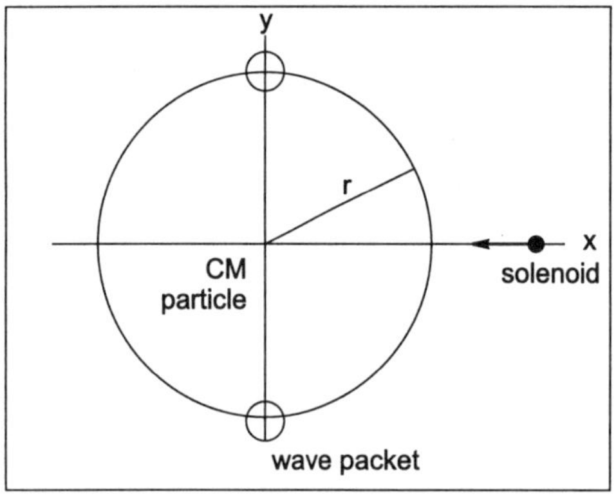

Figure 3. View from the reference frame where the center of mass of the charged particle is at rest: As the solenoid crosses the circle line of radius r, the modular angular velocity about the z axis of the charged particle changes.

particle in a superposition of two wave packets moving towards a solenoid. For the sake of simplicity we shall consider this case in the reference frame where the center of mass of the charged particle is at rest. We define the coordinate system thus: the origin coincides with the center of mass of the charged particle, the y axis passes through the centers of the two wave packets. The state of the charged particle is given by

$$\Psi = \Psi_1 + \Psi_2 \,, \tag{21}$$

where
$\Psi_1 = \Psi(\varphi - \frac{\pi}{2}, r)$ and $\Psi_2 = \Psi(\varphi + \frac{\pi}{2}, r)$. Ψ is a localized wave packet that is at rest, r is the distance of the center of the packet from the origin. The flux enclosed by the solenoid is Φ, with the magnetic field pointing in the -z direction. The solenoid moves towards the particle in the $-x$ direction (see fig. 3).

Consider the modular angular velocity about the z-axis. Let $\vec{A}$ be the vector potential surrounding the solenoid. Writing $\vec{A} = A_\varphi \hat{e}_\varphi + A_r \hat{e}_r$, we obtain,

$$e^{iIv_\varphi \pi} = e^{i(p_\varphi - \frac{e}{c} r A_\varphi)\pi} = e^{-i\frac{e}{c}\int_{\varphi'=\varphi, r'\equiv r}^{\varphi+\pi} \vec{A}\cdot d\vec{l}} e^{ip_\varphi} \,, \tag{22}$$

having used relation (20). p_φ and I are, respectively, the angular momentum and moment of inertia about the z axis.

We now calculate the expectation value of modular angular velocity. With the aid of (22) and using $e^{ip_\varphi \pi}\Psi_2 = \Psi_1$, $e^{ip_\varphi \pi}\Psi_1 = \Psi_2$, the result is,

$$\langle e^{iIv_\varphi \pi} \rangle = 1 \to \frac{1 + e^{i\alpha}}{2} \to \cos\alpha \,. \tag{23}$$

As the solenoid moves toward the charged particle and then away from it, the changes in the average angular modular velocity occur successively, and suddenly, on the solenoid's crossing the circle of radius r centered at the origin.

5 Conclusions

If one attempts to analyze the A-B effect in dynamical terms, then, because of its topological structure, it seems that there is an inherent freedom in the time when the effect occurs. We have shown that such freedom does not exist. As the charged particle passes by the solenoid, an exchange of a gauge invariant and otherwise conserved quantities takes place. Although it may appear that this exchange is non topological, any attempt to measure it by measuring the distribution of the velocity, to which we must resort since we are restricted to local measurements, will necessarily bring the two wave packets together and thus recover the topological nature of the effect. Finally, also the distribution of the energy changes abruptly. The time when this occurs will also be determined geometrically.

References:

[1] Y. Aharonov and D. Bohm, Phys. Rev. 115 (1959) 485.

[2] Y. Aharonov, H. Pendelton and A. Petersen, Int. J. Th. Phys. 2 (1969) 213.

THE WKB METHOD, COMPLEX-VALUED CLASSICAL TRAJECTORIES AND WEAK MEASUREMENTS

ATUSHI TANAKA
Department of Physics, Tokyo Metropolitan University,
Minami-Ohsawa, Hachioji, Tokyo, Japan 192-0397
email: tanaka@phys.metro-u.ac.jp

Within the semiclassical approximation, it is shown that the classical trajectories that appear in the semiclassical evaluations of Feynman kernels are *weak values* [Aharonov, Albert and Vaidman, *Phys. Rev. Lett.* **60** (1988) p. 1351] when the interference among the classical trajectories are negligible. The classical trajectories, including complex-valued ones, are accordingly observable by weak measurements, in principle. Furthermore, this explains why weak values can take "anomalous" values that lie outside the range of the eigenvalues of the corresponding operators, without relying on quantum coherence nor entanglement.

1 Introduction

Classical trajectories provide a lot of information about quantum phenomena through the WKB method[1]. In particular, the WKB method enable us to discuss the notion of "quantum chaos", by associating quantum wave propagations with classical trajectories, to which we can assign the notion of *chaos*[2]. Surprisingly, from the classical trajectories, we can learn a lot about the quantum phenomena that never occur in the corresponding classical systems, by extending the classical trajectories over the complex-valued phase space or time. There are many important examples: eigenenergy splittings caused by simple tunnellings[3,4], dynamics of chaotic tunnellings[5], nonadiabatic transitions[6,7], and coherent state path integrals[8,9]. In addition, the complex-valued classical trajectories are also employed in exact formulations of quantum theory[10]. The complex-valued classical trajectories thus provide a fundamental implement for the explanations of many physical phenomena.

The complex-valued classical trajectories, however, were presumed to be purely theoretical objects, since there was no known procedure to observe these trajectories directly. This point is completely opposite to the cases with real-valued trajectories: Ehrenfest's theorem tells us that the real-valued classical trajectories are certainly observable as expectation values for quantum ensembles, as long as wavepackets are not widely spread[11]. Hence we can accept real-valued trajectories as a reminiscent of classical mechanics in quantum theory. Complex-valued classical trajectories, however, are not subject to Ehrenfest's theorem, even in the simplest cases.

It is shown below that a some class of complex-valued classical trajectories are experimentally "observable" in principle. In order to do it, the notions of *weak measurement* and *weak values*[12] are employed. In the following, I elucidate the relationship between a class of both complex-valued and real-valued classical trajectories that contribute to semiclassical Feynman kernels, and *weak values*[12], which are experimentally observable[13] (see also, ref. [14]).

2 Complex-valued classical trajectories in the WKB method

In the semiclassical evaluation of Feynman kernels, here I focus on the coherent state representation of the Feynman kernels, since this is a good example in the sense that they are generically composed of complex-valued trajectories[8]. The coherent states[15] can be characterised with the help of a complex symplectic transformation[16,17] $(Q,P) = (q-ip, p-iq)/\sqrt{2}$, where q and p are the position and the momentum of the system, respectively. Operators $\hat{Q}$ and $\hat{P}$ are creation and annihilation operators, respectively. The coherent state $|q'p'\rangle$ is $\hat{P}$'s eigenstate whose eigenvalue is $(p'-iq')/\sqrt{2}$.

The Feynman kernel in the coherent state representation is $K(q''p''t''; q'p't') \equiv \langle q''p''|\hat{U}(t'',t')|q'p'\rangle$, where $\hat{U}(t'',t')$ is a time evolution operator, which is generated by a Hamiltonian $\hat{H}(t)$, for a time interval $[t',t'']$. The coherent state path integral representation[18] of K is evaluated by the stationary phase evaluation: $K \simeq \sum E\exp(iF/\hbar)$, where E and F are the amplitude factor and the action, respectively (the precise expressions are shown below), and the summation is taken over the classical trajectories $(Q(t), P(t))$ that satisfy both the Hamilton equation and Klauder's boundary condition $P(t') = P'(\equiv (p'-iq')/\sqrt{2})$ and $Q(t'') = Q''(\equiv (q''-ip'')/\sqrt{2})$ [8,9]. The residual boundary values of $Q' \equiv Q(t')$ and $P'' \equiv P(t'')$ are determined so as to satisfy both the Hamilton equation and Klauder's boundary condition. The classical trajectory $(q(t), p(t))$ is complex-valued[8], except for the case that the real-valued classical time evolution carries the point in the phase space (q', p') at time t' to (q'', p'') at time t''. It was presumed that such complex-valued trajectories did not have any counterpart in nature and were an artifact of the stationary phase evaluations of Feynman path integrals. The following argument shows that a class of the complex-valued trajectories are observable in principle.

3 Weak values

Aharonov, Albert and Vaidman proposed to consider the quantum ensemble that is specified not only by an initial state $|\psi'\rangle$ at time t' but also a final state $\langle\psi''|$ at time t'' $(> t')$[12]. The corresponding "expectation value" of an observable $\hat{A}$ at time t $(t' \le t \le t'')$ is called a *weak value* $\mathrm{W}(\hat{A},t)$[12], whose definition is

$$\mathrm{W}(\hat{A},t) \equiv \frac{\langle\psi''|\hat{U}(t'',t)\hat{A}\hat{U}(t,t')|\psi'\rangle}{\langle\psi''|\hat{U}(t'',t')|\psi'\rangle}. \quad (1)$$

Note that $\mathrm{W}(\hat{A},t)$ can take "anomalous" values that lie outside of the eigenvalues of $\hat{A}$. Actually, $\mathrm{W}(\hat{A},t)$ can be complex-valued even for an Hermite operator $\hat{A}$.

A weak value of a system can be measured by a kind of von Neumann type apparatus[19] whose "pointer" position has large quantum fluctuation. After the apparatus contact with the system weakly and the system is succeedingly postselected, we can gain the information of the corresponding weak value from the pointer: On one hand, the real part of the weak value is determined by the peak of the pointer position distribution; On the other hand, the imaginary part of the weak value is determined by the peak of the distribution of the conjugate quantity of the pointer position (e.g. momentum of the pointer). Complex-valued weak values are thus observable in principle by *weak measurements* explained above[12].

Although the postselection of quantum ensemble is the source of the possibility that we encounter anomalous weak values, there was no intuitive argument that explains such anomalous values. The following argument provides an explanation of this point.

4 Complex-valued classical trajectories as weak values

For a class of classical trajectories that compose semiclassical Feynman kernel, it is shown that the classical trajectories are, within the semiclassical approximation, weak values of the ensemble whose initial and final states are $\langle\psi''| = \langle q''p''|$ at $t = t''$ and $|\psi'\rangle = |q'p'\rangle$ at $t = t'$, respectively.

The weak values are evaluated with the

help of the following generating functional:

$$Z(\zeta(\cdot),\hat{A}) \equiv \left\langle \psi'' \middle| \overleftarrow{\exp}\left\{-\frac{i}{\hbar}\int_{t'}^{t''} \hat{H}_\zeta(t,\hat{A})dt\right\} \middle| \psi' \right\rangle \quad (2)$$

where $\overleftarrow{\exp}(\cdot)$ is the time-ordered exponential and $\hat{H}_\zeta(t,\hat{A},\zeta) \equiv \hat{H}(t) - \hat{A}\zeta(t)$. It is straightforward to show that

$$\mathrm{W}(\hat{A},t) = -\,i\hbar \left.\frac{\delta \ln Z(\zeta(\cdot),\hat{A})}{\delta\zeta(t)}\right|_{\zeta(\cdot)=0} \quad (3)$$

holds. This is evaluated by the WKB method in the following.

Let $A(q,p)$ be the classical counterpart of the operator $\hat{A}$. I ignore the operator ordering problem of $\hat{A}$, since it changes the result only $\mathcal{O}(\hbar)$, i.e., within the accuracy of the following semiclassical argument.

I assume that in the semiclassical evaluation of the Feynman kernel $Z(\zeta(\cdot),\hat{A})$ (2), only one classical trajectory $(Q(t),P(t))$ (or, equivalently $(q(t),p(t))$) is relevant for infinitesimally small values of $\zeta(\cdot)$. Namely, Z is assumed to be in a single-term form

$$Z \simeq E\exp(iF/\hbar) \quad (4)$$

where E and F are the amplitude factor and the action[8]:

$$E = (\partial Q''/\partial Q')^{-1/2} \quad (5)$$

$$F = \frac{1}{2}Q''\{P'' + i(Q'')^*\} + \frac{1}{2}\{Q' + i(P')^*\}P' + \int_{t'}^{t''}\Big\{\frac{1}{2}\{p(t)\dot{q}(t) - q(t)\dot{p}(t)\} - H(t) + A(q(t),p(t))\zeta(t)\Big\}dt. \quad (6)$$

For the semiclassical coherent state path integrals, the assumption (4) holds when the value of $\hbar$ is small, or, the time scale in question is short[20].

Applying the single-term condition (4) to (3), I obtain

$$\mathrm{W}(\hat{A},t) = \delta F/\delta\zeta(t)|_{\zeta(\cdot)\equiv 0} + \mathcal{O}(\hbar). \quad (7)$$

With the Klauder's boundary condition, which implies that $\delta Q'' = \delta P' = 0$ holds, the eq. (7) and (6) imply

$$\mathrm{W}(\hat{A},t) = A(q(t),p(t)) + \mathcal{O}(\hbar) \quad (8)$$

where $\zeta(\cdot) \equiv 0$ is imposed on $(q(t),p(t))$. Namely, the weak values are approximately determined, with an error of $\mathcal{O}(\hbar)$, by the classical trajectory $(q(t),p(t))$ that composes the semiclassical Feynman kernel $K = Z(\zeta(\cdot)=0)$, which is in the single-term form.

The application of (8) to the cases $\hat{A} = \hat{q}$ and $\hat{A} = \hat{p}$ shows that the complex-valued classical trajectory $(q(t),p(t))$ agrees with the time development of the pair of the weak values $(\mathrm{W}(\hat{q},t),\mathrm{W}(\hat{p},t))$, within the semiclassical approximation. Hence in the semiclassical case, when the small $\mathcal{O}(\hbar)$ errors can be ignored, the weak values of $(\hat{q},\hat{p})$ that are experimentally accessible through weak measurements agree with the complex-valued classical trajectory.

The same result (8) for different pairs of initial and final states (e.g. the eigenstates of the position or the momentum operators) can be obtained, when the single-term approximation holds for Z[14]. Accordingly, the classical trajectories that appears in the WKB method are weak values, with an error of $\mathcal{O}(\hbar)$, when the interference among multiple classical trajectories is absent. Thus I established above a relationship between weak measurements and classical trajectories in the WKB method. This is carried out without any discrimination between real and complex trajectories.

Note that the present evaluation of the weak values does not involve neither quantum interference nor quantum entanglement, which are typical origins of quantum phenomena.

5 Summary and Outlook

I elucidate that the complex-valued classical trajectories are observable as weak values. Since it is still unclear whether the weak values have physical reality or not, the present result does not need to imply that the complex trajectories have physical reality[14]. After all, what is reliable now is that *the "degrees" of the physical reality of real- and complex-valued classical trajectories are qualitatively the same*: In short time scales, their counterparts can be found in nature only approximately as well as statistically; For much longer time scales, we lost such a correspondence, as is concluded by the founders of quantum theory.

Although I consider above only the WKB method with the real-valued time, there are many applications of complex-valued time[4,7,21]. By developing an extension of the present WKB approach of the weak values, we may invent a way to observe the events that involve complex-valued time.

Acknowledgements

A part of this work is carried out at Institute of Physics, University of Tsukuba. This work is supported in part by University of Tsukuba Research Projects.

References

1. M. V. Berry and K. E. Mount, Rep. Prog. Phys. **35**, 315 (1972).
2. M. C. Gutzwiller, *Chaos in Classical and Quantum Mechanics* (Springer-Verlag, New York, 1990).
3. L. D. Landau and E. M. Lifshitz, *Quantum Mechanics (Non-relativistic Theory)* (Oxford University Press, New York, 1977), § 47.
4. L. S. Schulman, *Techniques and applications of path integration*, (Wiley, New York, 1981).
5. A. Shudo and K. S. Ikeda, Phys. Rev. Lett. **74**, 682 (1995).
6. L. Landau, Physik. Z. Sowjet. **2**, 46 (1932); E. C. G. Stueckelberg, Helv. Phys. Acta **5**, 369 (1932); J-T. Hwang and P. Pechukas, J. Chem. Phys. **67**, 4640 (1977).
7. W. H. Miller and T. F. George, J. Chem. Phys. **56**, 5668 (1972).
8. J. R. Klauder, in *Path Integrals*, edited by G. J. Papadopoulos and J. T. Devreese, NATO Advanced Summer Institute (Plenum, New York, 1978), p. 5; in *Random Media*, edited by G. Papanicolauou, (Springer-Verlag, New York 1987), p. 163.
9. S. Adachi, Ann. Phys. **195**, 45 (1989).
10. R. Balian and C. Bloch, Ann. Phys. (N. Y.) **85**, 514 (1974); M. S. Wang, Phys. Rev. Lett. **79**, 3319 (1997)
11. See, e.g., D. Bohm, *Quantum theory* (Dover, New York, 1989), § 9.26
12. Y. Aharonov, D. Z. Albert and L. Vaidman, Phys. Rev. Lett. **60**, 1351 (1988); Y. Aharonov and L. Vaidman, Phys. Rev. A **41**, 11 (1990); see also, Y. Aharonov, D. Z. Albert, A. Casher and L. Vaidman, Phys. Lett. A **124** 199 (1987)
13. N. W. M. Richie, J. G. Story and R. G. Hulet, Phys. Rev. Lett. **66**, 1107 (1991).
14. Atushi Tanaka, submitted.
15. J. R. Klauder and B.-S. Skagerstam, *Coherent states* (World Scientific, Singapore, 1985).
16. P. Kramer, M. Moshinsky and T. H. Seligman, in *Group Theory and Its Applications*, edited by E. M. Loebl (Academic Press, New York, 1975), Vol. III, p. 249.
17. Y. Weissman, J. Chem. Phys. **76**, 4067 (1982).
18. I. Daubechies and J. R. Klauder, J. Math. Phys. **26**, 2239 (1985).
19. J. von Neumann 1932, *Mathematische grundlagen der quantenmechanik* (Springer Berlag, Berlin, 1932), § VI.3.
20. Atushi Tanaka, (unpublished).
21. K. Takahashi and K. S. Ikeda, Ann. Phys. (N. Y.) **283**, 94 (2000)

DECOHERENT PARAMETERS IN CORRELATED TWO NEUTRAL KAON SYSTEMS

FUMIYO UCHIYAMA

Physics Division, Lawrence Berkeley National Laboratory, Berkeley, CA 94720, USA
and
Physics Department, Science University of Tokyo, Noda,Japan 278-8510
E-mail: uchiyama@thsrv.lbl.gov

We discuss the effects of decoherence parameters introduced in the correlated two neutral kaon system and show that their magnitudes are limited by the magnitudes of CP violation and of the strangeness non-conserving $\Delta S = \pm 2$ transitions.

1 Introduction

The neutral kaons played an important role in the history of quantum mechanics showing that the particle mixture[1] exhibits unusual and peculiar properties, the oscillating behaviors in the probabilities of finding a K^0 (or a $\bar{K^0}$) in a beam which is initially pure $\bar{K}^0$ (or K^0), as function of time. The oscillations stem from the interference terms in the particle mixture states. Based upon these striking quantum mechanical features of kaons, correlated two neutral kaons have been used in the proposals for testing EPR against Quantum Theory[2].

A decoherence parameter ζ parameterizing the deviation from Quantum Theory is introduced in an experimental proposal[3] which involves measurements of K_S and/or K_L particle states: The probability of finding two kaons in states, f_1 and f_2, in the decay products of Φ mesons, is written as

$$P_{decoh}(f_1, f_2) = \frac{1}{2}[|\langle f_1|K_S\rangle\langle f_2|K_L\rangle|^2 + |\langle f_1|K_L\rangle\langle f_2|K_S\rangle|^2 -(1-\zeta)(\langle f_1|K_S\rangle\langle f_2|K_L\rangle\langle f_1|K_L\rangle^*\langle f_2|K_S\rangle^* +\langle f_1|K_L\rangle\langle f_2|K_S\rangle\langle f_1|K_S\rangle^*\langle f_2|K_L\rangle^*)] \quad (1)$$

At $\zeta = 0$, $P_{decoh}(f_1, f_2)$ agrees with the result from a quantum mechanical calculation while $\zeta = 1$ gives zero interference terms in the above expression. Furry[4] in 1936, before the discovery of kaons, discussed theoretically in detail the degrees of agreement and disagreement between the results of quantum mechanical calculations and those to be expected on the assumption that a system once freed from dynamical interference can be regarded as possessing independently real properties. High energy accelerators made it possible to produce heavy flavored neutral mesons such as D and B, for which similar oscillations could be observed and neutrino oscillations among different flavored massive neutrinos, are predicted and many experimental investigation are underway.

We clarify that missing interference term ($\zeta = 1$) does not imply that the system is "'more classical" in an entangled state, and show that non zero decoherence parameter, $\zeta \neq 0$, introduces second order strangeness non-conserving ($|\Delta S| = 2$) transitions and a new CP violation, and therefore smallness of decoherence parameters are required.

2 Formulation

We use mass and strangeness eigenstates (K_S, K_L) and $(K^0, \bar{K}^0)$. This particular choice of base-sets enable us to derive the results by clear and definite arguments. The discussion and results can be easily generalized for any sets of two independent basis. The two bases are related at <u>all</u> time

$$|K_S\rangle = \frac{N}{\sqrt{2}}[(1+\epsilon)|K^0\rangle - (1-\epsilon)|\bar{K}^0\rangle] \quad (2)$$

$$|K_L\rangle = \frac{N}{\sqrt{2}}[(1+\epsilon)|K^0\rangle + (1-\epsilon)|\bar{K}^0\rangle] \quad (3)$$

where ϵ is the CP violation parameter ($|\epsilon| \sim 10^{-3}$) and N is a normalization constant, $\frac{1}{\sqrt{(1+|\epsilon|^2)}}$.

We consider the physical situations in which two neutral kaons are produced in correlated states of $J^{PC} = 1^{--}$. At time t after the productoion, the two kaon state is expressed in the strangeness basis as

$$\frac{1}{\sqrt{2}}(|K_0(t)\rangle_r|\bar{K}_0(t)\rangle_l - |\bar{K}_0(t)\rangle_r|K_0(t)\rangle_l) \quad (4)$$

where r and l stand for left and right to distinguish the two kaons. Or similarly in (K_S, K_L), the two kaon state is,

$$\frac{N'}{\sqrt{2}}(|K_S(t)\rangle_r|K_L(t)\rangle_l - |K_L(t)\rangle_r|K_S(t)\rangle_l) \quad (5)$$

where $N' = \frac{(1+|\epsilon|^2)}{(1-\epsilon^2)} = 1 + O(|\epsilon|^2)$ and

$$|K_{S,L}(t)\rangle = e^{-\frac{i}{2}\lambda_{S,L}}|K_{S,L}(0)\rangle$$

where $\lambda_{S,L}$ are complex mass eigenvalues of K_S and K_L respectively.

The probability of finding an f_1-type kaon to the right and an f_2-type kaon to the left at time t in the (K_S, K_L) basis is given by

$$P(f_1, f_2; t) = \frac{1}{2}|\langle f_1|K_S\rangle_r\langle f_2|K_L\rangle_l \quad (6)$$
$$-\langle f_1|K_L\rangle_r\langle f_2|K_S\rangle_l|^2 + O(\epsilon^2)$$

where time t is implicit in kaon states and the correction from the normalization of order $O(\epsilon^2)$ is separately written. Similarly in $(K^0, \bar{K}^0)$ bases,

$$P(f_1, f_2, t) = \frac{1}{2}|\langle f_1|K_0\rangle\langle f_2|\bar{K}_0\rangle$$
$$-\langle f_1|\bar{K}_0\rangle\langle f_2|K_0\rangle|^2 \quad (7)$$

where the subscripts of the first and second states in pairwise, r and l, are suppressed. Here we make further simplification taking $\epsilon = 0$ which makes it clear that the derivation of our conclusions is independent of weak interactions. The corrections from $\epsilon \neq 0$ will be considered perterbatively later.

Substituting $\epsilon = 0$ into equations (2, 3), we get two CP eigenstates which form another orthogonal basis set. They are eigenstates of the Hamiltonian, $H = H_{st} + H_{em}$ where H_{st} and H_{em} are strong and electro-magnetic interaction hamiltonians respectively. In this basis, the two kaon state is

$$\frac{1}{\sqrt{2}}(|K_S(t)\rangle_{\epsilon=0}|K_L(t)\rangle_{\epsilon=0}$$
$$-|K_L(t)\rangle_{\epsilon=0}|K_S(t)\rangle_{\epsilon=0}) \quad (8)$$

The probability in this basis is given by equation (6) with $\epsilon = 0$. Unless otherwise stated, we use the same notations, K_S and K_L, for CP+ and CP- eigenstates without subscript $>_{\epsilon=0}$ from now on. To transform the probability expressed in the set of bases $(K_0, \bar{K}_0)$, we define the following quantities;

$$A \equiv \langle f_1|K_0\rangle\langle f_2|K_0\rangle - \langle f_1|\bar{K}_0\rangle\langle f_2|\bar{K}_0\rangle \quad (9)$$
$$B \equiv \langle f_1|K_0\rangle\langle f_2|\bar{K}_0\rangle - \langle f_1|\bar{K}_0\rangle\langle f_2|K_0\rangle \quad (10)$$

where A has amplitudes of $\Delta S = \pm 2$ transitions while B is strangeness conserving amplitudes. We obtain for transformed probability in $(K_0, \bar{K}_0)$ bases:

$$P(f_1, f_2, t) = \frac{1}{2}BB^* = \frac{1}{2}|B|^2 \quad (11)$$

which is the equation(7) as we expect, showing the invariance of the probability $P(f_1, f_2; t)$ under base-transformations.

3 Decoherence Parameter and Invariance

Following Eberhard[3], let us assume that the physical processes of a particular interest might deviate from QM for some unknown reasons and that one of the conceivable parameterization for the probability in a particular experiment involving measurement on pairs of kaons in mass eigenstates K_S and K_L is given by equation (1). Transforming basis to K_0 and $\bar{K}_0$, we get

$$P_{decoh}(f_1, f_2; t) = \frac{1}{2}BB^* + (\frac{\zeta}{2})(AA^* - BB^*)$$

$$= \frac{1}{2}[|\langle f_1|K_0\rangle\langle f_2|\bar{K_0}\rangle|^2 + |\langle f_1|\bar{K_0}\rangle\langle f_2|K_0\rangle|^2$$
$$-(1-\zeta)(\langle f_1|K_0\rangle\langle f_2|\bar{K_0}\rangle\langle f_1|\bar{K_0}\rangle^*\langle f_2|K_0\rangle^*$$
$$+\langle f_1|\bar{K_0}\rangle\langle f_2|K_0\rangle\langle f_1|K_0\rangle^*\langle f_2|\bar{K_0}\rangle^*)$$
$$+\zeta \cdot T_{extra}\,] \qquad (12)$$

$\zeta = 0$ reproduces the eq.(7) as we expect. T_{extra} makes the form of $P_{decoh}(f_1, f_2; t)$ no-invariant under the transformation of bases. There is no apriori reason for invariance of the form of $P_{decoh}(f_1, f_2; t)$ under base transformations . (However the invariance requirement is important for spin-correlated two lepton case. See ref.(5)).

$$T_{extra} = \frac{1}{2}(|\langle f_1|K_0\rangle\langle f_2|\ K_0\rangle|^2$$
$$+|\langle f_1|\bar{K_0}\rangle\langle f_2|\bar{K_0}\rangle|^2$$
$$-\langle f_1|K_0\rangle\langle f_2|K_0\rangle\langle f_1|\bar{K_0}\rangle^*\langle f_2|\bar{K_0}\rangle^*$$
$$-\langle f_1|K_0\rangle^*\langle f_2|K_0\rangle^*\langle f_1|\bar{K_0}\rangle\langle f_2|\bar{K_0}\rangle$$
$$-|\langle f_1|K_0\rangle\langle f_2|\bar{K_0}\rangle|^2$$
$$-|\langle f_1|\bar{K_0}\rangle\langle f_2|K_0\rangle|^2$$
$$-\langle f_1|K_0\rangle\langle f_2|\bar{K_0}\rangle\langle f_1|\bar{K_0}\rangle^*\langle f_2|K_0\rangle^*$$
$$-\langle f_1|\bar{K_0}\rangle\langle f_2|K_0\rangle\langle f_1|K_0\rangle^*\langle f_2|\bar{K_0}\rangle^*)(13)$$

When $\zeta = 1$, the obvious interference terms [the terms proportional to $(1-\zeta)$] in the two probabilities in the two bases (K_S, K_L), eq.(2, 3 with $\epsilon = 0$), and $(K^0, \bar{K}^0)$, eq.(12) respectively vanish. But the interference terms in T_{extra} of equation(13), are nonzero and become the maximum. The first two terms in equation (13) of T_{extra} cause $|\Delta S| = 2$ transitions of order ζ even if the weak interactions has been turned off. Nonzero ζ breaks invariance of the probability under transformation of basis. For a different experiment involving measurements on eigenstates of strangeness, $(K_0, \bar{K_0})$, one may think to introduce a new decoherence parameter ζ' exactly in the same form for (K_S, K_L) bases as

$$P_{decoh}(f_1, f_2; t) = \qquad (14)$$
$$\frac{1}{2}[|\langle f_1|K_0\rangle\langle f_2|\bar{K_0}\rangle|^2 + |\langle f_1|\bar{K_0}\rangle\langle f_2|K_0\rangle|^2$$
$$-2(1-\zeta')Re(\langle f_1|K_0\rangle\langle f_2|\bar{K_0}\rangle\langle f_1|\bar{K_0}\rangle^*\langle f_2|K_0\rangle^*)]$$

The relationship between the two decoherence parameters, ζ and ζ' can be obtained by equating (12) and (15);

$$\zeta' = \zeta \cdot \qquad (15)$$
$$[1 + \frac{1}{2}\frac{T_{extra}}{Re(\langle f_1|K_0\rangle\langle f_2|\bar{K_0}\rangle\langle f_1|\bar{K_0}\rangle^*\langle f_2|K_0\rangle^*)}]$$

To see the magnitude of ζ', let us take $f_1 = \gamma(|K_0\rangle + \eta|\bar{K_0}\rangle)$ and $f_2 = \gamma(|\bar{K_0}\rangle - \eta|K_0\rangle)$ where η is a small number as an extreme example, and γ is a normalization. The denominator in equation(16) becomes as small as $\sim |\eta|^2$ while the numerator has a finite value, $\sim \frac{1}{2}$ meaning a large ζ'. This indicates that once ζ is nonzero, small deviation from quantum mechanics does not necessarily imply a small decoherence parameter in other basis.

One specific example of experimental analysis may be found in ref. [6].

4 Upper Bounds for the Magnitudes of Decoherence Parameters

It can be seen that a finite nonzero ζ induces nonzero probabilities for strangeness non-conserving , $\Delta S = \pm 2$ transitions of order ζ. Choosing $f_1 = K_0$ and $f_2 = K_0$, we get:

$$P_{decoh}(K_0, \bar{K_0}; t) =$$
$$\frac{\zeta}{4}(|\langle f_1|K_0\rangle\langle f_2|\ K_0\rangle|^2 \qquad (16)$$

This remains true at any arbitrary time t, as long as two sets of bases are related to each other by equations (2, 3). To our best knowledge, the $|\Delta S| = 2$ transitions are limited by order of the weak interaction and therefore the $|\Delta S| = 2$ transition probability is proportional to $\Delta m (\sim 10^{-6} ev)$ where Δm is the mass difference of K_L and K_S. To estimate the correction of week interaction on the results,

The corrections of weak interactions on the results may be estimated as follows: The eigenstates of effective hamiltonian

$$H_{ef} = H_{st} + H_{em} + H_{weak}$$

are give in equations (2, 3). Expanding the states in terms of ϵ, we obtain

$$|K_S\rangle = \frac{1}{\sqrt{2}}[|K_S\rangle_{\epsilon=0} + \epsilon|K_L\rangle_{\epsilon=0} + O(\epsilon^2)]$$
$$|K_L\rangle = \frac{1}{\sqrt{2}}[|K_L\rangle_{\epsilon=0} + \epsilon|K_S\rangle_{\epsilon=0} + O(\epsilon^2)]$$

Therefore the 1^{--} state of two neutral kaons can be written in terms of CP eigenstates as

$$\begin{aligned}(|K_S(t)\rangle|K_L(t)\rangle - |K_L(t)\rangle|K_S(t)\rangle) = \\ (|K_S(t)\rangle_{\epsilon=0}|K_L(t)\rangle_{\epsilon=0} \\ -|K_L(t)\rangle_{\epsilon=0}|K_S(t)\rangle_{\epsilon=0}) + O(\epsilon^2)\end{aligned} \quad (17)$$

Note there is no correction of order ϵ Therefore we can safely state

$$\frac{\zeta}{4} \leq |\epsilon|$$

Similarly, CP violating transition terms is

$$\frac{\zeta'}{4}(|\langle f_1|K_S\rangle\langle f_2|\ K_S\rangle|^2 + |\langle f_1|K_L\rangle\langle f_2|K_L\rangle|^2)$$

The derivation of this two terms is trivial because the mathematical procedures are exactly reversed but identical for the transformation from the strangeness basis $(K^0, \bar{K}^0)$ to CP basis $(K_S, K_L)_{\epsilon=0}$. As CP violating transitions are bounded by order of ϵ, we get

$$\frac{\zeta'}{4} \leq |\epsilon| \quad (18)$$

Therefore conservatively the decoherence parameters are limited by

$$\zeta',\ \zeta \leq 4|\epsilon| \quad (19)$$

The situation discussed above can be understood more clearly for two correlated electrons with total spin 0. The details of discussions for spin case can be found in ref.(5).

5 Concluding Remarks

Our naive intuition cultivated from interference phenomena of photons and electrons is not working for the correlated two neutral kaon case: Another example for which our cultivated intuition for quantum theory doesn't work is exhibited[7] in neutrino oscillations: Our naive intuition predicts that the smaller the mass difference(the gap of two energy levels) , the larger the rate of transition between them. However the probability of finding ν_l types neutrino in a beam which is initially pure ν neutrino, as function of time, is given by

$$P(\nu \to \nu_l) \sim sin^2(\frac{\Delta M^2 L}{4E})\ (\nu \neq \nu_l) \quad (20)$$

where $\Delta M^2 = m_\nu^2$ - $m_{\nu_l}^2$ and E and L are the energy and traveled distance of ν.

We have shown that the decoherence in correlated two neutral kaons in the form of modified interference terms is limited.

Acknowledgments

The author thanks P. Eberhard for discussion, S. Gopalakrishna for good questions and the theory group of Lawrence Berkeley National Laboratory for hospitality.

References

1. M.Gell-mann and A. Pais, Phys. Rev. 97 (1955) 1387; A. Pais and O. Piccioni, Phys. Rev. 100 (1955) 1487.
2. A. Einstein, B.Podolosky and N. Rosen, Phys. Rev. 47 (1935) 777; J.S.Bell, Speakable and Unspeakable in Quantum Mechanics, Cambridge University Press, Cambridge, 1987.
3. P.H.Eberhard, in The Second DaΦne Physics Handbook, ed. by L.Maiani, G Pancheri and N.Paver, (1994), Vol. I, p.99 and LBL-35983, UC-414.
4. W.H.Furry, Phys. Rev., 49 (1936) 393.
5. F.Uchiyama, LBLN-47187, HEP-PH-0012139.
6. R.A. Bertlemann, W. Grimus, and B.C. Hiesmayr, Phy. Rev. D60 (1999) 114032.
7. Thanks are due to H.J. Crawford for bringing up the subject to the author.

CLOSING ISQM-TOKYO '01

SADAO NAKAJIMA
*Superconductivity Research Laboratory, International Superconductivity Technology Center
Shinonome, Koto-ku, Tokyo 135-0062, Japan*

In place of his own closing speech as Chair, Professor Hidetoshi Fukuyama has kindly given me the chance to say a few words to you before I resign from the ISQM Organizing Committee, on which I am now getting too old to serve. As both he and Dr. Akira Tonomura have mentioned, I organized the first ISQM in the early 1980s with generous support from Hitachi. My decision was motivated by Dr. Tonomura's successful experimental use of electron beam holography to prove the existence of magnetic vector potential (the Aharonov-Bohm effect), which was undertaken at the Central Research Laboratory, Hitachi, Ltd. with the encouragement of Professor Chen Ning Yang. I was quite convinced that the time was ripe to organize such an interdisciplinary meeting as ISQM, but I did not expect this series of symposia to last for two decades. It was therefore especially movingt for me to listen to Dr. Tonomura's current presentation about his newly-developed 1-MeV electron microscope, in addition to being able to see his beautiful pictures of quantized vortices trapped in high-Tc superconductors.

Incidentally, in the early 1980s, I was serving as Director of ISSP (The Institute for Solid State Physics) at the University of Tokyo and asked professors there to help me plan the first ISQM, but they said that they were too busy to concern themselves with the *metaphysics* of quantum mechanical measurement. Clearly, mesoscopic physics was not yet their metier. (To restore the honor of ISSP, I should hastily add that the Institute now has a strong team of mesoscopic physicists headed by Professor Yasuhiro Iye.)

In contrast to this, at the second ISQM held in 1986, Processor Richard A. Webb spoke about the Aharonov-Bohm effect observed in a mesoscopic normal metal ring. At the same symposium, Professor Anthony J. Leggett even discussed the possibility of linear superpositions of macroscopically distinguishable states (Schrödinger's cat) in some detail, with particular reference to superconducting devices. This is of course directly connected with the fundamental concept of *"qubit"* in present-day quantum information theory. Indeed, at this seventh ISQM, we have frequently heard of *qubit*: how to set it up and how to "observe" it. We are now confronted by the challenge of understanding the physics of quantum mechanical measurement. It seems to me that we have reached the point where we should stop and think about how to continue this unique series of interdisciplinary symposia. I am confident that the answer will soon be forthcoming from Professor Fukuyama and the members of the Organizing Committee.

List of participants

Aikawa, Hisashi
Institute for Solid State Physics
University of Tokyo
5-1-5 Kashiwanoha, Kashiwa,
Chiba 277-8581
Japan
Phone:+81-471-36-3300
FAX:+81-471-36-3300
e-mail:astar@issp.u-tokyo.ac.jp

Aoki, Takatoshi
Faculty of Science and Technology
Science University of Tokyo
2641 Yamazaki, Noda-shi, Chiba 278-8510
Japan
Phone:+81-471-24-1501(Ext.3212)
FAX:+81-471-23-9361
e-mail:morinaga@ph.noda.sut.ac.jp

Astafiev, Oleg V.
Department of Basic Science
University of Tokyo
Building 16-622, Komaba 3-8-1, Meguro-ku, Tokyo
153-8902
Japan
Phone:+81-3-5454-6762
FAX:+81-3-5454-6762
e-mail:astf@mujin.c.u-tokyo.ac.jp

Barbara, Bernard A.
Laboratoire de Magnetisme Louis Neel
CNRS
25 Ave Des Martyrs, BP 166, 38042 Grenoble, Cedex
09
France
Phone:+33-4-76-88-11-92
FAX:+33-4-76-88-11-91
e-mail:barbara@labs.polycnrs-gre.fr

Cain, Paul A.
Cavendish Laboratory
Microelectronics research Centre, Cambridge
University
Madingley Road University of Cambridge
U.K.
Phone:+44-1223-337493
FAX:+44-1223-337706
e-mail:pac36@cam.ac.uk

Chatzidimitriou-Dreismann, C. A.
I-N-Stranski-Institute
Technical University of Berlin
Strasse des 17 Juni 112, D-10623 Berlin
Germany
Phone:+49-30-314-22692
FAX:+49-30-314-26602
e-mail:dreismann@chem.tu-berlin.de

Cheon, Taksu
Kochi University of Technology
Miyanokuchi, Tosa Yamada, Kochi 782-8502
Japan
Phone:+81-887-57-2302
FAX:+81-887-57-2320
e-mail:cheon@mech.kochi-tech.ac.jp

Cho, Alfred Yi
Vice President, Semiconductor Research
Bell Laboratories Lucent Technology
600 Mountain Ave. Murray Hill, NJ 07974-0636
U.S.A.
Phone:+1-908-582-2093
FAX:+1-908-582-2043
e-mail:ayc@lucent.com

Corbett, John V.
Mathematics Department,
Division of Information and Communication Sciences
Macquarie University
Sydney New South Wales 2109
Australia
Phone:+61-2-9850-8945
FAX:+61-2-9850-8114
e-mail:jvc@ics.mq.edu.au

Dalibard, Jean
Laboratory Kastler Brossel
Department de Physique de l'Ecole Normale Superieure
24 Rue Lhomond, 75231 Paris Cedex 05
France
Phone:+33-1-44-32-25-34
FAX:+33-1-44-32-34-34
e-mail:Jean.Dalibard@lkb.ens.fr

Das Sarma, Sankar
Department of Physics
University of Maryland
College Park, MD20742-4111
USA
Phone:+1-301-405-6145
FAX:+1-301-314-9465
e-mail:dassarma@physics.umd.edu

Ebisawa, Hiromichi
Graduate School of Information Sciences
Tohoku University
Aramaki-Aoba 04, Aoba-ku, Sendai 980-8579
Japan
Phone:+81-22-217-5846
FAX:+81-22-217-5851
e-mail:ebi@cmt.is.tohoku.ac.jp

Edamatsu, Keiichi
Graduate School of Engineering Science
Osaka University
1-3 Machikaneyama-cho, Toyonaka, Osaka 560-8531
Japan
Phone:+81-6-6850-6507
FAX:+81-6-6850-6509
e-mail:eda@mp.es.osaka-u.ac.jp

Ezawa, Hiroshi
Department of Physics
Gakushuin University
1-5-1 Mejiro, Toshima-ku, Tokyo 171-8588
Japan
Phone:+81-3-3941-3495
FAX:+81-3-3941-3883
e-mail:hiroshi.ezawa@gakushuin.ac.jp

Fujikawa, Kazuo
Department of Physics
University of Tokyo
7-3-1 Hongo, Bunkyo-ku, Tokyo 113-0033
Japan
Phone:+81-3-5841-4190
FAX:+81-3-5841-4224
e-mail:fujikawa@phys.su.u-tokyo.ac.jp

Fujimura, Touru
Central Research Laboratory, Hitachi Ltd.
Hatoyama, Saitama 350-0395
Japan
Phone:+81-49-296-6111
FAX:+81-49-296-6006
e-mail:toru-f@crl.hitachi.co.jp

Fujiwara, Akio
Department of Mathematics
Osaka University
1-16 Machikaneyama, Toyonaka, Osaka 560-0043
Japan
Phone:+81-6-6850-5721
FAX:+81-6-6850-5713
e-mail:fujiwara@math.wani.osaka-u.ac.jp

Fukuyama, Hidetoshi
Institute for Solid State Physics
University of Tokyo
5-1-5 Kashiwanoha, Kashiwa, Chiba 277-8581
Japan
Phone:+81-471-36-3250
FAX:+81-471-36-3250
e-mail:fukuyama@issp.u-tokyo.ac.jp

Funahashi, Haruhiko
Department of Physics
Kyoto University
Sakyo-ku, Kyoto 606-8502
Japan
Phone:+81-75-753-3842
FAX:+81-75-753-3887
e-mail:hal@nh.scphy.kyoto-u.ac.jp

Furusawa, Akira
Department of Applied Physics
University of Tokyo
7-3-1 Hongo, Bunkyo-ku, Tokyo 113-8656
Japan
Phone:+81-3-5841-6823
FAX:+81-3-5841-6823
e-mail:akiraf@ap.t.u-tokyo.ac.jp

Goto, Nobuharu
Institute for Solid State Physics
University of Tokyo
5-1-5 Kashiwanoha, Kashiwa,
Chiba 277-8581
Japan
Phone:+81-471-36-3300
FAX:+81-471-36-3300
e-mail:

Hara, Masahiro
Institute for Solid State Physics
University of Tokyo
5-1-5 Kashiwanoha, Kashiwa,
Chiba 277-8581
Japan
Phone:+81-471-36-3300
FAX:+81-471-36-3300
e-mail:

Harada, Ken
Advanced Research Laboratory, Hitachi, Ltd.
Hatoyama, Saitama 350-0395
Japan
Phone:+81-49-296-6111
FAX:+81-49-296-6006
e-mail:harada@harl.hitachi.co.jp

Harmans, Kees C.
Department of Applied Physics
Delft University of Technology
PO Box 5046, 2600 GA Delft
The Netherlands
Phone:+31-15-2785195
FAX:+31-15-2617868
e-mail:harmans@qt.tn.tudelft.nl

Haroche, Serge
Department de Physique
L'Ecole Normale Superieure
24, Rue Lhomond, 75231 Paris Cedex 05
France
Phone:+33-1-44 32 33 59
FAX:+33-1-45 35 00 76
e-mail:haroche@physique.ens.fr

Hashizume, Tomihiro
Advanced Research Laboratory, Hitachi, Ltd.
Hatoyama, Saitama 350-0395
Japan
Phone:+81-49-296-6111
FAX:+81-49-296-6006
e-mail:tomi@harl.hitachi.co.jp

Hasselbach, Franz
Institute fur Angewandte Physik der Universitat Tuebingen
Auf der Morgenstelle 10, D-72076 Tuebingen
Germany
Phone:+49-7071-2976328
FAX:+49-7071-295093
e-mail:franz.hasselbach@uni-tuebingen.de

Hayashi, Masahiko
Graduate School of Information Sciences
Tohoku University
Aramaki, Aoba-ku, Sendai 980-8579
Japan
Phone:+81-22-217-5847
FAX:+81-22-217-5851
e-mail:hayashi@cmt.is.tohoku.ac.jp

Hofmann, Holger F.
Research Institute for Electronic Science
Hokkaido University
Kita-12 Nishi-6, Kita-ku, Sapporo, Hokkaido 060-0812
Japan
Phone:+81-11-706-2648
FAX:+81-11-706-2648
e-mail:h.hofmann@osa.org

Hosoya, Akio
Department of Physics
Tokyo Institute of Technology
2-12-1 Oh-Okayama, Meguro-ku 152-8551
Japan
Phone:+81-3-5734-2463
FAX:+81-3-5734-2463
e-mail:ahosoya@th.phys.titech.ac.jp

Ichimura, Masahiko
Advanced Research Laboratory, Hitachi, Ltd.
Hatoyama, Saitama 350-0395
Japan
Phone:+81-492-96-6111
FAX:+81-492-96-5999
e-mail:ichimura@harl.hitachi.co.jp

Ide, Toshiki
Department of Physics, Faculty of Science
University of Tokyo
7-3-1 Hongo, Bunkyo-ku 113-0033
Japan
Phone:+81-3-5841-4228
FAX:+81-3-5841-4240
e-mail:ide@femto.phys.s.u-tokyo.ac.jp

Imamura, Hiroshi
Graduate School of Information Sciences
Tohoku University
Aramaki, Aoba-ku, Sendai 980-8579
Japan
Phone:+81-22-217-5849
FAX:+81-22-217-5849
e-mail:hima@cmt.is.tohoku.ac.jp

Imoto, Nobuyuki
School of Advanced Science
The Graduate University for Advanced Studies (SOKEN)
Shonan Village, Hayama, Kanagawa 240-0193
Japan
Phone:+81-468-58-1560
FAX:+81-468-58-1544
e-mail:imoto@soken.ac.jp

Inagaki, Satoru
Department of Physics
Meiji University
1-1-1 Higashi-Mita, Tama-ku, Kawasaki, Kanagawa 214-8571
Japan
Phone:+81-44-934-7432,7171
FAX:+81-44-934-7911
e-mail:inagaki@isc.meiji.ac.jp

Inouye, Shin
Department of Physics
Massachusetts Institute of Technology
77 Massachussetts Ave 26-255, Cambridge, MA 02139
USA
Phone:+1-617-253-2518
FAX:+1-617-253-4876
e-mail:sinouye@mit.edu

Ishibashi, Masayoshi
Advanced Research Laboratory, Hitachi, Ltd.
Hatoyama, Saitama 350-0395
Japan
Phone:+81-492-96-6111
FAX:+81-492-96-6006
e-mail:isibasi@harl.hitachi.co.jp

Iwasawa, Hiroshi
Department of Electronics
Hiroshima Kokusai-Gakuin University
3-3-40-304 Takasu, Nishi-ku, Hiroshima 733-0871
Japan
Phone:+81-82-274-2540
FAX:+81-82-274-2546
e-mail:iwasawa@es.hkg.ac.jp

Iye, Yasuhiro
Institute for Solid State Physics
University of Tokyo
5-1-5 Kashiwanoha, Kashiwa, Chiba 277-8581
Japan
Phone:+81-471-36-3300
FAX:+81-471-36-3300
e-mail:iye@issp.u-tokyo.ac.jp

Kamei, Osamu
EX. Ochanomizu University
2-58-2 Takinogawa, Kita-ku, Tokyo 114-0023
Japan
Phone:+81-3-3917-3160
FAX:
e-mail:

Karlsson, Erik B.
Department of Physics
Uppsala University
P.O.Box530, SE-75121 Uppsala
Sweden
Phone:+46-18-4713594
FAX:+46-18-4713524
e-mail:erk@fysik.uu.se

Kato, Takeo
Department of Applied Physics, Facility of Engineering
Osaka City University
3-3-138 Sugimoto, Sumiyoshi-ku, Osaka 558-8585
Japan
Phone:+81-6-6605-3090
FAX:+81-6-6605-2769
e-mail:kato@a-phys.eng.osaka-cu.ac.jp

Kaufherr, Tirzah
School of Physics and Astronomy
Tel Aviv University
P.O.Box 8604, Jerusalem 91086
Israel
Phone:+972-2-5662805
FAX:+972-3-6407932
e-mail:trka@post.tau.ac.il

Kawai, Tomoji
ISIR-Sanken
Osaka University
8-1 Mihogaoka, Ibaraki, Osaka 567-0047
Japan
Phone:+81-6-6879-8445
FAX:+81-6-6875-2440
e-mail:kawai@sanken.osaka-u.ac.jp

Kawamura, Kiyoshi
Department of Physics, Faculty of Science and Technology
Keio University
Hiyoshi 3-chome , Kohoku-ku, Yokohama 223-8522
Japan
Phone:+81-45-566-1678
FAX:+81-45-566-1672
e-mail:kawamura@rk.phys.keio.ac.jp

Kawasaki, Takeshi
Advanced Research Laboratory, Hitachi, Ltd.
Hatoyama, Saitama 350-0395
Japan
Phone:+81-49-296-6111
FAX:+81-49-296-6006
e-mail:tkawa@harl.hitachi.co.jp

Kitaguchi, Masaaki
Department of Physics
Kyoto University
Sakyo-ku, Kyoto 606-8502
Japan
Phone:+81-75-753-3871
FAX:+81-75-753-3887
e-mail:kitaguch@nh.scphys.kyoto-u.ac.jp

Koashi, Masato
The Graduate University for Advanced Studies
SOKEN
Shonan Village, Hayama, Kanagawa 240-0193
Japan
Phone:+81-468-58-1562
FAX:+81-468-58-1544
e-mail:koashi@soken.ac.jp

Kobayashi, Kensuke
Institute for Solid State Physics
University of Tokyo
5-1-5 Kashiwanoha, Kashiwa, Chiba 277-8581
Japan
Phone:+81-471-36-3301
FAX:+81-471-36-3301
e-mail:knsk@issp.u-tokyo.ac.jp

Kobayashi, Shun-ichi
Institute of Physical and Chemical Research
RIKEN
2-1 Wako, Saitama 351-0198
Japan
Phone:+81-48-462-1111
FAX:+81-48-462-4604
e-mail:skobaya@postman.riken.co.jp

Kobayashi, Takayoshi
Department of Physics
University of Tokyo
7-3-1 Hongo, Bunkyo-ku, Tokyo 113-0033
Japan
Phone:+81-3-5841-4227
FAX:+81-3-5841-4165
e-mail:kobayashi@phys.s.u-tokyo.ac.jp

Kodera, Katsuyoshi
Institute for Solid State Physics
University of Tokyo
5-1-5 Kashiwanoha, Kashiwa,
Chiba 277-8581
Japan
Phone:+81-471-36-3300
FAX:+81-471-36-3300
e-mail:

Koizumi, Hideaki
Central Research Laboratory
Hatoyama, Saitama 350-0395
Japan
Phone:+81-49-296-6111
FAX:+81-49-296-6005
e-mail:hkoizumi@harl.hitachi.co.jp

Kojima, Kunihiro
Section of Scientific Instrumentation and Control
Research Institute for Electronic Science, Hokkaido University
Kita-12 Nishi-6, Kita-ku, Sapporo, Hokkaido 060-0812
Japan
Phone:+81-11-706-2648
FAX:+81-11-706-2648
e-mail:kuni@es.hokudai.ac.jp

Komiyama, Susumu
Department of Basic Science
University of Tokyo
Building 16-622, Komaba 3-8-1, Meguro-ku, Tokyo 153-8902
Japan
Phone:+81-3-5454-6738
FAX:+81-3-5454-4327
e-mail:csusumu@asone.c.u-tokyo.ac.jp

Liu, Lerwen
Asian Technology Information Program
Harks Roppongi Bldg. 6-15-21 Roppongi, Minato-ku, Tokyo 106-0032
Japan
Phone:+81-3-5411-6670
FAX:+81-3-5411-6671
e-mail:lliu@atip.or.jp

Liu, Yu-xi
School of Advanced Science
The Graduate University for Advanced Studies (SOKEN)
Shonan Village, Hayama, Kanagawa 240-0193
Japan
Phone:+81-468-58-1579
FAX:+81-468-58-1544
e-mail:yxliu@koryuw01.soken.ac.jp

Majumdar, Archan S.
S. N. Bose National Centre for Basic Sciences
Block JD, Sector III, Salt Lake Calcutta 700098
India
Phone:+91-33-3355706
FAX:+91-33-3353477
e-mail:archan@boson.bose.res.in

Makhlin, Yuriy
Instut fur Theoretische Festkorperphysik
Universitat Karlsruhe
D-76128 Karlsruhe
Germany
Phone:+49-721-608-6054
FAX:+49-721-698150
e-mail:markhlin@tfp.physik.uni-karlsruhe.de

Maruyama, Eiichi
Frontier Research Sysmtem
RIKEN
2-1 Hirosawa, Wako, Saitama 351-0198
Japan
Phone:+81-48-467-8619
FAX:+81-48-465-8048
e-mail:emaru@postman.riken.go.jp

Matsuda, Ryouji
Ootsuka Laboratory, Institute of Physics
University of Tsukuba
1-1-1 Tennoudai, Tsukuba 305-8571
Japan
Phone:+81-298-53-4345
FAX:+81-298-53-4345
e-mail:matsuda@lt.px.tsukuba.ac.jp

Matsumoto, Takao
Advanced Research Laboratory, Hitachi Ltd.
Hatoyama, Saitama 350-0395
Japan
Phone:+81-49-296-6111
FAX:+81-49-296-6006
e-mail:matumoto@harl.hitachi.co.jp

Matsumoto, Takuya
Central Research Laboratory, Hitachi Ltd.
1-280 Higashi-koigakubo Kokubunji, Tokyo 185-8601
Japan
Phone:+81-42-323-1111
FAX:+81-42-327-7767
e-mail:m-takuya@crl.hitachi.co.jp

Mio, Norikatsu
Department of Advanced Materials Science
University of Tokyo
7-3-1 Hongo, Bunkyo-ku, Tokyo 113-0033
Japan
Phone:+81-3-5841-6871
FAX:+81-3-5841-8807
e-mail:mio@hagi.t.u-tokyo.ac.jp

Miyazaki, Hisao
Ootuka Laboratory, Institute of Physics
University of Tsukuba
1-1-1 Tennodai, Tsukuba 305-8571
Japan
Phone:+81-298-53-4345
FAX:+81-298-53-4345
e-mail:hmiya@lt.px.tsukuba.ac.jp

Morikoshi, Fumiaki
NTT Basic Research Laboratories
3-1 Morinosato-Wakamiya, Atsugi, Kanagawa 243-0198
Japan
Phone:+81-46-240-3483
FAX:+81-46-240-4726
e-mail:fumiaki@will.brl.ntt.co.jp

Morinaga, Atsuo
Deparment of Physics, Faculty of Science & Technology
Science Univ. of Tokyo
2641 Yamazaki, Noda-shi, Chiba 278-8510
Japan
Phone:+81-471-24-1501(Ex.3212)
FAX:+81-471-23-9361
e-mail:morinaga@ph.noda.sut.ac.jp

Motoyoshi, Akio
Deparment of Physics
Faculty of Science
Kumamoto University
2-39-1, Kurokami, Kumamoto 860-8555
Japan
Phone:+81-96-342-3363
FAX:+81-96-342-3320/3363
e-mail:motoyosi@aster.sci.kumamoto-u.ac.jp

Murayama, Yoshimasa
Department of Materials Science & Technology
Facility Engineering, Niigata University
Ikarashi-2-no-cho 8050, Niigata-shi, Niigata 950-2181
Japan
Phone:+81-25-262-6352
FAX:+81-25-262-6352
e-mail:murayama@eng.niigata-u.ac.jp

Nakajima, Sadao
ISTEC/SRC
2-44-1 Eifuku-cho, Suginami, Tokyo 168-0064
Japan
Phone:+81-3-3536-5700
FAX:+81-3-3536-5717
e-mail:

Nakamura, Kazuo
Fundamental Research Laboratories
NEC Corporation
34 Miyukigaoka, Tsukuba, Ibaraki 305-8501
Japan
Phone:+81-298-50-1192
FAX:+81-298-56-6136
e-mail:nakamura@frl.cl.nec.co.jp

Nakamura, Kenzo
High Energy Accelerator Research Organization (KEK)
Oho, Tsukuba, Ibaraki 305-0801
Japan
Phone:+81-298-64-5435
FAX:+81-298-64-7831
e-mail:kenzo.nakamura@kek.jp

Nakamura, Koichi
Division of Natural Science
Meiji University
1-9-1 Eifuku, Suginami-ku, Tokyo 168-8555
Japan
Phone:+81-3-5300-1252
FAX:+81-3-5300-1203
e-mail:knakam@isc.meiji.ac.jp

Nakamura, Michiharu
Research & Development Group
Hitachi Ltd.
New Marunouchi Building, 5-1 Marunouchi 1-Chome, Chiyoda-ku, Tokyo 100-8220
Japan
Phone:+81-3-3214-3101
FAX:+81-3-3212-3025
e-mail:nakamura@hitachi.co.jp

Nakamura, Yasunobu
Fundamental Research Laboratories
NEC Corporation
34 Miyukigaoka, Tsukuba, Ibaraki 305-8501
Japan
Phone:+81-298-50-1148
FAX:+81-298-56-6139
e-mail:yasunobu@frl.cl.nec.co.jp

Nakano, Hayato
Material Science Laboratory
NTT Basic Research Laboratories
3-1, Morinosato-Wakamiya, Atsugi-shi, Kanagawa 243-0198
Japan
Phone:+81-46-240-3528
FAX:+81-46-240-4722
e-mail:nakano@will.brl.ntt.co.jp

Obata, Shuji
Natural Sciences, Tokyo Denki University
Tokyo Denki University
Hatoyama, Hiki, Saitama 350-0394
Japan
Phone:+81-49-296-2911
FAX:+81-49-296-2915
e-mail:obata@u.dendai.ac.jp

Oh, Choo-Hiap
Physics Department
National University of Singapore
Lower Kent Ridge, Singapore 119260
Singapore
Phone:+65-8742603
FAX:+65-7776126
e-mail:phyohch@nus.edu.sg

Ohshima, Toshio
Nanotechnology Research Center
Fujitsu Laboratories Ltd.
10-1 Morinosato-Wakamiya, Atsugi-shi, Kanagawa 243-0197
Japan
Phone:+81-46-250-8234
FAX:+81-46-250-8844
e-mail:oshimato@flab.fujitsu.co.jp

Okamoto, Ryo
Research Institute for Electronic Science
Hokkaido University
Kita-12 Nishi-6, Kita-ku, Sapporo, Hokkaido 060-0812
Japan
Phone:+81-11-706-2648
FAX:+81-11-706-2648
e-mail:oka@es.hokudai.ac.jp

Ono, Teruo
Graduate School of Engineering Science
Osaka University
Toyonaka, Osaka 560-8531
Japan
Phone:+81-6-6850-6431
FAX:+81-6-6845-4632
e-mail:ono@mp.es.osaka-u.ac.jp

Ono, Yoshimasa A.
Advanced Research Laboratory, Hitachi, Ltd.
Hatoyama, Saitama 350-0395
Japan
Phone:+81-49-296-6111
FAX:+81-49-296-6005
e-mail:yaono@harl.hitachi.co.jp

Onogi, Toshiyuki
Advanced Research Laboratory, Hitachi, Ltd.
Hatoyama, Saitama 350-0395
Japan
Phone:+81-49-296-6111
FAX:+81-49-296-6006
e-mail:onogi@harl.hitachi.co.jp

Osakabe, Nobuyuki
Advanced Research Laboratory, Hitachi, Ltd.
Hatoyama, Saitama 350-0395
Japan
Phone:+81-49-296-6111
FAX:+81-49-296-6005
e-mail:osakabe@harl.hitachi.co.jp

Rikitake, Yoshiaki
Graduate School of Information Science
Tohoku University
Aramaki, Aoba-ku, Sendai, Miyagi 980-8579
Japan
Phone:+81-22-217-5850
FAX:+81-22-217-5851
e-mail:yoshiaki@cmt.is.tohoku.ac.jp

Sackett, Cass A.
UVA Physics Department
University of Virginia
382 McCormick Road, PO Box 400714,
Charlottesville, VA 22904-4714
USA
Phone:+1-434-924-6795
FAX:+1-434-924-4576
e-mail:sackett@virginia.edu

Saikawa, Kazuhiko
Dept. of Physics, Faculty of Science, Shinshu University
1-1-10 Igawajo, , Matsumoto, Nagano 390-0831
Japan
Phone:+81-263-27-2017
FAX:+81-263-37-2562
e-mail:ksaikaw@gipac.shinshu-u.ac.jp

Saito, Hiroki
Interactive Research Center of Science
Tokyo Institute of Technology
2-12-1 Ookayama, Meguro-ku, Tokyo 152-8551
Japan
Phone:+81-3-5734-3664
FAX:+81-3-5734-3664
e-mail:hsaito@stat.phys.titech.ac.jp

Saito, Shinichi
Central Research Laboratory, Hitachi Ltd.
1-280 Higashi-koigakubo Kokubunji, Tokyo 185-8601
Japan
Phone:+81-42-323-1111
FAX:+81-42-327-7773
e-mail:Shinichi-crl.Saito@c-net3.crl.hitachi.co.jp

Saitoh, Kazuo
Advanced Research Laboratory, Hitachi Ltd.
1-280 Higashi-koigakubo Kokubunji, Tokyo 185-8601
Japan
Phone:+81-42-323-1111
FAX:+81-42-327-7722
e-mail:

Sakurai, Akio
Dept. of Physics
Kyoto Sangyo University
Kamigamo Kita-ku, Kyoto 603-8555
Japan
Phone:+81-75-705-1606
FAX:+81-75-705-1640
e-mail:sakurai@cc.kyoto-su.ac.jp

Shibata, Junya
Department of Physics, Graduate School of Science
Osaka University
1-1 Machikaneyama, Toyonaka, Osaka 560-0043
Japan
Phone:+81-6-6850-5349
FAX:+81-6-6850-5351
e-mail:shibata@acty.phys.sci.osaka-u.ac.jp

Shimizu, Akira
Department of Basic Science
University of Tokyo
3-8-1 Komaba, Meguro-ku,
Tokyo 153-8902
Japan
Phone:+81-3-5454-6532
FAX:+81-3-5790-7229
e-mail:shmz@asone.c.u-tokyo.ac.jp

Shimizu, Fujio
University of Electro-Communications
1-5-1 Chofugaoka, Chofu, Tokyo 182-8585
Japan
Phone:+81-42-443-5701
FAX:+81-42-485-8960
e-mail:fshimizu@ils.uec.ac.jp

Shinba, Yutaka
Shizuoka Institute of Science and Technology
2200-2 Toyozawa, Fukuroi, Shizuoka 437
Japan
Phone:+81-538-45-0190
FAX:+81-538-45-0110
e-mail:shinba@ns.sist.ac.jp

Stopa, Michael
Tarucha Mesoscopic Correlation Project
ERATO-JST
45-3085 NTT Atsugi Research and Development
Laboratory 3-1 Morinosato Wakamiya, Atugi,
Kanagawa 243-0198
Japan
Phone:+81-46-248-4016
FAX:+81-46-248-4014
e-mail:stopa@tarucha.jst.go.jp

Sugano, Ryoko
Advanced Research Laboratory, Hitachi Ltd.
Hatoyama, Saitama 350-0395
Japan
Phone:+81-49-296-6111
FAX:+81-49-296-6006
e-mail:sugano@harl.hitachi.co.jp

Sugano, Takuo
Japan Science & Technology Corporation
CREST
Shibuya T. K. Building, Shibuya 3-13-11, Shibuya-ku,
Tokyo 150-0002
Japan
Phone:+81-3-5468-1491
FAX:+81-3-5468-1494
e-mail:sugano@hakusan.toyo.ac.jp

Takagi, Shin
Fuji Tokoha University
325 Obuchi, Fuji, Shizuoka 417-0801
Japan
Phone:+81-545-37-2033
FAX:+81-545-36-2651
e-mail:takagi@fuji-tokoha-u.ac.jp

Takayanagi, Kunio
Physics Department
Tokyo Institute of Technology
2-12-1 Oh-okayama, Meguro-ku, Tokyo 152-8550
Japan
Phone:+81-3-5734-2078
FAX:+81-3-5734-2079
e-mail:takayang@surface.phys.titech.ac.jp

Takeuchi, Shigeki
JST-Presto Project, Research Institute for Electronic Science
Hokkaido University
Kita-12 Nishi-6, Kita-ku, Sapporo, Hokkaido 060-0812
Japan
Phone:+81-11-706-2646
FAX:+81-11-706-4956
e-mail:takeuchi@es.hokudai.ac.jp

Tanaka, Atsushi
Department of Physics
Tokyo Metropolitan University
1-1 Minami Osawa, Hachioji, Tokyo 192-0397
Japan
Phone:+81-426-77-1111
FAX:+81-426-77-2483
e-mail:tanaka@phys.metro-u.ac.jp

Tanamoto, Tetsufumi
Advanced Materials & Devices Laboratory
Corporate Research and Development Center
Toshiba Corporation
1 Komukai Toshiba-cho, Saiwai-ku, Kawasaki, Kanagawa 212-8582
Japan
Phone:+81-44-549-2295
FAX:+81-44-520-1801
e-mail:tetsufumi.tanamoto@toshiba.co.jp

Terashima, Hiroaki
Physics Department
Tokyo Institute of Technology
152-8550
Japan
Phone:+81-3-5734-2486
FAX:
e-mail:terasima@stat.phys.titech.ac.jp

Tonomura, Akira
Advanced Research Laboratory, Hitachi, Ltd.
Hatoyama, Saitama 350-0395
Japan
Phone:+81-49-296-6111
FAX:+81-49-296-6005
e-mail:tonomura@harl.hitachi.co.jp

Torres Alonso, J. Antonio
Theory Group, JRCAT
1-1-1 Higashi Tsukuba, Ibaraki 305-0046
Japan
Phone:+81-298-542764
FAX:+81-298-542788
e-mail:jtorres@jrcat.or.jp

Toyoshima, Kouichi
Faculty of Science and Engineering
University of Saga
1 Honjo, Saga 839 0827
Japan
Phone:+81-952-28-8845
FAX:+81-952-28-8845
e-mail:toyo@cc.saga-u.ac.jp

Tsai, Jaw-Shen
Fundamental Research Laboratories
NEC Corporation
34 Miyukigaoka, Tsukuba, Ibaraki 305-8501
Japan
Phone:+81-298-50-1161
FAX:+81-298-56-6139
e-mail:tsai@cj.jp.nec.com

Tsai, Sheng-Yi
College of Science and Technology
Nihon University
1-8-14 Kanda Surugadai, Chiyoda-ku, Tokyo 101-8303
Japan
Phone:+81-3-3259-0867
FAX:+81-3-3293-8269
e-mail:tsai@phys.cst.nihon-u.ac.jp

Tsujimura, Tatsuya
Department of Science
Kyodo Tsushin
2-2-5 Toranomon, Minato-ku, Tokyo 105-8474
Japan
Phone:+81-3-5573-8151
FAX:+81-3-5573-8152
e-mail:utatsu@kyodonews.or.jp

Tsujino, Kenji
Research Institute for Electronic Science
Hokkaido University
Kita-12 Nishi-6, Kita-ku, Sapporo, Hokkaido 060-0812
Japan
Phone:+81-11-706-2648
FAX:+81-11-706-2648
e-mail:tsujino@es.hokudai.ac.jp

Tsutsui, Izumi
High Energy Accelerator Research Organization
KEK
1-1 Oho, Tsukuba, Ibaraki 305-0801
Japan
Phone:+81-298-79-6093
FAX:+81-298-79-6101
e-mail:izumi.tsutsui@kek.jp

Uchida, Fumihiko
Planning Office, Central Research Laboratory
1-280 Higashi-Koigakubo Kokubunji-shi, Tokyo
Japan
Phone:+42-327-7777
FAX:+42-327-7695
e-mail:fumihiko@crl.hitachi.co.jp

Uchiyama, Fumiyo
Nuclear Physics Division
MS-245, Lawrence Berkeley National Laboratory
Berkeley, CA 94720
USA
Phone:+1-510-495-2304
FAX:
e-mail:fuchiyam@gol.com

Udem, Thomas
Department of Laserspectroscopy
Max Planck Institut fur Quantenoptik
Hans-Kopfermann-Str. 1, 85748 Garching
Germany
Phone:+49-89-32905-262
FAX:+49-89-32905-200
e-mail:thu@mpq.mpg.de

Ueno, Kyu
Faculty of Science and Technology
Keio University
3-14-1 Hiyoshi, Kohoku-ku, Yokohama, Kanagawa 223-8522
Japan
Phone:+81-45-566-1676
FAX:+81-45-566-1672
e-mail:kueno@rk.phys.keio.ac.jp

Ulam-Orgikh, Duger
Graduate School of Engineering Science
Osaka University
1-3 Machikaneyama, Toyonaka, Osaka 560-8531
Japan
Phone:+81-6-6850-6336
FAX:+81-6-6850-6341
e-mail:uka@laser.ee.es.osaka-u.ac.jp

Utsumi, Yasuhiro
Graduate School of Information Sciences
Tohoku University
Aramaki, Aoba-ku, Sendai 980-8579
Japan
Phone:+81-22-217-5850
FAX:+81-22-217-5851
e-mail:utsumi@cmt.is.tohoku.ac.jp

Weihs, Gregor
Institute of Experimental Physics
University of Vienna
Boltzmanngasse 5, A-1090 Vienna
Austria
Phone:+43-1-4277-51207
FAX:+43-1-4277-9512
e-mail:gregor.weihs@univie.ac.at

Williams, David A.
Hitachi Cambridge Laboratory
Hitachi Europe Ltd.
Madingley Road, Cambridge CB3 0HE
UK
Phone:+44-1223-442902
FAX:+44-1223-467942
e-mail:williams@phy.cam.ac.uk

Yamada, Eizaburo
Faculty of Information Sciences
Meisei University
4-7-6, Renkoji, Tama-shi, Tokyo 206-0021
Japan
Phone:+81-42-375-8181
FAX:+81-42-375-8181
e-mail:

Yamaguchi, Takahide
Department of Physics
University of Tsukuba
1-1-1 Tennoudai, Tsukuba 305-8571
Japan
Phone:+81-298-53-4345
FAX:+81-298-53-4345
e-mail:yamaguchi@lt.px.tsukuba.ac.jp

Yamamoto, Yoshihisa
Ginzton Laboratory
Stanford University
Stanford, CA 94305-4085
USA
Phone:+1-650-725-3327
FAX:+1-650-723-5320
e-mail:yamamoto@loki.stanford.edu

Yanagimachi, Shinya
Department of Physics, Faculty of Science and Technology
Science Univerity of Tokyo
2641Yamazaki, Noda, Chiba 278-8510
Japan
Phone:+81-471-24-1501
FAX:+81-471-23-9361
e-mail:j6299706@ph.noda.sut.ac.jp

Yoneda, Tetsuya
College of Medical Science
Kumamoto University
4-24-1 Kuhonji, Kumamoto 862-0976
Japan
Phone:+81-96-373-5487
FAX:+81-96-373-5487
e-mail:tyoneda@cms.kumamoto-u.ac.jp

Zeilinger, Anton
Institute of Experimental Physics
University of Vienna
Boltzmanngasse 5, A-1090 Vienna
Austria
Phone:+43-1-4277-51201
FAX:+43-1-4277-9512
e-mail:anton.zeilinger@univie.ac.at

Author Index